Green Synthetic Processes and Procedures

Green Chemistry Series

Titles in the series:

1: The Future of Glycerol: New Uses of a Versatile Raw Material
2: Alternative Solvents for Green Chemistry
3: Eco-friendly Synthesis of Fine Chemicals
4: Sustainable Solutions for Modern Economies
5: Chemical Reactions and Processes under Flow Conditions
6: Radical Reactions in Aqueous Media
7: Aqueous Microwave Chemistry
8: The Future of Glycerol: 2nd Edition
9: Transportation Biofuels: Novel Pathways for the Production of Ethanol, Biogas and Biodiesel
10: Alternatives to Conventional Food Processing
11: Green Trends in Insect Control
12: A Handbook of Applied Biopolymer Technology: Synthesis, Degradation and Applications
13: Challenges in Green Analytical Chemistry
14: Advanced Oil Crop Biorefineries
15: Enantioselective Homogeneous Supported Catalysis
16: Natural Polymers Volume 1: Composites
17: Natural Polymers Volume 2: Nanocomposites
18: Integrated Forest Biorefineries
19: Sustainable Preparation of Metal Nanoparticles: Methods and Applications
20: Alternative Solvents for Green Chemistry: 2nd Edition
21: Natural Product Extraction: Principles and Applications
22: Element Recovery and Sustainability
23: Green Materials for Sustainable Water Remediation and Treatment
24: The Economic Utilisation of Food Co-products

25: Biomass for Sustainable Applications: Pollution Remediation and Energy
26: From C–H to C–C Bonds: Cross-dehydrogenative-coupling
27: Renewable Resources for Biorefineries
28: Transition Metal Catalysis in Aerobic Alcohol Oxidation
29: Green Materials from Plant Oils
30: Polyhydroxyalkanoates (PHAs) Based Blends, Composites and Nanocomposites
31: Ball Milling Towards Green Synthesis: Applications, Projects, Challenges
32: Porous Carbon Materials from Sustainable Precursors
33: Heterogeneous Catalysis for Today's Challenges: Synthesis, Characterization and Applications
34: Chemical Biotechnology and Bioengineering
35: Microwave-assisted Polymerization
36: Ionic Liquids in the Biorefinery Concept: Challenges and Perspectives
37: Starch-based Blends, Composites and Nanocomposites
38: Sustainable Catalysis: With Non-endangered Metals, Part 1
39: Sustainable Catalysis: With Non-endangered Metals, Part 2
40: Sustainable Catalysis: Without Metals or Other Endangered Elements, Part 1
41: Sustainable Catalysis: Without Metals or Other Endangered Elements, Part 2
42: Green Photo-active Nanomaterials
43: Commercializing Biobased Products: Opportunities, Challenges, Benefits, and Risks
44: Biomass Sugars for Non-fuel Applications
45: White Biotechnology for Sustainable Chemistry
46: Green and Sustainable Medicinal Chemistry: Methods, Tools and Strategies for the 21st Century Pharmaceutical Industry
47: Alternative Energy Sources for Green Chemistry
48: High Pressure Technologies in Biomass Conversion
49: Sustainable Solvents: Perspectives from Research, Business and International Policy
50: Fast Pyrolysis of Biomass: Advances in Science and Technology
51: Catalyst-free Organic Synthesis
52: Hazardous Reagent Substitution: A Pharmaceutical Perspective
53: Alternatives to Conventional Food Processing: 2nd Edition

54: Sustainable Synthesis of Pharmaceuticals: Using Transition Metal Complexes as Catalysts
55: Intensification of Biobased Processes
56: Sustainable Catalysis for Biorefineries
57: Supercritical and Other High-pressure Solvent Systems: For Extraction, Reaction and Material Processing
58: Biobased Aerogels: Polysaccharide and Protein-based Materials
59: Rubber Recycling: Challenges and Developments
60: Green Chemistry for Surface Coatings, Inks and Adhesives: Sustainable Applications
61: Green Synthetic Processes and Procedures

How to obtain future titles on publication:
A standing order plan is available for this series. A standing order will bring delivery of each new volume immediately on publication.

For further information please contact:
Book Sales Department, Royal Society of Chemistry, Thomas Graham House, Science Park, Milton Road, Cambridge, CB4 0WF, UK
Telephone: +44 (0)1223 420066, Fax: +44 (0)1223 420247
Email: booksales@rsc.org
Visit our website at www.rsc.org/books

Green Synthetic Processes and Procedures

Edited by

Roberto Ballini
University of Camerino, Italy
Email: roberto.ballini@unicam.it

ROYAL SOCIETY OF CHEMISTRY

Green Chemistry Series No. 61

Print ISBN: 978-1-78801-512-7
PDF ISBN: 978-1-78801-613-1
EPUB ISBN: 978-1-78801-839-5
Print ISSN: 1757-7039
Electronic ISSN: 1757-7047

A catalogue record for this book is available from the British Library

The Royal Society of Chemistry is a charity, registered in England and Wales, Number 207890, and a company incorporated in England by Royal Charter (Registered No. RC000524), registered office: Burlington House, Piccadilly, London W1J 0BA, UK, Telephone: +44 (0) 20 7437 8656.

For further information see our web site at www.rsc.org

Printed in the United Kingdom by CPI Group (UK) Ltd, Croydon, CR0 4YY, UK

Preface

The urgency for the publication of this new book relies greatly on the great growth of interest from the community towards sustainable chemistry and the enormous increase in publications in the field of green chemistry. In fact, the number of research groups dedicated to this field is increasing considerably, as well as the number of journals specialized on the topic and their Impact Factors.

The purpose of this book is to update the scientific community, in particular that interested in organic synthesis, on "green" synthetic processes and procedures whose literature has significantly expanded in recent years.

Roberto Ballini

Green Chemistry Series No. 61
Green Synthetic Processes and Procedures
Edited by Roberto Ballini

Published by the Royal Society of Chemistry, www.rsc.org

Contents

Green Chemistry Series No. 61
Green Synthetic Processes and Procedures
Edited by Roberto Ballini

Published by the Royal Society of Chemistry, www.rsc.org

CHAPTER 1

Sustainability of Green Synthetic Processes and Procedures†

ISTVÁN T. HORVÁTH*[a] AND EDIT CSÉFALVAY[b]

[a]City University of Hong Kong, Department of Chemistry, Tat Chee Avenue, Kowloon, Hong Kong; [b]Department of Energy Engineering, Faculty of Mechanical Engineering, Budapest University of Technology and Economics, Budapest, Hungary
*E-mail: istvan.thorvath@cityu.edu.hk

1.1 Development and Definition of Green Chemistry

Green chemistry emerged in the 1990s to address the increasing number of health and environmental issues caused by hazardous chemicals and materials.[1] Their toxicity was either not considered or underestimated and sometimes even just simply ignored by some of the stake holders. The eco-toxicology of chemicals had received even less attention until the negative environmental and health effects of dichlorodiphenyl-trichloroethane, DDT, were reported by Raphael Carson in 1962.[2] The number of serious environmental and health problems has rapidly increased in the second half of the last century, mostly due to the frantic expansion of the chemical, petrochemical, and pharmaceutical industry to supply all the goods

†This chapter is partly based on ref. 25 and 45 with permission from Elsevier and the American Chemical Society.

Green Chemistry Series No. 61
Green Synthetic Processes and Procedures
Edited by Roberto Ballini

Published by the Royal Society of Chemistry, www.rsc.org

and services for better quality of life of the growing population, as well as to generate more and more profits at the expense of the environment and population.[3] Some of the worst examples include the addition of tetraethyl lead to gasoline,[4] use of thalidomide by pregnant women,[3b] utilization of chlorofluorocarbons (CFCs) in refrigerators,[5] deadly accidents involving dioxins[3b] and methyl isocyanate,[3b] and contamination by crude oil,[6] dioxins,[3b] melamine,[7] and ammonia,[8] just to name a few. The emission of nitrogen oxides above the regulation level by hundred thousands of cheating cars for almost a decade[9] shows that profit making has remained more important than sustainability. While some of these health and/or environmental problems were the result of limited or lack of knowledge on toxicity, bioaccumulation, and ecotoxicity, others were simply due to fraudulent practices, individual or corporate greed, or both.

Environmental and health problems became so visible by the middle of the 1980s that the US Environmental Protection Agency switched the focus from 'end-of-the-pipe clean-up' approach to 'pollution prevention', which led to the enactment of the 1990 Pollution Prevention Act by the United States Congress.[10] In 1995, the traditional blue colour of the cover of *Chemical Reviews*, one of the flagship journals of the American Chemical Society, was changed to green for the one issue dedicated to environmental chemistry.[11] The editorial included one of the earliest published definitions of green chemistry: "It is no longer sufficient to make marvellous new molecules solely on the basis of their marketable properties. Although marketability is an appropriate goal, we, as scientists, must also be concerned with our creations' potential for environmental impact". The publication of the book entitled 'Green Chemistry: Theory and Practice' by Anastas and Warner in 1998 provided a carefully drafted definition: "*Green chemistry is the utilization of a set of principles that reduces or eliminates the use or generation of hazardous substances in the design, manufacture, and applications of chemical products*".[12] The prevention of the environmental and health impacts of hazardous chemicals, materials, and practices is addressed by the 12 principles of green chemistry (Box 1.1).

It should be emphasized that the simultaneous application of several principles of green chemistry for the design of a molecule, a material, a reaction, or a process has been mostly straightforward, but the utilization of all twelve of them is rather difficult, if not impossible. The definition and the twelve principles of green chemistry have not been challenged in the last thirty years, indicating their abiding nature.

1.2 Development and Definition of Sustainability

Sustainability has been the center of attention of humans for a long time, mostly focusing on how to ensure survival, individually or of smaller or larger communities, under sometimes dramatically changing social and environmental conditions. Although the overall achievements of the human race have been spectacular, the rapid development of modern

Box 1.1 The twelve principles of green chemistry[12]

1. It is better to prevent waste than to treat or clean up waste after it is formed.
2. Synthetic methods should be designed to maximize the incorporation of all materials used in the process into the final product.
3. Wherever practicable, synthetic methodologies should be designed to use and generate substances that process little or no toxicity to human health and the environment.
4. Chemical products should be designed to preserve efficacy of function while reducing toxicity.
5. The use of auxiliary substances (*e.g.*, solvents, separation agents, *etc.*) should be made unnecessary wherever possible and, innocuous when used.
6. Energy requirements should be recognized for their environmental and economic impacts and should be minimized. Synthetic methods should be conducted at ambient temperature and pressure.
7. A raw material of feedstock should be renewable rather than depleting wherever technically and economically practicable.
8. Unnecessary derivatization (blocking group, protection and deprotection, temporary modification of physical/chemical processes) should be avoided whenever possible.
9. Catalysts (as selective as possible) are superior to reagents.
10. Chemical products should be designed so that, at the end of their function, they do not persist in the environment and break down into innocuous degradation products.
11. Analytical methodologies need to be further developed to allow for real-time, in-process monitoring and control prior to the formation of hazardous substances.
12. Substances and the form of a substance used in a chemical process should be chosen so as to minimize the potential for chemical accidents, including releases, explosion, and fires.

society as well as globalization has had a very negative impact. It was already recognized in 1798 that the faster and faster growing population could surpass sufficient food production resulting in famine, plagues, and even war.[13] The environmental conditions and the future of the habitat of the Earth became worrisome to many by the last quarter of the last century.

In response to the increasing environmental and health concerns, the United Nations has created the World Commission on Environment and Development (or the Brundtland Commission)[14] to formulate a global agenda for change. Their report stated that "*humanity has the ability to make development sustainable to ensure that it meets the needs of the present without compromising the ability of future generations to meet their own needs*".[15] This definition has been used globally since 1987 and the overarching requisite that 'the needs of the future generations' can be predicted correctly has not been challenged in the last three decades. Although short-term needs can be identified with higher and higher certainty, long-term predictions are still unreliable due to the end of history illusion[16] and the extremely fast advances of science and technology.[17] Consequently, 'sustainability' has been regularly replaced with 'suitability' by many stake holders with vested or conflicts of interests.

It should be acknowledged that, despite the poor definition of sustainability, the worldwide recognition of several adverse environmental or health issues resulted in successful local and global responses. For example, the phasing-out of CFCs as refrigerants to eliminate the cause of ozone depletion or the ban of tetraethyl-lead as an antiknock agent was achieved efficiently. The recently discovered widespread pollution by micro- and nano-plastics is now spurring the global community including scientists and engineers to find efficient solutions.[18]

Although our track record for predicting economic changes or societal transformations has been unreliable, most next-generation definitions and metrics have combined ecological, economical, and societal components at different portions. For example, Robèrt *et al.* have suggested four system conditions as the first order principles for sustainability:[19] (1) *Substances from the lithosphere must not systematically increase in the ecosphere*, (2) *substances produced by society must not systematically increase in the ecosphere*, (3) *the physical basis for the productivity and diversity of Nature must not be systematically deteriorated*, and (4) *fair and efficient use of resources with respect to meeting human needs*. While the first three system conditions may be measured and analyzed, the fourth one, especially the 'fair use of resources' is subjective and probably impossible to define. Another approach proposed the use of the thermodynamic concept of '*lost work*' to address the sustainability of the processing industry: "*Sustainability in the ecological sense means that we do not place an intolerable load on the ecosphere and that we maintain the natural basis for life*".[20] In addition, "*sustainability implies utilization of renewable resources such as biomass or solar energy*". Since the processes converting natural resources to consumer materials and heat do not proceed ideally, part of the work-potential of primary materials is lost and may be the subject of exergy analysis. Another definition demanded that "*sustainable technology must be as efficient as possible and the production of harmless (waste) products may not exceed the assimilative uptake rate of these products in the ecosphere*".[21] The last two definitions have subjective terms such as 'intolerable load' or 'assimilative uptake rate', which would be difficult to define and measure. Sikdar's definition is even more complicated by combining environmental, economic, and societal aspects: "*Sustainability occurs when the material and social conditions for human health and the environment are maintained or improved over time without exceeding the ecological capabilities that support them*".[22] The term 'social conditions' could be rather fuzzy and its measurement could be very subjective again.

In 2015, even the United Nations acknowledged the problem of the ambiguous definitions of sustainability by identifying 17 sustainable development goals (SDGs) and 169 targets.[23] These were the result of the work of hundreds of organizations and thousands of individuals of many countries with significantly different environmental, societal, economic, and political conditions and, more importantly, vastly different short and long term needs and visions. Although the progress of the 17 SDGs will be monitored by the relevant 100 Potential and Indicative SDG Indicators and 152 Complementary National Indicators,[24] neither sustainable development nor its three dimensions were defined. Instead, circular reasoning was used repeatedly by beginning with sustainability to end up with sustainability. For example, "We envisage a world in which every country enjoys *sustained*, inclusive, and *sustainable* economic growth and decent work for all. A world in which consumption and production patterns and use of all natural resources – from air to land, from rivers, lakes, and aquifers to oceans and seas – are *sustainable*".[23b] Thus, the UN's 17 SDGs and the associated targets may serve as a 'roadmap to happiness', but not as a definition to develop metrics to set deliverables and achieve accountability and credibility.

Owing to said ambiguous definitions, 'sustainability' has been regularly replaced with 'suitability' by many stake holders with vested or conflicts of interests to make profits for businesses, secure financial sponsorships for non-governmental organizations and environmentalists, and attract support for politicians and political organizations to be elected or re-elected. In order to prevent the possibility to call 'suitable development' sustainable, sustainability must be defined as an intrinsic property of a molecule, a material, a reaction, a process, a technology, a house, a city, a country, a planet, *etc.* We have recently suggested an alternative simple definition, which is based on the intrinsic evolutionary rules of Nature and independent of the randomly changing economical and societal structures: *Nature's resources, including energy, should be used at a rate at which they can be replaced naturally, and the generation of wastes cannot be faster than the rate of their remediation by Nature.*[25] It is important to recognize that Nature sets the upper limits for the replacement of resources and remediation of wastes, which of course could depend on the geological location and season of the year. This definition is similar to the first two principles, or system conditions, of the four sustainability principles by Robèrt *et al.*[19] It should also be noted that sustainability can be increased by using sustainable repairing, reconditioning, remanufacturing or recycling methods, processes, and technologies. The integration of the use of energy, chemicals, and materials could increase the overall sustainability locally and globally, as both the resource demand and proportional waste generation could be reduced.

1.2.1 Definition of Sustainable Chemistry

Since some parts of green chemistry may not be sustainable at all, a definition of sustainable chemistry should be established and applied for the assessment of each chemical reaction or process. Based on the definition of sustainability,[25] it was proposed that "*sustainable chemistry should use resources, including energy, at a rate at which they can be replaced naturally, and the generation of waste cannot be faster than the rate of their remediation*".[26] It should be emphasized that, according to this definition, all fossil resources cannot be sustainable! Natural gas, crude oil, and coal have been used at a rate much faster than their natural replacement.[27] While their formation requires hundred millions of years, fossil resources will be exhausted in a few hundreds of years. Furthermore, the photosynthesis (Scheme 1.1) is slower than the generation of carbon dioxide and water evidenced by the increasing carbon dioxide levels in the atmosphere from 315 ppm in 1955 to 411 ppm in 2019.

Consequently, a sustainable future requires all carbon chemicals to be produced from carbon dioxide or biomass to achieve a net zero carbon footprint. The easy conversion of polysaccharides to monosaccharides has opened the door for the fermentation to bioethanol (Scheme 1.2, step 1), which can be used as a sustainable resource to produce basic and intermediate chemicals (Scheme 1.2, steps 2–17).

Finally, it should be noted that not all sustainable chemicals or reactions/processes can be green. Therefore, the selection of chemicals, reactions, and processes sustainable and green at the same time should be preferred or the target of future design and innovation.

$$6\,CO_2 + 6\,H_2O \xrightarrow{h\nu} C_6H_{12}O_6 + 6\,O_2$$

Scheme 1.1 Photosynthesis of C_6-monosaccharides and oxygen.

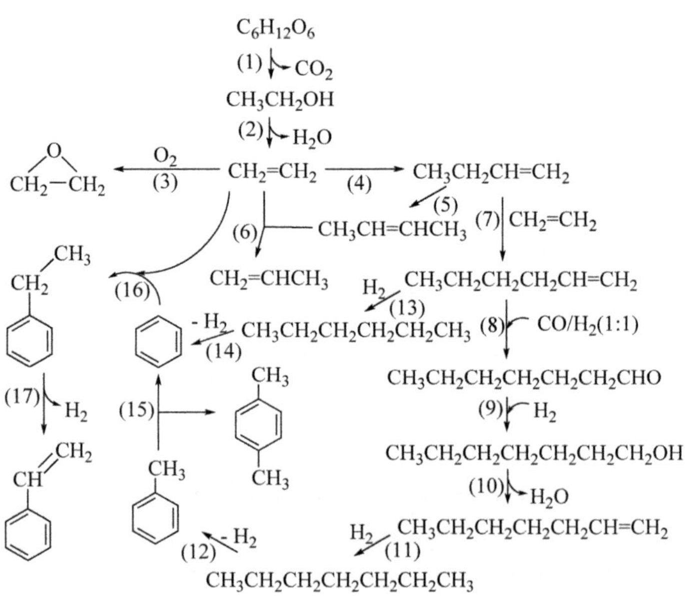

Scheme 1.2 Biomass-based production of basic carbon chemicals.

1.3 Metrics for Chemistry

The simplest measure of a chemical reaction is the *conversion* of the reactant(s), which shows the level of transformation of the reactant(s) but does not reveal the amount of product(s) and side product(s). The ideal performance of a chemical reaction can be described by the *theoretical yield*, which is the formation of the maximum amount of product(s) in a chemical reaction or process according to the corresponding stoichiometry. The latter is based on the law of conservation of mass, *e.g.*, the total mass of the reactants must equal the total mass of the products and side products. While some reactions can be described theoretically by a chemical equation leading to a single product (*e.g.*, isomerization and addition reactions), others result in two or even more products (*e.g.*, substitution and elimination reactions). It is important to emphasize that the *theoretical yield* requires 100% *conversion* of the starting substrate(s) to the product(s), which is only possible when the *selectivity* of the reaction, the ratio of the desired product(s) to the converted reactants, is 100%. The actual *yield* of a reaction is generally lower than the *theoretical yield* due to low reaction rates (*e.g.*, incomplete reaction) and the simultaneous occurrence of side reactions resulting in side products formed from the starting substrate(s) or secondary products formed from the desired products or side products. The *yield* of product(s) can thus be simply calculated using the *conversion* and *selectivity* (eqn (1.1)):

$$\text{Conversion}(\%) = \frac{\text{Converted reactants}(\text{mol})}{\text{Reactants}(\text{mol})} \times 100$$

$$\text{Selectivity}(\%) = \frac{\text{Desired product (mol)}}{\text{Converted reactants (mol)}} \times 100 \qquad (1.1)$$

$$\text{Yield}(\%) = \frac{\text{Conversion}(\%) \times \text{Selectivity}(\%)}{100}$$

A much more realistic measure is the *isolated yield*, which accounts for all the losses during isolation procedures and only considers the isolated moles of the desired products. While these metrics are simple and have been used for a long time, they do not account for all the required chemicals/materials and make it difficult to design and select greener chemical systems.

1.4 Metrics for Green Chemistry

The concept of ecological footprint[28] was developed to understand the human impact on Earth by measuring the consumption of the resources provided by the ecosystem, including the production of goods, services, and some waste remediation capabilities as well. Although determination of the carbon footprint is a well-accepted metric in chemical enterprises, it has been rarely applied for the characterization of chemical reactions, processes, products, and materials. The two most popular metrics, Trost's *atom economy*[29] (which is identical to Sheldon's *atom utilization*[30]) and Sheldon's *environmental factor*,[31] have been developed for green chemistry to provide measurable and verifiable mass balances of chemical reactions and processes at the molecular level. By using the same chemical equations, those needed to calculate the theoretical yield, one can also calculate the *atom economy* or *atom utilization* of the reaction, *e.g.*, how many of all the atoms in the reaction are part of the product(s) (eqn (1.2)). While the *atom economy* of isomerization and addition reactions is 100% theoretically, substitution and elimination reactions are intrinsically dependent on the size of the side products. The *atom economy* or *atom utilization* is of course very useful at the design stage in the development of new chemical processes to select the greenest reactions and minimize waste.

$$\text{Atom economy}(\%) = \frac{\text{Molar mass of products}}{\text{Molar mass of all reactants}} \times 100 \tag{1.2}$$

The major shortcoming of the *atom economy* is that it does not show all the chemicals and materials required to perform a chemical reaction or process. Various auxiliary substances (solvents, additives, reagents, *etc.*) can be used during the workup of the reactions to ensure the highest yield of the product(s), which in general will become un-recyclable waste. The *Environmental* or *E-factor* was defined to account for all chemicals and materials involved by calculating the ratio of the mass of all wastes per mass of the product(s) (eqn (1.3)).

$$\text{E} - \text{factor} = \frac{\text{Mass of all wastes}}{\text{Mass of products}} \tag{1.3}$$

For example, Jamison *et al.*[32] demonstrated that E-factors can be significantly reduced by using continuous-flow chemistry,[33] computational calculations, and solvent minimization. The E-factor for the synthesis of atropine was lowered from 2245 to 24, and that for the synthesis of diazepam was reduced from 36 to 9. Several other green metrics have been critically reviewed recently,[34,35] including the carbon efficiency, effective mass yield, mass intensity, reaction mass efficiency, stoichiometric factor, and solvent and catalyst environmental impact parameter.

1.5 Metrics for Sustainability

The exact meaning of the definition of sustainable development by the World Commission on Environment and Development is very subjective and the corresponding goals and deliverables could be markedly different from person to person, city to city, and country to country. Consequently, more than five hundred approaches have been suggested for the measurement of sustainability,[36] most of which were based on the three major dimensions of sustainability at different ratios.[21,37–39] The one-dimensional metrics can be classified as environmental, economic, and social indicators. Two-dimensional metrics combine environmental-social, environmental-economic, and social-economic indicators, and three-dimensional metrics can be obtained from the intersection of the environmental, economic, and social dimensions. Even four indicators, including the material intensity, energy intensity, potential chemical risk, and potential environmental impact can be used for a wide range of process systems.[40]

1.5.1 Ethanol Equivalent

The cost of fossil resources may become so high in the future that carbon chemicals and renewable energy will be produced from biomass. A new metrics called the Ethanol Equivalent (EE) was recently defined to evaluate the sustainability of biomass-based chemicals and energy.[25] The EE is "the mass of ethanol, expressed in million tons ethanol equivalent or MtEE, needed to deliver the equivalent amount of energy from a given feedstock using energy equivalency or produce the equivalent amount of mass of a carbon chemical using molar equivalency". Since the required energy for transportation, storage, mixing, heating, cooling, *etc.* of a given process can also be calculated, the EE of a complete production process or even a total technology can be estimated. The EE has been successfully used as a translational tool between fossil- and biomass-based feedstocks, products, processes, and technologies. Since the EE can be produced by different biomass-based technologies, the required mass of a possible biomass feedstock, the size of the corresponding land, and even the required volume of water can be calculated.

EE calculations are based on first-generation corn-based bioethanol technology commercially practiced in the USA in 2008,[41] including the well-known overall equation of photosynthesis for the production of corn (Scheme 1.1),[42] followed by its conversion into bioethanol by fermentation (Scheme 1.2, step 1) using established technologies. The reason for selecting corn as the biomass, bioethanol as the biomass-based feedstock, and the year 2008 as the reference year was to allow the use of reliable data supported by long-term commercial experience.[43] The used and available resources in the USA in 2008 are listed in Table 1.1.

It is important to note that, in a sustainable world, the energy requirements for the production of ethanol from biomass should also be produced from ethanol. Therefore, the real ethanol equivalent (EE_x) should be used, which also includes the use of one unit of bioethanol to produce x units of bioethanol. For example, when one unit of ethanol was used up for fertilizers,

Table 1.1 Used and available resources in the USA in 2008.

	Units	2008
Crude oil consumption	Million tons oil equivalent	884.5
Natural gas consumption	Million tons oil equivalent	600.7
Coal consumption	Million tons oil equivalent	565
Total fossil resource consumption	Million tons oil equivalent	2050.2
	Exajoules	86.1
Bioethanol production	Million tons	27.5
Corn production	Million tons	308
Total cornfield (planted corn)	Million hectares	78.6
Total planted land	Million hectares	132
Total land	Million hectares	916
Actual irrigated land	Million hectares	22

powering the farming machinery, fermentation, distillation *etc.*, the estimated return was 2.3 units of ethanol in 2008. Thus, the abbreviation $EE_{2.3}$ means a 2.3 output/input bioethanol ratio or efficiency.

The $EE_{2.3}$ was first used for the evaluation of biomass-based production of gasoline, jet fuel, and basic carbon chemicals, including ethylene, propylene, toluene, xylenes, styrene, and ethylene oxides. The role of $EE_{2.3}$ was similar to the use of million tons of oil equivalents (Mtoe) for coal, crude oil, and natural gas. Accordingly, the $EE_{2.3}$ values of fossil resource-based energy products and carbon-based basic chemicals are summarized in Table 1.2. The $EE_{2.3}$ values of the reaction enthalpies for the production of carbon-based chemicals are based on the heat of formation of the components of the overall reactions from ethanol to the product.

The replacement of 387 million tons of gasoline used in 2008 would have required the production of 808 million tons of $EE_{2.3}$. The proportional corn demand would have been 2739 million tons, about ten-times greater than the actually harvested 308 million tons of corn. Since almost 31% of the total land or 214% of the farmed land of the US would have been needed to grow such 2739 million tons of corn, the use of biomass-based gasoline cannot be sustainable. The replacement of 73.4 million tons jet-fuel with 152 million tons bioethanol on the other hand would only require about half of the currently farmed land, which could be reduced to a sustainable level by the development of more efficient technologies. It should be emphasized that the sustainability of biofuels depends on 'resource availability' since the rate of natural replacement of the resource (*e.g.*, corn *via* photosynthesis) is about a year and the rate of depletion (*e.g.*, burning the biofuel to produce energy, carbon dioxide, and water) is in the range of days to months. Theoretically, the catalytic dehydration of ethanol to ethylene could be sustainable; 54.91 million tons bioethanol could cover the production of 22.5 million tons of ethylene. The production of 14.8 million

Table 1.2 Required $EE_{2.3}$, corn, and land to supply the total demand of gasoline, jet fuel, and several carbon-based basic chemicals from bioethanol in 2008.

	Used in 2008	$EE_{2.3}$	$EE_{2.3}$ of the reaction enthalpy of production	Sum of $EE_{2.3}$	Required corn to produce sum of $EE_{2.3}$		Required land to produce corn by first-generation bioethanol technology in the US in 2008		
								% of land	
Units	Million tons	Million tons	Million tons	Million tons	Million tons	% of total corn	Million hectares	Total	Farmed
Gasoline	387	808	n/a	808	2739	890	283	30.9	214
Jet fuel	73.4	152	n/a	152	514	167	53.2	5.8	40.7
Ethylene[a]	22.5	53.2	1.71	54.91	187.0	60.77	19.36	2.11	14.81
Propylene[b]	14.8	34.8	0.13	34.93	118.3	38.44	12.25	1.34	9.37
Toluene[c]	3.13	7.86	−0.02	7.84	26.51	8.62	2.75	0.30	2.10
Xylenes[d]	5.41	13.47	−0.1	13.37	45.21	14.69	4.68	0.51	3.58
Styrene[e]	4.1	10.42	0.13	10.55	43.40	14.10	4.49	0.49	3.44
Ethylene oxide[f]	2.9	4.36	0.71	5.07	17.69	5.75	1.83	0.20	1.40

[a]Ethanol → ethylene + H_2O.
[b]3 Ethanol → 2 propylene + $3H_2O$.
[c]7 Ethanol → 2 toluene + $7H_2O$ + $6H_2$.
[d]4 Ethanol → 1 xylene + $4H_2O$ + $3H_2$.
[e]4 Ethanol → 1 styrene + $4H_2O$ + $4H_2$.
[f]2 Ethanol + O_2 → 2 ethylene oxide + $2H_2O$.

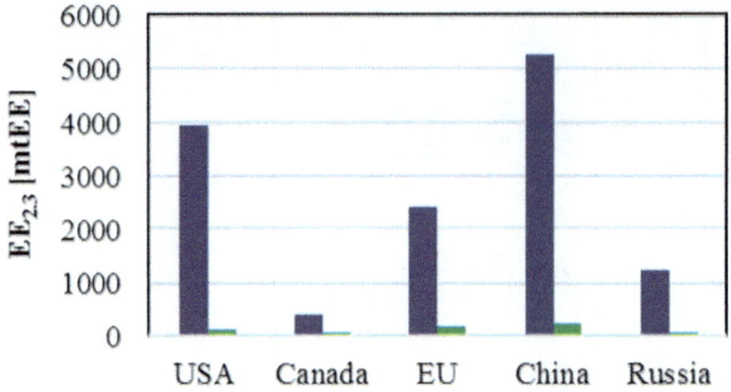

Figure 1.1 $EE_{2.3}$ of total fossil (blue) and renewable (green) energy in 2014. Reproduced from ref. 44 with permission from American Chemical Society, Copyright 2018.

tons propylene, the second largest volume basic chemical, requires 34.9 million tons bioethanol. The volumes of the rest of the basic chemicals in Table 1.2 are much lower.

The sustainability of the replacement of natural gas, crude oil, and coal with corn-based bioethanol was assessed by using ethanol equivalents of fossil resources used in the USA, Canada, EU, China, and Russian Federation in 2008–2014 (Figure 1.1).[44] Based on the $EE_{2.3}$ values, the required volume of corn and the corresponding size of land were calculated, and compared to the actual land used for corn production, which was only enough to replace one sixth of fossil resources in the USA, EU, and China, and practically insufficient in Canada and the Russian Federation. Until the utilization of electricity becomes practical and economical in aviation, biomass-based liquid fuels could be a sustainable alternative.

The 'ethanol equivalent' concept was also used in the development of a novel metric to evaluate the sustainability of carbon-based chemicals.[45] The *sustainability value of resource replacement* (SV_{rep}) and *sustainability value of the fate of waste* (SV_{waste}) were defined and used to establish a *sustainability indicator* (SUS_{ind}) to assess the overall sustainability.

The sustainability value of resource replacement (SV_{rep}) was defined to calculate how much of the necessary resources ($EE_{necessary\ resource}$) consumed during a given time ($t_{consumption}$) can be replaced in a given time ($t_{replacement}$) by using available biomass-based resources ($EE_{available\ resource}$) and the best replacement technologies, which have measurable effectiveness values (ERoE) (eqn (1.4)). While the $EE_{necessary\ resource}$ is the sum of the EE based on carbon-atom equivalency according to the overall yield of production from bioethanol and the EE of the standard enthalpy of the reaction ($EE_{overall\ standard\ enthalpy\ of\ reaction}$), the $EE_{available\ resource}$ is the total amount of bioethanol available on the market. The measurable effectiveness of the reference technology is characterized by its ethanol return of ethanol or ERoE, which is equal to 2.3 indicating that, out of the total 3.3 units of bioethanol, one unit is required for the production of 2.3 units of bioethanol.[46] Thus, the total $EE_{available\ resource}$

has to be multiplied by 2.3/(1 + 2.3) = 0.7 to secure the sustainable production of the resource.

$$SV_{rep} = \frac{\frac{\text{ERoE}}{1+\text{ERoE}} \times \frac{\text{ERoE}}{2.3} \times EE_{\text{necessary resource}} + EE_{\text{secondary resource}}}{t_{\text{replacement}}} \Bigg/ \frac{EE_{\text{necessary resource}}}{t_{\text{consumption}}} \quad (1.4)$$

Ideally, the available resource should be at least equal to the necessary resource and the amount required to produce the necessary resource during the same time period to avoid either resource shortage or over production leading to storage issues. In general, the amount of resources could be increased by using the same chemical as a secondary resource ($EE_{\text{secondary resource}}$) from another biomass-based technology.

The sustainability value of the fate of waste (SV_{waste}) is defined as one when the continuously produced generated waste ($EE_{\text{generated waste}}$) is equal to the continuously treated waste ($EE_{\text{treated waste}}$) in the same time period ($t_{\text{waste generation}} = t_{\text{waste treatment}}$) and no waste is released to the environment ($EE_{\text{untreated waste}} = 0$ and $t_{\text{waste natural decomposition}} = 0$) (eqn (1.5)). The treatment methods could include incineration, chemical and biological treatments, and disposal to official waste storage sites including landfills. All waste released to the environment must be considered as untreated waste.

$$SV_{\text{waste}} = \frac{EE_{\text{treated waste}} + EE_{\text{untreated waste}}}{t_{\text{waste treatment}} + t_{\text{waste natural decomposition}}} \Bigg/ \frac{EE_{\text{generated waste}}}{t_{\text{waste generation}}} \quad (1.5)$$

The following assumptions have been made: (a) if no waste treatment is applied, the associated EE and time must be equal to zero; (b) the time of natural decomposition of untreated waste should include one year (if waste generation time is assumed to be one year) and the time needed to reach the local government's regulation level in the environment based on its natural half-life; (c) in the case of multi-step technologies with the potential to generate waste in each step, the longest time of natural decomposition should be used for the overall process; and (d) in the case of contamination of air, water, and/or soil, the longest time of natural decomposition should be used. If a side or minor product can be utilized in another technology, it should be considered as a secondary resource and not as a waste. Since the sum of $EE_{\text{treated waste}}$ and $EE_{\text{untreated waste}}$ must be equal to $EE_{\text{generated waste}}$, all waste streams are accounted for. By selecting the longest time for the natural decomposition of untreated waste, the persistent component of the waste with the longest half-life will significantly lower SV_{waste}. In other words, prevention of persistent waste formation by the development of efficient processes or integrated valorization of the waste is the preferred path to reach sustainability.

The sustainability indicator (SUS_{ind}) is based on the sustainability values of resource replacement (SV_{rep}) and sustainability values of fate of waste (SV_{waste}) by merging a two-body issue into a single one (eqn (1.6)).[45]

$$SUS_{ind} = \frac{SV_{rep} \times SV_{waste}}{SV_{rep} + SV_{waste}} \tag{1.6}$$

Consequently, sustainability requires that all resources must be replaced (SV_{rep} = 1) and all wastes can be recycled or the remaining parts treated (SV_{waste} = 1) within a specific time. While $SUS_{ind} \geq 0.5$ indicates equal or better than the required sustainability, $SUS_{ind} < 0.5$ denotes an unsustainable situation.

In order to demonstrate the applicability of these sustainability metrics, the SV_{rep}, SV_{waste}, and SUS_{ind} values for ethylene, propylene, toluene, *p*-xylene, styrene, and ethylene oxide were calculated as if they were produced from bioethanol by known reactions and processes (Scheme 1.2) for the consumed quantities in the USA in 2008 (Table 1.3). If these basic chemicals had to be produced from biomass, the total amount of required bioethanol would have been 151.93 million tons, which is significantly over the total bioethanol production of 27.8 million tons. The limited amount of bioethanol was used proportionally for each chemical.

The SV_{rep} values for these chemicals are between 0.128–0.625 (Figure 1.2) and, therefore, none of them are sustainable. The SV_{rep} of styrene is much higher than the others due to the utilization of benzene, produced as a side product in the production of *p*-xylenes as a secondary resource. This result indicates that integration of side and waste products could significantly lower the resource demand and increase the sustainability. The SV_{rep} for these chemicals would be much smaller if we had taken out all the bioethanol burned as gasoline additive.

In order to achieve sustainability, the total available bioethanol volume should be equal or higher than 151.93 million tons. This would require the production of 515 million tons of corn, which would be 1.67 times the 308 million tons of total corn produced in the US in 2008. Subsequently, 53 million hectare land or 40.2% of the total 132 million hectares of farmed land should be used for corn production. Of course, the demand for bioethanol, corn, and arable land would be significantly higher if bioethanol was used as fuel at the same time (since the calculations in this paper used every drop of bioethanol for the production of carbon-based chemicals). One possible approach to secure the required 151.93 million tons bioethanol is to improve the ERoE from 2.3 to at least 18, a formidable challenge on production technologies. Alternatively, the valorization of agricultural residues and food wastes could also contribute significant amounts of bioethanol, not to mention the beneficial effect on waste management in highly populated areas. The $EE_{generated\ waste}$ for each chemical is based on the EE of wastes in each step shown in Scheme 1.2. The SV_{waste} values for these chemicals are between 0.48–0.96 (Figure 1.2) and therefore none of them are sustainable. While waste prevention is the best for the dehydration of ethanol to ethylene (SV_{waste} = 0.96), all the other technologies produce significant amounts of waste.

Table 1.3 Sustainability analysis of basic chemicals in the USA in 2008.

	Basic chemicals						
	Ethylene	Propylene	Toluene	*p*-Xylene	Styrene	Ethylene oxide	Σ Chemicals
Produced amount [mt year^{-1}][a,b]	22.5	14.8	3.13	5.41	4.1	2.9	52.84
$EE_{required\ resource}$ [mt][c]	37.47	47.26	8.76	49.37	5.62	3.40	151.87
$EE_{overall\ standard\ enthalpy\ of\ reaction}$ [mt][d]	1.206	0.177	−0.078	−1.287	0.191	−0.151	0.058
$EE_{neccessary\ resource}$ [mt]	38.68	47.43	8.68	48.08	5.81	3.25	151.93
$EE_{proportional\ necessary\ resource}$ [%][e]	25.46	31.22	5.71	31.65	3.82	2.14	100
$EE_{secondary\ resource}$ of benzene[f] [mt]	0.00	0.00	0.00	0.00	16.21	0.00	16.21
$EE_{proportional\ secondary\ resource}$ [mt]	0.00	0.00	0.00	0.00	2.89	0.00	2.89
$EE_{available\ resource}$ [mt year^{-1}][g]	27.8						
$EE_{proportional\ available\ resource}$ [mt year^{-1}][h]	7.08	8.68	1.59	8.80	1.06	0.59	27.8
$SV_{rep2.3}$	0.128	0.128	0.128	0.128	0.625	0.128	0.147
$EE_{generated\ waste}$ [mt year^{-1}]	0.37	7.62	2.08	12.02	0.17	0.37	22.63
$EE_{recycled\ waste}$ [mt year^{-1}]	0.00	0.00	0.00	0.00	0.00	0.00	0.00
$EE_{treated\ waste}$ [mt year^{-1}]	0.00	2.70	2.02	11.61	0.00	0.00	16.33
$EE_{untreated\ waste}$ [mt]	0.37	4.92	0.06	0.41	0.17	0.37	6.30
$1 + t_{waste\ natural\ decomposition}$ [year]	1.044	1.099	1.044	1.044	1.137	1.601	n/a
SV_{waste}	0.96	0.48	0.49	0.49	0.88	0.62	n/a
SUS_{ind}	0.113	0.101	0.101	0.101	0.365	0.106	n/a

[a] mt = million tons.
[b] The chemical is completely consumed within one year.
[c] Based on carbon-atom molar equivalency of the chemicals and combined yields (conversion × selectivity) of all reactions from ethanol to the chemical.
[d] Overall standard enthalpy of reaction: the sum of each reaction's enthalpy.
[e] $(EE_{neccessary\ resource}/EE_{\Sigma Chemicals}) \times 100$.
[f] The amount of benzene available from xylene and styrene production.
[g] Bioethanol production in the USA in 2018.
[h] $(EE_{neccessary\ resource}/EE_{\Sigma Chemicals}) \times EE_{available\ resource}$.

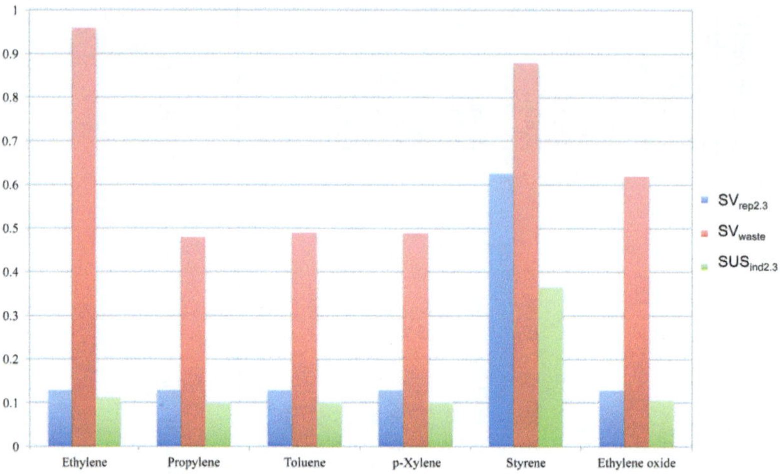

Figure 1.2 Sustainability metrics of basic chemicals in the USA in 2008.

Since the values of SV_{rep} and SV_{waste} are below one, the SUS_{ind} index lies between 0.1 and 0.365 for all these chemicals (Figure 1.2), indicating that none of them are sustainable. Therefore, bioethanol-based carbon products should be labeled 'sustainable' only when the necessary land is available, independently of social and economic changes.

In conclusion, these six basic chemicals cannot be produced sustainably due to the limited availability of bioethanol. However, a ten-fold increase of the volume of bioethanol could easily result in a completely biomass-based production of carbon-based chemicals. The SV_{waste} of the processes is higher than SV_{rep} because of the successful waste prevention and process integration of contemporary petrochemical technologies.

The results suggest that chemical companies could achieve sustainability based on the molar equivalency of the total carbon atoms used by simply owning enough land to produce renewable resources. Of course, the position and size of the lands and the crops produced should be carefully selected to secure sustainable farming and production technologies. Although sustainability may be reached on a molar basis, an additional challenge is to cover the associated energy demand of production processes, storage, and transportation by renewable energy.

1.5.2 Atom Equivalent

The ethanol equivalent-based sustainability metrics are limited to energy and carbon chemicals. In order to expand the applicability of SV_{rep} and SV_{waste} beyond carbon-chemicals, the replacement of EE with the Atom Equivalent (AE) was proposed to represent any elements of the periodic table as constituents of chemicals and materials.[47] The AE is defined as the equivalent mass of one of the constituent elements of a chemical or a material based on molar equivalency. For example, in the case of lithium batteries, the lithium equivalent (LiE) is the mass of all lithium atoms present in a resource, a product, a side-product, or a waste in the form of LiCl, $LiCO_3$, $LiCoO_2$, $LiFePO_4$, $LiMn_2O_4$, Li_2MnO_3, $LiNiMnCoO_2$, $LiNiCoAlO_2$, or $Li_4Ti_5O_{12}$. The energy vector of any given technology must be an intrinsic part of the calculation of SV_{rep}, SV_{waste}, and SUS_{ind}. The calculations will require reference technologies commercially used for years. It should be emphasized that the available resources for any element must be accessible by ecologically sound processes and technologies.

References

1. P. T. Anastas, Origins and Early History of Green Chemistry, Series on Chemistry, Energy and the Environment, in *Advanced Green Chemistry, Part 1: Greener Organic Reactions and Processes*, ed. I. T. Horvath and M. Malacria, World Scientific, Singapore, 2018.
2. R. Carson, *Silent Spring*, Houghton Mifflin Harcourt, Boston, 1968.
3. (a) D. H. Meadows, D. L. Meadows, J. Randers, *et al.*, *The Limits to Growth*, Universe Books, New York, 1972; (b) A. Heaton, *The Chemical Industry*, Chapman & Hall, London, 1994.
4. D. Seyferth, *Organometallics*, 2003, **22**, 2346; D. Seyferth, *Organometallics*, 2003, **22**, 5154.
5. Japanese Ministry of Economy, Trade and Industry, http://www.meti.go.jp/policy/chemical_management/ozone/files/pamplet/panel/08e_basic.pdf, accessed on January 26, 2019.
6. ITOPF, http://www.itopf.com, accessed on January 26, 2019.
7. WHO, Expert meeting to review toxicological aspects of melamine and cyanuric acid, December 1–4, 2008, https://www.who.int/foodsafety/fs_management/conclusions_recommendations.pdf, accessed on January 26, 2019.
8. J. Muller, *CNN*, September 5, 2013, http://edition.cnn.com/2013/09/05/world/asia/china-river-dead-fish/, last accessed on January 26, 2019.
9. G. Gates, J. Ewing, K. Russell and D. Watkins, *New York Times*, March 16, 2017, https://www.nytimes.com/interactive/2015/business/international/vw-diesel-emissions-scandal-explained.html, accessed on January 26, 2019.
10. Pollution Prevention Act of 1990, http://www.epa.gov/p2/pubs/p2policy/act1990.htm, accessed on January 26, 2019.
11. I. T. Horváth, *Chem. Rev.*, 1995, **95**, 1.
12. P. T. Anastas and J. C. Warner, *Green Chemistry: Theory and Practice*, Oxford University Press, Oxford, 1998.
13. T. Malthus, *An Essay on the Principle of Population*, Printed for J. Johnson, in St. Paul's Church-Yard, London, 1798, http://www.esp.org/books/malthus/population/malthus.pdf, accessed on January 26, 2019.

14. Named after Ms. G. H. Brundtland, the Chair of the World Commission on Environment and Development, https://en.wikipedia.org/wiki/Gro_Harlem_Brundtland, accessed on January 26, 2019.
15. World Commission on Environment and Development, *Our Common Future*, Oxford University Press, Oxford, 1987.
16. J. Quoidbach, D. T. Gilbert and T. D. Wilson, *Science*, 2013, **339**, 96.
17. The Emerging Future, LLC, http://theemergingfuture.com/speed-technological-advancement.htm, accessed on January 26, 2019.
18. C. B. Crawford and B. Quinn, *Microplastic Pollutants*, Elsevier, 2016.
19. K.-H. Robèrt, H. Daly, P. Hawken and J. Holmberg, *Int. J. Sustainable Dev. World Ecol.*, 1997, **4**, 79.
20. P. Hinderink, H. J. van der Kooi and J. de Swaan Arons, On the efficiency and sustainability of the process industry, *Green Chem.*, 1999, **1**, G176–G180.
21. J. Dewulf, H. Van Langenhove, J. Mulder, M. M. D. van den Berg, H. J. van der Kooi and J. de Swaan Arons, *Green Chem.*, 2000, **2**, 108.
22. S. K. Sikdar, *AIChE J.*, 2003, **49**, 1928.
23. (a) Resolution adopted by the General Assembly on 25 September 2015 (A/RES/70/1): Transforming our world: the 2030 Agenda for Sustainable Development; (b) Declaration/Our vision/Paragraph 9.
24. Sustainable Development Solutions Network, Indicators and a Monitoring Framework for the Sustainable Development Goals - Launching a data revolution for the SDGs, Report, June 12, 2015, https://sustainabledevelopment.un.org/content/documents/2013150612-FINAL-SDSN-Indicator-Report1.pdf, accessed on January 26, 2019.
25. E. Cséfalvay, G. R. Akien, Q. Long and I. T. Horváth, *Catal. Today*, 2015, **239**, 50.
26. I. T. Horváth, *Chem. Rev.*, 2018, **118**, 369.
27. K. S. Deffeyes, *Beyond Oil: The View from Hubbert's Peak*, Farrar, Straus and Giroux, New York, 2005.
28. M. Wackernagel and W. E. Rees, *Our Ecological Foot-print: Reducing Human Impact on the Earth*, Gabriola Press, New Society Publishing, B.C, 1996.
29. B. M. Trost, *Science*, 1991, **254**, 1471.
30. (a) R. A. Sheldon, 13th Seminar on Frontier Technology, Fuji Hakone, Japan, July 23–26, 1991; (b) 3rd International Conference on Catalytic Chemistry for Global Environment, Hokkaido University, Japan, July 27, 1991; (c) Interview with R. A. Sheldon, *Chem. Eng. Prog.*, 1991, **87**, 11.
31. R. A. Sheldon, *Chem. Ind.*, 1992, 903.
32. A.-C. Bédard, A. R. Longstreet, J. Britton, Y. Wang, H. Moriguchi, R. W. Hicklin, W. H. Green and T. F. Jamison, *Bioorg. Med. Chem.*, 2017, **25**, 6233.
33. K. Hellgardta and K. K. Hii, Continuous Flow Technologies in the Development of "Green" Organic Reactions and Processes, in *Advanced Green Chemistry, Part 1: Greener Organic Reactions and Processes*, ed. I. T. Horvath and M. Malacria, Series on Chemistry, Energy and the Environment, World Scientific, Singapore, 2018.

34. M. Tobiszewski, M. Marć, A. Gałuszka and J. Namieśnik, *Molecules*, 2015, **20**, 10928.
35. R. A. Sheldon and J. P. M. Sanders, *Catal. Today*, 2015, **239**, 3.
36. C. Böhringer and P. E. P. Jochem, *Ecol. Econ.*, 2007, **63**, 1.
37. P. Hinderink, H. J. van der Kooi and J. de Swaan Arons, *Green Chem.*, 1999, **1**, G176.
38. S. Lems, H. J. van der Kooi and J. de Swaan Arons, *Green Chem.*, 2002, **4**, 308.
39. J. G. Moretz-Sohn Monteiro, O. de Queiroz Fernandes Araujo and J. L. de Medeiros, *Clean Technol. Environ. Policy*, 2009, **11**, 209.
40. A. A. Martins, T. M. Mata, C. A. V. Costa and S. K. Sikdar, *Ind. Eng. Chem. Res.*, 2007, **46**, 2962.
41. H. Shapouri, P. W. Gallagher, W. Nefstead, R. Schwartz, S. Noe and R. Conway, *Energy Balance for the Corn-Ethanol Industry*, USDA, 2008, Agricultural Economic Report Number 846, June, 2010.
42. D. O. Hall and K. Rao, *Photosynthesis*, Cambridge University Press, Cambridge, 2009.
43. C. Wyman, *Handbook on Bioethanol: Production and Utilization*, Applied Energy Technology Series, CRC Press, Taylor & Francis Group, 1996.
44. E. Cséfalvay and I. T. Horváth, *ACS Sustainable Chem. Eng.*, 2018, **6**, 8868.
45. I. T. Horváth, E. Cséfalvay, L. T. Mika and M. Debreczeni, *ACS Sustainable Chem. Eng.*, 2017, **5**, 2734.
46. D. J. Murphy and C. A. S. Hall, *Ann. N. Y. Acad. Sci.*, 2010, **1185**, 102.
47. I. T. Horvath, Atom Equivalents Based Metrics to Bridge Green Chemistry and Sustainability, 7th Asia-Oceania Conference on Green and Sustainable Chemistry (AOC7-GSC), Singapore, November 19–21, 2018.

CHAPTER 2

One-pot Organic Reactions

WEI ZHANG*

University of Massachusetts Boston, Department of Chemistry,
100 Morrissey Boulevard, Boston, MA 02125, USA
*E-mail: wei2.zhang@umb.edu

2.1 Introduction

Green chemistry aims to maintain the benefits of chemical products and chemistry processes and minimize their side-effect to the environment and human health. It is a new theory and practice for chemistry and chemical engineering. The safety and toxicity of chemical products, and sustainability and efficiency of chemical processes are the major concerns of the 12 principles of green chemistry.[1] Since organic synthesis plays a critical role in the preparation of functionalized products with special physical, chemical, and biological utilities, it is an important part of green synthesis, covering the topics of waste prevention, renewable feedstock, catalysis, alternative solvents, less toxic and safer reaction conditions, energy efficiency, and synthetic efficiency.[2] Among them, synthetic efficiency is particularly important because it is not only driven by the green chemistry considerations, but may also have huge economic benefits. Presented in this chapter are one-pot organic reactions as important methods to achieve high synthetic efficiency.

Conducting reactions in a single vessel (one-pot) to avoid the need for intermediate isolation has intrinsic efficiency in terms of pot, atom, and step

Green Chemistry Series No. 61
Green Synthetic Processes and Procedures
Edited by Roberto Ballini

Published by the Royal Society of Chemistry, www.rsc.org

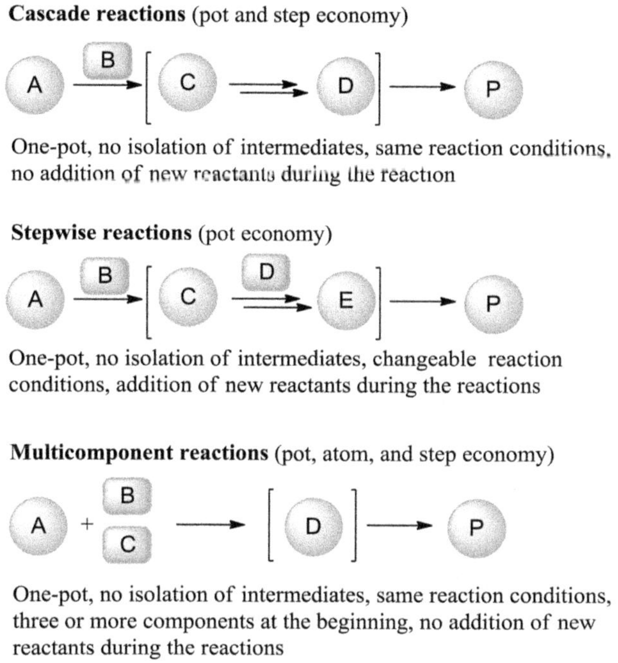

Scheme 2.1 The three types of one-pot reactions.

economy (PASE).[3] One-pot reactions offer the advantages of simple operation and reduced time, energy consumption, and amount of waste. Three one-pot reactions are here discussed: (1) cascade reactions, (2) stepwise reactions, and (3) multicomponent reactions (Scheme 2.1). The efficiency of such one-pot reactions can be evaluated by performing green chemistry metrics analysis.

Cascade reactions are also called domino or tandem reactions,[4–6] in which, after initiation of the reaction, sequential transformations proceed spontaneously under the same conditions. The intermediates are usually not isolable but might be trapped, such as with the radical trapping agents used in radical reactions. Their great operational simplicity allows cascade reactions to present superior pot and step economy. However, the reactions have limited scope because they are only applicable to specially designed substrates.

In *stepwise reactions*, multistep reactions are carried out in the same vessel. Each reaction step generates a new intermediate, which may but not necessarily be isolated. The reaction process can be closely monitored, and the reaction conditions can be changed at the different steps by addition of new reagents, changing the temperature, and even switching the solvent. This type of reactions present good pot economy and operational flexibility for the incorporation of reaction conditions and the design of new reaction sequences.

In *multicomponent reactions* (MCRs), three or more reactants are added together at the beginning of the reaction.[7–10] Most components of the reactants end up in the structure of the product. MCRs are carried out under the same conditions and no additional reactants or catalysts are introduced during the process. MCRs are convergent and offer good atom, step, and pot economy. However, only a dozen or so MCRs have been reported in the literature.

Over the years, several monographies and many review articles have been published on one-pot syntheses, especially on MCRs.[4–10] This chapter is intended to offer clear definitions of stepwise, cascade, and multicomponent reactions. Since there is a huge body of publications on one-pot reactions, only selective papers from recent years are highlighted in this chapter to exemplify different one-pot reactions and their utilities.

2.2 Cascade Reactions

In cascade reactions, once the initial reaction starts, the sequential transformation proceeds spontaneously under the same conditions. No additional reagents and catalysts are introduced during the reactions. Depending on the substrates and conditions, the reactions proceed *via* an ionic, radical, or pericyclic mechanism. The mechanism can alternate during the reactions, but the reaction conditions are not supposed to be changed.

2.2.1 Cascade Ionic Reactions

Ionic reactions initiated by heterolytic bond cleavage are an important class of organic reactions. Since many intermediates are involved in cascade reactions, challenges such as functional group tolerance, regio- and stereoselectivities, and competition between kinetic and thermodynamic processes have to be suitably addressed. In this section, three examples of nucleophilic addition-initiated cascade reactions are described.

Cationic cascade reactions initiated by nucleophilic addition for the synthesis of functionalized carbazoles **3** are shown in Scheme 2.2.[11] In this sequence, vinyl-substituted α-hydroxy silyl enol ether **1** is readily ionized with catalytic camphorsulfonic acid (CSA) to form 2-(silyloxy)pentadienyl cations **4a** and **4b** as *Z*- and *E*-conformational isomers. The addition of substituted indole **2** to cation **4a** gives **5**, followed by acidic desilyation of TBSO affording cationic intermediate **6**. Cyclization of **6** and dehydrative aromatization furnish carbazole products **3**. In this reaction process, double-bond isomerization to form (*Z*)-silyldienol ethers **4a** and regioselective addition of indoles to form **5** are the critical steps for the cascade sequence. Over a dozen analogs with different R^1–R^4 groups and ring size were synthesized in 48–94% yield.

Another cationic cascade reaction initiated by nucleophilic addition of CF_3^- to carbonyls has been introduced for the synthesis of trifluoromethylated

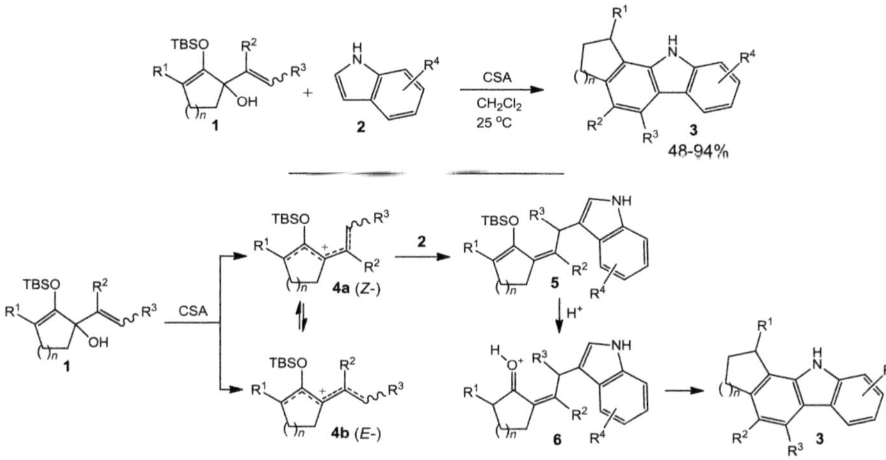

Scheme 2.2 Nucleophilic addition-based cascade reactions for carbazoles **3**.

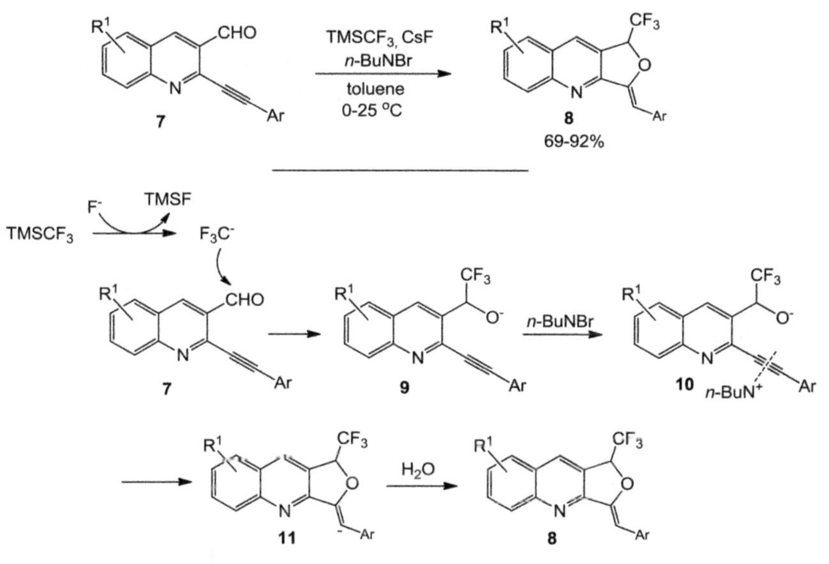

Scheme 2.3 Nucleophilic addition-based cascade reactions for 1,3-dihydrofuro [3,4-*b*]quinolines **8**.

1,3-dihydrofuro[3,4-*b*]quinolines **8** (Scheme 2.3).[12] In this reaction process, the CF_3^- generated from $TMSCF_3$ adds to the carbonyl of 2-arylalkynylquinoline aldehyde **7** to form intermediate **9**. Under catalysis of *n*-Bu_4NBr, intermediate **10** undergoes 5-*exo*-dig cyclization to form **11** and then product **8** after protonation.

Base-promoted reactions of 2-fluorobenzaldehyde with heterocyclic ketene aminals for the synthesis of 1,3-diazaheterocycle-fused [1,2-*a*]quinolines **14** are shown in Scheme 2.4.[13] The α-position of ketene aminal **12** with

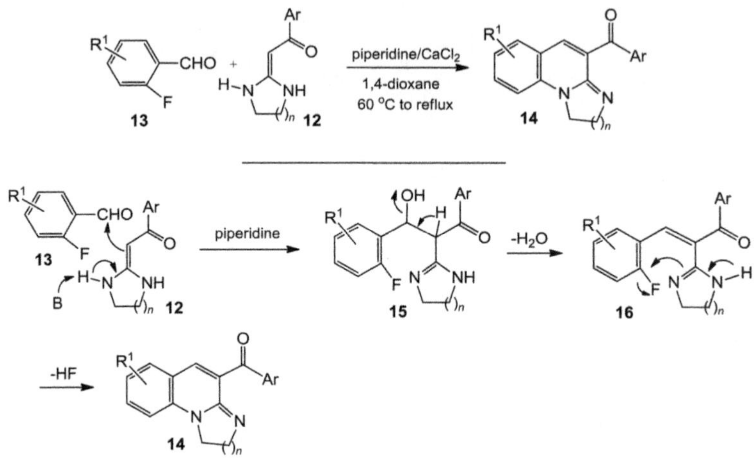

Scheme 2.4 Nucleophilic addition-based cascade reactions for 1,3-dihydrofuro-[3,4-*b*]quinolines **14**.

conjugated ketocarbonyl and two electron-donating diamino groups serve as a nucleophile to react with the formyl group of 2-fluorobenzaldehyde **13** to form intermediate **15** *via* aza-ene addition. Dehydrative aromatization of **15** affords **16**, which then undergoes cyclative nucleophilic substitution to produce product **14**.

2.2.2 Cascade Transition Metal-catalyzed Reactions

In transition metal-catalyzed cascade reactions, once a reaction is initiated with a catalyst, the remaining steps are processed under the same conditions until formation of the final product. Highlighted in this section are two examples of palladium-catalyzed cascade reactions.

The Pd-catalyzed cascade reaction for the synthesis of substituted cyclooctatetraenes **18** is shown in Scheme 2.5.[14] The Pd species is inserted in the C–Br bond of **17** forming **19** for further 4-*exo*-dig cyclocarbopalladation to afford intermediate **20**. It then undergoes a 5-*exo*-dig cyclocarbopalladation to give **21**, followed by an 8π-electrocyclization pathway to form **22**. A 1,3-π-allyl Pd shift occurs to give **23**, followed by a Stille cross-coupling to give product **18**. The key step in this unique process is the 8π-electrocyclization of ene triyne **21**.

Pd-catalyzed cascade Heck–Sonogashira reactions in water have been developed for the synthesis of 3,3-disubstituted oxindoles **25** (Scheme 2.6).[15] The reaction process is initiated with the oxidative insertion of Pd(0) to the carbon–iodine bond of acrylamide **24** to form intermediate **26**. The intramolecular Heck reaction of **29** affords primary alkylpalladium complex **27**. In the following transformations, one of the ligands in complex **27** is replaced

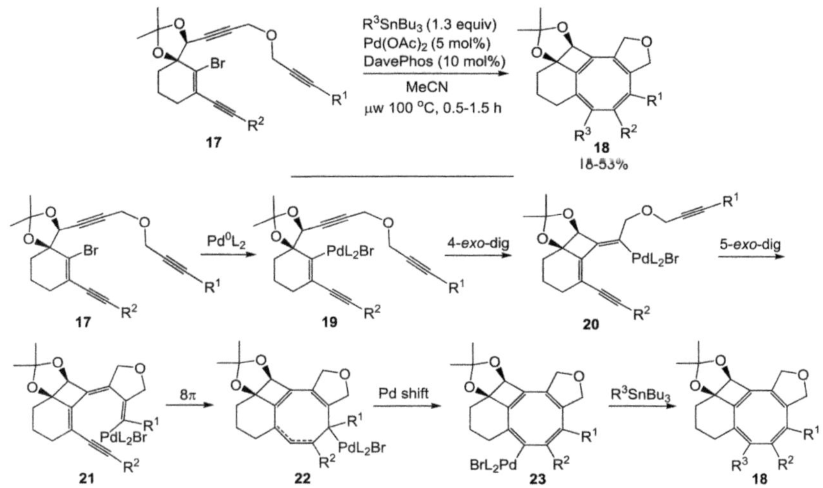

Scheme 2.5 Pd-catalyzed cascade reaction for cyclooctatetraenes **18**.

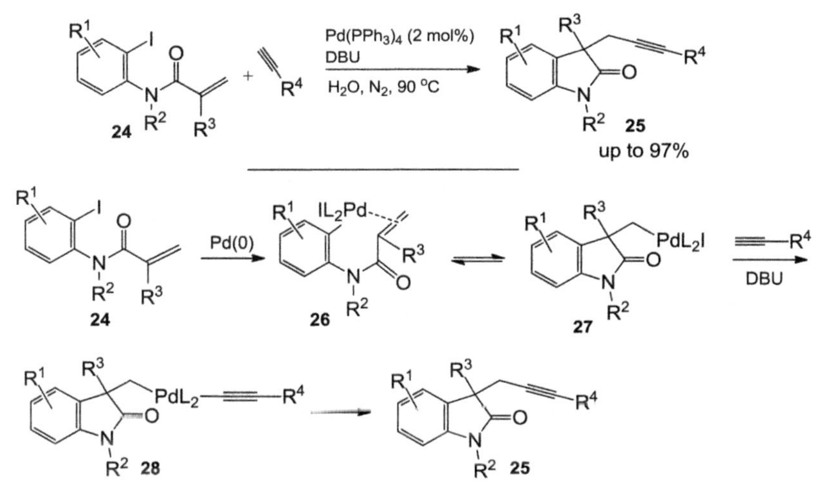

Scheme 2.6 Pd-catalyzed cascade reactions for 3,3-disubstituted oxindoles **25**.

by an alkyne, followed by dehydroiodination in the presence of DBU to give intermediate **28**. The reductive elimination of **28** affords the desired product **25** and also regenerates the Pd(0) catalyst.

2.2.3 Cascade Radical Reactions

Radical cascade reactions are well known in the literature. This section describes three examples of different methods to initiate radical cascade reactions.

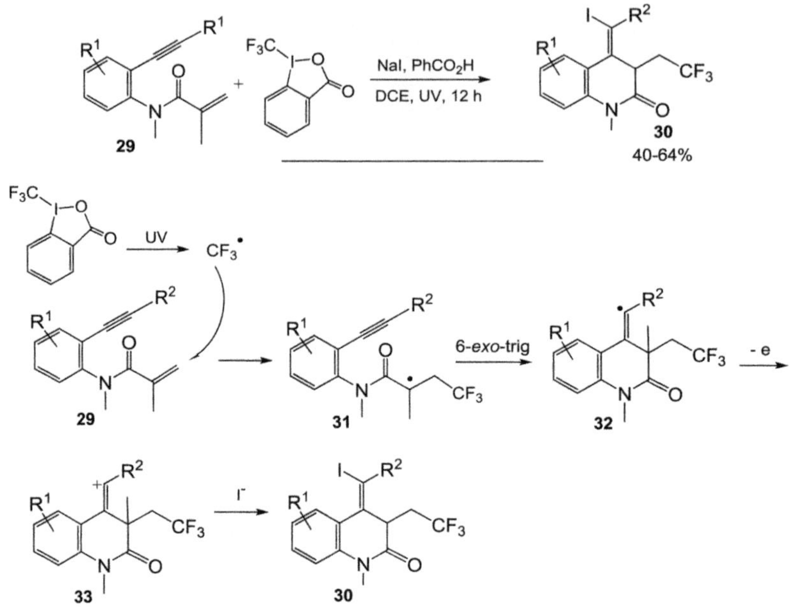

Scheme 2.7 Radical-initiated cyclative difunctionalization for 3,4-dihydroquinolinones **30**.

The photolytic generation of CF_3 radical from the Togni reagent for cyclative difunctionalization during the synthesis of 3,4-dihydroquinolinones **30** is shown in Scheme 2.7.[16] At the first step, the CF_3 radical adds to the double bond of *N*-methyl-*N*-(2-(phenylethynyl)phenyl)methacrylamide **29** to form **31**. 6-*Exo*-trig cyclization of radical **31**, followed by oxidation of radical **32**, furnishes cation **33**. Finally, iodination of **33** affords trifluoromethylated 3,4-dihydroquinolinone products **30**.

Other than the generation of π-complexes mentioned in the previous section, transition metals can also be used to generate radicals. Ni(II)-catalyzed cascade radical reactions of alkylidenecyclopropanes **34** for the synthesis of benzo[*b*]naphtho[1,2-*d*]azepines **35** have been reported in the literature (Scheme 2.8).[17] In this reaction process, LNi(0) generated from a Ni catalyst and 1,10-phen ligand reacts with **34** to form the initial radical **36**, which is cyclized into alkylidenecyclopropane to form radical **37** and then cyclopropane ring-opened radical **38**. Cyclization of **38** on the benzene ring, followed by single electron transfer and deprotonative aromatization, provides the final products **35**.

Increasing numbers of publications exist on photoredox-based radical reactions. Shown in Scheme 2.9 is an example of the synthesis of fused pyran derivatives **41**.[18] An acyl radical, generated from acyl chloride under photoredox conditions using Ir^{III} (ppy)$_3$, adds to 1,7-enyne **40** to form radical **42**,

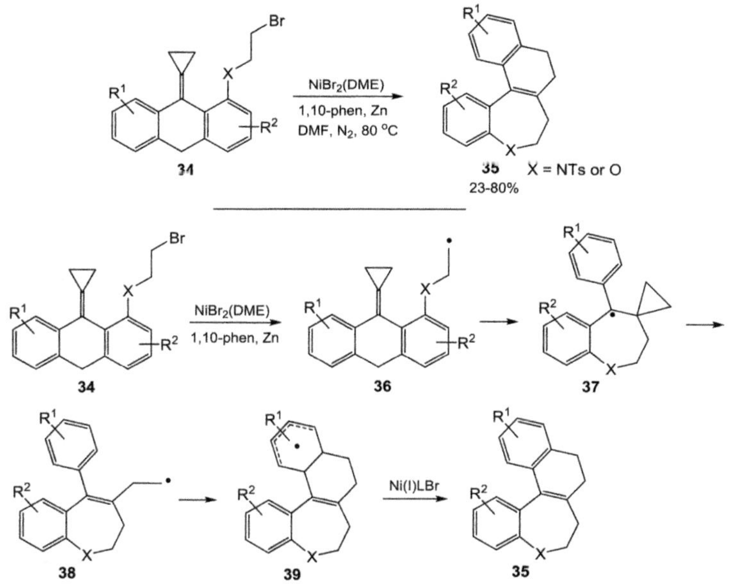

Scheme 2.8 Radical cascade reactions for benzo[*b*]naphtho[1,2-*d*]azepines **35**.

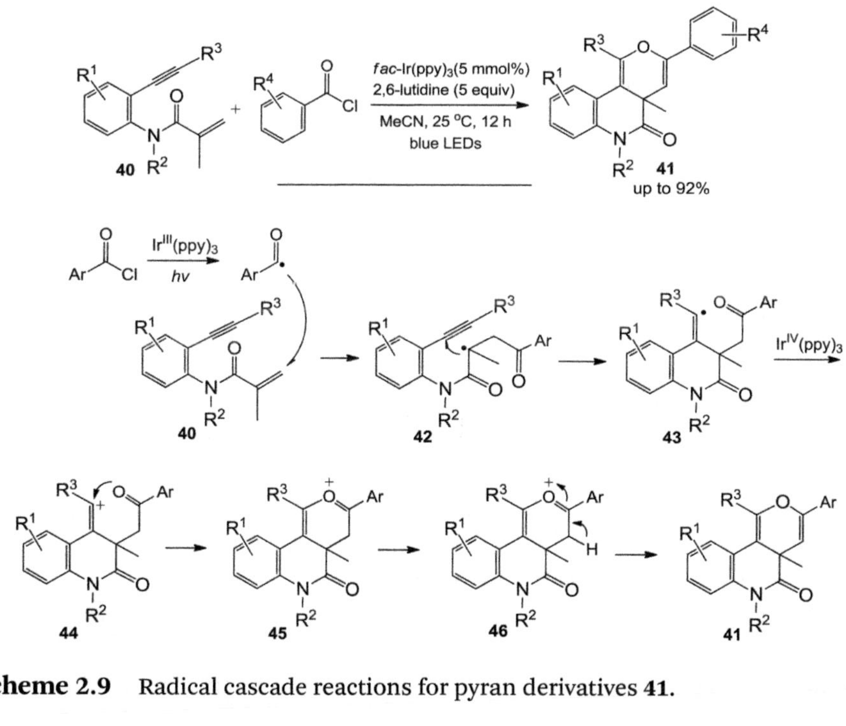

Scheme 2.9 Radical cascade reactions for pyran derivatives **41**.

which then cyclizes on the triple bond to form **43**. Single electron oxidation gives cation **44**, followed by cyclization, affording **45** and finally leading to products **41** after deprotonation.

2.2.4 Cascade Pericyclic Reactions

Pericyclic reactions are powerful to prepare complex molecular structures. The combination of two or more pericyclic reactions in one-pot makes the method even more powerful. However, pericyclic reactions demand high stereo chemistry requirements on the substrates. Highlighted in this section are two examples of pericyclic cascade reactions for the construction of polycyclic scaffolds.

Using a carefully designed substrate of tertiary propargyl vinyl ethers (PVEs) **47**, a Claisen rearrangement and intramolecular Diels–Alder reaction were integrated in the synthesis of compounds **48** bearing a 2-oxabicyclo[2.2.2]oct-5-ene core (Scheme 2.10).[19] The cascade sequence starts with the Claisen rearrangement of **47** to form β-alleneal **49**, which undergoes imidazole-catalyzed 1,3-H transfer to afford 1-oxatriene **50**. Oxa-6π-electrocyclization of **50** affords 2*H*-pyran **51**, which is then subjected to an intramolecular Diels–Alder reaction to give product **48**. This highly efficient one-pot reaction sequence provides access to polycyclic scaffolds through the formation of three new rings with six stereogenic centers.

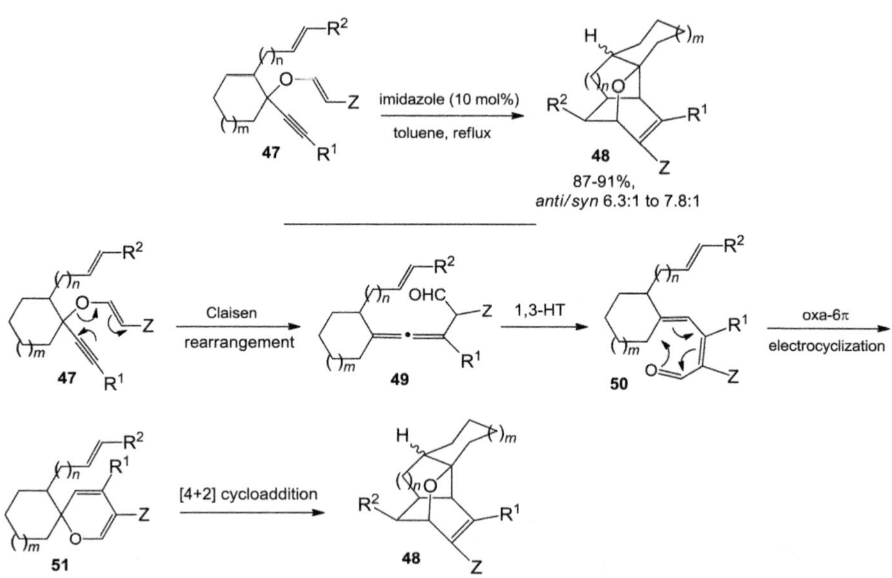

Scheme 2.10 Pericyclic cascade reactions for polycyclic compounds **48**.

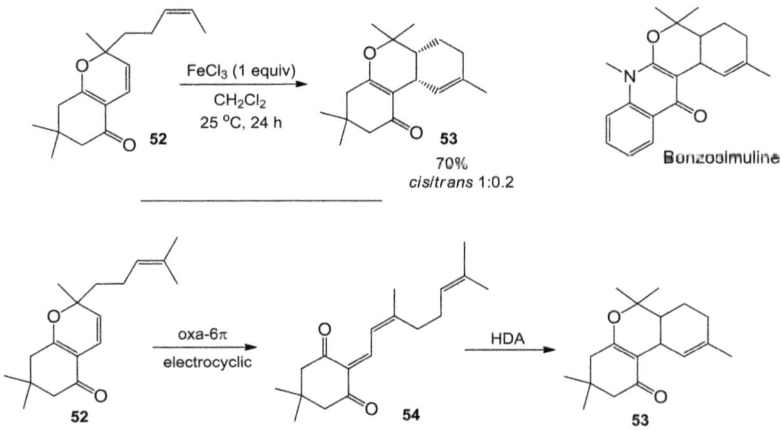

Scheme 2.11 Pericyclic cascade reactions for 2*H*-pyran scaffold **53**.

Shown in Scheme 2.11 is another example of pericyclic cascade reactions for the synthesis of 2*H*-pyran scaffold **53**.[20] The reaction sequence involves an oxa-6π electrocyclic ring opening for intermediate **54**, which then undergoes a hetero-Diels–Alder reaction to give 2*H*-pyran scaffold **53**. An analog prepared by this method was used for the synthesis of alkaloid benzosimuline.

2.3 Stepwise Reactions

Different from cascade reactions, which are conducted under the same reaction conditions, one-pot stepwise reactions can be carried under different conditions and are thus more flexible for the design of new reaction sequences in the synthesis of complicated structures.

2.3.1 Stepwise Non-catalytic Reactions

A stepwise reaction sequence comprising the [3 + 2] cycloaddition of azomethine ylide, Staudinger–aza-Wittig reaction of an azide, formation of isothiocyanate, and cyclization of an amine to isothiocyanate has been reported for the synthesis of dihydroquinazolinethiones **56** (Scheme 2.12).[21] The initial three-component [3 + 2] cycloaddition involving 2-azidobenzaldehyde **55** gives diastereoselective adduct **57**, which is subsequently treated with PPh_3 for the Staudinger reaction to form iminophosphorane **58** and then converted into isothiocyanate **59** *via* the aza-Wittig reaction with CS_2. In the last step, cyclization of the amine leads to the formation of dihydroquinazolinethione **56**. During the four-step reaction sequence, new reagents are added and the heating method is also changed.

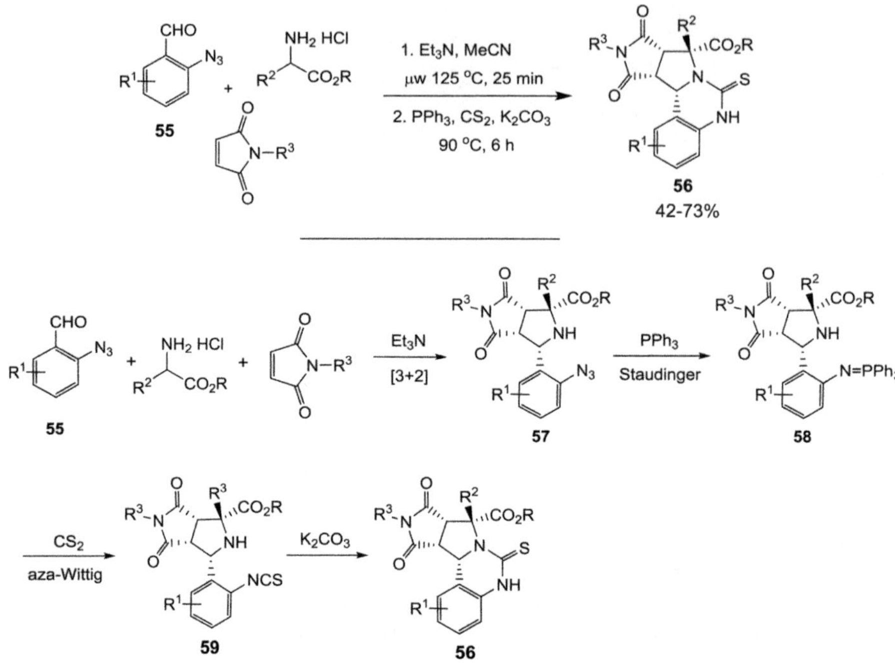

55 56 57 58 59

Scheme 2.12 One-pot stepwise reaction for dihydroquinazolinethiones **56**.

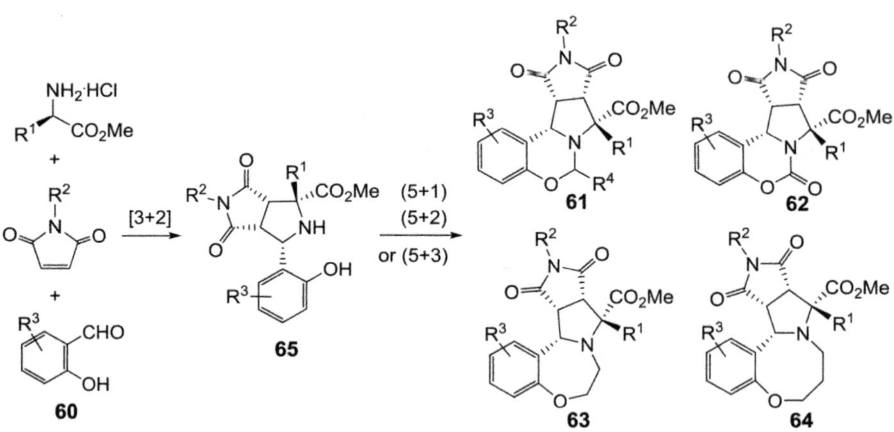

60 65 61 62 63 64

Scheme 2.13 Modular synthesis of **61–64** *via* [3 + 2] cycloaddition and (5 + *n*) annulation.

Shown in Scheme 2.13 is a modular approach for the synthesis of dihydrobenzoxazines **61** and **62**, tetrahydrobenzoxazepines **63**, and tetrahydrobenzoxazocines **64** through one-pot reactions involving a [3 + 2] cycloaddition for **65** and then (5 + *n*) annulation of **65** with bifunctional reagents for double nucleophilic substitutions.[22] The one-pot reactions of the [3 + 2] cycloaddition and (5 + 2) annulation sequence for tetrahydrobenzoxazepines **63** are shown in Scheme 2.14. Upon completion of the [3 + 2] cycloaddition involving 2-hydroxy benzaldehyde **60**, adduct **65** is reacted with 1,2-dibromoethane for (5 + 2) annulation to form **63**.

2.3.2 Stepwise Catalytic Reactions

Through careful design of a catalytic system, one-pot and stepwise reactions can be accomplished to make complicated scaffolds in a stereoselective fashion. Highlighted in this section are two examples of one-pot asymmetric syntheses promoted by recyclable organocatalysts. The bifunctional cinchona alkaloid/thiourea catalyst **cat-1** bearing a fluorous tag (C_8F_{17}) can be readily isolated from the reaction mixture by fluorous solid-phase extraction (F-SPE).

The one-pot asymmetric synthesis of spiro-γ-lactam-oxindole **67** through sequential thiol-Michael/Mannich/lactamization reactions was conducted in two operation steps (Scheme 2.15).[23] Isatin derivative **66** reacts with arylthiol

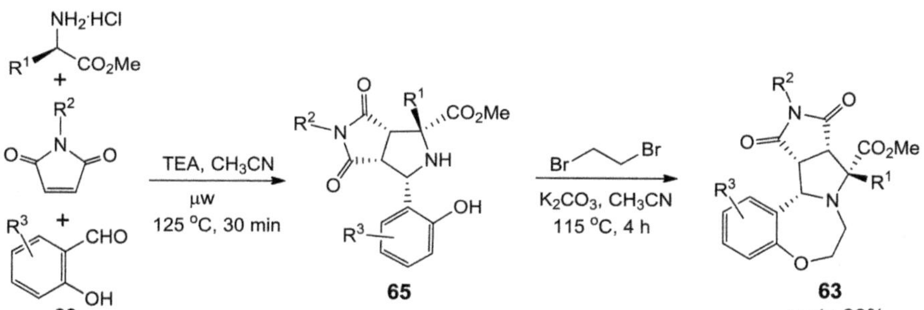

Scheme 2.14 One-pot synthesis of tetrahydrobenzoxazepines **63**.

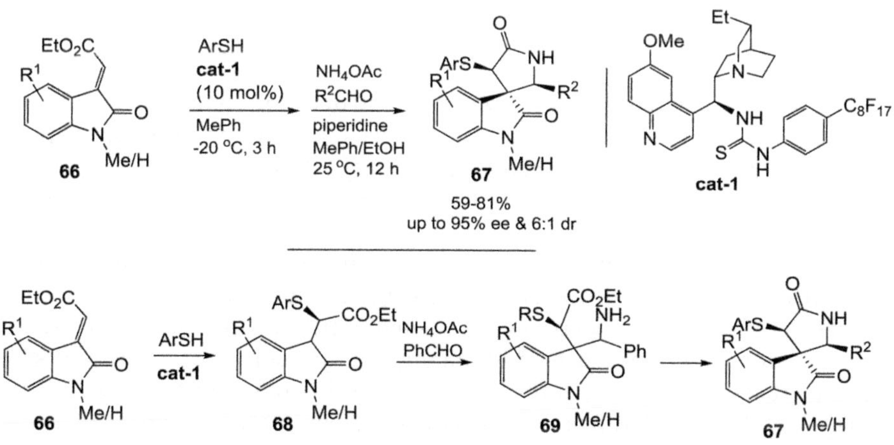

Scheme 2.15 Organocatalytic synthesis of spiro-γ-lactam-oxindoles **67**.

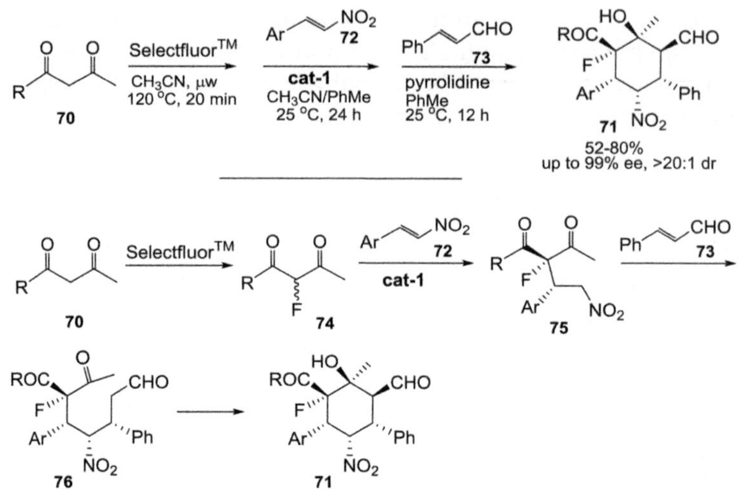

Scheme 2.16 One-pot organocatalytic synthesis of 2-fluorocyclohexanols **71**.

under the catalysis of **cat-1** for thiol Michael addition to form **68**. The addition of NH_4OAc and aldehyde to the reaction mixture for the Mannich reaction affords **69**, which is then cyclized to form spiro product **67** bearing three stereocenters.

In the second example of one-pot organocatalysis reactions, a sequence of fluorination/Michael/Michael/aldol reactions was developed for the synthesis of fully-substituted cyclohexanols **71** bearing six stereocenters with excellent ee and dr values (Scheme 2.16).[24] 1,3-Dicarbonyl **70** is treated with Selectfluor™ to form *a*-fluorinated intermediate **74**. Under the catalysis of **cat-1**, Michael reaction of **74** with **72** affords **75**, which is used for another Michael addition with **73** to afford **76**. Intramolecular aldol reaction of **76** gives 2-fluorocyclohexanol **71**.

2.3.3 Stepwise Cycloaddition Reactions

Cycloadditions are powerful methods to construct functionalized cyclic compounds. The combination of two or more cycloadditions in one-pot is powerful to access highly condensed ring systems. Highlighted in this section are two examples of [3 + 2] cycloaddition-initiated reaction sequences for the synthesis of polycyclic scaffolds.

1,3-Dipolar [3 + 2] cycloaddition of azomethine ylides is a good method for the synthesis of pyrrolidine-containing ring systems. Scheme 2.17 shows a double [3 + 2] cycloaddition sequence involving the reaction of unsubstituted *a*-aminoester **77** with aldehyde and maleimide for the diastereoselective synthesis of fused-pyrrolidine compound **78**.[25] In this reaction process, azomethine ylide **79** generated from the reaction of **77** with an aldehyde undergoes first a [3 + 2] cycloaddition reaction with maleimide to form **80**. It is then

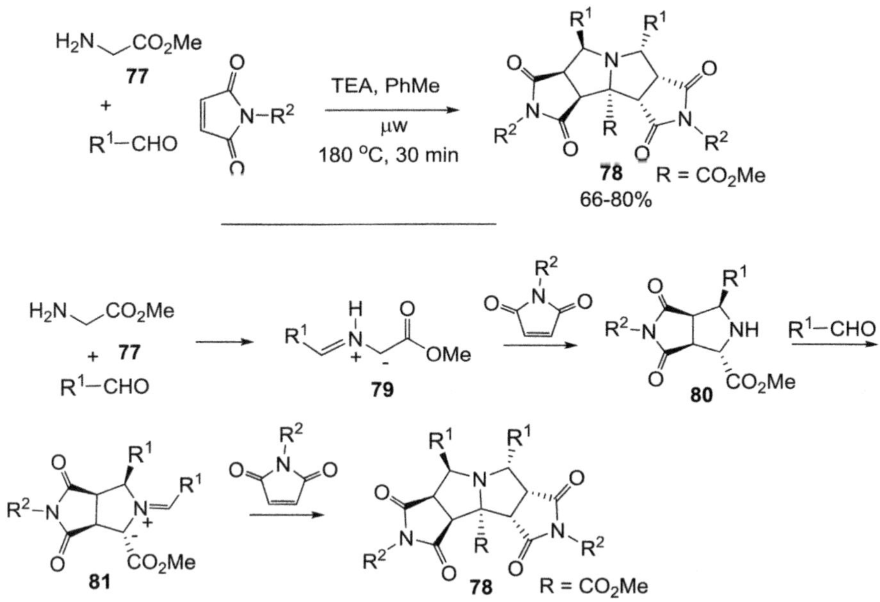

Scheme 2.17 Double [3 + 2] cycloaddition for the synthesis of fused pyrrolidine compounds **78**.

reacted with another equivalent of aldehyde to generate azomethine ylide **81** for a second [3 + 2] cycloaddition reaction to diastereoselectively furnish products **78**.

Another pyrrolidine-containing polycyclic system **82** was synthesized by sequential intermolecular and intramolecular [3 + 2] cycloadditions involving propargyl aldehyde **83** (Scheme 2.18). After the intermolecular [3 + 2] cycloaddition of *a*-aminoester **77** with aldehyde and maleimide to form intermediate **80**, propargyl aldehyde **83** is added for the generation of azomethine ylide **84** for intramolecular [3 + 2] cycloaddition to form products **82** diastereoselectively.

2.4 Multicomponent Reactions (MCRs)

Multicomponent reactions (MCRs) employ three or more reactants, and all the reactants are introduced together at the beginning of the reaction. MCRs are pot, atom, and step economic in the generation of complex molecular structures. MCRs have inter- and intramolecular versions. For example, the intermolecular [3 + 2] cycloaddition of azomethine ylides involves three components, while the intramolecular version of this reaction has only two components (Scheme 2.19). Thus, the intermolecular one is denoted as 3C3C-MCR (three-center and three-component reaction), while the intramolecular one is referred to as 3C2C-MCR (three-center and two-component reaction).

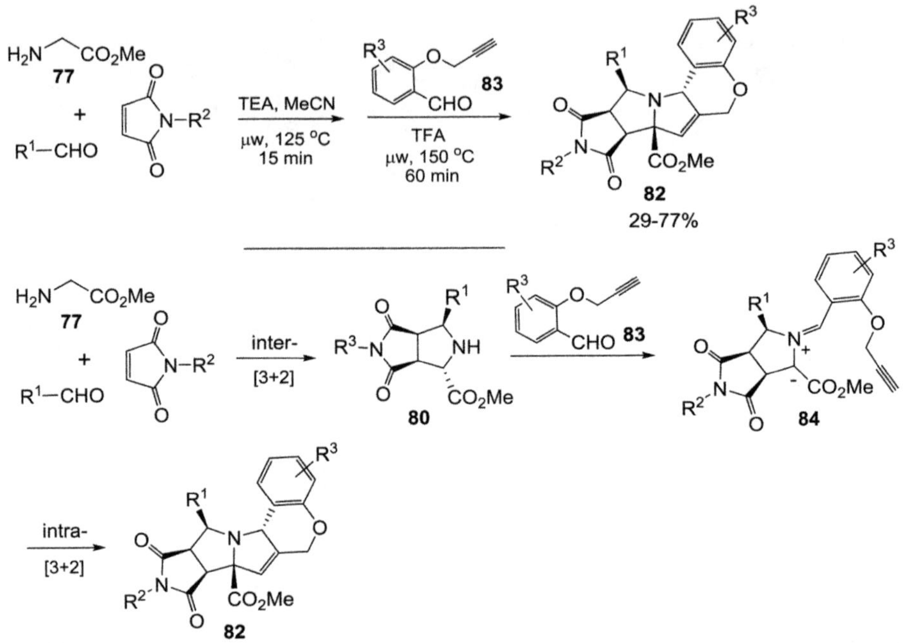

Scheme 2.18 Inter- and intramolecular [3 + 2] cycloadditions for polycyclic systems **82.**

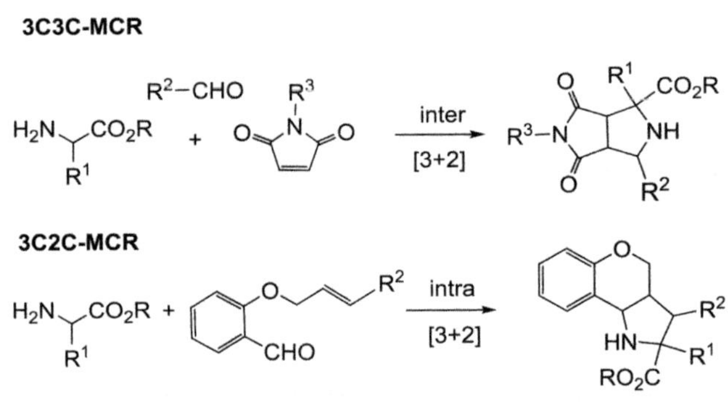

Scheme 2.19 Inter- and intramolecular [3 + 2] cycloadditions.

Reactants such as isonitriles/isocyanides, aldehydes, amines, and carboxylic acids are commonly used for MCRs. Several reactions, including Ugi, Biginelli, Huisgen, Mannich, GBB, and Petasis protocols, are well-known MCRs.[7–10] Many reports exist on the applications of these reactions in the literature. Only selective examples of high-order MCRs and integrated MCRs are discussed in this section.

2.4.1 High-order MCRs

MCRs with five or more reaction centers are called high-order MCRs.[26] They are more synthetically efficient, but their number is very limited because it is more difficult to incorporate such a large number of centers/components in an orderly manner to form desirable structures. The combination of two MCRs with greater than five components should be called consecutive MCRs instead of high-order MCRs.

An Ugi-5CR involving an isonitrile, aldehyde, amine, carboxylic acid, and thiol has been developed for the synthesis of methionine-type peptides **85** (Scheme 2.20).[27] The reaction sequence begins with a thiol Michael addition to the aldehyde to form **86**, which then reacts with the amine to give **87**, with the acid to give **88**, with the isonitrile to give **89**, and then with the acid anion to form **90**. An acyl shift from the ester to form an amide gives the final products **85**. It is nice to see a mixture of five components reacting in such an orderly manner to afford thiolated dipeptide products in this 5C5C-MCR.

A 5C4C-MCR involving an amino acid, isonitrile, ketone, and alcohol for the synthesis of dipeptides **91** has also been developed (Scheme 2.21).[28] The reaction is diastereoselective, and the condensation product **91** was treated with BF_3 for the synthesis of 2,6-diketopiperazines **92**.

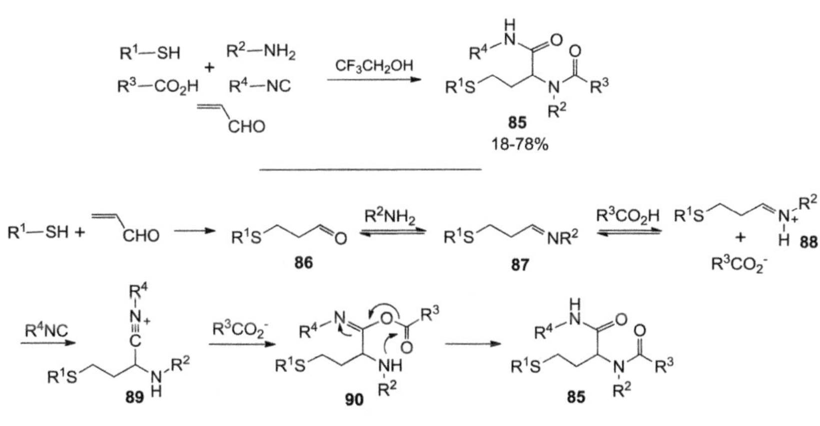

Scheme 2.20 5C5C-MCR for the synthesis of methionine-type peptides **85**.

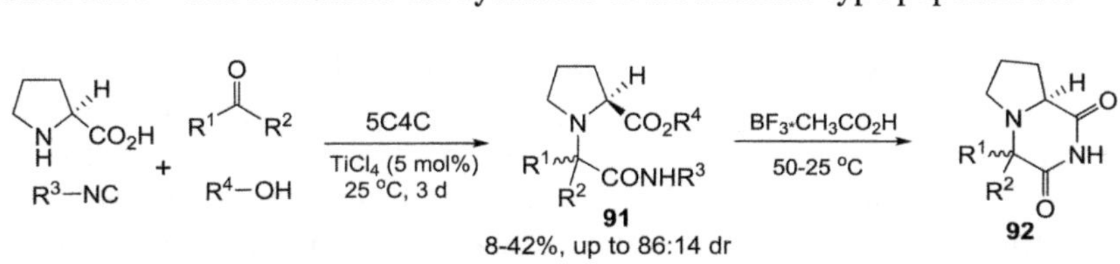

Scheme 2.21 5C4C-MCR for the synthesis of dipeptides **91** and 2,6-diketopiperazines **92**.

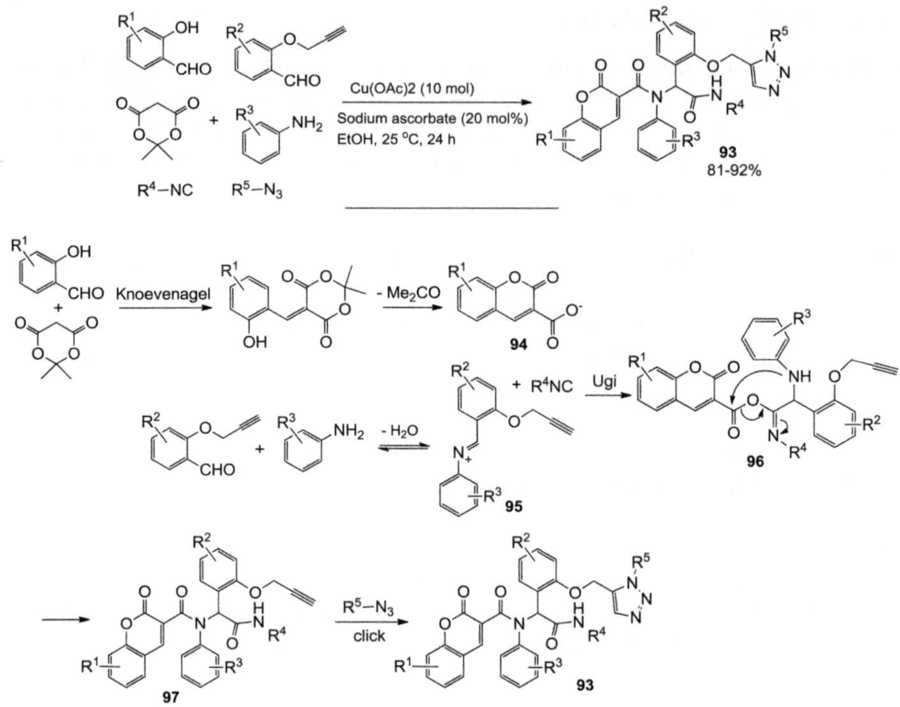

Scheme 2.22 Sequential Knoevenagel/Ugi/click reactions for compounds **93**.

2.4.2 Integrated MCRs

Since the number of reported MCRs is limited, the integration of MCR with other reactions is a good strategy to synthesize novel and complex structures. A sequence of Knoevenagel/Ugi/click reactions involving six components was developed for the synthesis of triazole- and coumarin-containing carboxamides **93** (Scheme 2.22).[29] The Knoevenagel reaction of 2-hydroxy aldehyde and Meldrum's acid gives intermediate **94.** Intermediate **95** is generated from the reaction of propargyl aldehyde with aniline. The mixture of these two intermediates through an Ugi reaction affords **97** after the rearrangement of **96.** The reaction sequence finishes with a click reaction of **97** with an azide to give products **93.**

Conducting consecutive MCRs is powerful to quickly assemble complicated molecular scaffolds. The Pictet–Spengler reaction is an intramolecular version of a three-component Mannich reaction, which is useful for the construction of aromatic ring-fused heterocyclic systems such as tetrahydropyridoindoles and tetrahydroisoquinolines. Shown in Scheme 2.23 are sequential Ugi and Pictet–Spengler (PS) reactions for the synthesis of polycyclic rings **98.**[30] Bifunctional 2-formyl benzoic acid is reacted with 2-isocyanoethylindole and carbonyl-protected aminoacetaldehyde for an Ugi-3CR reaction to form **99,** which is then treated with formic acid to free the carbonyl group for the

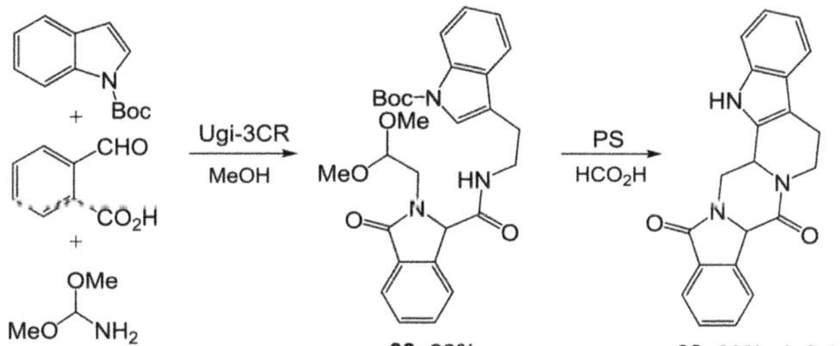

Scheme 2.23 Ugi-3CR and PS reactions for polycyclic compound **98**.

PS reaction to form product **98** as a mixture of diastereomers. Different polycyclic indole alkaloid-type compounds can be prepared by using alternative cyclic keto carboxylic acids as starting materials.

2.5 Summary

Three different types of one-pot reactions, including cascade, stepwise, and multicomponent reactions, are described in this chapter. These reactions have intrinsic pot, atom, and step economy. They are powerful for the synthesis of complex molecules with a broad range of applications. These one-pot reactions thus deserve more attention for the design and synthesis of organic molecules.

References

1. P. T. Anastas and J. C. Warner, *Green Chemistry: Theory and Practice*, Oxford University Press, Oxford, 1998.
2. *Green Techniques for Organic Synthesis and Medicinal Chemistry*, ed. W. Zhang and B. W. Cue, Wiley, Chichester, 2nd edn, 2018.
3. P. A. Clarke, S. Santos and W. H. C. Martin, *Green Chem.*, 2007, **9**, 438.
4. L. F. Tietze, G. Brasche and K. M. Gericke, *Domino Reactions in Organic Synthesis*, Wiley-VCH, Weinheim, 2006.
5. *Science of Synthesis: Applications of Domino Transformations in Organic Synthesis 1 & 2*, ed. S. A. Snyder, Thieme, Stuttgart, 2016.
6. H. Pellisser, *Asymmetric Domino Reactions*, The Royal Society of Chemistry, Cambridge, 2013.
7. *Multicomponent Reactions*, ed. J. Zhu and H. Bienayme, Wiley-VCH, Weinheim, 2005.
8. J. Zhu, Q. Wang and M.-X. Wang, *Multicomponent Reactions in Organic Synthesis*, Wiley-VCH, Weinheim, 2015.
9. *Multicomponent Reactions – Concepts and Applications for Design and Synthesis*, ed. R. P. Herrera and E. Marques-Lopez, Wiley, Hoboken, 2015.
10. *Multicomponent Reactions – Synthesis of Bioactive Heterocycles*, ed. K. L. Ameta and A. Dandia, CRC Press, London, 2017.

11. J. R. Stepherson, C. E. Ayala, T. H. Tugwell, J. L. Henry, F. R. Fronczek and R. Kartika, *Org. Lett.*, 2016, **18**, 2002.
12. K. Mishra, J. B. Singh, T. Gupta and R. M. Singh, *Org. Chem. Front.*, 2017, **4**, 1926.
13. L. Chen, R. Huang, L.-B. Kong, J. Lin and S.-J. Yan, *ACS Omega*, 2018, **3**, 1126.
14. S. Blouin, V. Gandon, G. Blond and J. Suffert, *Angew. Chem., Int. Ed.*, 2016, **55**, 7208.
15. D.-C. Wang, H.-X. Wang, E.-J. Hao, X.-H. Jiang, M.-S. Xie, G.-R. Qu and H.-M. Guo, *Adv. Synth. Catal.*, 2016, **358**, 494.
16. Y. An, Y. Kuang and J. Wu, *Org. Chem. Front.*, 2016, **3**, 994.
17. B. Jiang, J.-X. Liu, Y. Wei and M. Shi, *Org. Lett.*, 2018, **20**, 6229.
18. C.-G. Li, G.-Q. Xu and P.-F. Xu, *Org. Lett.*, 2017, **19**, 512.
19. D. Tejedor, S. Delgado-Hernandez, J. Peyrac, J. Gonzalez-Platas and F. Garcia-Tellado, *Chem. - Eur. J.*, 2017, **23**, 10048.
20. M. J. Riveira, A. La-Venia and M. P. Mischne, *J. Org. Chem.*, 2016, **81**, 7977.
21. W. Zhang, X. Zhang, X. Ma and W. Zhang, *Tetrahedron Lett.*, 2018, **59**, 3845.
22. A. Muthengi, X. Zhang, G. Dhawan, W. Zhang, F. Corsini and W. Zhang, *Green Chem.*, 2018, **20**, 3134.
23. X. Huang, M. Liu, K. Pham, X. Zhang, W.-B. Yi, J. P. Jasinski and W. Zhang, *J. Org. Chem.*, 2016, **81**, 5362.
24. X. Huang, M. Liu, J. P. Jasinski, B. Peng and W. Zhang, *Adv. Synth. Catal.*, 2017, **359**, 1919.
25. Q. Lu, G. Song, J. P. Jasinski, A. C. Keeley and W. Zhang, *Green Chem.*, 2012, **14**, 3010.
26. S. Brauch, S. S. van Berkel and B. Westermann, *Chem. Soc. Rev.*, 2013, **42**, 4948.
27. G. N. Kaluđerović, M. Abbas, H. C. Kautz, M. A. M. Wadaan, C. Lennicke, B. Seligerc and L. A. Wessjohann, *Chem. Commun.*, 2017, **53**, 3777.
28. M. Dawidowski, S. Sobczak, M. Wilczek, A. Kulesza and J. Turło, *Mol. Diversity*, 2014, **18**, 61.
29. S. Shaabani, A. Shaabani and S. W. Ng, *ACS Comb. Sci.*, 2014, **16**, 176.
30. W. Wang, E. Herdtweck and A. Domling, *Chem. Commun.*, 2010, **46**, 770.

CHAPTER 3

Application of Step, Cumulative, and Global E-factor and Process Mass Intensity Metrics to Gauge Synthesis Efficiency: L-DOPA and Apixaban Pharmaceutical Examples

JOHN ANDRAOS*

CareerChem, Research and Development, 504-1129 Don Mills Road, Toronto, ON M3B 2W4, Canada
*E-mail: c1000@careerchem.com

3.1 Introduction

The related quantities E-factor (environmental factor)[1] and PMI (process mass intensity)[2] are now well-established metrics[3,4] that are routinely used, particularly in the process pharmaceutical industry, to assess the material efficiencies for chemical reactions and synthesis plans. The E-factor refers to the ratio of the mass of waste produced from all sources to the mass of target product collected in a chemical process. On the other hand, process mass

Green Chemistry Series No. 61
Green Synthetic Processes and Procedures
Edited by Roberto Ballini

Published by the Royal Society of Chemistry, www.rsc.org

intensity refers to the ratio of the total mass of input materials used to the mass of target product collected. For a balanced chemical equation, where a mass balance is applied to all input and output materials so that the mass of input materials equals the mass of waste produced plus the mass of target product, it is obvious that E and PMI are related by the simple relationship PMI = E + 1. With respect to the goal of reaction or synthesis optimization, it is desired to minimize both PMI and E values in an effort to simultaneously minimize waste production and maximize the input material consumption. In any synthesis plan, the E and PMI metrics may be determined at each reaction step and are therefore called step E-factor and step PMI, respectively. When these metrics are applied to an entire synthesis plan covering all reaction steps, they are called global or overall E-factor and global or overall PMI. For linear synthesis plans consisting of N reaction steps, the mathematical relationships connecting the step metrics to the global metrics are given by eqn (3.1) and (3.2) for PMI and E, respectively:[5–8]

$$(\mathrm{PMI})_{1\to N} = \frac{1}{m_{\mathrm{P}_N}}\left[m_{\mathrm{P}_N}(\mathrm{PMI})_N + \sum_{j=1}^{N-1} m_{\mathrm{P}_j}\left((\mathrm{PMI})_j - 1\right)\right] \quad (3.1)$$

$$\mathrm{E}_{1\to N} = \mathrm{E}_N + \frac{1}{m_{\mathrm{P}_N}}\sum_{j=1}^{N-1} m_{\mathrm{P}_j}\mathrm{E}_j \quad (3.2)$$

where m_{P_j} and m_{P_N} refer to the masses of intermediate product P_j and final product P_N pertaining to step j and step N, respectively. The reliability and speed of computing the step and global metrics for individual reactions and synthesis plans is significantly enhanced using radial pentagons[9] and synthesis tree diagrams[10] as visual aids to guide the accuracy of the computations and to pinpoint the strengths and weaknesses in the material efficiency of a given process. A radial diagram decomposes the step E and step PMI for an individual reaction into its kernel, by-product, solvent, and catalyst contributions according to the reaction yield, atom economy, excess reagent consumption, and auxiliary material consumption. Synthesis tree diagrams depict reaction schemes composed of sequenced reaction steps in a compact way so that all input materials connected to a given intermediate made in a specified step are tracked according to the reaction yield, molecular weight, and appropriate stoichiometric coefficients. In parallel with step and global E and PMI metrics parameters, corresponding cumulative metrics[5–8] may also be determined, as shown by the recursive formulae in eqn (3.3) and (3.4):

$$(\mathrm{cPMI})_{1\to i} = \frac{m_{\mathrm{P}_{i-1}}}{m_{\mathrm{P}_i}}\left[(\mathrm{cPMI})_{1\to i-1} - 1\right] + (\mathrm{PMI})_i \quad (3.3)$$

$$(\mathrm{cE})_{1\to i} = \frac{m_{\mathrm{P}_{i-1}}}{m_{\mathrm{P}_i}}(\mathrm{cE})_{1\to i-1} + (\mathrm{E})_i \quad (3.4)$$

where $\frac{m_{P_{i-1}}}{m_{P_i}} = \frac{MW_{P_{i-1}}}{MW_{P_i}\varepsilon_i}$, MW refers to the molecular weight of intermediates multiplied by the stoichiometric *coefficients* as appropriate from balanced chemical equations, ε_i is the reaction yield of step i, the $1 \rightarrow i$ notation means that the cumulative PMI quantity extends from step 1 to step i, and the counting index goes from $i = 2, 3, \ldots$, *etc.* Clearly, when $i = N$, the cumulative metrics appearing on the left-hand sides of eqn (3.3) and (3.4) correspond to the global metrics found on the left-hand sides of eqn (3.1) and (3.2).

In the next sections, we demonstrate the determination of the step, global, and cumulative E-factor and PMI values for industrial syntheses of two well-known pharmaceuticals, L-DOPA and apixaban. L-DOPA is a neurotransmitter used to treat Parkinson's disease and apixaban is a blood anti-coagulant.

3.2 L-DOPA

The following three synthesis plans to manufacture L-DOPA are considered: the Hofmann-La Roche process[11,12] employing a classical resolution step to obtain the desired enantiomer, the Monsanto process[12] employing a stereoselective hydrogenation, and the Ajinomoto process[12] employing a one-step bio-fermentation process from catechol, sodium pyruvate, and sodium chloride. Figures 3.1–3.3 show the complete mass inventories for all three processes to manufacture 100 kg of L-DOPA. Based on these findings, we found that the relative ranking of these plans based on material efficiency is as follows: Ajinomoto (PMI = 36.49) > Monsanto (PMI = 44.46) > Hofmann-La Roche (PMI = 419.36).

3.2.1 Hofmann-La Roche Process

From the mass balance inventory shown in Figure 3.1, we can obtain the following step E-factors: (a) step 1, $E_1 = 9433.7/710.3 = 13.28$; (b) step 2, $E_2 = 24\,908.6/479.5 = 51.95$; (c) step 3, $E_3 = 3204.6/411.0 = 7.80$; and (d) step 4, $E_4 = 4288.8/100 = 42.89$. Since PMI = E + 1, we can then immediately obtain the following step PMIs: (a) step 1, $(PMI)_1 = 14.28$; (b) step 2, $(PMI)_2 = 52.95$; (c) step 3, $(PMI)_3 = 8.80$; and (d) step 4, $(PMI)_4 = 43.89$. The global E-factor is found by adding up all masses of waste from all steps and dividing by the mass of target product. Hence, we have $(9433.7 + 24\,908.6 + 3204.6 + 4288.8)/100 = 418.36$. Again, using the relation PMI = E + 1, the global PMI for the synthesis is 419.36. We can verify the relationship given in eqn (3.1) connecting the global PMI to the step PMIs as follows:

$$\begin{aligned}&\left[m_{P_1}\left((PMI)_1 - 1\right) + m_{P_2}\left((PMI)_2 - 1\right) + m_{P_3}\left((PMI)_3 - 1\right) + m_{P_4}(PMI)_4\right] / m_{P_4}\\&= [710.3(14.28-1) + 479.5(52.95-1) + 411.0(8.80-1) + 100(43.89)]/100\\&= 419.38\end{aligned}$$

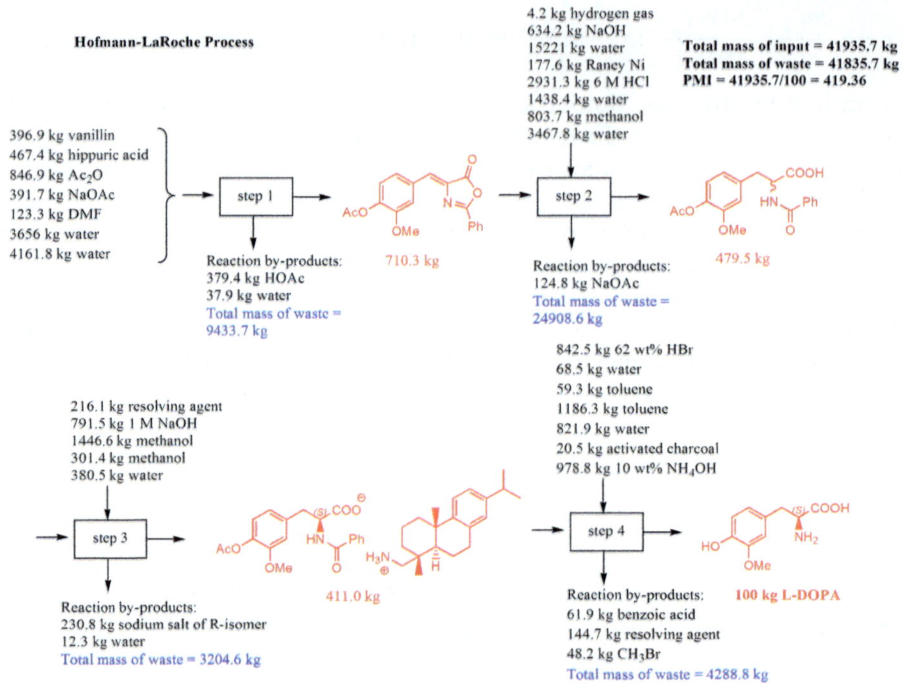

Figure 3.1 Hofmann-La Roche process to manufacture 100 kg of L-DOPA.

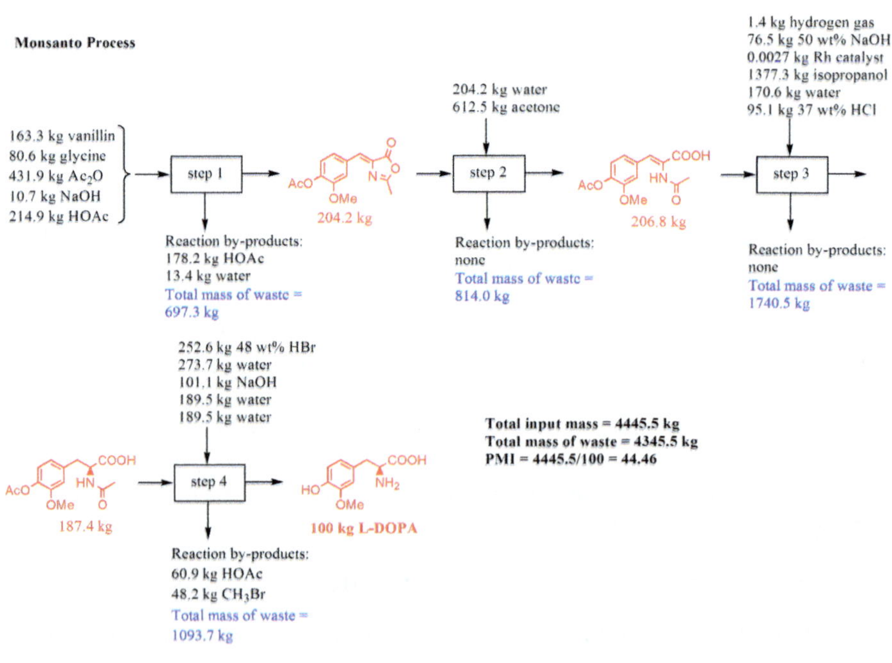

Figure 3.2 Monsanto process to manufacture 100 kg of L-DOPA.

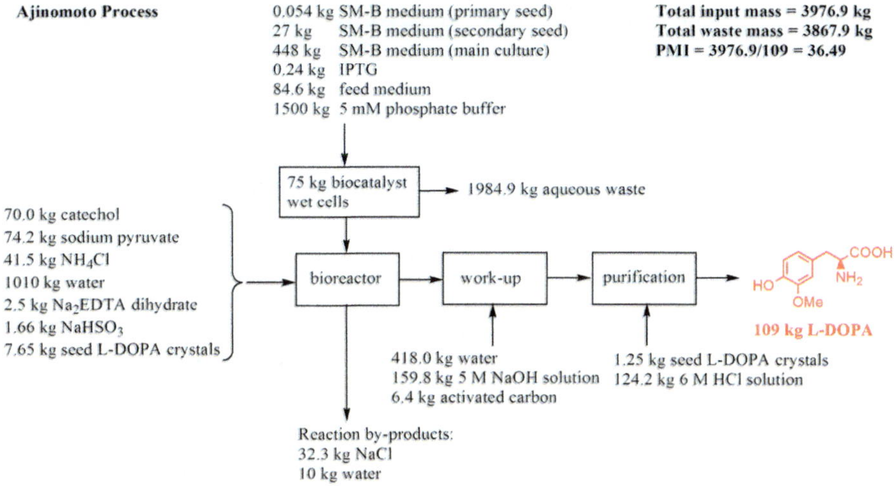

Figure 3.3 Ajinomoto process to manufacture 109 kg of L-DOPA.

Similarly, eqn (3.2) connecting the global E to the step E-factors can also be verified:

$$\left[m_{\mathrm{P}_1}\mathrm{E}_1 + m_{\mathrm{P}_2}\mathrm{E}_2 + m_{\mathrm{P}_3}\mathrm{E}_3\right]\left(1/m_{\mathrm{P}_4}\right) + \mathrm{E}_4$$
$$= [710.3(13.28) + 479.5(51.95) + 411.0(7.80)](1/100) + 42.89$$
$$= 418.38$$

It is important to point out here that the global E-factor is *not* the sum of the step E-factors. In the Appendix (Section 3.5), a simple mathematical analysis demonstrates the precise connection between the true global E-factor and the sum of the step E-factors. In this particular example, if the step E-factors are added, we obtain 13.28 + 51.95 + 7.80 + 42.89 = 115.92, compared to the actual global E-factor of 418.38. This example represents a direct numerical verification of the above statement.

The cumulative PMI recursive formula given in eqn (3.3) is verified as follows:

$$(\mathrm{cPMI})_{1\to1} = (\mathrm{PMI})_1 = 14.28$$

$$(\mathrm{cPMI})_{1\to2} = \frac{m_{\mathrm{P}_1}}{m_{\mathrm{P}_2}}[(\mathrm{cPMI})_{1\to1} - 1] + (\mathrm{PMI})_2 = \frac{710.3}{479.5}[14.28 - 1] + 52.95 = 72.62$$

$$(\mathrm{cPMI})_{1\to3} = \frac{m_{\mathrm{P}_2}}{m_{\mathrm{P}_3}}[(\mathrm{cPMI})_{1\to2} - 1] + (\mathrm{PMI})_3 = \frac{479.5}{411.0}[72.62 - 1] + 8.80 = 92.36$$

$$(\mathrm{cPMI})_{1\to4} = \frac{m_{\mathrm{P}_3}}{m_{\mathrm{P}_4}}[(\mathrm{cPMI})_{1\to3} - 1] + (\mathrm{PMI})_4 = \frac{411.0}{100}[92.36 - 1] + 43.89 = 419.38$$

Similarly, the cumulative E-factor recursive formula given in eqn (3.4) is verified as follows:

$$(\mathrm{cE})_{1\to 1} = (\mathrm{E})_1 = 13.28$$

$$(\mathrm{cE})_{1\to 2} = \frac{m_{\mathrm{P}_1}}{m_{\mathrm{P}_2}}(\mathrm{cE})_{1\to 1} + (\mathrm{E})_2 = \frac{710.3}{479.5}\times 13.28 + 51.95 = 71.62$$

$$(\mathrm{cE})_{1\to 3} = \frac{m_{\mathrm{P}_2}}{m_{\mathrm{P}_3}}(\mathrm{cE})_{1\to 2} + (\mathrm{E})_3 = \frac{479.5}{411.0}\times 71.62 + 7.80 = 91.36$$

$$(\mathrm{cE})_{1\to 4} = \frac{m_{\mathrm{P}_3}}{m_{\mathrm{P}_4}}(\mathrm{cE})_{1\to 3} + (\mathrm{E})_4 = \frac{411.0}{100}\times 91.36 + 42.89 = 418.38$$

It is clear that the bottlenecks in this plan are the excessive use of auxiliary solvents in the hydrogenation step (step 2) and the loss of half the racemic mixture in step 4 when the diastereomeric salt is treated with acid. The last step also has the lowest atom economy at 28% due to the loss of the high molecular weight chiral auxiliary as a by-product. However, in the actual manufacturing process, this resolving agent can be recovered and re-used for further batches producing L-DOPA.

3.2.2 Monsanto Process

The step E-factors are as follows: (a) azlactone synthesis (E = 697.3/204.2 = 3.41), (b) azlactone hydrolysis (E = 814.0/206.8 = 3.94), (c) stereoselective hydrogenation (E = 1740.5/187.4 = 9.29), and (d) hydrolysis of two acetyl groups (E = 1093.7/100 = 10.94). The expressions in eqn (3.1)–(3.4) can be verified in the same way as shown for the Hofmann-La Roche process from the data shown in Figure 3.2. The bottlenecks in the plan are the 50% atom economies in steps 1 and 4, the excess reagent contribution in step 3, and the high auxiliary material contribution in step 4.

3.2.3 Ajinomoto Process

It is evident that the one-step bio-fermentation process is the most material efficient without having to perform any computations of E or PMI metrics. The extra 9 kg of L-DOPA arise due to the use of seed crystals of the product as an input material used in the process. The PMI value of 36.49 shown in Figure 3.3 for the process includes the preparation of the biocatalyst. The PMI value excluding the preparation of the biocatalyst is ((3867.9 − 1984.9)/109) + 1 = 18.28. Since the process involves one step, there are no cumulative metrics to consider.

3.3 Apixaban

Six synthesis plans are considered, including two from Bristol-Myers Squibb (BMS),[13,14] one from an academic lab in China,[15] and three from generic drug companies.[16–18] Figure 3.4 shows a reaction network summarizing all synthesis routes to apixaban.

We observe from this network that all synthesis plans to apixaban are convergent, where the common intermediate **X** is made in a parallel single-step branch that is then used as input material in a subsequent convergent step. For convergent plans, the computation of the global E-factor and global PMI is carried out by applying eqn (3.1) and (3.2) in a piece-wise manner over appropriate branches of the plan. We illustrate the computation for a convergent plan consisting of two branches, as shown in Figure 3.5. Intermediate P_r is the end product obtained over r steps along the branch from starting material S_2, intermediate P_n is the end product obtained over n steps along the branch from starting material S_1, intermediate P^* is the end product obtained in the convergent step from starting materials P_r and P_n, and the final target product P_N is the end product obtained over q steps along the branch from starting material P^*. The total number of steps in the plan along the longest branch is $N = n + q + 1$.

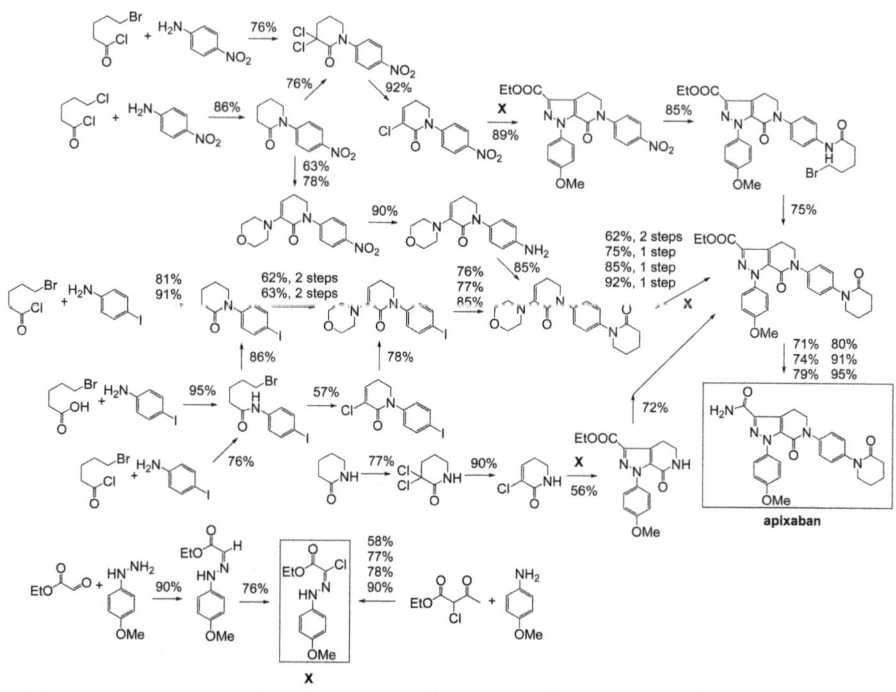

Figure 3.4 Reaction network for the synthesis of apixaban.[19]

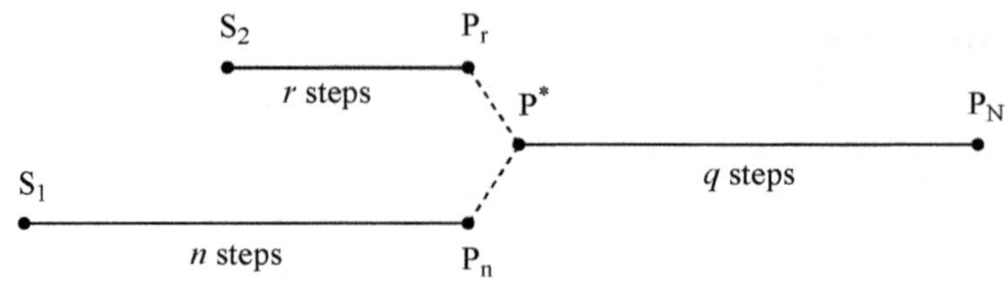

Figure 3.5 Generic convergent synthesis plan showing two branches.

The global E-factor is given by eqn (3.5):

$$E_{1\to N} = \frac{1}{m_{P_N}}\left[m_{P_m} E_{S_2\to P_m} + m_{P_n} E_{S_1\to P_n} + m_{P_N} E_{P^*\to P_N} + m_{P^*} E^*\right] \quad (3.5)$$

where $E_{S_2\to P_m} = E_m + \frac{1}{m_{P_m}} \sum_{j=1}^{m-1} m_{P_j} E_j$, $E_{S_1\to P_n} = E_n + \frac{1}{m_{P_n}} \sum_{j=1}^{n-1} m_{P_j} E_j$, $E_{P^*\to P_N} = E_N + \frac{1}{m_{P_N}} \sum_{j=1}^{q-1} m_{P_j} E_j$, E^* is the E-factor for the convergent step, and the m parameters refer to the mass of intermediates. The expression for global PMI is found by replacing each E parameter appearing in eqn (3.5) by PMI − 1, as shown in eqn (3.6):

$$(\mathrm{PMI})_{1\to N} = \frac{1}{m_{P_N}}[m_{P_m}((\mathrm{PMI})_{S_2\to P_m} - 1) + m_{P_n}((\mathrm{PMI})_{S_1\to P_n} - 1) + m_{P_N}((\mathrm{PMI})_{P^*\to P_N} - 1) + m_{P^*}((\mathrm{PMI})^* - 1)] + 1$$

where $(\mathrm{PMI})_{S_2\to P_m} = (\mathrm{PMI})_m + \frac{1}{m_{P_m}} \sum_{j=1}^{m-1} m_{P_j}\left((\mathrm{PMI})_j - 1\right)$,

$$(\mathrm{PMI})_{S_1\to P_n} = (\mathrm{PMI})_n + \frac{1}{m_{P_n}} \sum_{j=1}^{n-1} m_{P_j}\left((\mathrm{PMI})_j - 1\right), \text{ and}$$

$$(\mathrm{PMI})_{P^*\to P_N} = (\mathrm{PMI})_N + \frac{1}{m_{P_N}} \sum_{j=1}^{q-1} m_{P_j}\left((\mathrm{PMI})_j - 1\right) \quad (3.6)$$

Table 3.1 summarizes the essential material efficiency metrics calculated for the six plans to apixaban arranged in ascending order from lowest global PMI to highest global PMI. Entries in bold show the best performing values and entries in italic show the worst ones. Also included is a breakdown of the E-factor contributions for each plan according to waste produced from by-products (E-kernel), excess reagent consumption (E-excess), reaction solvents (E-solv), catalyst (E-cat), workup material consumption

Table 3.1 Summary of green metrics for various synthesis plans to apixaban.[19]

Plan	# Steps	Overall Y (%)[a]	Overall AE (%)	E-kernel	E-excess	E-solvent	E-cat	E-workup	E-purif	E-total	PMI-total
Bristol-Myers 2006	7	**37.4**	**20.8**	**10.32**	**7.72**	**80.65**	1.12	**116.85**	27.2	**243.86**	**244.86**
Optimus 2017	9	16.7	18.4	18.63	15.90	117.40	2.46	214.10	9.41	377.90	378.90
Jiang-Ji 2013	7	26.8	18.4	11.15	16.84	104.15	**0.05**	220.37	29.53	382.09	383.09
MSN 2015	8	16.3	19.2	24.29	28.9	126.45	0.63	308.40	8.78	497.45	498.45
Bristol-Myers 2003	7	25.6	18.5	16.06	19.15	128.22	0.80	460.73	**0**	624.96	625.96
DrReddy 2017	7	22.2	19.8	19.86	24.24	239.37	0.86	442.27	129.36	855.96	856.96

[a]Calculated along the longest branch.

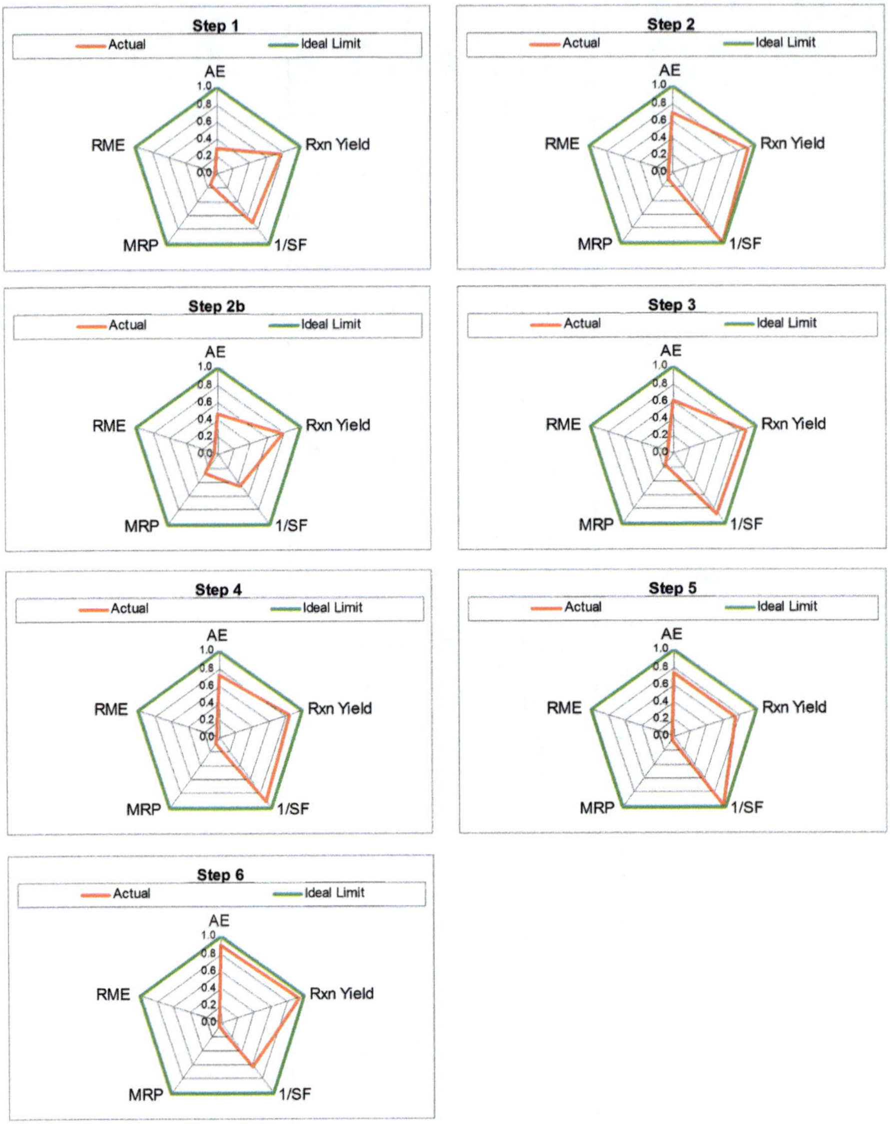

Figure 3.6 Radial pentagons for the best performing BMS 2006 plan to apixaban.[14,19]

(E-workup), and purification material consumption (E-purif). For comparison, the sum of the step E-factors for each plan are as follows: 197.81 (Bristol-Myers, 2006), 208.72 (Optimus), 338.30 (Jiang-Ji), 274.18 (MSN), 419.51 (Bristol-Myers, 2003), and 688.55 (DrReddy).[19] We can see that, in all cases, the values obtained by summing the step E-factors are noticeably lower than the true values of E-total shown in the 11th column of Table 3.1.[19] The additive values are found to be between 12% and 45% lower than the true overall E-factor values. These results reinforce once again that adding step E-factors in a synthesis plan does not lead to the global E-factor. We can see from Table 3.1 that the Bristol-Myers 2006 plan ranks the highest in all categories except for E-cat. Figure 3.6 shows the radial pentagons for each step in that plan according to the reaction yield, atom economy (AE), excess reagent consumption (SF = stoichiometric factor), auxiliary material consumption (MRP = material recovery parameter), and reaction mass efficiency (RME, which is equal to 1/PMI). From these pictures, we observe the following bottlenecks that can be further improved: (a) the low AE for steps 1 (amidation-α,α′-dichlorination sequence) and 2b (diazotization-imination sequence) and (b) the high SF for steps 2b and 6 (final carboxyamination).

We can use Figure 3.5 to guide the computation of cumulative PMI metrics for convergent plans as follows. From S_1 to P_n, we have the following sequence of recursive formulas given by eqn (3.7)–(3.9):

$$(\text{cPMI})_{1\to 1} = (\text{PMI})_1 \tag{3.7}$$

$$(\text{cPMI})_{1\to 2} = \frac{m_{\text{P}_1}}{m_{\text{P}_2}}\left[(\text{cPMI})_{1\to 1} - 1\right] + (\text{PMI})_2 \tag{3.8}$$

$$(\text{cPMI})_{1\to n} = \frac{m_{\text{P}_{n-1}}}{m_{\text{P}_n}}\left[(\text{cPMI})_{1\to n-1} - 1\right] + (\text{PMI})_n \tag{3.9}$$

From S_2 to P_r, we have the following sequence of recursive formulas given by eqn (3.10)–(3.12):

$$(\text{cPMI})_{1^*\to 1^*} = (\text{PMI})_{1^*} \tag{3.10}$$

$$(\text{cPMI})_{1^*\to 2^*} = \frac{m_{\text{P}_{1^*}}}{m_{\text{P}_{2^*}}}\left[(\text{cPMI})_{1^*\to 1^*} - 1\right] + (\text{PMI})_{2^*} \tag{3.11}$$

$$(\text{cPMI})_{1^*\to m} = \frac{m_{\text{P}_{m-1}}}{m_{\text{P}_m}}\left[(\text{cPMI})_{1\to m-1} - 1\right] + (\text{PMI})_m \tag{3.12}$$

At the convergent step, we have the following recursive formula given by eqn (3.13):

$$(\text{cPMI})_{1\to n+1} = \frac{m_{\text{P}_{n-1}}}{m_{\text{P}^*}}\left[(\text{cPMI})_{1\to n} - 1\right] + (\text{PMI})_{n+1} + \frac{m_{\text{P}_m}}{m_{\text{P}^*}}\left[(\text{cPMI})_{1^*\to m} - 1\right] \tag{3.13}$$

Continuing from P_n to P_N we have the following sequence of recursive formulas given by eqn (3.14) and (3.15):

Table 3.2 Summary of cumulative complete PMI values for various synthesis plans to apixaban.[19]

Step j	BM-2006	Optimus-2017	Jiang-Ji-2013	MSN-2015	BM-2003	DrReddy-2017
1	45.92	7.62	36.93	10.35	70.23	65.82
2	75.31	35.40	101.29	107.46	167.16	289.54
3	78.81	78.61	151.55	124.80	288.44	399.80
4	95.52	106.71	214.04	190.78	354.82	579.92
5	183.54	189.87	308.68	224.37	**625.64**	**857.06**
6	**245.01**	161.64	**382.83**	336.47		
7		269.01		**498.35**		
8		**379.30**				

Table 3.3 Summary of cumulative PMI ranks for various synthesis plans to apixaban.[19]

Plan	Complete cPMI	Rank (complete)	Kernel cPMI	Rank (kernel)	Kernel + excess cPMI	Rank (kernel + excess)
BM-2006	245.01	1	11.17	1	19.04	1
Optimus-2017	379.30	2	19.63	5	35.56	3
Jiang-Ji-2013	382.83	3	12.05	2	28.95	2
MSN-2015	498.35	4	24.73	6	54.19	6
BM-2003	625.64	5	16.56	3	36.22	4
DrReddy-2017	857.06	6	18.65	4	45.12	5

$$(\text{cPMI})_{1\to n+2} = \frac{m_{\text{P}_{n+1}}}{m_{\text{P}_{n+2}}}\left[(\text{cPMI})_{1\to n+1} - 1\right] + (\text{PMI})_{n+2} \tag{3.14}$$

$$(\text{cPMI})_{1\to N} = \frac{m_{\text{P}_{N-1}}}{m_{\text{P}_N}}\left[(\text{cPMI})_{1\to N-1} - 1\right] + (\text{PMI})_N = (\text{PMI})_{1\to N}\text{, where } N = n + q + 1 \tag{3.15}$$

Table 3.2 summarizes the cumulative step PMIs for all six apixaban plans determined according to the protocol given by eqn (3.7)–(3.15). The final cumulative values shown in bold agree with the global PMIs shown in Table 3.1. Table 3.3 summarizes the rankings of the six plans according to various degrees of PMI calculations. The kernel PMI values account for by-product formation only, kernel plus excess PMI values account for by-product formation as well as excess reagent consumption, and the complete PMI values account for all waste produced from all sources. This last column of entries is identical to the global PMIs determined in Table 3.1.

We observe from the rankings shown in Table 3.3 that the BMS 2006 plan ranks first in all PMI categories. If a plan consistently ranks first under these three regimes, then the probability that it represents the "greenest" material efficient plan to a given target is very high and consequently this criterion can be considered an excellent test for robustness of "green" material efficiency for a synthesis plan.[19] If a synthesis plan is ranked first under the kernel regime, then this means that such a plan likely has the highest atom economy and overall yield; if it is ranked first under the kernel plus excess regime, then this means that it also consumes the least excess reagents; and finally, if it is ranked first under the complete regime, then this means that it also consumes the least auxiliary materials in terms of reaction solvents, workup materials, and purification materials.[19]

3.4 Conclusion

In this chapter, we have demonstrated the computation of step, global, and cumulative E-factors and PMIs for linear and convergent synthesis plans using the industrial syntheses of L-DOPA and apixaban as representative examples. The mathematical relationships given in all equations are applicable universally to any synthesis plan. The results show that the cumulative metrics covering all reaction steps in a plan are identically equal to the global metrics. Furthermore, it is shown that the sums of step metrics do not yield the global metrics. Most material efficient plans are those that rank the highest when computations are performed under both the kernel and complete regimes, where only by-product waste materials and all waste materials are considered, respectively.

3.5 Appendix

The expression given in eqn (3.2) relating the global E-factor to the step E-factors for a linear synthesis plan can be manipulated, as shown below in eqn (3.16):

$$
\begin{aligned}
E_{1\to N} &= E_N + \frac{1}{m_{P_N}}\left[m_{P_1}E_1 + m_{P_2}E_2 + m_{P_3}E_3 + \ldots + m_{P_{N-1}}E_{N-1}\right] \\
&= E_N + \left(E_1 - E_1 + \frac{m_{P_1}}{m_{P_N}}E_1\right) + \left(E_2 - E_2 + \frac{m_{P_2}}{m_{P_N}}E_2\right) \\
&\quad + \left(E_3 - E_3 + \frac{m_{P_3}}{m_{P_N}}E_3\right) + \ldots + \left(E_{N-1} - E_{N-1} + \frac{m_{P_{N-1}}}{m_{P_N}}E_{N-1}\right) \\
&= E_N + E_1 + E_2 + E_3 + \ldots + E_{N-1} + E_1\left(\frac{m_{P_1}}{m_{P_N}} - 1\right) + E_2\left(\frac{m_{P_2}}{m_{P_N}} - 1\right) \\
&\quad + E_3\left(\frac{m_{P_3}}{m_{P_N}} - 1\right) + \ldots + E_{N-1}\left(\frac{m_{P_{N-1}}}{m_{P_N}} - 1\right) \\
&= \sum_{j=1}^{N} E_j + \sum_{j=1}^{N-1} E_j\left(\frac{m_{P_j}}{m_{P_N}} - 1\right)
\end{aligned}
\tag{3.16}
$$

We observe that the first summed term corresponds to the sum of the step E-factors and the second summed term accounts for the variation of masses of intermediate products along the way with respect to the mass of the target product collected at the end of the synthesis plan. There are three main scenario cases that can be explored. If $\frac{m_{P_j}}{m_{P_N}} = 1$ for each intermediate, as would be the case for a sequence of rearrangements with no loss of material (*i.e.*, 100% yield in each reaction step), then $E_{1\to N} = \sum_{j=1}^{N} E_j$. If $\frac{m_{P_j}}{m_{P_N}} < 1$ for each intermediate, then this implies that the second summed term has a negative value. Hence, the magnitude of the overall or global E-factor of a linear synthesis is *lower* than the sum of the reaction or step E-factors. If $\frac{m_{P_j}}{m_{P_N}} > 1$ for each intermediate, then this implies that the second summed term has a positive value. Hence, the magnitude of the overall or global E-factor of a linear synthesis is higher than the sum of the reaction or step E-factors. This scenario corresponds to the results of all synthesis plans analyzed for L-DOPA and apixaban.

References

1. R. A. Sheldon, *CHEMTECH*, 1994, **24**(3), 38.
2. C. Jiménez-González, C. S. Ponder, Q. B. Broxterman and J. B. Manley, *Org. Process Res. Dev.*, 2011, **15**, 912.
3. *Green Chemistry Metrics – Measuring and Monitoring Sustainable Processes*, ed. A. Lapkin and D. Constable, Wiley, Chichester, 2008.
4. *Handbook of Green Chemistry – Green Metrics*, ed. D. J. C. Constable and C. Jiménez-González, Wiley-VCH, Weinheim, 2018, vol. 11.
5. J. Andraos and A. Hent, *J. Chem. Educ.*, 2015, **92**, 1820.
6. J. Andraos and A. Hent, *J. Chem. Educ.*, 2015, **92**, 1831.
7. J. Andraos, *ACS Sustainable Chem. Eng.*, 2016, **4**, 1917.
8. J. Andraos, *ACS Sustainable Chem. Eng.*, 2018, **6**, 3206.
9. J. Andraos and M. Sayed, *J. Chem. Educ.*, 2007, **84**, 1004.
10. J. Andraos, *Org. Process Res. Dev.*, 2006, **10**, 212.
11. G. M. Jaffe and W. R. Rehl, *US Pat.*, US3714242, Hofmann-La Roche, 1973.
12. T. Li and X. Li, *Green Chem.*, 2014, **16**, 4241.
13. J. Zhou, L. M. Oh, P. Ma, H. Y. Li and P. Confalone, *US Pat.*, US2003181466, Bristol-Myers Squibb Company, 2003.
14. R. Shapiro, L. T. Rossano, B. M. Mudryk, N. Cuniere, M. Obelholzer, H. Zhang and B. C. Chen, *US Pat.*, US2006069258, Bristol-Myers Squibb Company, 2006.
15. J. Jiang and Y. Ji, *Synth. Commun.*, 2013, **43**, 72.
16. S. T. Rajan, S. Eswaraiah and M. Venkatesh, *US Pat.*, US2015353541, MSN Laboratories Limited, 2015.
17. N. R. Nevuluri, R. K. Rapolu, J. Iqbal, B. Kandagatla, S. Sen, V. H. Dahanukar and S. Oruganti, *Monatsh. Chem.*, 2017, **148**, 1477.
18. D. Reddy, S. Reddy, D. R. Rane and S. R. Velivela, *World Pat.*, WO2017187245, Optimus Drugs (P) Limited, 2017.
19. J. Andraos, *Green Process. Synth.*, 2019, **8**, 324.

CHAPTER 4

Flow Chemistry in Drug Discovery

J. ALCAZAR*[a], A. DE LA HOZ[b] AND A. DÍAZ-ORTIZ[b]

[a]Discovery Sciences, Janssen Research and Development, Janssen-Cilag, S.A., Jarama 75A, 45007 Toledo, Spain; [b]Facultad de Ciencias Químicas, Universidad de Castilla-La Mancha, Av. Camilo José Cela 10, 13071, Ciudad Real, Spain
*E-mail: jalcazar@its.jnj.com

4.1 Introduction

The pharmaceutical industry is continuously focusing on productivity improvement, waste reduction, as well as quality improvement and control in both research-and-development (R&D) and manufacturing areas.[1]

In 2005, the American Chemical Society (ACS), Green Chemistry Institute (GCI), and several global pharmaceutical corporations founded the ACS GCI Pharmaceutical Roundtable. In 2007, they defined the main key green chemistry and engineering areas and, among the top ten research areas, continuous processing and manufacturing were the first in priority.[2]

First let us define what continuous processing and manufacturing means. Continuous processing represents a single operation in flow or a complex assembly of operations or synthetic steps. Continuous manufacturing refers

Green Chemistry Series No. 61
Green Synthetic Processes and Procedures
Edited by Roberto Ballini

Published by the Royal Society of Chemistry, www.rsc.org

to the integration of unit operations linked in a continuous train to deliver the drug substance (DS) or drug product (DP) under Good Manufacturing Practices (GMP) conditions.[3]

Flow chemistry is known for almost a century and used routinely in other industries such as petroleum refineries. In this industry, continuous processing is the key of process intensification. The main objective of process intensification is to reduce costs, reduce the size of process equipment, improve product quality, and reduce the energy consumption, solvent utilization, and generated waste.[1] Many of these objectives are closely related to the Green Chemistry principles.[4]

Green chemistry was defined by Anastas and Warner as the design of chemical products and processes that reduce or eliminate the generation of hazardous substances, which they articulated in the 12 principles of Green Chemistry. More recently, the 12 principles of Green Engineering have been proposed.[5]

Continuous flow chemistry is based on the inherent properties of the corresponding reactors that affect fluid dynamics and the heat and mass transfer for streams of chemicals. It is considered that flow methodologies are only constrained by the limits of chemistry and physics, whereas batch processing is by the required equipment. The major characteristics of flow chemistry that can be used advantageously can be classified as:[6–8]

- Better control of the reaction conditions in comparison with batch processes, *i.e.*, rapid and efficient mixing of reactants, even with immiscible phases, short residence times with precise control over it and the local stoichiometry, and efficient heat and mass transfer with low operating volumes. This results in an improvement of the yield and selectivity and reduces the process variance. The 'steady-state' operation under flow conditions is clearly superior to batch-to-batch variability.
- High temperature and pressure conditions can be applied, with the subsequent acceleration of reaction rates and reduction of reaction times. Exothermic reactions are perfectly controlled due to the high mass- and heat transfer rates.[1]
- Reduction of solvent consumption can be achieved by applying high concentrations or a single solvent during the whole process. Solventless conditions are also possible using neat or molten reagents, providing higher reactivity. In this regard, clogging, an important problem in flow chemistry, can be prevented by careful choice of the solvents and reagents, using wide bore channels or tubing, applying agitation or sonic irradiation, and controlling the fluid velocity. Solvent usage and emissions are also reduced due to the lower frequency of cleaning compared to batch production.[7]
- Flow chemistry is an inherently safer, faster, and cleaner technology than batch processes with reduced footprint. Hazards can be diminished due to the short residence times and better control of the temperature. The

size of the reactor is very small and, in consequence, the amount of hazardous reagents or intermediates is very small, reducing the possibilities of thermal runaway or explosions. The use of corrosive reagents can be prevented by employing novel process windows and gaseous reagents can be safely used and precisely controlled by tube-in-tube reactors based on semipermeable Teflon.[7]

The combination of incompatible reactions is possible by the so-called telescoped flow chemistry. Multistep flow chemical sequences can be optimized using the smallest possible number of reactions, work-up procedures, and equipment (pumps, reactors, mixers, *etc.*) to minimize the complexity of the system and improve the reproducibility.[6,9]

Reactions can be scaled-up by parallel processing or process intensification. Process intensification has been defined as the use of new equipment or techniques that promote a significant improvement of production processes at large scale.[6]

Enabling technologies as flow chemistry have enabled the realization of reactions considered impracticable in the laboratory, especially at industrial scale. This approach has evolved onto the concept of Novel Process Windows (NPWs), which was defined as the use of new equipment, micro/macro reactors, and reaction conditions to perform chemical transformations in a vigorous but completely safe and scalable manner.[6,10]

In situ analysis of flow reactions can be performed by in-line or on-line methodologies. In this regard, analytical techniques such as FT-IR, NIR, Raman, NMR spectroscopy and MS have been used. In-line analysis allows precise control of the reaction conditions (concentration of the reactants, addition of reagents, *etc.*) and full control of the end of the reaction, avoiding the decomposition of the reaction product(s) and energy waste.[7]

Purification can be performed also in flow. Isolation can be achieved by in-line filtration of solid co-products, liquid–liquid phase separation, gas–liquid phase separation, or through the use of solid phase-supported scavengers. Finally, purification can be performed by continuous crystallization, semi-batch recrystallization, neutralization, catch-and-release chromatography, and continuously operated simulated moving bed (SMB) chromatography.[7,9,11]

The small size of continuous flow reactors makes photo- and electrochemical reactors feasible. Photochemical transformations have regained new interest in organic synthesis since they began to be performed under continuous flow conditions. The efficiency of these reactions has been significantly improved under continuous flow compared to batch conditions due to the homogeneous light irradiation enabled by the use of polymer tube reactors (PFA, FEP). Under batch conditions, a number of limitations prevent efficient photochemical reactions.[7,12]

Quantification of the green character of a process can be performed using different metrics, such as Atom Economy (AE), E-factor, Mass

Intensity (MI), and Life Cycle Assessment (LCA). The E-factor, defined by Sheldon as the ratio between the mass of residues and mass of desired product, is especially high in the pharmaceutical industry compared to others such as the petrochemical industry. The use of multistep sequences and the preparation of very complex products, often enantiomerically pure, that require solvents for reaction and purification steps, and sophisticated reagents and catalysts, produce often secondary and concomitant products that should be removed or better avoided. However, the E-factor does not consider the large production in the bulk chemical industry, which generates larger residue volumes than the more sophisticated fine chemical industry.[13]

Process Mass Intensity (PMI), the ratio of the total mass of material to the mass of the target isolated product, and Molar Residue (MR) are more precise parameters to account for the sustainability of a flow process.[13]

The AE, the ratio of the molecular weight of the product to the sum of the molecular weights of the reagents or products, is an essential parameter, such as the yield and selectivity, to determine the efficiency of a process, in this case in terms of its Green character.[13]

The LCA is described as the compilation and evaluation of the ingredients (inputs) and results (outputs) and the environmental impacts of a process or product through its whole life cycle, in our case the pharmaceutical production. It compiles all the impacts associated with the process from the acquisition of the starting materials to the delivery (from cradle to gate) and even to the elimination after use (from cradle to grave).[13] In this regard, Janssen, S.A., evaluated the LCA of the production of an anti-Alzheimer drug, galantamine. They found that replacement of a batch step by a flow process afforded a reduction of the resource requirements.[14]

Finally, a green process needs to be scalable, if possible to industrial scale, without extra environmental impact, which means reducing the need for reoptimizing the reaction conditions at every scale. Continuous flow processes may fulfil these reaction condition requirements in terms of the solvent, concentration, reagents, product quality, energy consumption, equipment cost, waste generation, and safety.[6,8,13]

All these benefits are required for the manufacturing of Active Pharmaceutical Ingredients (APIs) in the pharmaceutical industry.

Continuous flow was introduced in the pharmaceutical industry in process chemistry as an excellent technology for complete integration of the early stages of drug discovery, since scale-up to medium scale can be performed without modification of the reaction conditions by just extending the process time.[11]

An efficient cycle design–synthesis–screening is essential to improve the success and reduce the costs in the pharmaceutical industry. Synthesis and biological evaluation are usually the limiting steps in the process of lead development; in consequence, the design of new and efficient methodologies that provide large libraries in shorter times is a priority objective.[9,11]

In continuous flow chemistry, the screening can be integrated into a single continuous platform with in-line or on-line analysis of the new compounds or libraries, thus reducing the time of analysis and biological evaluation and allowing the rational design of new libraries.[9,11]

Considering the production scale, in batch production, normally only one unit operation is conducted at a time (reaction, separation, purification, *etc.*), while continuous processing can be designed to run a number of reactions and separation unit operations simultaneously with a constant flow of materials and a footprint that is constant while the steady-state is maintained, something rarely achieved from batch to batch.[6,8] In contrast, in batch processes the equipment utilization is reduced and significant periods of time with little activity are observed.

However, although many efforts and advances have been undertaken in the past few years, some challenges remain, for example:[11]

- The possibility of performing flow reactions in medium to high throughput parallel synthesis.
- Control over API polymorph formation in flow.
- Control of residual solvents, impurities, inorganics, metals, *etc.*
- Automatization of the reaction optimization and library design using machine learning neural networks.
- Design and optimization of flow systems using machine learning and 3D printing.

Continuous flow chemistry requires a thorough redesign of conventional chemical batch processes and new equipment and technology. In developed countries, the batch manufacturing capacity is very important and they are reluctant to change unless the advantages of flow technology are evident, both in efficiency and cost.[12] In less developed countries, the existing infrastructure is smaller, and they have the freedom to introduce continuous manufacturing if the cost and performance are favorable.[15]

Introduction of continuous manufacture could be very important for realizing the United Nations Millennium Development Goals; specially for improving the access to essential medicines for the world's poor.[16]

Recently, drug manufacturing machines have been introduced for the *in situ* preparation of several drugs. This 'pharmacy on demand' could be interesting to empower local populations in the preparation of required medicines and avoid the suspicions observed when they are supplied by foreign aid organizations.[17]

Continuous manufacturing can be leveraged to lower drug costs, create resilient supply chains, assure product quality, and empower local populations.

In this way, the choice of drug candidates for continuous manufacturing in developing economies is critical and should be based on different aspects, such as the preparation of high-value molecules and the benefit compared to batch processes, the volume of drug needed to meet the market demand, the availability of starting materials, the number of steps translated to continuous manufacturing, and the possibility to telescope the synthesis.[15]

4.2 Flow Chemistry in Drug Discovery and Development

Before starting the discussion on the applications of flow chemistry in drug discovery, it is important to explain how this process works.[3,18] From the recognition of a suitable protein target to the identification of a potential medicine, the process is divided in four main stages:

- Target validation and hit identification. This is the step where a suitable protein and a first set of molecules are identified. Different approaches can be used, being the most common the High Throughput Screening (HTS) due to its capacity to test up to several hundred thousand compounds for campaign.[19] Fragment-based drug design is another approach that requires having the structure of the protein previously defined, *i.e.*, determined by X-ray diffraction.[20] With the defined structure in hand, millions of molecules can be docked *in silico* to identify the most suitable ones, later confirmed by biochemical assays against the protein.
- Hit to Lead. The active molecules in the previous step are called hits. They are usually active in the target protein but only in the micromolar range. On top of that, they are not usually selective, as most of them were initially designed and synthetized to interact with a different target. At this stage, these initial hits are modified synthetically to get more active and selective structures. The goal is obtaining compounds that interact with the target in the nanomolar range and can validate the target protein *in vivo*. The molecules that reach this stage are called lead compounds.[21]
- Lead Optimization. Even though lead compounds are active *in vivo*, they are usually far from being suitable for humans. Toxicity, metabolism, absorption, undesired side effects, as well as other parameters, need to be modulated to combine all required features in one structure. This is the goal at this stage, getting a compound with a potential therapeutic index to be used in humans (therapeutic index: window between the dose required for desired effect and the dose where a tolerably side reaction starts to show up).
- Late Lead Optimization is the stage where a very limited set of candidates for clinical trials need to be prepared in multigram quantities (40–100 g) to support preclinical studies and complete the dossier for presentation to the regulatory authorities.

The role of chemistry changes through all this process. From the preparation of diverse libraries required in the first two steps, to more focused approaches once the lead compounds are identified or multi-gram scale-up campaigns for the most suitable clinical candidates. This part of the chapter discloses how flow chemistry can be used to aid each phase of this process.

4.2.1 Hit Finding and Hit to Lead

Early stages of drug discovery require the fast exploration of the chemical space to identify the main binding features of the target protein. This provides the key structural elements that interact with the protein. Combination of these elements allows access to compounds with nanomolar potency that may demonstrate the relation of the target protein with the target disease. Library synthesis,[22] diversity-oriented approaches,[23] and access to new chemical space[24] are the usual tools here that can also benefit from flow chemistry.

4.2.1.1 Library Synthesis

Segmented flow is one of the most common tools used for library synthesis in flow. The fact that a single line must be used to prepare several compounds in an efficient manner has made chemists consider dividing the flow stream in small reactor fragments, or plugs, separated by an inert solvent. Each plug can be considered as an individual reactor. The collection of the different plugs provides the reaction products after purification by preparative HPLC-MS in line. Tens of milligrams are usually obtained, which perfectly fits the requirements for fast *in vitro* target screening, avoiding wasting or accumulating useless materials. The inert solvent carrier, typically a fluorous solvent, is recovered, reducing also the amount of waste produced. An example of this approach was performed at Abbott.[25,26] They validated the approach with the synthesis of two small pyrazole and thiazole-containing libraries (Scheme 4.1). Both libraries were prepared in half the time required by conventional microwave synthesis in yields ranging 9–79%.

Further evolution of this platform was performed in a collaboration between MIT and Novartis.[27] They developed an automated platform that can even handle multi-phase and multi-step chemistries. Argon gas was used as the carrier to eliminate any potential miscibility with the organic solvent. The presence of on-line preparative HPLC/MS/ELSD allowed integration of the synthesis, purification, and analysis at microliter scale (Figure 4.1). The product quantities varied between 50 and 500 µg, fitting perfectly the amounts required for initial biochemical assays. The authors showed the flexibility of the design by carrying out five different reactions: nucleophilic aromatic substitution, sulfonylation, acylation,

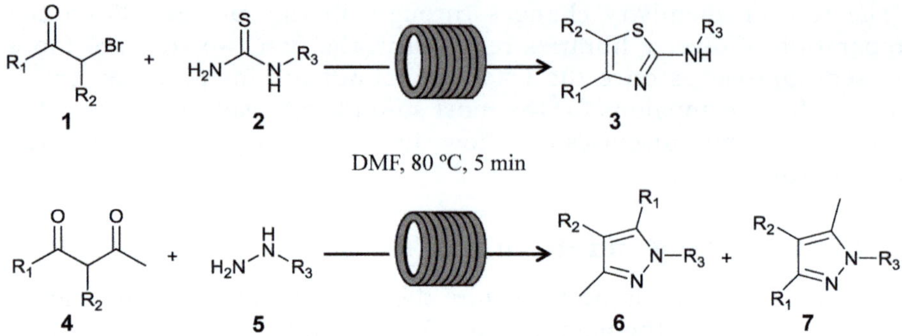

Scheme 4.1 Flow synthesis of thiazole and pyrazole libraries performed at Abbot.

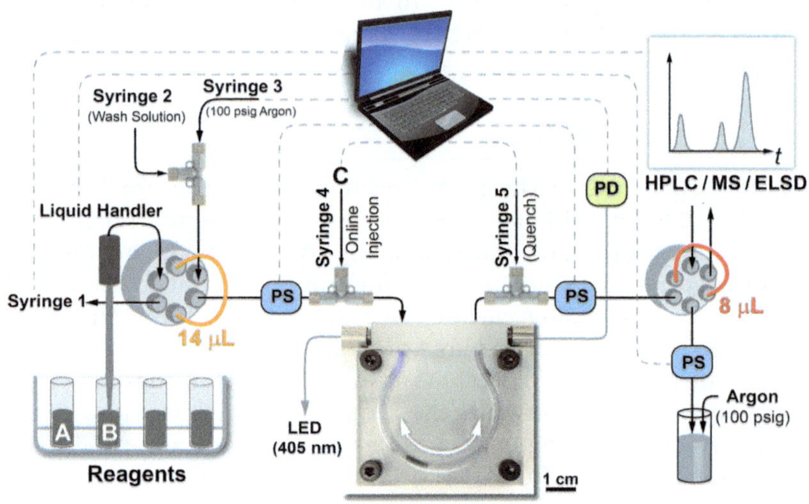

Figure 4.1 Droplet-based platform. PS: phase sensor; PD: photodetector. Reproduced from ref. 27 with permission from the Royal Society of Chemistry.

reductive amination, and Suzuki and Buchwald cross-couplings. To demonstrate the value of the platform, a 36-member compound library was synthesized.

Scientist at Pfizer went a step further in this approach and developed a new flow-based synthesis platform designed from commercially available components, which integrated both rapid nanomole-scale reaction screening and micromole-scale synthesis in a single modular unit avoiding the use of a carrier inert solvent.[28] The reactants are directly injected in the flow stream and mixed by passive diffusion of the system under suitable control of the residence time. The segments are properly separated to avoid cross-contamination between them (Figure 4.2). Once detected, they are directed alternatively to two LC-MS instruments to perform the analysis. The

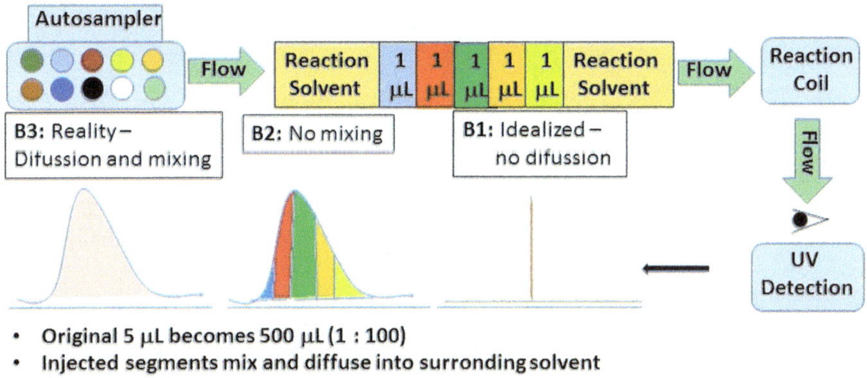

Figure 4.2 Injected segments and its diffusion with the surrounding solvent to get a reaction droplet.

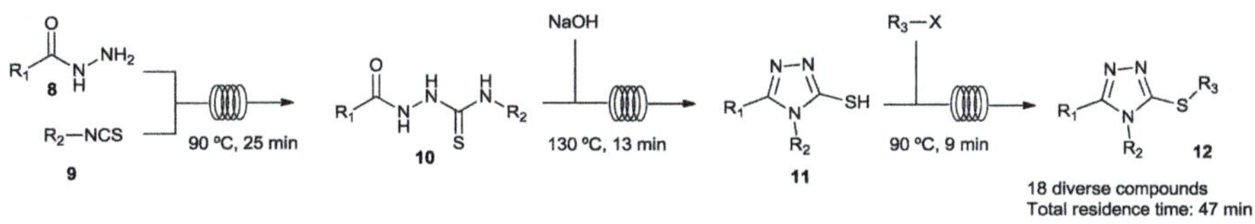

Scheme 4.2 Telescoped synthesis of an 18-member library of 1,2,4-triazoles.

throughput of the system is unparalleled, being able to run ~1500 reactions in 24 h. Although this design is primarily devoted to reaction screening, its applicability for synthesis is clear.

Telescoping reactions in segmented flow is also another interesting approach to library production, as the amount of waste is reduced considerably. Several steps can be combined in a single process without the need for intermediate work-up procedures. To exemplify this approach, Pastre and co-workers developed the preparation of an 18-member library of substituted 3-thio-1,2,4-triazoles.[29] The protocol included the condensation of isothiocyanates with hydrazides, followed by cyclization in basic media to obtain the triazole core and alkylation of the thiol (Scheme 4.2). This protocol allows modification at three different positions of the heterocycle. The compounds were obtained in higher yields and shorter times when compared to batch conditions.

4.2.1.2 Access to New Chemical Space

The development of novel reactions and transformations is clearly the preferred approach toward accessing new chemical space. Two technologies, photochemistry,[30,31] and electrochemistry,[32] are currently facilitating unsuitable transformations and providing new insight into drug discovery for the

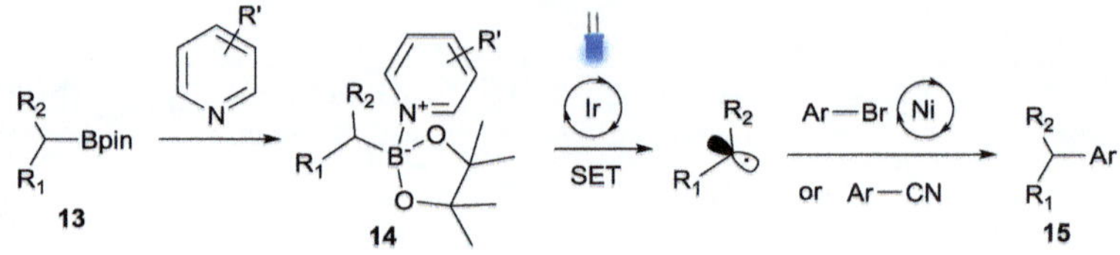

Scheme 4.3 Generation of C(sp^3) radicals through the combination of an iridium photocatalyst with a Lewis base. SET: Single Electron Transfer.

preparation of new bioactive compounds. These two technologies can clearly benefit from their combination with flow methods. Flow approaches can overcome their intrinsic limitations in terms of reproducibility and scalability, the main reasons why they have been traditionally overlooked by medicinal chemists.

One of the current challenges in drug discovery is increasing the complexity of the molecules by increasing the fraction of carbons with sp^3 hybridization. It has been demonstrated that highly saturated molecules are more likely to become suitable clinical candidates. Photochemistry has become an important tool to achieve this goal.

Recently, Ley and co-workers have developed a flow protocol to generate radicals from readily available boronates by combination of a Lewis base catalyst and an iridium-based photocatalyst.[33] The radicals for this process can react with cyanoheteroarenes or follow a nickel catalytic cycle in the presence of suitable bromoarenes to form new C(sp^3)–C(sp^2) bonds (Scheme 4.3).

One of the main limitations of most new photoredox chemistry developed recently from a pharma perspective is the need to use photocatalysts based on expensive and rare metals, such as iridium and ruthenium. This may compromise the future applications of this chemistry in the production of the active molecules. Janssen and researchers from Universidad de Castilla-La Mancha have developed alternative methodologies based on irradiation with visible light of commonly used metals. They demonstrated that saturated carbons can be introduced in a single step starting from readily available alkyl halides by formation of the organozinc reagent in flow, followed by an accelerated cross-coupling with aryl halides in the presence of nickel[34] or palladium[35] catalysts. The procedure is based in the well-known Negishi cross-coupling, but the effect of light allows to expand the scope of the reaction to substrates that are challenging in the absence of light. In this way, electron-rich aryl or heteroaryl derivatives, as well as aryl chlorides, can be coupled in a few minutes in high yield. The light energy is absorbed by different bimetallic interactions between organozinc derivatives and nickel or palladium complexes to accelerate the key step of the catalytic cycle: the reductive elimination in the case of nickel and the oxidative addition in the case of palladium (Scheme 4.4).

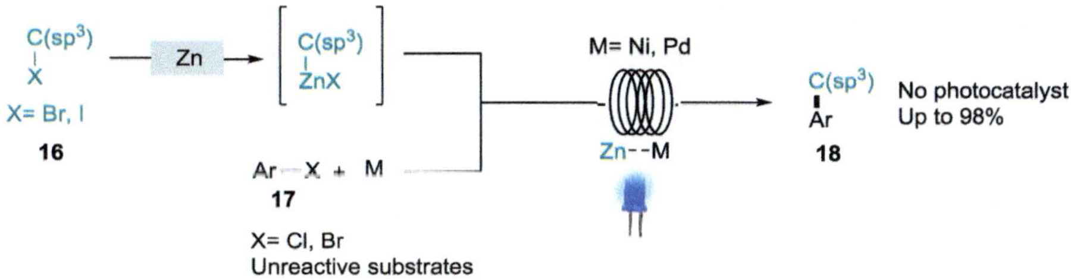

Scheme 4.4 Flow procedure for photo-Negishi reactions.

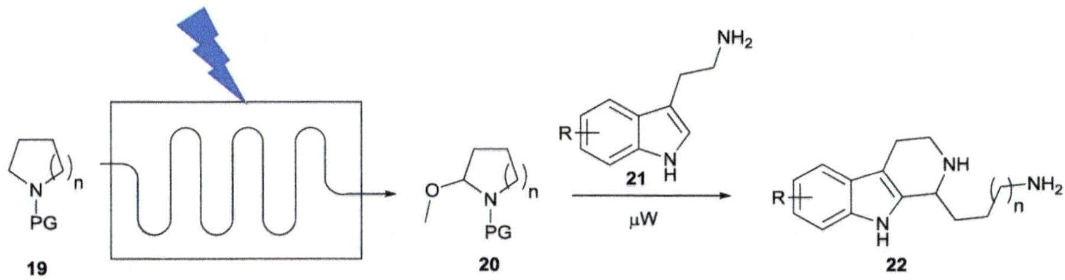

Scheme 4.5 Application of electrochemistry in flow to produce a small library of indole alkaloids.

More recently, the integration of electrochemistry in flow has received significant attention.[32] The large surface-to volume ratio achieved in flow microreactors allows much better control of the electrochemical parameters, improving its reproducibility and scalability. Ley and co-workers developed a key electrochemical Shono oxidation in flow of cyclic amines to produce α-methoxyamine intermediates that, after a Pictet–Spengler reaction under microwave irradiation, produced a small library of indole alkaloids (Scheme 4.5).[36]

4.2.1.3 *Diversity Oriented Synthesis (DOS)*

In early hit-to-lead projects, it is important to start with a diverse set of compounds to cover the largest chemical space possible. At this stage, diversity-oriented approaches are important to obtain such coverage. DOS has become an important tool to achieve this goal. In a nutshell, this approach consists of getting the maximum diversity starting from simple starting materials in few steps.

An example of this approach was described by Seeberger and co-workers. They designed different interchangeable modules to access three different scaffolds depending on their combination.[37] For instance, starting from ketones and alcohols, different combinations can provide either β-amino acids, γ-lactams, or γ-amino acids. This example represents a proof-of-concept, demonstrating how a variety of molecular frameworks can be produced in flow in a modular way.

Baumann and Baxendale in contrast focused on the diversification of 1,2,4-thiadiazoles, an important core for medicinal chemistry.[38] The preparation of the core was achieved by the reaction of benzamidines with trichloromethane sulfenyl chloride, a versatile but hazardous reagent that can be suitably handled in flow by quenching in-line to avoid malodorous and corrosive by-products. The chlorine atom of the core was then reacted with different oxygen and nitrogen atoms of sulfur-containing nucleophiles (Scheme 4.6). Following this procedure, three different modifications were introduced at the 1,2,4-thiadiazole core: an aryl group at position 5, a linker atom, and a substituent at position 3.

One of the main advantages of flow chemistry is the preparation of reactive intermediates and their use *in situ*. This can be employed for DOS purposes as the same reagent can provide different compounds by simply varying the substrate used in the subsequent reaction. As examples of this approach, organozinc and Grignard reagents were prepared by flowing the halogenated derivatives through a column containing the corresponding metal, and these reactive species were then reacted with different electrophiles to obtain a diverse set of products.[39,40] In an initial report, the organozinc reagent was quenched as it was formed in the column with ketones or nitriles to produce β-hydroxy esters or β-ketoesters as primary compounds. These primary compounds can be collected in batch over different reagents to produce saturated or unsaturated, monocyclic or bicyclic heterocycles in a single experiment (Scheme 4.7).

In a following work, Grignard reagents were prepared following the same procedure but, in this case, diversity was achieved by quenching the

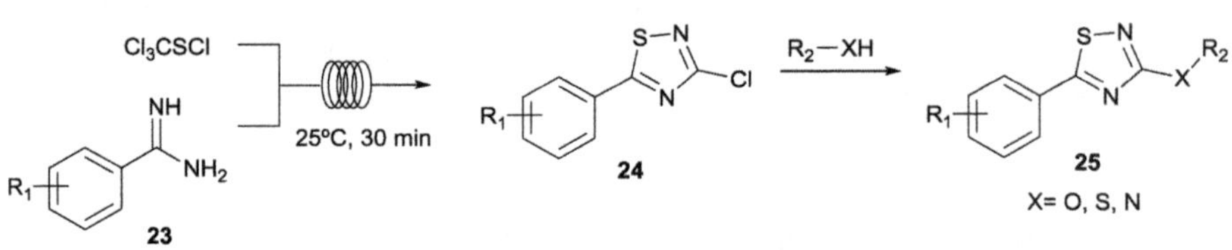

Scheme 4.6 Diversification of the 1,2,4-thiadiazole core.

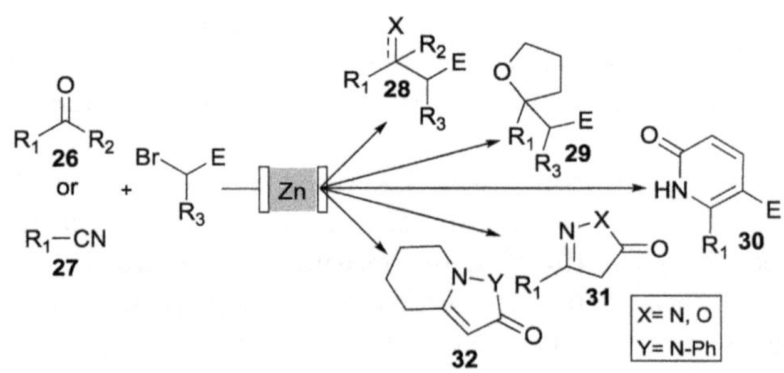

Scheme 4.7 Organozinc reagents in DOS.

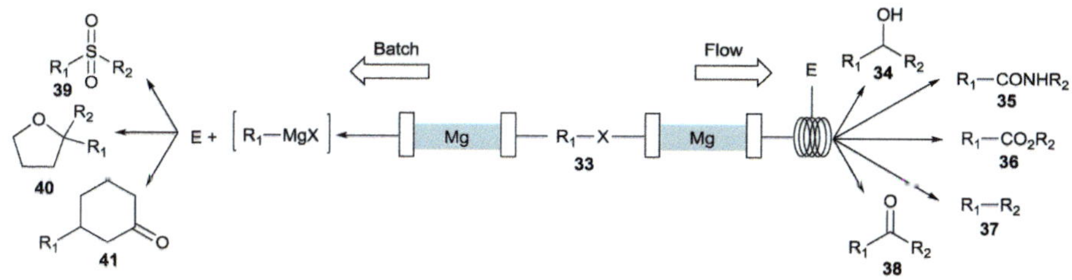

Scheme 4.8 Grignard reagents in DOS.

highly reactive intermediate with different electrophiles, either in flow or batch when the electrophiles were not soluble (Scheme 4.8). From this approach, up to eight different compounds were obtained from the same starting material.

4.2.2 Lead Optimization

The hit-to-lead and lead optimization stages are critical and rate-limiting steps in the drug discovery process and thus present great potential for new enabling technologies and innovations. The introduction of deadlines for the preparation of new entities makes medicinal chemists to continuously balance speed and optimization. The efficiency of the classical and iterative design–synthesize–screen cycle is determined by key factors including time, success rate, and cost.

Continuous flow has emerged as a novel chemical tool that can help synthetic chemists to combine efficiency and sustainability for complete integration of the early stages of drug discovery. The recent and significant advances in both chemical synthesis in flow and biological screening in flow, and the advantages of continuous flow have allowed new and important advances especially in the field of lead optimization.[11]

4.2.2.1 Integrated Platforms

Despite the new synthetic methodologies, HTS techniques, and robust structure–activity relationship (SAR) models, the major bottleneck in the design–synthesize–screen cycles is the interaction between biological evaluation and chemical synthesis.

The most usual solutions proposed to improve and optimize drug lead development are parallelization and miniaturization. Parallelization and the integration of synthesis, purification, and biological screening is the most common approach to reduce the cycle time per synthesis–assay run and enable real-time optimization of the next generation of chemical candidates, reducing the preparation of irrelevant chemical analogs and waste production and avoiding unwanted losses of time and money.

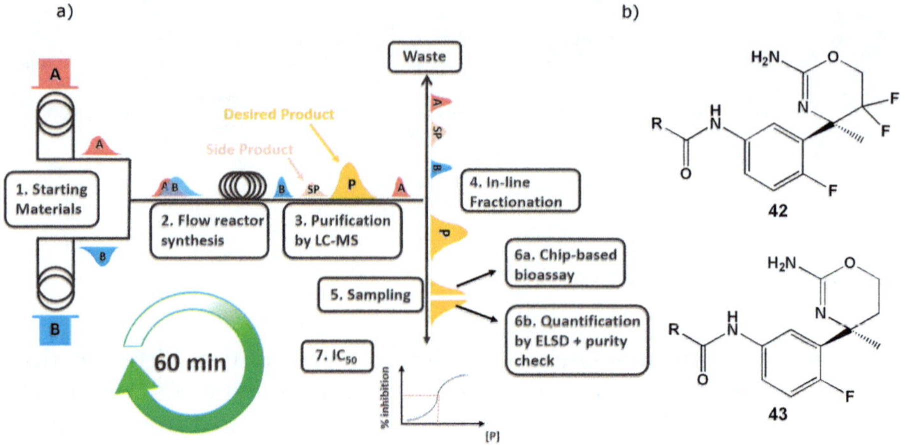

Figure 4.3 (a) Integrated system, including compound synthesis, purification, in-line UV-detected fractionation, and fractionation for bioassay and quantitation. (b) Chemical structure of the two libraries of BACE1 inhibitors. Reproduced from ref. 43 with permission from John Wiley and Sons, Copyright © 2014 WILEY-VCH Verlag GmbH & Co. KGaA, Weinheim.

From the seminal work of Hawkes in 2005 at GlaxoSmithKline,[41] there has been significant progress in the adaptation of microfluidic devices for chemical biology applications, including biological screening of compounds.[42] These applications can be benefited by flow methodologies since they require only picoliter volumes of compound solutions, resulting in reduced waste, improved reproducibility, and shorter analysis times.

Werner and co-workers reported a flow-based biochemical assay and its seamless integration in a fully automated system comprising flow chemical synthesis, purification, and in-line quantification of the compound concentration.[43] This synthesis-screening integrated platform allowed obtaining SAR data from the half maximal inhibitory concentration (IC_{50}) of β-secretase (BACE1) inhibitors at a cycle time of only 1 h instead of the several days usually needed (Figure 4.3). Since this platform led to productivity gains and cost reductions, this example represents a more efficient and sustainable alternative for drug discovery. The assay was validated using known BACE1 inhibitors by comparing the IC_{50} values using the flow bioassay and traditional well plate-based assays.

An alternative approach implies introducing a screening stage at the end of the synthesis stream with the aim of increasing the biological information with minimal amount of compound. An interesting possibility is Frontal Affinity Chromatography (FAC). In FAC, the molecule to be evaluated is continuously infused through a column containing an immobilized target biomolecule and retained in relation to its affinity (Figure 4.4). As a proof of concept of this integrated approach reported by Ley and co-workers, a set of imidazo[1,2-*a*]pyridines was prepared by

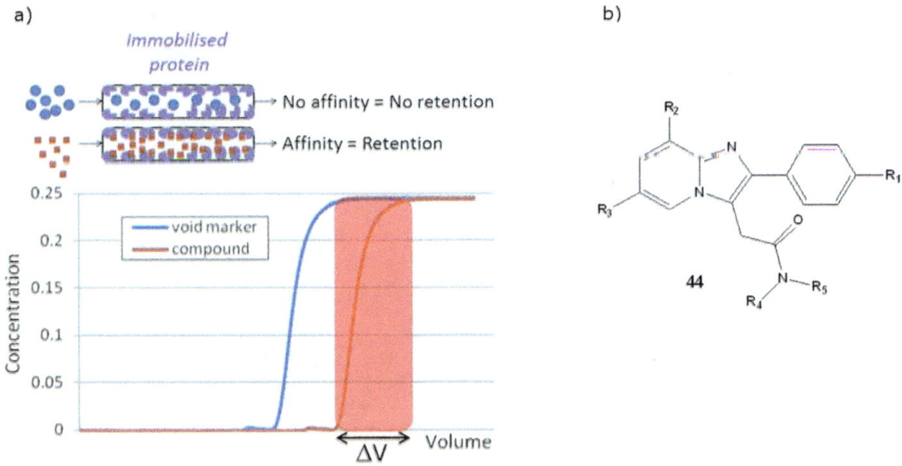

Figure 4.4 (a) Principle of Frontal Affinity Procedure. (b) Structure of the imidazo[1,2-*a*]pyridines evaluated as $GABA_A$ agonists. Reproduced from ref. 44 with permission from the Royal Society of Chemistry.

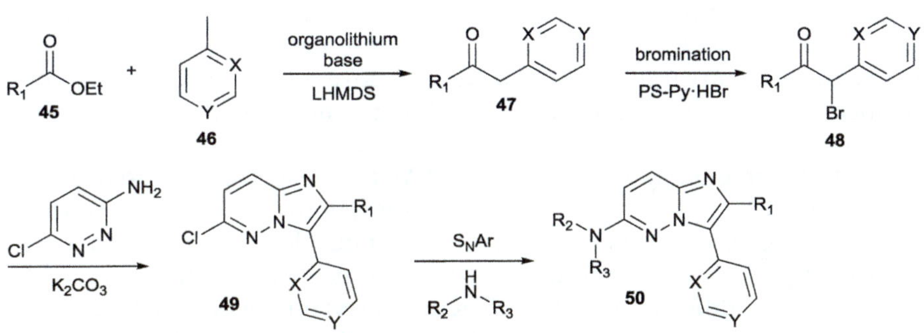

Scheme 4.9 General route for the scale up of casein kinase I inhibitors.

flow synthesis and evaluated by FAC using human serum albumin (HSA) as potential $GABA_A$ agonists.[44]

4.2.2.2 *Multigram Synthesis for* in vivo *Evaluation*

The process of lead optimization of the most promising candidates includes their preparation in multimilligram scale for appropriate biological evaluation.

In this context, Ley and co-workers demonstrated how flow-chemistry methods can facilitate the scale up of casein kinase I inhibitors using a four-step sequence of reactions (Scheme 4.9)[45] including: (i) the use of organometallic

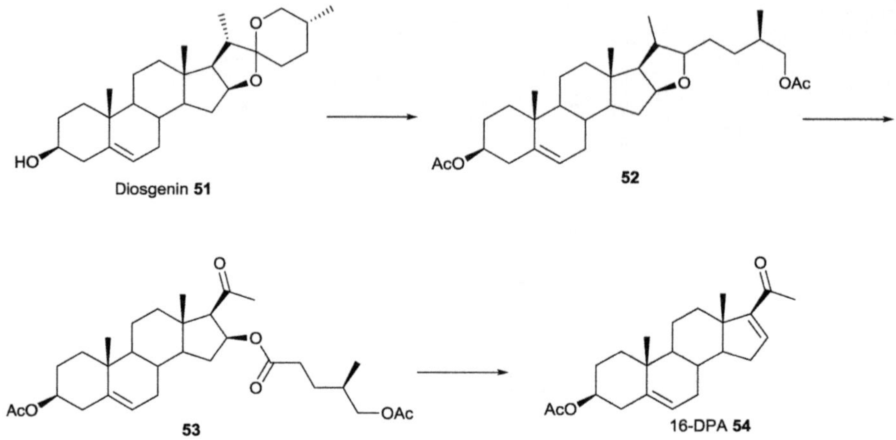

Scheme 4.10 Synthetic route from diosgenin to 16-DPA.

bases at low temperature, (ii) the application of solid-supported reagents to avoid the need for hazardous materials and facilitate the reaction work up of unstable monobrominated products, (iii) the use of an automated flow device to quickly optimize the reaction conditions and facilitate the scale up process, and (iv) an autosampler to perform automated S_NAr reactions leading to a set of twenty imidazopyridazine derivatives. The final products were obtained with excellent purity, thus avoiding the need for chromatographic purification.

16-Dehydropregnenolone acetate (16-DPA), the key compound to produce most steroidal drugs, can be prepared from diosgenin in a three-step process (Scheme 4.10). Recently, the telescoped continuous flow synthesis of 16-DPA has been reported with a productivity of 2.1 g per day for single reactor.[46] The process, in addition to higher efficiency and yield under milder reaction conditions, includes the use of lower amounts of reagents, reduced manual handling, and simpler downstream procedures. The result is high-quality products and safety standards.

4.2.3 New Molecular Entities and Active Pharmaceutical Ingredients

The adoption of continuous flow technologies for the preparation of New Molecular Entities (NMEs) and APIs is now strongly recommended by regulatory authorities such as the US Food and Drug Administration (FDA) and the European Medicines Agency (EMA). Continuous-flow chemistry approaches to synthesize NMEs and APIs offers great potential to maximize the efficiency, thus reducing significantly the overall manufacturing costs and environmental impact. These facts present added importance in developing economies,[47] where the basic chemical industry is under-developed and sometimes scarce, and new synthetic technologies are being implemented. Problems with the availability and costs of intermediates or oscillations in the amount of drug sold can leave the market unattended, and telescoped flow synthesis in that country can offer substantial benefits.[15]

The importance and incidence of flow chemistry in the preparation of NMEs and APIs has been widely reported in recent years. For example, Greiner and co-workers[7] have reviewed the challenges of flow chemistry involved in chemical transformations and the subsequent analysis and purification process, whereas De Souza and co-workers[6] focused on a comparison between batch and flow modes of production for APIs. Monbaliu and co-workers[12] focused their review in the advances related to the multistep continuous flow preparation of drugs currently in shortage, orphan drugs, and drugs listed on the World Health Organization (WHO) list of essential medicines.

4.2.3.1 *Compounds for Clinical Trials*

Prexasertib is an inhibitor of checkpoint kinase 1 (CHK1) with potential antineoplastic activity. Upon administration, prexasertib selectively binds to CHK1 preventing its activity and avoiding the repair of damaged DNA. The result is an accumulation of damaged DNA that may promote genomic instability and apoptosis.[48] Cole and co-workers have developed a multistep continuous flow CGMP process that produced 24 kg of prexasertib monolactate monohydrate for use in human clinical trials.[49] The NME is produced in laboratory fume hoods in eight continuous unit operations at approximately 3 kg per day, employing small continuous reactors, extractors, evaporators, crystallizers, and filters with higher purity than the compound obtained in batch processes (Scheme 4.11). Continuous manufacturing offers remarkable advantages, such as

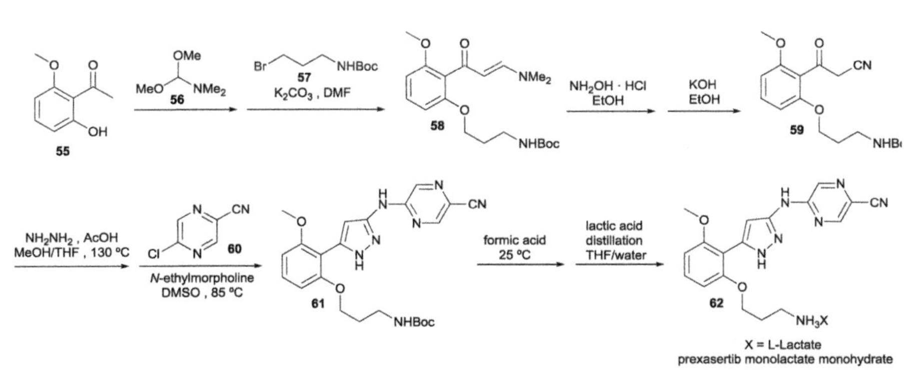

Scheme 4.11 Continuous manufacturing preparation of prexasertib monolactate monohydrate.

working at high temperature in a low-boiling solvent, improved safety for a hazardous reaction, better yield and purity *via* countercurrent multistage extraction, the use of dedicated and disposable equipment set in transparent vessels, and the elimination of steps for the isolation and elimination of solids.

AZD7594 is a non-steroidal inhaled selective glucocorticoid receptor modulator with promising prospects for the treatment of asthma and chronic obstructive pulmonary disease. Shibasaki and co-workers have reported a scalable flow synthesis of a key intermediate of AZD7594 *via* an *anti*-selective nitroaldol process (Scheme 4.12).[50] The solid-phase catalyst turnover number exceeded 1600 with high stereoselectivity, it was prepared with inexpensive metal salts, and afforded 8.89 g of the product after 398 h (81% yield, *anti*/*syn* > 20 : 1, 95% ee *anti*). In addition, the column volume for cryogenic conditions (−40 °C) was only 1.66 cm^3, considerably lower than that required under batch conditions (372 cm^3).

4.2.3.2 Making Green APIs

From an economic and environmental point of view, minimizing the waste stream associated with the production of APIs is highly interesting. Recently, Jamison and co-workers[51] reported a highly optimized and environmentally conscious continuous-flow synthesis of atropine and diazepam, two APIs identified as essential medicines by the WHO (Figure 4.5).

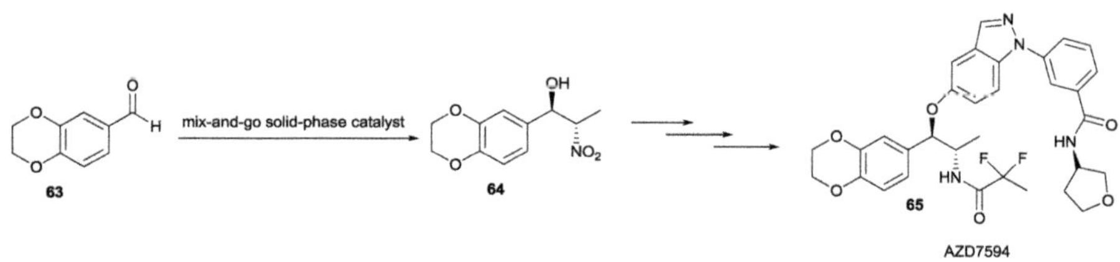

Scheme 4.12 Preparation of a key intermediate of AZD7594 in a continuous flow platform *via* a highly stereoselective nitroaldol reaction.

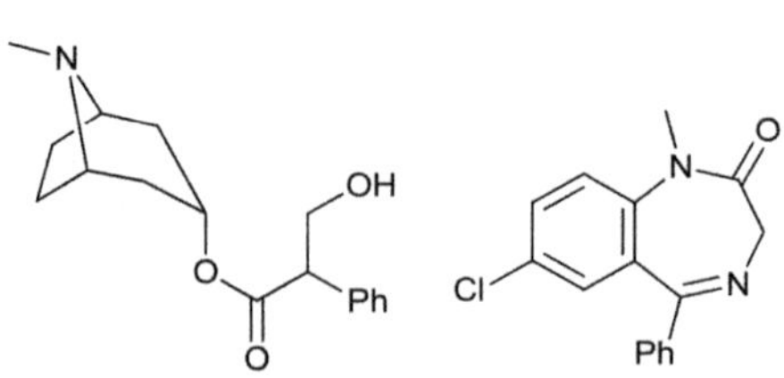

Figure 4.5 Structure of APIs atropine and diazepam.

The new procedure significantly reduced the E-factor (eqn (4.1)) of the previously reported synthesis using continuous-flow chemistry, computational calculations, and solvent minimization. These authors reduced the E-factor associated with the synthesis of atropine from 2245 to 24, about two orders of magnitude. In the case of diazepam, the E-factor was reduced four-fold, from 36 to 9.

$$\text{E-factor} = \text{kg waste/kg product} \tag{4.1}$$

In 2015, the continuous-flow synthesis of ibuprofen from its elementary building blocks was reported with a total residence time of 3 min and an average yield of over 90% for each step (Figure 4.6).[52] The authors scaled up the five-stage process to obtain ibuprofen at a rate of 8.09 g h^{-1}. Other benefits of the procedure included a Friedel–Crafts acylation run under neat conditions promoted by $AlCl_3$, liquid–liquid separations performed at 200 psi to provide solvent-free products, and the safe use of highly aggressive oxidants, such as iodine monochloride.

In a recent communication, a strategy for the telescoped flow synthesis of anti-HIV pharmaceutical drug Dolutegravir (DTG) was reported in seven steps in three separate flow operations in 24% overall yield (Scheme 4.13).[53] DTG was produced in 4.5 h from commercially available materials, avoiding the isolation of intermediates and the use of silica gel chromatography. Moreover, this synthesis should be adaptable to produce analog drugs such as Cabotegravir and Bictegravir.

4.2.3.3 *Pharmacy on Demand*

Most chemical and pharmaceutical manufacturers are located in Europe and America with an excess of manufacturing capacity, the majority of them produced by batch methods. Less developed countries present fewer manufacturers and little infrastructure to produce chemicals. Currently, these countries import APIs principally from India and China; however, prices are still high since they are conditioned by the use of equipment and raw materials.

Three approaches have been envisaged to increase the local availability of APIs in developing countries, where there is a lack of infrastructure, and especially in places where a particular API is required due to an epidemic crisis.

The first approach is research and processing in Universities of these developing countries. This approach can be exemplified by the study toward the synthesis of emtricitabine (Figure 4.7), a crucial drug in antiretroviral therapy (HAART) developed in the Nelson Mandela University in South Africa, a country with a high impact of human immunodeficiency virus-1 (HIV-1) and immunodeficiency syndrome (AIDS).[54]

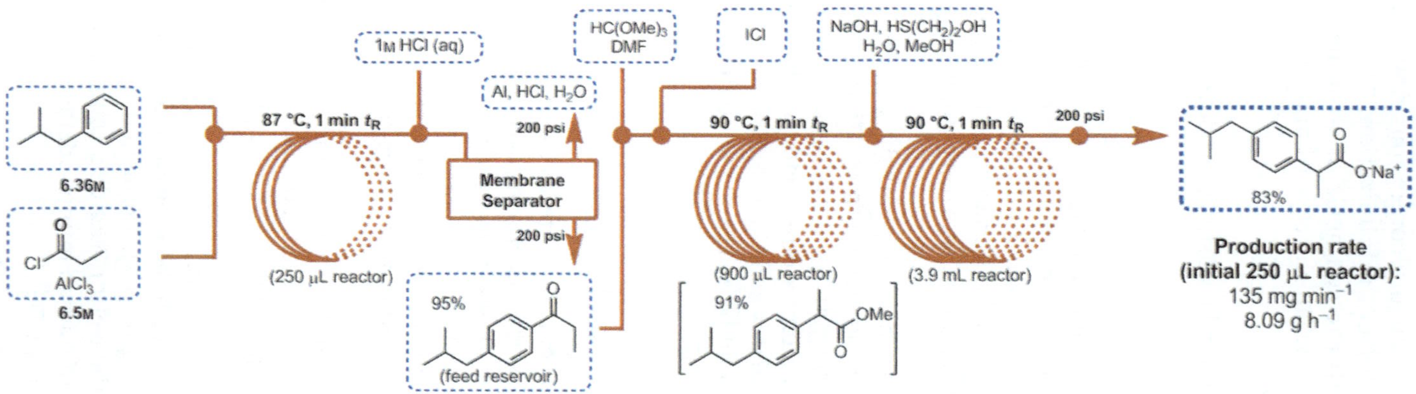

Figure 4.6 Scale-up of the three-minute ibuprofen continuous-flow synthesis. Reproduced from ref. 52 with permission from John Wiley and Sons, © 2015 WILEY-VCH Verlag GmbH & Co. KGaA, Weinheim.

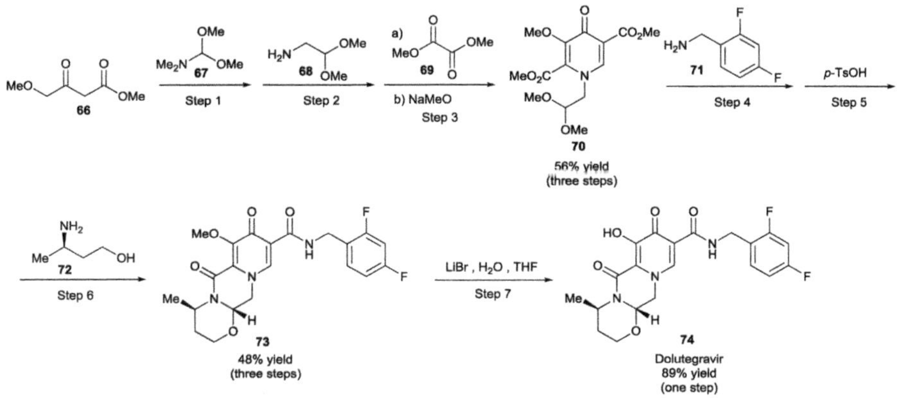

Scheme 4.13 Telescoped flow synthesis of Dolutegravir.

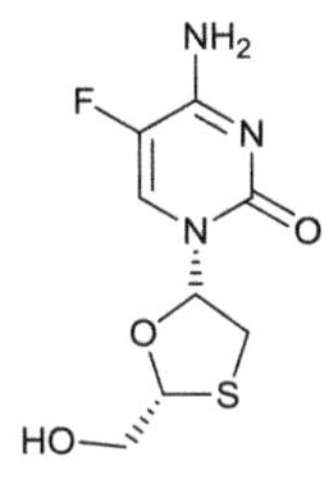

Figure 4.7 Structure of Emtricitabine 1.

The authors optimized the procedure, telescoped two steps of the synthesis and scale-up using a Uniqsis FlowSyn platform with an overall yield of 37% with 99.75% chiral purity. The University is currently working with a South African partner to establish commercial manufacturing.

A second approach is the reported 'pharmacy on demand' concept. This approach uses a compact and reconfigurable manufacturing platform that permits the in-site continuous flow synthesis and formulation of a large variety of APIs. The concept of end-to-end continuous manufacturing of pharmaceuticals was developed by Trout with the flow synthesis of Aliskiren hemifumarate (Figure 4.8).[55] The entire process, from chemical synthesis to tableting, runs continuously and is controlled automatically to check the quality of the final product. They produce 45 g h^{-1}, which corresponds to 2.7 $\times 10^6$ tablets per year (112 mg of Aliskiren per tablet).

The authors reported an economic analysis of these integrated processes and estimated savings ranging from 9% to 40%. The overall cost can be reduced by improving the recycling and direct tablet formation, and also by developing more efficient synthetic routes, separation, dosage, and plant modeling.[56]

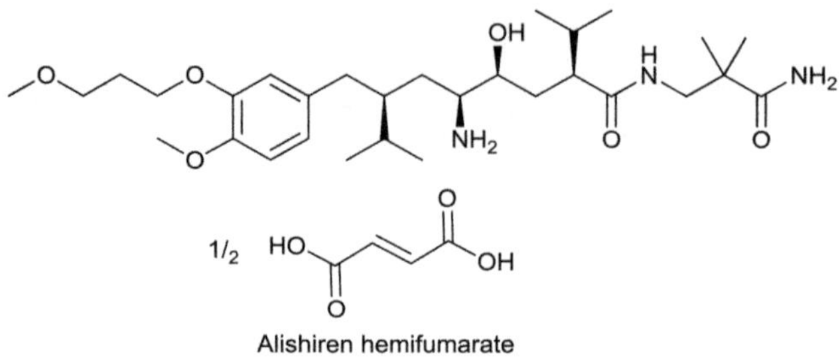

Figure 4.8 Structure of Alishiren hemifumarate.

This approach has been optimized by Behnam, Zhang, and co-workers.[17] They developed a continuous end-to-end synthesizer with a refrigerator-sized system [1.0 m (width) × 0.7 m (length) × 1.8 m (height)]. With this size and versatility, the system enabled production on demand in a simpler and cheaper way than a batch plant *in situ* for small populations or to cover the needs of an epidemic outbreak.

The system is capable of multiple-steps synthesis, in-line purification, and work-up (precipitation, filtration, crystallization), real-time monitoring (ATR, FTIR, *etc.*), and automated formulation. The authors tested the system for the preparation of diphenylhydramine, lidocaine, diazepam, and fluoxetine. In the last example, four reactors and four in-line separation units were used. Formulation was centered on liquid oral or topical dosage, but the system should be further developed to allow for the production of tablets (by 3D-printing), capsules, and sterile injection solutions.

The third approach is the use of the so-called Cloud Chemistry to perform remote-controlled experiments.[57] Cloud chemistry offers the advantages of exploiting infrastructures not available in developing countries and the facilities of the web. In this way, the experiments can be conducted on-line at any time from any location with internet access. The hosting laboratory may provide the equipment, technicians, and products to perform the synthesis. Recently, the across-the-world automated synthesis of tramadol, lidocaine, and bupropion has been reported.[58] The report used cloud chemistry employing servers located in Japan, managed in Los Angeles, and with the equipment located in Cambridge (UK). The authors described the synthesis and optimization of these three APIs.

This approach is very related to the idea of Chemical Leasing, although still some questions need to be considered in this shared relation; for example, financial support, intellectual property, and cybersecurity must be perfectly clear and secured. The idea is to maximize the access to flow instruments, promote industry–academia collaborations, and promote new business models and opportunities.

Cloud chemistry will contribute to improve reaction optimizations, machine learning, and share synthetic procedures around the world, both in industry and academia.

4.3 Conclusions and Outlook

Pharma companies are always under pressure to produce compounds in the most efficient way, and here flow chemistry can be an excellent platform to build in efficiency as well as sustainability. In this chapter, the added value that this technology offers at different stages of the drug discovery process has been presented.

Although most pharmaceutical companies are considering flow chemistry for the production and commercialization of final APIs, there is room for the application at the discovery stage of the business. Clearly, flow technologies can be adapted to the preparation of libraries followed by direct screening of the compounds, thus reducing the cycle times for the discovery of new leads. When combined with batch methods, it allows access to a diverse set of compounds that will support the exploration of the chemical space. One of the key flow chemistry advantages is the increase in the surface-to-volume ratio. This is clearly beneficial for technologies with low penetration depth, typically underused in the pharma sector due to their low reproducibility and scalability. The combination with flow makes them suitable for the exploration of new transformations and hence new chemistries.

The excellent reproducibility of the chemistry performed in flow enables easy access to multigram quantities for *in vivo* evaluation and, once the molecule is selected for human studies, to the preparation of GMP batches required for this purpose. At this stage, it should also be considered the great applicability of flow for the optimization of APIs and the possibility to redesign synthetic routes to further improve the preparation of active ingredients, thus reducing costs and waste at the same time.

As exemplified in last part of the chapter, the future of this technology is linked to its potential for automation. Clearly, this will reduce the need for large stocks of intermediates and the preparation of drugs on demand. This will have outstanding impact on the reduction of waste, not only in terms of the generated residues, but also in the reduction of time-expired medicines by adapting the production to the real market.

References

1. C. Jiménez-González, P. Poechlauer, Q. B. Broxterman, B.-S. Yang, D. am Ende, J. Baird, C. Bertsch, R. E. Hannah, P. Dell'Orco, H. Noorman, S. Yee, R. Reintjens, A. Wells, V. Massonneau and J. Manley, *Org. Process Res. Dev.*, 2011, **15**, 900.
2. D. J. C. Constable, P. J. Dunn, J. D. Hayler, G. R. Humphrey, J. L. Leazer, R. J. Linderman, K. Lorenz, J. Manley, B. A. Pearlman, A. Wells, A. Zaks and T. Zhang, *Green Chem.*, 2007, **9**, 411.
3. S. A. May, *J. Flow Chem.*, 2017, **7**, 137.

4. P. T. Anastas and J. Warner, *Green Chemistry. Theory and Practice*, Oxford University Press, New York, 1998.
5. P. T. Anastas and J. B. Zimmerman, *Environ. Sci. Technol.*, 2003, **37**, 94A.
6. J. M. De Souza, R. Galaverna, A. A. N. De Souza, T. J. Brocksom, J. C. Pastre, R. O. M. A. De Souza and K. T. De Oliveira, *An. Acad. Bras. Cienc.*, 2018, **90**, 1131.
7. P. Bana, R. Örkényi, K. Lövei, Á. Lakó, G. I. Túrós, J. Éles, F. Faigl and I. Greiner, *Bioorg. Med. Chem.*, 2017, **25**, 6180.
8. K. P. Cole and M. D. Johnson, *Expert Rev. Clin. Pharmacol.*, 2018, 5.
9. M. Baumann, *Org. Biomol. Chem.*, 2018, **16**, 5946.
10. (a) V. Hessel, D. Kralisch, N. Kockmann, T. Nöel and Q. Wang, *ChemSusChem*, 2013, **6**, 746; (b) S. C. Stouten, T. Nöel, Q. Wang and V. Hessel, *Aust. J. Chem.*, 2013, **66**, 121.
11. G. S. Fleming and A. B. Beeler, *J. Flow Chem.*, 2017, **7**, 124.
12. R. Gerardy, N. Emmanuel, T. Toupy, V.-E. Kassin, N. N. Tshibalonza, M. Schmitz and J.-C. M. Monbaliu, *Eur. J. Org. Chem.*, 2018, 2301.
13. D. Dallinger and C. O. Kappe, *Curr. Opin. Green Sustain. Chem.*, 2017, **7**, 6.
14. G. Van der Vorst, W. Aelterman, B. De Witte, B. Heirman, H. Van Langenhove and J. Dewulf, *Green Chem.*, 2013, **15**, 744.
15. R. O. M. A. de Souza and P. Watts, *J. Flow Chem.*, 2017, **7**, 146.
16. M. M. Bio, *Pharma Horizon*, 2018, **2**, 38.
17. A. Adamo, R. L. Beingessner, M. Behnam, J. Chen, T. F. Jamison, K. F. Jensen, J.-C. M. Monbaliu, A. S. Myerson, E. M. Revalor, D. R. Snead, T. Stelzer, N. Weeranoppanant, S. Y. Wong and P. Zhang, *Science*, 2016, **352**, 61.
18. L. Malet-Sanz and F. Susanne, *J. Med. Chem.*, 2012, **55**, 4062.
19. R. Macarron, M. N. Banks, D. Bojanic, D. J. Burns, D. A. Cirovic, T. Garyantes, D. V. S. Green, R. P. Hertzberg, W. P. Janzen, J. W. Paslay, U. Schopfer and G. S. Sittampalam, *Nat. Rev. Drug Discovery*, 2011, **10**, 188.
20. D. C. Rees, M. Congreve, C. W. Murray and R. Carr, *Nat. Rev. Drug Discovery*, 2004, **3**, 660.
21. K. H. Bleicher, H.-J. Böhm, K. Müller and A. I. Alanine, *Nat. Rev. Drug Discovery*, 2003, **2**, 369.
22. L. A. Thompson and J. A. Ellman, *Chem. Rev.*, 1996, **96**, 555.
23. W. R. J. D. Galloway, A. Isidro-Llobet and D. R. Spring, *Nat. Commun.*, 2010, **1**, 1–13.
24. A. Karawajczyk, F. Giordanetto, J. Benningshof, D. Hamza, T. Kalliokoski, K. Pouwer, R. Morgentin, A. Nelson, G. Müller, A. Piechot and D. Tzalis, *Drug Discovery Today*, 2015, **20**, 1310.
25. C. M. Thompson, J. L. Poole, J. L. Cross, I. Akritopoulou-Zanze and S. W. Djuric, *Molecules*, 2011, **16**, 9161.
26. J. E. Hochowski, P. A. Searle, N. P. Tu, J. Y. Pan, S. G. Spanton and S. W. Djuric, *J. Flow Chem.*, 2011, **2**, 56.

27. Y.-J. Hwang, C. W. Coley, M. Abolhasani, A. L. Marzinzik, G. Koch, C. Spanka, H. Lehmannd and K. F. Jensen, *Chem. Commun.*, 2017, **53**, 6649.
28. D. Perera, J. W. Tucker, S. Brahmbhatt, C. J. Helal, A. Chong, W. Farrell, P. Richardson and N. W. Sach, *Science*, 2018, **359**, 429.
29. M. C. F. C. B. Damião, R. Galaverna, A. P. Kozikowski, J. Eubanks and J. C. Pastre, *React. Chem. Eng.*, 2017, **2**, 896.
30. Y. Su, N. J. W. Straathof, V. Hessel and T. Noël, *Chem. - Eur. J.*, 2014, **20**, 10562.
31. D. Cambié, C. Bottecchia, N. J. W. Straathof, V. Hessel and T. Noël, *Chem. Rev.*, 2016, **116**, 10276.
32. M. Atobe, H. Tateno and Y. Matsumura, *Chem. Rev.*, 2018, **118**, 4541.
33. F. Lima, M. A. Kabeshov, D. N. Tran, C. Battilocchio, J. Sedelmeier, G. Sedelmeier, B. Schenkel and S. V. Ley, *Angew. Chem., Int. Ed.*, 2016, **55**, 14085.
34. I. Abdiaj, A. Fontana, M. V. Gomez, A. de la Hoz and J. Alcázar, *Angew. Chem., Int. Ed.*, 2018, **57**, 8473.
35. I. Abdiaj, L. Huck, J. M. Mateo, A. de la Hoz, M. V. Gomez, A. Díaz-Ortiz and J. Alcázar, *Angew. Chem., Int. Ed.*, 2018, **57**, 13231.
36. M. A. Kabeshov, B. Musio, P. R. D. Murray, D. L. Browne and S. V. Ley, *Org. Lett.*, 2014, **16**, 4618.
37. D. Ghislieri, K. Gilmore and P. H. Seeberger, *Angew. Chem., Int. Ed.*, 2015, **54**, 678.
38. M. Baumann and I. R. Baxendale, *Bioorg. Med. Chem.*, 2017, **25**, 6218.
39. L. Huck, M. Berton, A. de la Hoz, A. Díaz-Ortiz and J. Alcázar, *Green Chem.*, 2017, **19**, 1420.
40. L. Huck, A. de la Hoz, A. Díaz-Ortiz and J. Alcázar, *Org. Lett.*, 2017, **19**, 3747.
41. S. Y. F. W. Hawkes, M. J. V. Chapella and M. Montembault, *QSAR Comb. Sci.*, 2005, **24**, 712.
42. J. Alcázar, in *Sustainable Flow Chemistry: Methods and Applications*, ed. L. Vaccaro, Wiley-VCH, Wienheim, 2017, p. 135.
43. M. Werner, C. Kuratli, R. E. Martin, R. Hochstrasser, D. Wechsler, T. Enderle, A. I. Alanine and H. Vogel, *Angew. Chem., Int. Ed.*, 2014, **53**, 1704.
44. L. Guetzoyan, N. Nikbin, I. R. Baxendale and S. V. Ley, *Chem. Sci.*, 2013, **4**, 764.
45. F. Venturoni, N. Nikbin, S. V. Ley and I. R. Baxendale, *Org. Biomol. Chem.*, 2010, **8**, 1798.
46. V. Mancino, B. Cerra, A. Piccinno and A. Gioiello, *Org. Process Res. Dev.*, 2018, **22**, 600.
47. J. Verghese, C. J. Kong, D. Rivalti, E. C. Yu, R. Krack, J. Alcázar, J. B. Manley, D. T. McQuade, S. Ahmad, K. Belecki and B. F. Gupton, *Green Chem.*, 2017, **19**, 2986.

48. F. S. Farouz, R. C. Holcomb, R. Kasar and S. S. Myers, *WO Pat.*, WO2010077758A1, 2010.
49. K. P. Cole, J. M. Groh, M. D. Johnson, C. L. Burcham, B. M. Campbell, W. D. Diseroad, M. R. Heller, J. R. Howell, N. J. Kallman, T. M. Koenig, S. A. May, R. D. Miller, D. Mitchell, D. P. Myers, S. S. Myers, J. L. Phillips, C. S. Polster, T. D. White, J. Cashman, D. Hurley, R. Moylan, P. Sheehan, R. D. Spencer, K. Desmond, P. Desmond and O. Gowran, *Science*, 2017, **356**, 1144.
50. A. Nonoyama, N. Kumagai and M. Shibasaki, *Tetrahedron*, 2017, **73**, 1517.
51. A.-C. Bédard, A. R. Longstreet, J. Britton, Y. Wang, H. Moriguchi, R. W. Hicklin, W. H. Green and T. F. Jamison, *Bioorg. Med. Chem.*, 2017, **25**, 6233.
52. D. R. Snead and T. F. Jamison, *Angew. Chem., Int. Ed.*, 2015, **54**, 983.
53. R. E. Ziegler, B. K. Desai, J.-A. Jee, B. F. Gupton, T. D. Roper and T. F. Jamison, *Angew. Chem., Int. Ed.*, 2018, **57**, 7181.
54. D. Mandala, S. Chada and P. Watts, *Chem. Today*, 2018, **36**, 28.
55. S. Mascia, P. L. Heider, H. Zhang, R. Lakerveld, B. Benyahia, P. I. Barton, R. D. Braatz, C. L. Cooney, J. M. B. Evans, T. F. Jamison, K. F. Jensen, A. S. Myerson and B. L. Trout, *Angew. Chem., Int. Ed.*, 2013, **52**, 12359.
56. S. D. Schaber, D. I. Gerogiorgis, R. Ramachandran, J. M. B. Evans, P. I. Barton and B. L. Trout, *Ind. Eng. Chem. Res.*, 2011, **50**, 10083.
57. R. A. Skilton, R. A. Bourne, Z. Amara, R. Horvath, J. Jin, M. J. Scully, E. Streng, S. L. Y. Tang, P. A. Summers, J. Wang, E. Pérez, N. Asfaw, G. L. P. Aydos, J. Dupont, G. Comak, M. W. George and M. Poliakoff, *Nat. Chem.*, 2015, **7**, 1.
58. D. E. Fitzpatrick, T. Maujean, A. Evans and S. V. Ley, *Angew. Chem., Int. Ed.*, 2018, **57**(46), 15128–15132.

CHAPTER 5

Sustainable Batch or Continuous-flow Preparation of Biomass-derived Fuels Using Sulfonated Organic Polymers

VALERIA TROMBETTONI, FILIPPO CAMPANA, ASSUNTA MARROCCHI AND LUIGI VACCARO*

Laboratory of Green Synthetic Organic Chemistry, Dipartimento di Chimica, Biologia e Biotecnologie, Università degli Studi di Perugia, via Elce di Sotto 8, 06100 Perugia, Italy
*E-mail: luigi.vaccaro@unipg.it

5.1 Introduction

Research and development directed toward building up innovation around the production processes of biomass-derived fuels still stands at the forefront of the liquid fuel industry sector. This is strongly tied to the search for process intensification and improved product purity later on, as well as reduced power consumption and manufacturing costs. The development of innovative bio-products with excellent fuel qualities is also the subject of further research.

Many national and state governments in Europe and worldwide have established high targets in terms of the use of biofuels in the next few years, and have introduced financial incentives to promote their diffusion.

Green Chemistry Series No. 61
Green Synthetic Processes and Procedures
Edited by Roberto Ballini

Published by the Royal Society of Chemistry, www.rsc.org

For instance, according to the EU Renewable Energy Directive set aside in 2009,[1] the mandate for renewable fuels production is at least 10% of all road transport fuels by 2020. In addition, a revised Fuel Quality Directive requires a 6% cut in carbon intensity on the fossil diesel and gasoline levels in the road transport fuel mix in the EU by 2020.[2]

Thus, a real market opportunity exists that could drive the biofuel production and distribution further. Most certainly, the incentive programs supporting the companies include a series of requirements to ensure that biofuels are produced through sustainable, safe and cost-efficient manufacturing processes.

This section is organized as follows. First, the basic concepts underlying biodiesel fuel are introduced, with a reference to the pro's & con's of the traditional industrial process. Next, we focus on two main key technologies that, particularly when combined each other, play a key role in the realization of sustainable chemical production, *i.e.*, heterogeneous catalysis and flow chemistry. Thus, a brief discussion of heterogeneous catalysis, focusing specifically on organic polymer-supported acid catalysts is provided. Moreover, a survey of the principles and operating characteristics of continuous flow technologies is given. The aim of the above is to provide the essential background for the subsequent sections, which give an overview of the recent advances in the field.

5.1.1 Biodiesel

Of the biomass-derived diesel fuels, biodiesel has gained momentum as a complement or substitute of its traditional counterpart, with production reaching ~33 billion liters in 2016.[3]

Biodiesel is made up of a mixture of alkyl esters of C_{10}–C_{22} fatty acids (FAs), typically methyl esters (FAMEs, Figure 5.1). It is a low-toxic, clean-burning, biodegradable fuel that can be derived from a variety of feedstocks, including (non-food) plant oils or animal fats; although the latter are limited in volume; microalgae are an emerging feedstock, albeit the associated biodiesel production technology is still in its infancy.

A large amount of low-cost, not feed–food competitive raw materials, such as non-edible oils or restaurant waste oils, can be in principle converted into biodiesel. The problem with processing these feedstocks is that they often contain a large amount of free fatty acids (FFAs) along with water, as well as traces of metal salts, that may interfere with the typical biodiesel production process (*vide infra*).

Biodiesel is produced from the triglycerides (Figure 5.1) that make up the vegetal oils or animal fats and a short-chain alcohol (namely methanol or ethanol) through a transesterification reaction. Crude glycerol is the main by-product of biodiesel production,[4,5] accounting for ~10% by weight of the starting material.

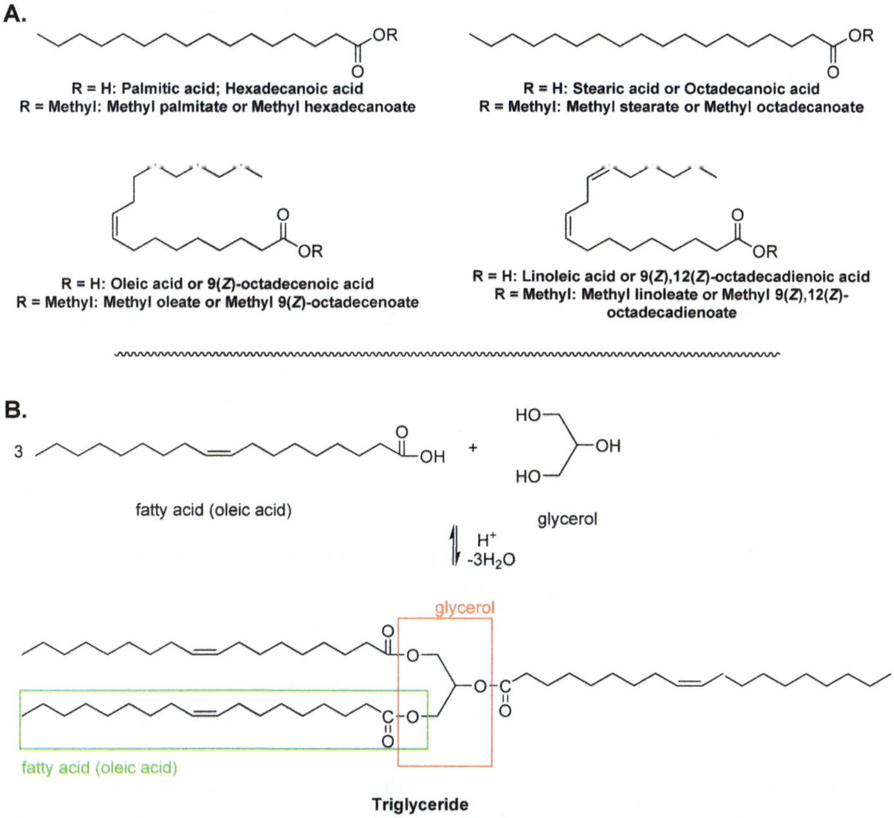

Figure 5.1 (a) Chemical structure of common long chain fatty acids and corresponding methyl esters (FAMEs), and (b) chemical structure of a common triglyceride. Reproduced from ref. 65 with permission from Elsevier, Copyright 2018.

The transesterification process can be promoted in principle by a wide range of catalysts, including homogeneous systems (*i.e.*, bases, acids, metal complexes), heterogeneous catalysts, phase-transfer catalysts, and biocatalysts (enzymes).[4,6–10]

For the industrial production process, however, triglycerides are typically converted into biodiesel esters with the help of homogeneous strong base catalysts such as alkali metal hydroxides and alkoxides. Thus, when using feedstocks with high FFA content (see above), the alkali catalyst will react undesirably with the latter, thereby forming significant amounts of soap, which contaminate the fuel, thereby lowering its yield and ultimately rising its production costs. Therefore, a refining process to remove FFAs from raw oil to less than 3 wt% is normally required for biodiesel fuel production.[11–13] A well-consolidated strategy envisages FFA esterification with methyl alcohol typically catalyzed by strong homogeneous acids (*e.g.*, H_2SO_4). The major drawbacks to the use of such feedstock 'pre-treatment' are that the catalyst must be subsequently neutralized, and the resulting salts disposed; moreover, high temperatures are simultaneously required, which are associated with equipment corrosion issues. Therefore, the development of efficient, selective, and stable acid catalysts that can make viable the cost-effective production of biodiesel remains a great challenge.

5.1.2 Heterogeneous Organic Polymer-based Acid Catalysts

For several industrial processes, both petro- and bio-based homogeneous catalysts are still preferred, despite their most frequent time-consuming and costly removal from the reaction mixture. However, heterogeneous catalysts are considered more appealing, which is motivated by the fact that they can be, in principle, easily separated from the reaction mixture by filtration or decantation and recovered, and they may be reused for several times, which enable to set waste-minimized protocols. In this context, a heterogeneous catalyst is a key tool for a sustainable approach to biofuels, which is the topic of this chapter. Moreover, replacing conventional homogeneous acid catalysts with their heterogeneous counterparts can overcome the equipment corrosion (see above).[6,7] Heterogeneous catalysts also open the possibility to convert batch operations, which are typical of reactions promoted by homogeneous systems, into continuous processing through fixed-bed reactors, this aspect leading to further reduction of the waste associated with a given chemical process and ultimately to process intensification.[14]

Briefly, a heterogeneous catalyst consists, in most cases, in a solid support on which given active sites should be stably anchored. In order to enable a reaction to occur, first the reactants must be adsorbed on the catalyst surface and, thereafter, the obtained products should be easily released (the so-called Langmuir–Hinshelwood model).[6,7] With this concept in mind, it becomes clear the crucial role played by the support in the catalyst activity and/or selectivity, and the consequent need to tailor the support based on its affinity to the structural features of the reactant(s)/solvent/product(s) to limit, *e.g.*, the effect of mass transfer between the different phases of all the reaction components and the accessibility of the catalytic sites by the reactants in solution, so as to make heterogeneous systems competitive with their homogeneous counterparts.

In the context of acid-catalyzed biodiesel preparation, there is a difference in the mechanism of the esterification reaction when homogeneous or heterogeneous catalysts are used, which correlates with the degree of hydrophobicity of the heterogeneous catalyst surface.[15] More specifically, with the increasing number of (hydrophilic) Brönsted acid sites, surrounded by a hydrophobic environment, the lipophilic tail of a fatty acid likely approaches and adsorbs in a parallel (Figure 5.2, left)

Figure 5.2 Cartoon representing the influence of the hydrophobicity of the catalyst surface on the activity.

to perpendicular (Figure 5.2, middle) fashion with respect to the hydrophobic surface; in this latter case, a locally hydrophobic environment is formed. However, if there are many adjacent (hydrophilic) acidic sites, water, which is the by-product of the esterification reaction, will be preferentially adsorbed on the catalyst surface (Figure 5.2, right), forming a layer that hinders the access of the substrates (*i.e.*, acid and alcohol). In other words, water causes the deactivation of the catalyst. On the other hand, an excess of the alcohol may avoid the catalyst deactivation, since it mixes preferentially with water than with the fatty acid, and the reaction mixture remains 'monophasic' until equilibrium conversion is reached; the use of a large excess of alcohol, however, is not desirable from the point of view of the sustainable production of biofuel, unless it can be conveniently recycled.

A variety of different heterogeneous acid catalysts have been used in recent times both for FFA esterification and simultaneous esterification/transesterification reactions, *e.g.*, those based on inorganic materials such as zeolites, metal oxides and silicas, or organic insoluble polymers, nanoparticles, carbonaceous materials, including graphene and carbon nanotubes, to name a few.[6,7,16–18]

Generally speaking, organic polymer-supported systems are among the most versatile heterogeneous catalysts, enabling a wide range of organic fundamental transformations, including the (trans)esterification reaction to produce biofuels. Another key attraction is the tunability of the physicochemical and morphological properties of the organic polymer structure, as well as the availability of efficient, functional group-tolerant synthetic protocols allowing their controlled preparation, as well as the appropriate incorporation of catalytic centers.[19–22]

Cation-exchange resins are by far the most widely used organic polymer-supported heterogeneous acid catalysts, the sulfonated styrene (S)-divinylbenzene (DVB) copolymer species finding the majority of applications (Figure 5.3, left).[23–32] The S-DVB copolymer support can be readily obtained in the form of spherical beads of suitable size by suspension polymerization, in which the organic phase containing the monomer(s) and free radical initiator is dispersed as droplets through efficient stirring in an aqueous phase,

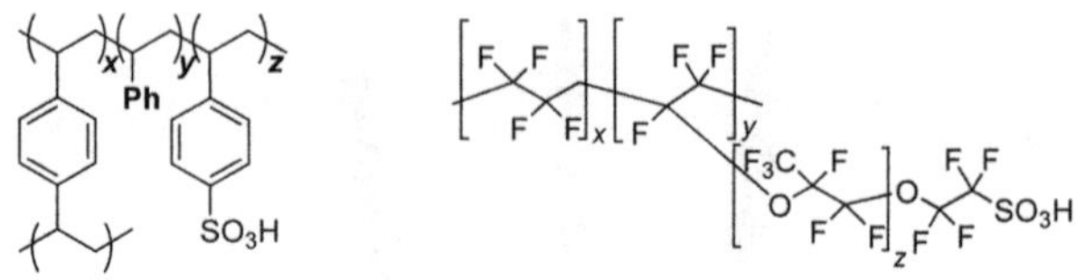

Figure 5.3 Structure of sulfonated crosslinked polystyrene (left) and perfluorinated sulfonic acid resin (right). Reproduced from ref. 65 with permission from Elsevier, Copyright 2018.

often containing a dissolved polymer (*e.g.*, polyvinyl alcohol) to help stabilize such droplets. Polymerization is initiated by heating the mixture typically at 70–90 °C and, during the reaction, the monomer droplets are converted into solid spherical polymer beads.

So-called **gel-type** resins are obtained from, *e.g.*, styrene monomer and low levels of a crosslinker (*e.g.*, DVB), typically <5 mol%; no other components except the free radical initiator are involved.[33] The resulting materials consist typically of hard transparent beads, featuring a surface area that is low in the dry state (typically <10 m^2 g^{-1} as per the Brunauer–Emmet–Teller (BET) model). However, gel-type resins can swell in solvents with associated solubility parameters similar to those of the co-polymer (so-called 'good' solvents), originating therefore a porous structure and increasing the penetrability in the network segments. Not unexpectedly, the swelling degree relates inversely to the crosslinker content.

The so-called **macroporous** (**macroreticular**) resins are typically prepared from a vinyl monomer (*e.g.*, styrene) and significant amounts of DVB (10–80 mol%). It is worth-mentioning here that 'macroporous' in this context indicates a class of resins exhibiting a permanently porous morphology, even in the dry state.

The morphology of these resins is generated through the introduction of a suitable organic solvent (porogen) in the polymerizing mixture, which induces the growing polymer network to phase separate/precipitate during the suspension polymerization. At the end of polymerization, after removal of the porogen, hard, opaque resin beads with a permanent porous structure are formed. Macroporous resins typically feature a surface area in the dry state much higher (50–100 m^2 g^{-1}) than gel-type resins and exhibit a pore size distribution ranging from micropores (<2 nm) to mesopores (2–50 nm) to macropores (>50 nm). To finely control this morphology, careful selection of the nature and the amount of both the porogen and crosslinker employed is required. Amberlyst® 15 is a well-known commercially available example of macroreticular styrene-based sulfonic acid.[34–36]

Another group of resins with unique morphology are the **hyper-crosslinked** species. In their most interesting form, a conventional gel-type copolymer (S-DVB), lightly crosslinked, is subjected to extensive crosslinking in a post-polymerization Friedel–Crafts reaction involving a bis-alkylating agent (*e.g.*, di-halogen methyl hydrocarbons) and a Lewis acid such as Al_2Cl_3. Permanent microporous structures with surface areas up to 1000 m^2 g^{-1} can be obtained.[37]

Perfluorinated resins represent another well-known class of (non-porous) sulfonated cation exchangers that have been demonstrated to have an acidity as high as that of 100% H_2SO_4.[38–40] In particular, Figure 5.3 (right) shows a Nafion® resin, which is a sulfonated tetrafluoroethylene-based fluoropolymer-copolymer. Nafion® resins (*e.g.*, Nafion® NR50) are commercially available as millimeter-sized beads and their dry-state surface area in this form is very low.

5.1.3 Flow Technology

Flow technology[41–52] is widely recognized as a fundamental tool for use on laboratory as well as industrial scale when manufacturing large quantities of a given material, and to develop waste-minimized and sustainable chemical processes.[51–62] A flow set-up is typically made up of several devices connected one to each other that allow performing synthetic processes in continuous regime; the design of the reactors, tailored on the basis of the protocol to be developed, is a crucial aspect of flow technology. Typically, flow reactors incorporate a heterogeneous catalysts (and/or heterogeneous reactants), whereas the homogeneous reactants and, when applicable, the reaction medium are fed to the reactor by means of a suitable pump.

A broad list of advantages in the use of flow reactors compared to traditional stirred tank batch reactors can be identified; for example, reactant mixing can be achieved within seconds, the reaction temperature can be raised above the boiling point of the reaction components since the volume of the devices is typically small, heat transfer is intensified, higher yield and selectivity are typically achieved, and less manpower is needed. Among others, typical flow drivers offer higher safety levels than conventional batch equipment. When used in combination with a heterogeneous catalyst, flow conditions prevent the breakdown to fines occurring under stirring in batch mode, and the substrate is forced into an excess of catalyst, allowing for significant minimization of the reaction time. Moreover, continuous flow reactors operate under steady-state conditions and can therefore be automated with far less effort than the batch counterpart; this automation allows also to sequentially investigate a range of possible reaction parameters (*e.g.*, reactants stoichiometry, residence time, temperature). In addition to the above advantages, continuous flow technology enables the arrangement of multi-step reactions in a continuous sequence, which is beneficial especially when the generated intermediates are, *e.g.*, unstable. It is also possible to straightforwardly perform multi-phase reactions, and readily scale up flow reactors with little process development work, by either changing the reactor size or connecting several reactors in series (the so-called 'numbering up').

5.2 Batch Production of FAMEs *via* (Trans) esterification Mediated by Sulfonated Cation-exchange Resins

In this section, we summarize the most representative studies on the production of biodiesel fuel through (trans)esterification reactions mediated by organic polymer-supported acid catalysts under traditional batch conditions. We focus the discussion on sulfonated cation-exchange resin-mediated (trans)esterification to FAMEs, which stands as the most widely explored route in the field to date, although several interesting studies in the area of supported acidic ionic liquids, ionomeric membranes, and hybrid systems have been carried out.[63–67] The influence of parameters, such as the catalyst morphology, reactant molar ratio, and FFA and impurity (including water) contents in the oil feedstock, on the activity of the catalysts themselves, as well as on their recyclability will be critically discussed.

Rothenberg and co-workers[68] explored the activity of two different commercially available benchmarks, *i.e.*, Amberlyst® 15 and Nafion® NR50 (see above) in the esterification reaction of lauric acid (a C_{12} fatty acid) with 2-ethylhexanol, taking into account that macroreticular Amberlyst® 15 exhibits a large pore volume and BET surface area (4.8 $cm^3\ g^{-1}$ and ~50 $m^2\ g^{-1}$, respectively), whereas Nafion® NR50 is a non-porous resin with a BET surface area of ~0.02 $m^2\ g^{-1}$. Both catalysts were employed at 3 wt% loading in the presence of an acid/alcohol molar ratio of 1 : 1 and 150 °C, whereby Amberlyst® 15 showed the highest activity, reaching 96% conversion to the target ester. The lower acid conversion achieved with Nafion® NR50 (78%) was attributed in part to the low swelling degree of such a (non-porous) catalyst at the low alcohol content employed.[74] The reaction proved to be selective in both cases, with no side reactions such as ether formation or 2-ethylhexanol dehydration taking place. Despite the good esterification results, however, Amberlyst® 15 and Nafion® NR50 deactivated after short times under the above conditions (*i.e.*, 2 and 4.5 h, respectively). The faster deactivation of porous Amberlyst® 15 was ascribed to solvation of the catalytic sites by the esterification by-product water, which makes the protons less available to assist the ester formation, as well as to the preferential adsorption of the water itself on the catalyst inner/outer surfaces, leading to limited accessibility of the (hydrophobic) lauric acid.[69] Marchetti and co-workers used Amberlyst® 15 for the esterification of oleic acid with ethanol.[70] The authors found that, by employing 20 wt% of catalyst and an optimal 1 : 1 reactants molar ratio at 60 °C, the oleic acid conversion into the corresponding ethyl ester reached 53% after 6 h. It is noteworthy to mention here that a study on the influence of the alcohol-to-acid molar ratio on the catalyst activity was also carried out, and it was demonstrated that in the presence of excess alcohol, the oleic acid conversion decreased, which appears to be somehow counter-intuitive, the esterification reaction being an equilibrium. The authors, however, claimed that the observed effect can be explained by the fact that the reaction was considered not at equilibrium, but at an initial stage. In other words, a 1 : 1 alcohol/acid molar ratio enabled the highest initial reaction rate under the investigated conditions. The possibility of recycling the catalyst was also investigated, and it was observed a catalyst activity loss after a single re-use of the system, which was ascribed to the solvation of active sites by the esterification by-product water.

Amberlyst® 15 was also used by Paterson *et al.*[71] in the transesterification reaction of triolein featuring an acid value of 2.5 mg KOH g^{-1} with methanol. Under the optimized reaction conditions of a 24 : 1 alcohol/oil molar ratio, 13 wt% catalyst loading, and 200 °C, a 97% conversion of triolein was achieved after 10 h. Recyclability studies showed that the activity of Amberlyst® 15 decreased by *ca.* 17% compared to that of the fresh catalyst after three runs. Ilgen and co-workers[72,73] investigated the activity of commercially available macroreticular Amberlyst® 46 in the esterification reaction of oleic acid with methanol. Under optimized conditions (3 : 1 alcohol excess, 15 wt% catalyst, 100 °C, 2 h), a nearly quantitative (98%) conversion of oleic acid was achieved.

Note that the use of a higher methanol excess (*i.e.*, 9 : 1) was found to reduce the acid conversion, which was attributed to preferential adsorption of the alcohol on the active sites, ultimately hindering the appropriate interaction between acid and catalyst. Notably, after the tenth reuse of the catalyst, a negligible loss of its activity (2%) was observed. The same authors carried out the simultaneous esterification–transesterification of a model sunflower oil containing 6% oleic acid, mediated by Amberlyst® 46 (20 wt%), by employing a 6 : 1 methanol/acid molar ratio at 130 °C. A conversion as high as 76% was reached after 10 h. Furthermore, Amberlyst® 46 showed 15% activity loss after the fourth reaction cycle, which was attributed in part to incomplete mass recovery upon filtration.

Park *et al.* explored the catalytic activity of Amberlyst® 15 and BD20 in the esterification reaction of FFA-enriched (50% and 99%) oils with methanol.[74,75] Amberlyst® BD20 is a gel-type sulfonated polystyrene-based resin (Dow Chemicals) featuring an average particle diameter of 750 μm, BET surface area <0.1 m^2 g^{-1}, and acidity of 5.10 eq kg^{-1}. When using 20 wt% catalyst and 6 : 1 methanol excess at 80 °C, both catalysts showed high activity, with negligible residual FFA content in the oils (0.2–3%). However, it was observed that the activity of Amberlyst® 15 decreased with the increasing amount of water co-produced during the esterification process. Moreover, a progressive loss of catalyst activity was observed (1) upon contamination of the oil feedstocks with increasing amounts of water and (2) during recycling of the catalyst. All these findings were explained by the negative effect of water solvation on the active sites, along with the difficult water removal from the catalyst pores, which ultimately leads to limited accessibility of reactants to the catalytic sites within the pores. On the other hand, Amberlyst® BD20 exhibited good water tolerability. The authors tentatively explained this phenomenon by considering that a non-porous structure such as that of Amberlyst® BD20 implies poorer water adsorption on the inner and/or outer catalyst surface, compared to porous Amberlyst® 15. This hypothesis was supported by the fact that the activity of both catalysts was not hindered when oils with low FFA contents (*i.e.*, 2.5%) were used as the feedstock, since only small amounts of water were produced in this case.

Fu *et al.* recently studied the influence of impurities such as water, metal ions, and organic acids other than FFAs and water, in the biodiesel production from waste cooking oil over Amberlyst® BD20 and 15. When using 10 wt% catalyst loading at 70–100 °C, conversions in the ranges of 5–14% and 1–14% were obtained for Amberlyst® BD20 and 15, respectively.[76,77] The results from a blank reaction (no catalyst) indicated that the metal ions behave as Lewis-acid catalysts in the esterification reaction, but at the same time they induce resin catalyst deactivation through metal exchange. This phenomenon was particularly evident when employing a large methanol excess, since the reaction mixture is more viscous and the dispersion of metal ions in the alcohol phase is higher, which ultimately promotes the metal ion exchange. Regarding the water deactivation effect, the authors considered that solvation of the active sites occurred, which could be limited by increasing the reaction temperature, but not the methanol concentration. Finally, the presence of organic acids other than FFAs was found to efficiently promote the esterification reaction, showing a synergy with the Amberlyst® resins.

Hoelderich and co-workers studied the pre-esterification reaction of diverse FFAs containing vegetable oils with methanol, over three different EBD resins (EBD-100, EBD-200, and EBD-300).[78] Such EBD resins are sulfonated styrene/DVB co-polymers, with gel-type (EBD 100) or macroporous (EBD-200 and EBD-300) morphology. The gel-type EBD resin exhibited the highest ion-exchange capacity (IEC, 5.37 mol kg^{-1}) while the macroporous resins had IEC values of 2.81 and 3.33 mol kg^{-1} for EBD-200 and EBD-300, respectively. The catalytic activity of gel-type EBD-100 was found to be superior compared to those of macroporous EBD-200 and EBD-300, affording 100% FFA conversion. These findings were explained based on the low crosslinking degree of gel-type EBD-100, which resulted in high swelling and alcohol uptake. Furthermore, the authors claimed that the active sites in EBD-100 were mostly accessible to the reactants, whereas in the macroporous resins with a higher crosslinking degree, the acidic sites were accessible only to a small extent *via* the surface of the catalyst, most of them being embedded in the dense polymer network. Recycling studies were also carried out and it was found that fast catalyst deactivation occurred after four cycles, probably due to progressive ion-exchange with traces of cations present in the oil feedstock as contaminants. However, the deactivation process was reversible, the original catalyst activity being restored by washing with hydrochloric acid.

Hidaka and co-workers[79] investigated the performance of macroreticular Amberlyst® 15 and gel-type Amberlyst® 31 resins for the esterification of oleic acid with methanol. By using a 70 : 1 methanol/acid molar ratio, and 100 wt% catalyst loading at 60 °C, the gel-type resin led to 82% FA conversion after 1.5 h and, consistently with other reports, it revealed to be more active than macroreticular Amberlyst® 15, which in turn led to 69% FA conversion. The authors ascribed this behavior not only to the higher accessibility of the catalytic sites in Amberlyst® 31, but also to the greater hydrogen exchange capacity of the latter *vs.* Amberlyst® 15 (7.1 *vs.* 5.4 mmol g^{-1}). Furthermore, Amberlyst® 31 experienced a loss of activity of only *ca.* 15% after four reaction cycles.

Özbay *et al.*[80] studied the esterification reaction of FFAs contained in used oil by comparing the performance of macroreticular resins Amberlyst® 15, 35, and 16, along with the gel-type Dowex® HCR W2. They observed that the macroreticular resins were more effective than the gel-type system. More specifically, they found that the higher average pore diameter and BET surface area of the former were far more effective in promoting the esterification reaction compared to the higher swelling capacity of the latter. The catalytic activity followed the order: Amberlyst® 15 > Amberlyst® 35 > Amberlyst® 16 > Dowex® HCR W2. The apparent discrepancy with the results obtained by Park,[74,75] Hoelderich,[78] and Hidaka[79] might be attributed to the fact that the FFA content in the employed oil feedstock was very low (*ca.* 0.5%) and, as a consequence, the amount of water produced during the esterification reaction was small.

Fu *et al.*[81] studied the esterification reaction of FFA-enriched rapeseed oil (50 wt%) and methanol by employing a self-developed macroreticular sulfonated polystyrene-based resin, namely CLD-50. Such resin features a degree of crosslinking of 50%, IEC of 3.4 mmol g^{-1}, average pore diameter of 97 nm, and BET surface area of 185 m^2 g^{-1}. The optimal conditions were found to be a 15 : 1 excess of methanol, 10 wt% catalyst loading, and 100 °C. After a short time (3 h), the CLD-50 resin allowed nearly quantitative FFA conversion (98%). Next, the effect of water contamination (0–5 wt%) of the oil feedstock on the activity of the CLD-50 resin, along with the effect of water produced during the esterification reaction was studied. For comparison purposes, the same experiments were performed over a commercial gel-type resin – name not specified. CLD-50 was found to be negatively affected by both the water contaminating the oil feedstock and the water esterification by-product. Particularly, the authors demonstrated that the activity of macroporous resin CLD-50 was more affected by the amount of water contaminating the feedstock than that produced during FFA esterification. On these bases, the authors speculated that, in the presence of high FFA contents, the FFAs themselves compete more for the acid sites with the water esterification by-product, thereby limiting the deactivation effect of the latter. In contrast, the reference gel-type resin was found to be less affected by the water contaminating the oil feedstock, which was attributed to its low surface area (<0.1 m^2 g^{-1}). However, the authors pointed out that such resin possesses a limited number of active sites, which restricts its performance with high FFA-content oil feedstocks. Finally, the recyclability of macroporous CLD-50 was explored, and it was observed a *ca.* 7% loss of activity after the tenth reaction cycle with catalyst washing/drying between runs. The reduction in performance was attributed to the not excellent mechanical resistance of the resin beads, which made difficult their complete recovery upon filtration.

The commercially available microporous sulfonated hyper-crosslinked polystyrene resin Purolite® C275 was employed by Papayannakos and co-workers in the esterification of vegetable oil with a 38 wt% FFA content with methanol.[82] The main resin characteristics were the following: acidity of 5.2 eq H^+ g^{-1}, pore volume of 0.3 cm^3 g^{-1}, BET surface area of 31×10^{-3} m^2 kg^{-1}. By employing a 6.6 : 1 methanol-to-acid molar ratio and 2 wt% catalyst loading at 100 °C, quantitative conversion of FFAs into the corresponding FAMEs was observed, although a substantial loss of catalytic activity was observed after the first reaction cycle.

Two sulfonated hyper-crosslinked polystyrene resins, namely Purolite® D5081 and D5082, have also been employed successfully in the esterification reaction of oleic acid with methanol.[83] From the physico-chemical and morphological points of view, Purolite® D5081 features a much higher BET surface area (700 *vs.* 380 m^2 g^{-1}), half the acidic sites (1.0 *vs.* 2.0 mol SO_3H g^{-1}), and smaller pore volume and average pore diameter (0.2 *vs.* 0.4 cm^3 g^{-1} and 19 *vs.* 27 nm, respectively) compared to D5082. By setting the reaction conditions at 44 : 1 alcohol-to-acid molar ratio, 5 wt% catalyst loading, and 65 °C, Purolite® D5081 was much more active than D5082 (turnover frequency number – TOF – of 65 h^{-1} and 26 h^{-1}, respectively) despite exhibiting a smaller number of acidic sites. The authors explained the better performance of Purolite® D5081 on the basis of its higher surface area. For comparison purposes, the reaction was carried out over macroreticular Amberlyst® 15, under the same reaction conditions. In this case, a TOF of only 2 h^{-1} was observed. However, Purolite® D5081 significantly lost its original activity after the first reaction cycle. Interestingly, by using higher reaction temperatures (*i.e.*, 85 °C), it was possible to maintain the catalytic activity up to the fourth reaction cycle. This was attributed to the faster diffusion of FFA molecules through the resin micropores, which somehow prevented their blockage.

The scope of Purolite® D5081 and D5082 resins was extended by Saha and co-workers,[84] who explored the esterification of waste cooking oil with an acid value of 12.2 mg KOH g^{-1} with methanol. Purolite® D5081 gave better results compared to D5082, reaching 88% *vs.* 78% FFA conversion by employing a 6 : 1 methanol-to-acid molar ratio and 1 wt% catalyst loading at 65 °C (8 h). These findings highlight once again the predominant influence of the BET surface area on the catalyst efficiency. Recycling studies demonstrated that ~10% loss of catalyst activity occurred per reaction cycle (in a total of four cycles), which was ascribed to fouling phenomena and –SO_3H leaching.

Bhaumik and co-workers reported self-developed sulfonated hyper-crosslinked exchangers for the esterification of a wide set of FFAs with methanol.[85–87] In a first study, the authors developed the HMP-1-SO_3H system

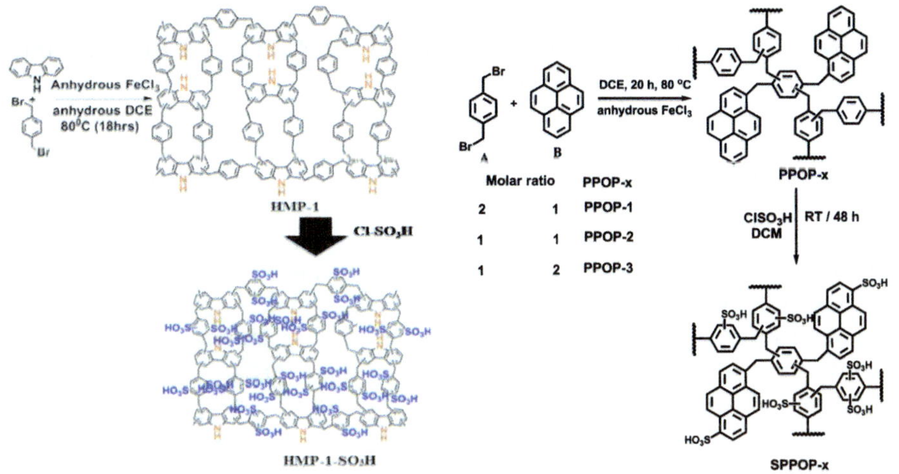

Figure 5.4 (Left) Friedel–Crafts alkylation reaction between carbazole and α,α′-dibromo-*p*-xylene, followed by sulfonation yielding HMP-1-SO_3H. Reproduced from ref. 65 with permission from Elsevier, Copyright 2018, and from ref. 85 with permission from the Royal Society of Chemistry. (Right) Synthesis of sulfonic acid-functionalized pyrene-based hyper-crosslinked SPPOP-*x*. Reproduced from ref. 65 with permission from Elsevier, Copyright 2018, and from ref. 87 with permission from American Chemical Society, Copyright 2015.

(Figure 5.4, left), which was prepared through a $FeCl_3$-promoted Friedel–Crafts reaction between α,α′-*p*-bromoxylene and carbazole, followed by post-polymerization sulfonation with chlorosulfonic acid.[85] HMP-1-SO_3H features an acidity of 3.7 mmol g^{-1}, average pore size distribution of 1.54 nm, BET surface area of 346 m^2 g^{-1}, and pore volume of ~0.2 cm^3 g^{-1}.

The catalyst activity was explored in the esterification of a range of FFAs, *i.e.*, lauric, mirystic, palmitic, stearic, oleic, adipic, and sebacic acids, using a methanol excess of 50 : 1 (or 100 : 1) and 20 mg of HMP-1-SO_3H mmol FFA^{-1} at room temperature (10–12 h). Excellent conversions were achieved in all cases (93–99%). Interestingly, the catalytic systems maintained their original activity for four reaction runs, the catalyst being washed/dried between runs. In a subsequent study, the same group developed a novel class of sulfonated pyrene-based hyper-crosslinked catalysts, namely SPPOP-*x* (Figure 5.4, right) by means of $FeCl_3$-mediated Friedel–Crafts reaction between pyrene and bis(1,4-dibromomethyl)benzene at molar ratios of 2 : 1, 1 : 1, and 1 : 2 (SPPOP-1, SPPOP-2, and SPPOP-3, respectively), followed by sulfonation of the resulting polymer with chlorosulfonic acid.[86] All these polymers were thermally stable and exhibited high acidity (2.2–3.2 mmol SO_3H g^{-1}), while their BET surface areas followed the order: SPPOP-1 > SPPOP-2 > SPPOP-3 (280 m^2 g^{-1}, 244 m^2 g^{-1}, and 205 m^2 g^{-1}, respectively), which is the opposite trend to the sulfonation degree. The activity of the catalysts was tested in the esterification of different FFAs with methanol. It was found that, by using a methanol excess of 50 : 1 and 0.054 mmol $-SO_3H$ per mmol FFA (10 h, room temperature), FFA conversions up to 90% were obtained in all cases. Next, recyclability studies demonstrated no substantial activity losses after six reaction cycles, the catalyst being washed/dried between cycles. Notably, SPPOP-3 revealed to be a promising solution for the transesterification reaction of olive oil (94% conversion). Very recently, the same group[91] has developed and investigated the activity in FFA esterification with methanol of a different type of microporous catalysts (*i.e.*, PVDTA-SO_3H) synthesized through post-polymerization functionalization with chlorosulfonic acid of a poly-DVB-*co*-triallyl amine polymer. The catalyst exhibited a SO_3H loading of 2.3 mmol g^{-1}, good thermal stability (4–10% weight loss at 200 °C), BET surface area of 410 m^2 g^{-1}, and average micropore diameter of 1.4 nm. Under the conditions of 120 : 1 methanol excess, 10 mg PVDTA-SO_3H per mmol FFA, at 25 °C (12 h), FFA conversions of 92–98% were achieved. Notably, the catalyst substantially retained its original activity after six reaction cycles.

Liu *et al.*[88–90] developed sulfonated mesoporous polydivinylbenzenes (PDVBs) for the esterification of FFAs with diverse alcohols, and the simultaneous esterification/transesterification of FFA-enriched oil to biodiesel fuel. First, the authors synthesized a set of PDVB-SO_3H catalysts through hydrothermal polymerization of divinyl benzene monomer, followed by sulfonation with chlorosulfonic acid, by varying the conditions such as the porogen solvent, the monomer/porogen ratio, and the reaction time. The catalysts revealed excellent activities in the esterification of lauric acid with ethanol (89–99% FFA conversions) when employing a 20 : 1 alcohol excess and 0.187 : 1 catalyst/acid mass ratio (70 °C, 5 h). For comparison purposes, the reaction was carried out over the Amberlyst® 15 benchmark under the same conditions, with which a much lower FFA conversion was obtained (75%), partially ascribed to the higher BET surface area, larger pore volume, and more hydrophobic nature of the PDVB network. The newly developed catalysts substantially retained their activity after four reaction cycles, although the catalyst required a cumbersome regeneration protocol between cycles. The PDVB-SO_3H-type catalysts showed good performance also in the transesterification of tripalmitin contaminated with both FFAs and water impurities. Particularly, by employing a 60 : 1 metanol/tripalmitin molar ratio and 0.12 : 1 catalyst/tripalmitin mass ratio (65 °C), conversions of up to 76% were achieved. The optimal PDVB-SO_3H-type catalyst substantially retained its original activity after three reaction cycles.

More recently, the same group developed a novel set of mesoporous catalysts, namely H-PDVB-SO_3H, by solvothermal co-polymerization between sodium *p*-styrene sulfonate and divinyl benzene under a variety of conditions. The catalytic activity was tested by employing the reaction between acetic acid and butanol (cyclohexanol) as a model esterification reaction. When using a 1:1.5 (1:2.6) alcohol/acid molar ratio and 0.20 or 0.65 g catalyst per mol acid (90–100 °C), acid conversions in the range of 78–90% were obtained, 1-butanol giving the optimal results compared to cyclohexanol based on its smaller steric hindrance, enabling ready access to the catalyst active sites. The catalytic activity of the optimal H-PDVB-SO_3H-type catalyst was substantially maintained after five reaction cycles[90] (Table 5.1).

5.3 Biodiesel Production over Sulfonated Cation-exchange Resins under Flow Regime

As already mentioned, the use of heterogeneous catalysis has the potential to significantly benefit from flow chemistry. Despite this fact, the synergy between these two crucial technologies for the sustainable production of biofuels remains poorly explored. In this section, we give an overview of the advances in the use of sulfonated ion-exchange resins to produce biodiesel fuel in continuous mode. Trends are identified between the catalyst physico-chemical and morphological properties and the performance, and a comparison with the batch counterpart is described when provided in the original papers. Catalyst stability aspects are also considered.

In 2011, Kusakabe and co-workers[91] performed the esterification reaction of oleic acid with methanol using Amberlyst® 15 packed in a three-phase fixed-bed reactor. An optimal methanol/acid molar ratio of 6:1 was used and three different catalysts loading were investigated (*i.e.*, 0.5, 1.0, 2.0, and 5.0 g). Oleic acid and the alcohol were pumped through the reactor, the feed rate of the oily phase being fixed at 9.0 mL h^{-1} and the methanol feed rate varying from 3.2 to 8.6 mL h^{-1}. An optimal 95% FAME yield was achieved at 100 °C for methanol and acid feed rates of 8.6 and 9.0 mL h^{-1}, respectively. The possibility to set the temperature of the system at the boiling point of water enabled its evaporation, thereby shifting the reaction equilibrium toward the product. Additionally, no catalyst activity losses were observed during the first 5 h. In the same study, it was demonstrated that, by adding sunflower oil to oleic acid, esterification and transesterification occurred simultaneously, though lower conversions to FAME were observed, *i.e.*, 65% and 55% with 75 and 50 wt% FFA, respectively. The authors also investigated the effect of water on the catalyst performance. To this end, methanol in water solution was fed into the reactor. Interestingly, no significant change was observed due to the continuous removal of water from the reaction environment. Moreover, the obtained results were compared to those for the same reaction carried out in a batch reactor at 60 °C. For water contents in oleic acid of 0 and 20 wt%, the methyl oleate yields in the batch reactor were found to be 70% and 15%, respectively. The yield drop was explained on the basis of catalyst poisoning by water, as well as

Table 5.1 Results of cation-exchange sulfonic acid resins for FFA esterification reactions under batch conditions.[a]

		Characterization							Reaction conditions						
Catalyst	Preparation	SA (m^2 g^{-1})	APD (nm)	APV (cm^3 g^{-1})	CD (%)	IEC/acidity (mmol g^{-1})	Others	Feedstock	Alcohol/FFA (oil) molar ratio	Cat. (wt%)	T (°C)	Time (h)	Conversion (%)	% cat. activity (cycle nr.)	Reference
A 15	Commercial grade	—	—	—	—	—	—	LA	1:1 (Ethylhexanol)	3	150	2	96	Desact. after 2 h	68
N NR50	Commercial grade												78	Desact. after 4.5 h	
A 15	1) 5× washing with 1:1 v/v $EtOH/H_2O$ and acetone + drying at 70 °C (72 h)	—						OA	1:1 (EtOH)	20	60	6	50	Desact.	70
A15 (wet)	Commercial grade	43	23	0.25	—	4.4[b]	4–10 wt% loss at 200 °C	Triolein (2.5 mg KOH g^{-1})	24:1 (MeOH)	13	200	10	97	17 (3)	71
A 46	Commercial grade	75	24	—	—	0.4[b]	Macro-reticular	OA	3:1 (MeOH)	15	100	2	98	98 (10)	72
A 46	Commercial grade	—	—	—	—	—	—	SFO (6 wt% oleic acid)	6:1 (MeOH)	20	130	10	76	85 (4)	73

A 15	Commercial grade	40	—	—	—	4.7[b]	Porous	OA	6:1 (EtOH)	20	80	—	97[c] 92[c] 94[c]	Low (5) 100 (5) Low (5)	74, 75
A BD20 (wet)	Washing with 1:1 v/v EtOH/H_2O, acetone and petroleum ether + drying (72 h)	<0.1	—	—	—	5.1[b]	Non-porous, gel type	OA/SBO (2.5 wt% FFA)					98[c] 93[c] 96[c]	~100 (5)	
A BD20	Dried overnight under vacuum (80 °C)	—	—	—	—	—	—	WCO (1.6 mmol $H^+ g^{-1}$)	15:1	10	100	3	97	—	76, 77
EBD100	Washed 3× with MeOH and dried (1 h)	—	—	—	—	5.4	Gel type, 175 kg m^{-3} (wet)	Stearic acid (10 wt%)/SFO (0.6% FFA)	20:1 (SFO, MeOH)	1	120	24	100 (TOF 1.5 h^{-1})	TOF 0.5 h^{-1} (4)	78
EBD200						2.8	Macroporous, 318 kg m^{-3} (wet)	WCO (15.7% FFA)	12:1 (WFO, MeOH)				100 (TOF 1.4 h^{-1})		
EBD300						3.3	Macroporous, 377 kg m^{-3} (wet)						100 (TOF 0.6 h^{-1})		
A 31 (wet) A 15	Washed with MeOH and dried (100 °C, 24 h)	—	—	—	4 20	7.1 5.4	Gel type Macroreticular	OA	70:1 (MeOH)	100	60	1.5	82 69	85 (4) —	79

(*continued*)

Table 5.1 (*continued*)

		Characterization							Reaction conditions						
Catalyst	Preparation	SA (m^2 g^{-1})	APD (nm)	APV (cm^3 g^{-1})	CD (%)	IEC/acidity (mmol g^{-1})	Others	Feedstock	Alcohol/FFA (oil) molar ratio	Cat. (wt%)	T (°C)	Time (h)	Conversion (%)	% cat. activity (cycle nr.)	Reference
A 15	Washed with MeOH and dried (12 h)	53	30	—	—	5.2	Macroreticular MeOH swelling volume: 10–20% (50–60 °C)	WCO (0.4–0.5 wt% FFAs)	20^e (MeOH)	2	60	3	46	—	80
A 35		50	30	—	—	5.2	MacroreticularMeOH swelling volume: 20–30% (50–60 °C)						40	—	
A 16		30	25	—	—	4.8	Macroreticular MeOH swelling volume: 30–40% (50–60 °C)						42	—	
Dowex® HCR W2		—	—	—	—	4.8	Gel type MeOH swelling vol.: 40–60% (50–60 °C)						38	—	

CLD-50	1) S co-polymerized with DVB in presence of porogen + BPO + gelatin at 82–95 °C (2–3 h) 2) Sulfonation with 93% H_2SO_4, 82 °C, 6 h. Pre-swollen with CH_2Cl_2 (60 °C).	185	97	0.45	50	3.4	Macroporous	RO/OA (65 mg KOH g^{-1})	15:1 (MeOH)	10	100	3	98	93 (10)	81
Purolite® CT-275	Commercial grade	30	—	0.29		5.2	HC T_{max} 150 °C	VO (acidity 38 wt%)	6.6:1 (MeOH)	2	100	50	1.2[b]	Lower (1 × 12 h) Stable (2–7 × 12 h)	82
Purolite® D5081	Grounded beads cooled in liquid N_2	700	19	0.39		0.8[g]	HC	OA	44:1 (MeOH)	5	55	—	TOF 65 h^{-1}	Lower (1)	83
Purolite® D5082	Activated at 80 °C (1 h)	380	25	0.24		1.2[g]	PS < 25 μm						TOF 26 h^{-1}	Stable at 85 °C (4)	
Purolite® D5081	In MeOH under ultrasonic conditions and dried at 100 °C, 6 h	510	3.7	0.47		1.6	HC sulfur 4.1% PS 280–800 μm	WCO (12.2 mg KOH g^{-1})	6:1 (MeOH)	1	60	8	88	60–68 (4)	84
Purolite® D5082		460	3.1	0.36		1.8	HC sulfur 5.9% PS 375–900 μm						78	—	
HMP-1-SO_3H	1) Friedel–Crafts alkylation: 2 α,α′-dibromo-*p*-xylene + 1 carbazole ($FeCl_3$) 2) Sulfonation	350	1.5	0.16		3.7	HC	LA, MA, PA, STA, OA, AA, SEA	50:1 100:1 (MeOH)	7–14	25	10–12	93–99	LA: ~100 (4)	85

(*continued*)

Table 5.1 *(continued)*

Catalyst	Preparation	Characterization: SA (m^2 g^{-1})	APD (nm)	APV (cm^3 g^{-1})	CD (%)	IEC/acidity (mmol g^{-1})	Others	Feedstock	Reaction conditions: Alcohol/FFA (oil) molar ratio	Cat. (wt%)	T (°C)	Time (h)	Conversion (%)	% cat. activity (cycle nr.)	Reference
SPPOP-1	1) Friedel–Crafts alkylation: 2 bis(1,4-bromomethyl) benzene + 1 pyrene ($FeCl_3$) 2) Sulfonation	280	1.7	0.18		2.2	HC nanofiber morphology TGA; 25 wt% loss up to 250 °C	LA, MA, PA, STA, OA, AA, SEA, GA	50:1 (MeOH)	0.05[f]	25	10	~90	LA: ~100 (6)	86
PDVTA-SO_3H	1) Solvothermal polymerization: 4 DVB + 1 triallylamine (in acetone 120 °C, 24 h) 2) Sulfonation	410	1.4	0.22		2.3	TGA: 5–10 wt% loss up to 200 °C TEM: 1.4 nm micropores FE-SEM: 25 nm particles	LA, MA, PA, STA, OA, AA, SEA	1:5[d] (MeOH)	4–7	25	10–12	92–98	LA/AA: ~100 (6)	87
PDVB-xs-SO_3H	1) Hydrothermal polymerization with DVB/AIBN/THF/H_2O molar ratio of 1:0.02:16.1:0 (7.2) at 100 °C (48 h) 2) Sulfonation	380	23.5	0.90		4.1	Sulfur 11.9%	LA	20:1 (EtOH)	19	70	5	99	~100 (4)	88

PDVB-SO_3H-xs	1) Hydrothermal polymerization with DVB/AIBN/EA molar ratio of 1:0.02:13.5 at 100 °C (24 h) 2) Sulfonation	360	29.4	0.44	3.4	Sulfur 10.0%	TP	60:1 (MeOH)	12	65	—	76	~95 (3)	89
H-PDVB-$x$$SO_3H$	1) Solvothermal co-polymerization with DVB/sodium *p*-styrene sulfonate/AIBN/THF:H_2O molar ratio of 1:x:0.01:16.1:7.2 2) Ion exchange with 1 M H_2SO_4 (25 °C, 24 h)	380	7.3	0.46	1.0	Sulfur 4.2% Monolithic morphology x = 0.33 (styrene:DVB molar ratio) TEM: wormhole 3–20 nm mesopores	AA	1:1.5 (BuOH)	0.3	90–100	4	90 (TOF 1102 h^{-1})	~100 (5) when x = 0.33	90
						TGA: decomposition of SO_3H groups at T > 245 °C		1:2.6 (Cyclohexanol)	1.1			78 (TOF 421 h^{-1})		

[a]Abbreviations: A (Amberlyst®); APD (average pore diameter); BPO (benzoyl peroxide); CD (crosslinking degree); N (Nafion®); SA (surface area); OA (oleic acid); SO (soybean oil); S (styrene); RO (rapeseed oil); SFO (sunflower oil); VO (vegetal oil); TPV (total pore volume); WCO (waste cooking oil); AA (adipic acid); GA(glutaric acid); HC (hyper-crosslinked); LA (lauric acid); MA (myristic acid); PA (palmitic acid); PS (particle size); SEA (sebacic acid); STA (stearic acid); AIBN (azobisisobutyronitrile); EA (ethylacetate); THF (tetrahydrofuran); TP (tripalmitin).
[b]Expressed as meq g^{-1}.
[c]Conversion refers to FFAs.
[d]Expressed as w v^{-1} ratio.
[e]Expressed as vol%.
[f]Expressed in mmol SO_3H per mmol FFA.
[g]Based on NH_3 adsorption.

the limited accessibility of reactants to the active acidic sites; thus, the use of flow technology enabled a longer lifetime of the catalytic system and, additionally, the ready isolation of methyl oleate from methanol and water by vaporization. Also, separation of the product from the heterogeneous catalyst was avoided.

Santacesaria and co-workers[92–96] compared the performance of four different sulfonated polystyrene-based exchangers crosslinked with DVB, *i.e.*, Amberlyst® 15, Amberlyst® 16, Amberlyst® 131, and Relite CFS. The esterification reaction of a mixture of FFAs as well as the simultaneous esterification/transesterification of oleic acid-enriched soybean oil with methanol were investigated under batch and continuous modes. All the catalysts featured a degree of crosslinking in the range of 20–25%, acidity of 4.7 meq g^{-1}, and a particle average diameter of 0.7 mm. Among the catalysts tested, Relite CFS in particular exhibited high activity under batch conditions, leading to 80% FFA conversion in a short time (2 h). The optimal conditions were as follows: 8 : 1 methanol/FFA molar ratio, 2.5 wt% catalyst loading, and 85 °C. Subsequently, the authors switched from batch to flow mode, transferring the results obtained in the former case. It is worth-noting here that the authors realized tailor-made flow reactors to overcome the limitations associated with catalyst 'swelling', *i.e.*, excessive pressure drops along the catalytic bed; this is an important aspect highlighting the versatility of flow technology. Thus, different flow reactor configurations were investigated, and the optimal results were obtained by using a fixed-bed tubular reactor (50 cm × 8.2 cm i.d.; Figure 5.5, left) containing *ca.* 1 kg of Relite CFS, which was able to

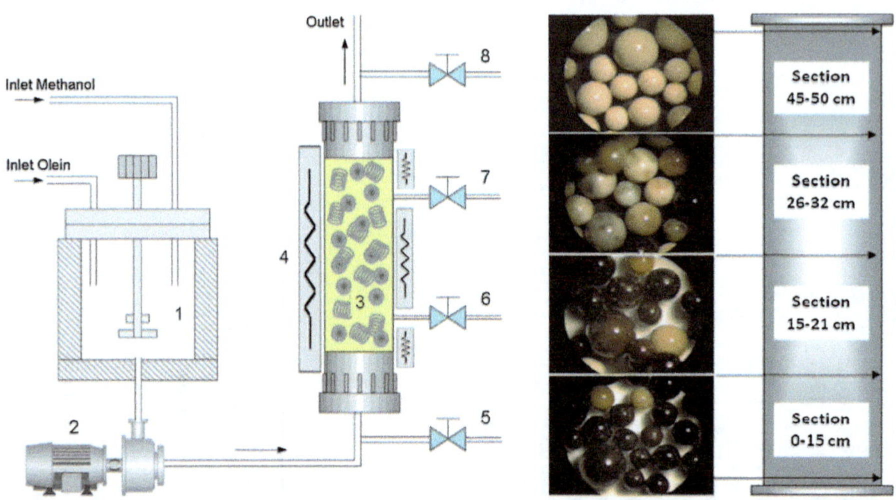

Figure 5.5 (Left) Continuous reactor experimental apparatus (1, feeding reservoir containing free fatty acids and methanol; 2, feeding pump; 3, packed bed tubular reactor; 4, heating system). (Right) Microphotographs of resin particles discharged from various reactor sections after consecutive continuous runs. Reproduced from ref. 65 and 96 with permission from Elsevier, Copyright 2018, 2010.

operate on stream for 242 h, producing biodiesel with conversions over 95%. FFAs and methanol (0.72 : 1 weight ratio) were pre-mixed in a reservoir thank, then pumped through to the reactor at a flow rate of 1 L h^{-1}; the reactor was kept at 100 °C and 6 bar, so that methanol remained in the liquid state. However, Relite CFS deactivated progressively *via* both fouling and ion exchange with metals present in the raw feed. This was also supported by observation of a visible discoloration of the resin beads during continuous operation (Figure 5.5, right). The original activity of the resin could be restored, although a time-consuming protocol was needed.

Jeřábek *et al.* led a study on the performance of sulfonated polystyrene-based resins, featuring different crosslinking degrees, on the esterification of stearic acid-enriched vegetal oil (5 wt%) with methanol using flow-through fixed-bed reactors.[97,98]

The proposed flow system was based on a reactor made of a glass column packed with the dry resin, and connected with the other components of the system through polytetrafluoroethylene (PTFE) end pieces.

Firstly, gel-type Amberlyst® BD20 was employed, and it was demonstrated that two separate phases formed at methanol concentrations higher than 20%, *i.e.*, an oil phase saturated with methanol (predominant) and a methanol-rich phase, with conversions of ~50%. For methanol concentrations below 20%, a homogeneous mixture (not alcohol-saturated) was observed and the conversion started to decrease. Notably, it was found that, by lowering the methanol content, the specific volume of the Amberlyst® BD20 bed decreased, which indicated a shrinking of the 'swelled' resin under those conditions, ultimately leading to poorer access of the reaction mixture to the acid sites. These findings suggested that, besides lower alcohol excesses, also changes in the resin morphology may influence the FFA conversion. For comparison, three sulfonated macroreticular polystyrene-based resins with varying crosslinking degrees (DVB) from 15% to 40% to 60%, namely M-15, M-40, and M-60, respectively, were investigated. In this case, the lowest FFA conversion was achieved for the methanol-saturated reaction mixture. Additionally, the macroreticular resins achieved a maximum FFA conversion of 35%, significantly lower than the results for Amberlyst® BD20. The authors ascribed this behavior to the fact that, under alcohol-saturated conditions, the $-SO_3H$ groups become less available for interaction with the carboxylic groups of the FFA. By decreasing the methanol content in the reaction mixture, the FA starts to compete with the alcohol for the $-SO_3H$ groups. Additionally, the authors observed that the effect of the decrease in methanol content on the reaction kinetics was more pronounced for the macroreticular resin exhibiting the lower the crosslinking degree, since in this case the difference between swollen and shrunk morphologies is higher; this correlates well with the results obtained for Amberlyst® BD20.

Liu *et al.*[99] carried out the esterification of rapeseed oil deodorizer distillate, with ~50 wt% FFA content, over a macroreticular polystyrene resin, namely D002, under continuous flow using a fixed-bed reactor. The D002 resin featured a 32% crosslinker degree (DVB), an IEC of 4.0 mmol g^{-1}, and a particle size of 0.05 nm. The flow reactor comprised a water-jacketed glass column (30 cm × 11 cm i.d.) packed with the resin in question and placed in a thermostatic water bath (40–80 °C). The reaction solution was pre-mixed and pumped through the reactor at a flow rate of 1 mL min^{-1}. The optimal conditions (18 wt% catalyst loading, 9 : 1 methanol/oil ratio, 60 °C, 4 h) enabled conversions of up to 96%. When the same reaction was performed using the gel-type resins 002 CR and 732 with crosslinking degrees and IEC values comparable to those of the macroreticular one (35–38% and 4.5–5.0% mmol g^{-1}, respectively) but with bigger particle sizes (1.18 and 1.28 nm), the maximum conversion obtained was only 74%. The authors also demonstrated that, during continuous operation, the by-product water formed a layer on the catalyst surface preventing the access of the (hydrophobic) reactants (see above), leading to a progressively decreasing FFA conversion. They addressed the issue by regenerating the catalyst trough a washing procedure that envisaged the use of diethyl ether and absolute ethanol as 'drying agents'. Further, the authors emphasized the gain in sustainability by replacing the batch protocol for the flow counterpart, although the former used a homogeneous catalyst (H_2SO_4), which prevents any direct comparison.

Feng *et al.* compared the activity of three different polystyrene-based cation-exchange resins, *i.e.*, NKC-9, 001x7, and D-61, to realize the continuous flow esterification of high acid value (13.7 mg KOH g^{-1}) using cooking oil with methanol.[100–102]

First, the three catalysts were studied in batch mode in order to identify the optimal system. The macroreticular NKC-9 resin gave the best results under the following conditions: 6 : 1 methanol/oil molar ratio, 20 wt% catalyst loading, 64 °C, and 2 h. An FFA conversion of *ca.* 80% was obtained. The high average pore diameter of NKC-9 (56 nm) favor the access of the reactants to the active sites. Macroreticular resin D61 has indeed a similar BET surface area (84 *vs.* 77 m^2 g^{-1}) but a much smaller average pore diameter (*ca.* 11 nm), giving the lowest conversion (10%). Finally, the gel-type 001x7 resin enabled conversions up to 32%.

Recycling studies on the best performing NKC-9 resin demonstrated that the catalyst substantially retained its original activity after ten reaction cycles, but poor mass recovery was observed, due to extensive breakdown of the catalyst particles under mechanical stirring. Thus, in order to solve this problem, the same group developed an FFA esterification protocol under continuous flow, setting up a fixed-bed reactor packed with NKC-9 (catalyst bed of 450 mm × 25 mm i.d.). The reactor comprised a water-jacketed stainless-steel column with side outlets. A pre-mixed solution of the reactants (1.25 : 1 methanol/acid molar ratio) was fed to the reactor at a flow rate of 0.8 mL min^{-1}, leading to 96% conversion of FFAs in the waste cooking oil at 65 °C. No significant catalyst activity losses were observed after 500 h on stream.

Zecca and co-workers recently developed a set of novel gel-type sulfonated polystyrene-based resins embedding acyl groups (*i.e.*, acetyl-, butyroyl-, and caproyl-) to tune the hydrophobic character of the resulting catalysts.[103] The acylation degree of the resins varied in the range of 50–80%, and the IEC values ranged from 2.8 to 4.2 mmol g^{-1}. The catalyst activities were tested in the esterification of stearic acid-enriched oil (5 wt%) with methanol using a flow-through fixed-bed reactor at 85 °C. It was found that, for high methanol contents (*i.e.*, 20 wt%), the influence of the degree of swelling was predominant on the catalyst activity, meaning that the best performing catalyst was the less hydrophobic one (*i.e.*, the non-acylated blank material). Interestingly, however, the (more hydrophobic) caproyl-functionalized catalyst showed comparable conversion (~75%) despite its lower degree of swelling (4.5 *vs.* 6.6 cm^3 g^{-1}); this was attributed to the enhanced adsorption of stearic acid on the catalytic sites, helped by the hydrophobic caproyl groups. When a lower amount of methanol was used (5 wt%), all the catalysts exhibited a comparable degree of swelling, but it was observed that the activity was low and, as a general trend, it increased with the hydrophobicity of the catalyst, the caproyl-functionalized system being once again the best performing one (Table 5.2).

5.4 Batch and Continuous Flow Production of Levulinates over Resin-supported Acid Catalysts

In this section, we give a brief overview of representative papers related to the performance of cation-exchange resins for the synthesis of alkyl levulinate esters in batch and continuous flow modes by converting the biomass-derived levulinic acid (LA) platform molecule, which is readily accessed from lignocellulose feedstocks.[104] Indeed, levulinates are identifiable as structurally related to biodiesel esters (Scheme 5.1) and are prospect oxygenated additives, as well as blending components in conventional fuels.[105–108]

They can be obtained in good yield by different routes,[108] one of the most straightforward one being the abovementioned direct esterification reaction of levulinic acid with alcohols, typically mediated by homogeneous Brönsted acid catalysts (*e.g.*, H_2SO_4) (Scheme 5.1). In recent times, diverse heterogeneous acid catalysts have been employed, mostly zeolites, supported heteropolyacids, hybrid catalysts, sulfonated carbon-based materials, sulfated metal oxides, and silicas.[108]

However, only a handful of studies were undertaken to explore the potential of resin-supported acid catalysts.

Thus, in regard to the synthesis of levulinates over sulfonated cation-exchange resins in batch mode, Teixeira da Silva and co-workers investigated the esterification reaction of LA with ethanol mediated by Amberlyst® 15. Under the conditions of 5 : 1 excess ethanol, 2.5 wt% catalyst loading, and 70 °C, LA conversion of 55% was achieved in a relatively short time (5 h).[108] Recycling studies demonstrated that Amberlyst® 15 substantially retained its original activity after the fourth reaction cycle,

Table 5.2 Results of cation-exchange sulfonic acid resins for FFA esterification reactions under flow conditions.[a]

		Characterization							Reaction conditions						
Catalyst	Preparation	SA (m^2 g^{-1})	APD (nm)	APV (cm^3 g^{-1})	CD (%)	IEC/Acidity (mmol g^{-1})	Others	Feedstock	Alcohol: FFA (oil) molar ratio	Cat. (wt%)	T (°C)	Time (h)	Conversion (%)	% cat. Activity (cycle nr.)	Reference
A 15	Commercial grade	—	—	—	—	—	—	OA	6:1 (MeOH)	CF (5 cm bed height)	100	6.4 mL h^{-1}	93	Stable (5 h)	91
								50 wt% OA/SFO					55	Desact.	
								75 wt% OA/SFO					65		
								SFO					35		
A 15, A 16, A 131	Dried at 100 °C (24 h)	—	—	—	—	—	—	OA/SBO (50% acidity)	8:1 (MeOH)	2.5	100	—	85–90	—	92–96
Relite CFS			0.7			5.2		Oleins (95% acidity)	0.7:1[b] (MeOH)	CF	100	1 L h^{-1}	95	242 h on stream	
A BD20	Washed with 10% HCl and dH_2O; dried	—	—	—	—	—	Gel type	Stearic acid (5 wt%)/VO (0.04% acidity)	0.25:1[b] (MeOH)	CF	75	55[c]	50	—	97, 98
M15	overnight (110 °C)	15			15		Macro-reticular		0.05:1[b] (MeOH)	CF	85	83[c]	25		
M40	and stored over	31			40								40		
M60	P_2O_5	233			60								35		
D002	Immersed in 5% HCl/EtOH (30 min),	—	—	—	32	4.0	Macro-porous	RO deodorizer distillate (~50 wt% FFAs)	8:1 (MeOH)	15	60	4	94	92 (10)	99
002CR	eluted with EtOH				38	5.0	Gel type						44	—	
732	until neutral pH + dried (70 °C, 2 h)				35	4.5	Gel type						74	—	

NKC-9	Swollen in EtOH (30 min) and dried at 90 °C (24 h)	—	—	—	—	—	—	WCO (36 mg g^{-1} KOH) OA/SBO (198 mg g^{-1} KOH)	1.25–2.50:1 (MeOH)	CF (2.5 × 44 cm)	65	0.8 mL h^{-1}	96	No des-act. (500 h on stream)	102
C2A(B) C4A(B) C8A(B)	1) Commercial grade S-*co*-DVB swelled with acyl chloride + $AlCl_3$ (50–75 wt%) (50 °C, 3 h), followed by washing with 1:1 v/v HCl/D, 1:1 v/v dH_2O/D and MeOH + drying 2) Sulfonation with 96% H_2SO_4 (4:1 v/w, 84 °C, 3 h)	—	—	—	2	2.8–4.2	Gel type Acylation degree: 0.5–0.8 Sulphonation degree: 0.8–0.9	Edible oil (5 wt% stearic acid)	5–20[d] (MeOH)	CF	85	13[e]	75	—	103

[a]Abbreviations: A (Amberlyst®); APD (average pore diameter); CD (crosslinking degree); CF (continuous flow); D (dioxane); SA (surface area); OA (oleic acid); SBO (soybean oil); S (styrene); RO (rapeseed oil); SFO (sunflower oil); VO (vegetal oil); TPV (total pore volume); WCO (waste cooking oil).
[b]Mass ratio.
[c]Expressed as g h mol^{-1}.
[d]Expressed as wt%.
[e]Expressed as g h kg^{-1}.

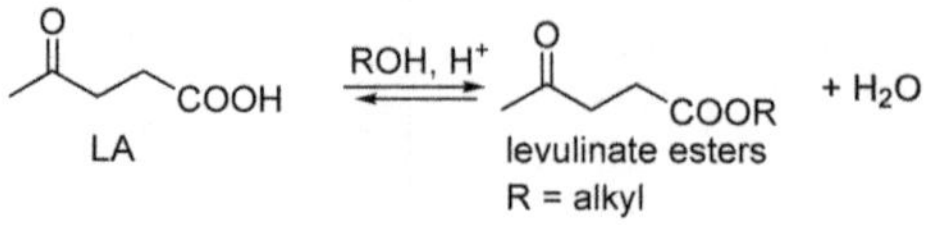

Scheme 5.1 Synthesis of alkyl levulinates by direct esterification of levulinic acid with alcohols.

provided that the catalyst underwent washing/drying between the cycles. By employing analogous reaction conditions, but not the catalyst loading (0.8 mol%), Melero *et al.* obtained only 5% LA conversion after 3 h reaction time.[109] A similar behavior for Nafion® NR50 was demonstrated (8% conversion). The two benchmarks Amberlyst® 15 and Nafion® NR50 were also tested by Maggi and coworkers[110] in the esterification reaction of LA with 1-pentanol. The activity of a fluorinated Nafion® NR50 congener (Aquivion®) was also explored. When using a 1:1 alcohol/LA molar ratio and 1 mol% catalyst loading at 100 °C (2 h), Aquivion® led to the maximum conversion (yield) of 84% (80%). On the other hand, macroreticular Amberlyst® 15 gave a much lower conversion and yield of 52% and 31%, respectively, thereby suggesting the better activity of fluorinated catalysts. A much better selectivity toward the target levulinate ester was also observed in the latter case (94% *vs.* 60%, respectively).

Tejero and coworkers[111] demonstrated the potential of a set of sulfonated polystyrene-based exchangers in mediating the esterification of LA with 1-butanol. Indeed, yields in the range of 64–94% were obtained under the optimal conditions of 3:1 excess of butanol, 0.8% catalyst loading, 80 °C, and 2.5 MPa. The general trend observed was that gel-type resins featuring low crosslinking degrees (*e.g.*, Dowex® 50Wx2, Dowex® 50Wx4, Purolite® CT224) afforded better conversions than macroreticular ones (*e.g.*, Amberlyst® 15, 16, 35, and 46), thereby highlighting the predominant influence of the polymer matrix swelling degree on the catalyst activity. Moreover, within the gel-type series, Purolite® CT224 – over sulfonated – resulted less active than Dowex® 50Wx2 – conventionally sulfonated – at the same crosslinking degree (4%). The authors ascribed this behavior to a possible higher stiffness of Purolite® CT224 conferred by the associated over sulfonation, which ultimately limited its swelling.

Very recently,[112] some of us have investigated for the first time the continuous flow esterification of LA with structurally diverse alcohols using sulfonated exchangers with different morphologies, *i.e.*, the macroreticular Amberlyst® 15 benchmark, macroporous polystyrene-supported *p*-toluenesulfonic acid (PS-*p*TsOH), and a micronized form of non-porous Aquivion® (Aquivion mP98, average particle size ~0.7 mm). The model esterification of LA with 1-pentanol was first investigated under classical batch conditions (10:1 excess of 1-pentanol, 10% mol catalyst loading, 90 °C), with Aquivion® mP98 giving the best results (92% conversion). On the opposite, conversions of up to 83% and 76% were possible with PS-pTsOH and Amberlyst® 15, respectively. This

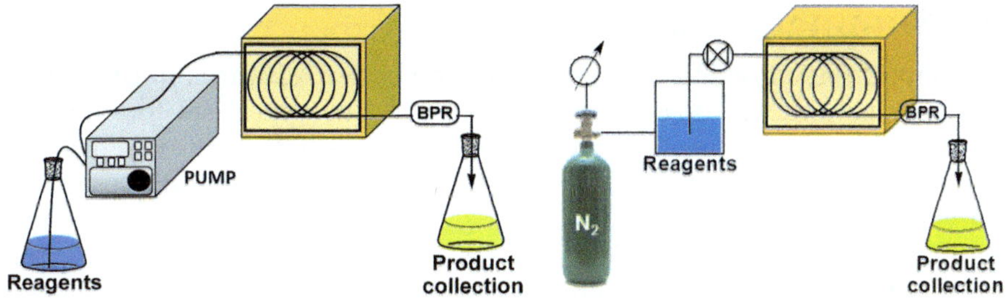

Figure 5.6 (Left) Continuous flow reactor system composed of a feed HPLC pump, a heated PTFE coil reactor, and back-pressure regulator (BPR) set at 40 psi. Reproduced from ref. 113 with permission from the Royal Society of Chemistry. (Right) Schematic of the pump-free reactor system developed in this study with the BPR set at 40 psi.

behavior can be partially explained based on the differences in the porous structure between the three catalysts, in line with the trends identified by, *e.g.*, Park,[74,75] Hoelderich,[78] and Hidaka[79] (see above) for the esterification of FFAs. Notably, the use of an excess of alcohol was not a problem in terms of minimization of the waste associated with the process, since nearly complete recovery of unreacted 1-pentanol by distillation was possible for subsequent reuse. The protocol was successfully extended to a set of aliphatic alcohols featuring short-, moderate-, or branched-alkyl chains. Recycling studies demonstrated that all the three catalysts substantially maintained their original activity after five reaction cycles, though only non-porous Aquivion mP98 was fully recoverable in mass, featuring superior mechanical strength. The use of flow reactors (Figure 5.6, left) revealed to be an extremely efficient solution to preserve the physical integrity of the catalysts (see above), thus enabling their better reusability. Additionally, by transferring the optimal batch conditions in flow mode, it was possible to enlarge the scale of levulinate production while intensifying the process. Indeed, the production of levulinates under flow conditions was in the range of ~1 to ~5 g h^{-1} (Amberlyst® 15, TOF 0.88 h^{-1}; Aquivion mP96, TOF 4.68 h^{-1}).

In this context, in a further contribution, the same group[113] exploited the features of a newly developed gel-type sulfonated polystyrene-type catalytic system in combination with those of flow technology to define a versatile protocol for the direct esterification of levulinic acid with a set of structurally diverse alcohols with short, moderate, and branched alkyl chains.

The protocol featured a minimal cost in terms of waste disposal and enabled the production of levulinates on (multi)gram per hour. From the observation that the catalyst could be reused after easy regeneration by nitrogen flow at moderately low temperatures, an unconventional pump-free flow protocol (Figure 5.6, right) was successfully set, in which the reactants are fed to the reactor by a drying nitrogen stream to discard the by-product water from the microenvironment; the catalyst holds therefore the promise to work indefinitely under such conditions.

5.5 Summary and Outlook

This chapter has provided an overview of the recent advances in the use of sulfonated cation-exchange resins as heterogeneous catalysts for the efficient overall production of biodiesel fuel from cost-effective FFA-enriched biomass feedstocks, as well as of alkyl levulinates – biofuel additives structurally related.

It is now clear that, in the last few years, excellent results holding the promise of sustainability and technical viability for this approach have been achieved, especially when exploiting the synergy between such catalytic systems and flow technology.

Most sulfonated cation exchangers, however, still appear to be unsatisfactory in terms of reactivity toward the transesterification reaction, especially from an industrial point of view, but it must be taken into account that the currently available data on this topic are very limited. Some trends in the resin catalyst activity may be identified, although the experimental results are often contradictory, which makes it difficult to define at this stage the systems appropriate for biofuel production. For example, large surface areas may help promoting the esterification reaction; on the other hand, the polymer matrix swelling ability (inverse to the crosslinking degree) may be predominant in influencing the catalyst activity when working in the presence of large alcohol excesses. Also, the hydrophobic catalyst surfaces appear to favor the adsorption of (hydrophobic) FFAs and this may limit the poisoning of active sites by the water by-product. Resins with a non-porous or gel-type structure (*e.g.*, the Amberlyst® BD20 or Nafion® NR50 benchmarks) seem to deactivate to a minor extent compared to those showing a permanent structure, as the latter exhibit greater capacity to adsorb the water. Finally, catalyst deactivation through metal exchange may occur when producing biodiesel fuel from metal impurity-containing waste cooking oil, particularly when working with large alcohol excesses.

The utmost aim of this overview was to point out the need for further research toward the development of selective, mechanically robust, and stable organic polymer-based heterogeneous acid catalysts suitable for continuous processing able to efficiently improve the overall sustainable production of biofuel esters dramatically.

References

1. Directive 2009/28/EC of the European Parliament and of the Council of 23 April 2009 on the promotion of the use of energy from renewable sources and amending and subsequently repealing Directives 2001/77/EC and 2003/30/EC, https://eur-lex.europa.eu/legal-content/EN/ALL/?uri=celex%3A32009L0028, last accessed October 2018.
2. Directive 2009/30/EC of the European Parliament and of the Council of 23 April 2009 amending Directive 98/70/EC as regards the specification

of petrol, diesel and gas-oil and introducing a mechanism to monitor and reduce greenhouse gas emissions and amending Council Directive 1999/32/EC as regards the specification of fuel used by inland waterway vessels and repealing Directive 93/12/EEC, https://eur-lex.europa.eu/legal-content/EN/TXT/?uri=CELEX:32009L0030, last accessed October 2018.

3. https://www.spglobal.com/platts/en/market-insights/latest-news/agriculture/070516-world-biodiesel-productionconsumption-to-rise-14-by-2020-oecdfao, last accessed October 2018.
4. (a) R. Luque, J. C. Lovett, B. Datta, J. Clancy, J. M. Campelo and A. A. Romero, *Energy Environ. Sci.*, 2010, **3**, 1706; (b) D. Sun, Y. Yamada, S. Sato and W. Ueda, *Green Chem.*, 2017, **19**, 3186.
5. S. Veluturla, N. Archna, D. Subba Rao, N. Hezil, I. S. Indraja and S. Spoorthi, *Biofuels*, 2018, **9**, 305.
6. M. R. Avhad and M. J. Marchetti, *Catal. Rev.*, 2016, **58**, 157.
7. S. M. Coman and V. I. Parvulescu, Heterogeneous catalysis for biodiesel production, in *The Role of Catalysis for the Sustainable Production of Bio-fuels and Bio-Chemicals*, ed. K. Triantafylidis, A. Lappas and M. Stocker, Elsevier, Oxford, 1st edn, 2013.
8. E. Gutierrez-Arnillas, M. S. Alvarez, F. J. Deive, A. Rodriguez and M. Á. Sanromán, *Renewable Energy*, 2016, **98**, 92.
9. L. J. Konwar, J. Boro and D. Deka, *Renewable Sustainable Energy Rev.*, 2014, **29**, 546.
10. H. Li, Z. Fang, R. L. Smith Jr and S. Yang, *Prog. Energy Combust. Sci.*, 2016, **55**, 98.
11. *Handbook of Biofuels Production. Processes and Technologies*, ed. R. Luque, J. Campelo and J. Clark, Woodhead, Cambridge, UK, 2010.
12. K. A. Rogers and Y. Zheng, *ChemSusChem*, 2016, **9**, 1750.
13. W. Li, Y. Gao, S. Yao, D. Ma and N. Yan, *Green Chem.*, 2015, **17**, 4198.
14. L. Vaccaro, D. Lanari, A. Marrocchi and G. Strappaveccia, *Green Chem.*, 2014, **16**, 3680.
15. A. A. Kiss, A. C. Dimian and G. Rothenberg, *Adv. Synth. Catal.*, 2006, **348**, 75.
16. F. Su and Y. Guo, *Green Chem.*, 2014, **16**, 2934.
17. M. J. Climent, A. Corma and S. Iborra, *Green Chem.*, 2014, **16**, 516.
18. Z. Hu, Y. Peng, Y. Gao, Y. Qian, S. Ying, D. Yuan, S. Horike, N. Ogiwara, R. Babarao, Y. Wang, N. Yan and D. Zhao, *Chem. Mater.*, 2016, **28**, 2659.
19. Y. Chang and C. Bae, *Curr. Org. Chem.*, 2011, **8**, 208.
20. A. Marrocchi, P. Adriaensens, E. Bartollini, B. Barkakaty, R. Carleer, J. Chen, D. K. Hensley, C. Petrucci, M. Tassi and L. Vaccaro, *Eur. Polym. J.*, 2015, **73**, 391.
21. L. Bianchi, E. Ballerini, M. Curini, D. Lanari, A. Marrocchi, C. Oldani and L. Vaccaro, *ACS Sustainable Chem. Eng.*, 2015, **3**, 1873.

22. M. Alonzi, M. P. Bracciale, A. Broggi, D. Lanari, A. Marrocchi, M. L. Santarelli and L. Vaccaro, *J. Catal.*, 2014, **309**, 260.
23. A. Izci and F. Bodur, *React. Funct. Polym.*, 2007, **67**, 1458.
24. A. P. Toor, M. Sharma, G. Kumar and R. Wanchoo, *Bull. Chem. React. Eng. Catal.*, 2011, **6**, 23.
25. (a) A. K. Kolah, N. S. Asthana, D. T. Vu, C. T. Lira and D. J. Miller, *Ind. Eng. Chem. Res.*, 2008, **47**, 5313; (b) A. K. Kolah, N. S. Asthana, D. T. Vu, C. T. Lira and D. J. Miller, *Ind. Eng. Chem. Res.*, 2007, **46**, 3180.
26. Y. T. Tsai, H. M. Lin and M. J. Lee, *Chem. Eng. J.*, 2011, **171**, 1367.
27. E. O. Akbay and M. Altiokka, *Appl. Catal., A*, 2011, **396**, 14.
28. I. B. Jua, H. W. Limb, W. Jeona, D. J. Suha, M. J. Park and Y. W. Suh, *Chem. Eng. J.*, 2011, **168**, 293.
29. I. B. Jua, H. W. Limb, W. Jeona, D. J. Suha, M. J. Park and Y. W. Suh, *Chem. Eng. J.*, 2011, **168**, 293.
30. V. K. S. Pappu, V. Kanyi, A. Santhanakrishnan, C. T. Lira and D. J. Miller, *Bioresour. Technol.*, 2013, **130**, 793.
31. G. A. Olah, P. S. Iyer and G. K. S. Prakash, *Synthesis*, 1986, **7**, 513.
32. L. Castrica, F. Fringuelli, L. Gregoli, F. Pizzo and L. Vaccaro, *J. Org. Chem.*, 2006, **71**, 9536.
33. D. C. Sherrington, *Chem. Commun.*, 1998, 2275.
34. T. A. Peters, N. E. Benes, A. Holmen and J. T. F. Keurentjes, *Appl. Catal., A*, 2006, **297**, 182.
35. M. Sharma, R. K. Wanchoo and A. P. Toor, *Ind. Eng. Chem. Res.*, 2014, **53**, 2167.
36. M. Sharma, A. P. Toor and R. K. Wanchoo, *Int. J. Chem. React. Eng.*, 2014, **12**, 451.
37. M. P. Tsyurupa and V. A. Davankov, *React. Funct. Polym.*, 2006, **66**, 768.
38. M. A. Harmer and Q. Sun, *Appl. Catal., A*, 2001, **221**, 45.
39. M. A. Harmer, W. E. Farneth and Q. Sun, *J. Am. Chem. Soc.*, 1996, **118**, 7708.
40. M. A. Harmer, Q. Sun, A. J. Vega, W. E. Farneth, A. Heidekumb and W. Hoelderich, *Green Chem.*, 2000, **2**, 7.
41. D. C. Fabry, E. Sugiono and M. Rueping, *Isr. J. Chem.*, 2014, **54**, 341.
42. K. Geyer, J. D. Codee and P. H. Seeberger, *Chem. - Eur. J.*, 2006, **12**, 8434.
43. B. Gutmann, D. Cantillo and C. O. Kappe, *Angew. Chem., Int. Ed.*, 2015, **54**, 6688.
44. L. Vaccaro, D. Lanari, A. Marrocchi and G. Strappaveccia, *Green Chem.*, 2014, **16**, 6688.
45. P. Watts and S. J. Haswell, *Chem. Soc. Rev.*, 2005, **34**, 235.
46. L. Vaccaro, *Sustainable Flow Chemistry*, Wiley-VHC, Weinheim, Germany, 2016.
47. I. Davies and W. Welch, *Science*, 2009, **325**, 701.

48. V. Hessel, D. Kralisch, N. Kockmann, T. Noël and Q. Wang, *ChemSusChem*, 2013, **6**, 746.
49. J. Yoshida, Y. Takahashi and A. Nagaki, *Chem. Commun.*, 2013, **49**, 9896.
50. M. Oelgemoller, *Chem. Rec.*, 2011, **35**, 1144.
51. D. Cantillo, C. Mateos, J. A. Rincon, O. De Frutos and C. O. Kappe, *Chem. - Eur. J.*, 2015, **21**, 12894.
52. D. Ghislieri, K. Gilmore and P. H. Seeberger, *Angew. Chem., Int. Ed.*, 2015, **54**, 678.
53. A. Adamo, R. L. Beinges sner, M. Behnam, J. Chen, T. F. Jamison, K. F. Jensen, J. C. M. Monbaliu, A. S. Myerson, E. M. Revalor, D. R. Snead, T. Stelzer, N. Weeranoppanant, S. Yee Wong and P. Zhang, *Science*, 2016, **352**, 61.
54. P. Zhang, N. Weeranoppanant, D. A. Thomas, K. Tahara, T. Stelzer, M. G. Russell, M. O'Mahony, A. S. Myerson, H. Lin, L. P. Kelly, K. F. Jensen, T. F. Jamison, C. Dai, Y. Cui, N. Briggs, R. L. Beingessner and A. Adamo, *Chem. - Eur. J.*, 2018, **24**, 2776.
55. K. P. Cole, J. M. Groh, M. D. Johnson, C. L. Burcham, B. M. Campbell, W. D. Diseroad, M. R. Heller, J. R. Howell, N. J. Kallman, T. M. Koenig, S. A. May, R. D. Miller, D. Mitchell, D. P. Myers, S. S. Myers, J. L. Phillips, C. S. Polster, T. D. White, J. Cashman, D. Hurley, R. Moylan, P. Sheehan, R. D. Spencer, K. Desmond, P. Desmond and O. Gowran, *Science*, 2017, **356**, 1144.
56. D. Ghislieri, K. Gilmore and P. H. Seeberger, *Angew. Chem., Int. Ed.*, 2015, **127**, 688.
57. P. Bhanuchander, S. S. Priya, V. P. Kumar, S. Hussain, N. Pethane Rajan, S. K. Bhargava and K. V. R. Chary, *Catal. Lett.*, 2017, **147**, 845.
58. R. B. Mane and C. V. Rode, *Green Chem.*, 2012, **14**, 2780.
59. C. Aellig, F. Jenny, D. Scholz, P. Wolf, I. Giovinazzo, F. Kollhoff and I. Hermans, *Catal. Sci. Technol.*, 2014, **4**, 2326.
60. J. D. Lewis, S. Van De Vyver, A. J. Crisci, W. R. Gunther, V. K. Michaelis, R. G. Griffin and Y. Román-Leshkov, *ChemSusChem*, 2014, **7**, 2255.
61. J. Luo, J. Yu, R. J. Gorte, E. Mahmoud, D. G. Vlachos and M. A. Smith, *Catal. Sci. Technol.*, 2014, **4**, 3074.
62. J. Luo, M. Monai, C. Wang, J. D. Lee, T. Duchoň, F. Dvořák, V. Matolín, C. B. Murray, P. Fornasiero and R. J. Gorte, *Catal. Sci. Technol.*, 2017, **7**, 1735.
63. F. Liu, L. Wang, Q. Sun, L. Zhu, X. Meng and F.-S. Xiao, *J. Am. Chem. Soc.*, 2012, **134**, 16948.
64. Z. Wu, C. Chen, Q. Guo, B. Li, Y. Que, L. Wang, H. Wan and G. Guan, *Fuel*, 2016, **184**, 128.
65. V. Trombettoni, D. Lanari, P. Prinsen, R. Luque, A. Marrocchi and L. Vaccaro, *Prog. Energy Combust. Sci.*, 2018, **65**, 136.

66. X. Zhang, Y. Zhao, S. Xu, Y. Yang, J. Liu, Y. Wei and Q. Yang, *Nat. Commun.*, 2014, **5**, 3170.
67. X. Zhang, L. Zhang and Q. Yang, *J. Mater. Chem. A*, 2014, **2**, 7546.
68. A. A. Kiss, A. C. Dimian and G. Rothenberg, *Adv. Synth. Catal.*, 2006, **348**, 75.
69. J.-Y. Park, Z.-M. Wang, D.-K. Kim and J.-S. Lee, *Renewable Energy*, 2010, **35**, 614.
70. A. Hykkerud and J. M. Marchetti, *Biomass Bioenergy*, 2016, **95**, 340.
71. G. Paterson, T. Issariyakul, C. Baroi, A. Bassi and A. Dalai, *Catal. Today*, 2013, **212**, 157.
72. O. Ilgen, *Fuel Process. Technol.*, 2014, **124**, 134.
73. M. I. Senoymak and O. Ilgen, *Energy Sources, Part A*, 2016, **38**, 3139.
74. J.-Y. Park, D.-K. Kim and J.-S. Lee, *Bioresour. Technol.*, 2010, **101**(Suppl 1), S62.
75. J.-Y. Park, Z.-M. Wang, D.-K. Kim and J.-S. Lee, *Renewable Energy*, 2010, **35**, 614.
76. J. Fu, L. Chen, P. Lv, L. Yang and Z. Yuan, *Fuel*, 2015, **154**, 1.
77. J. Fu, L. Zhibing, X. Shiyou, W. Zhiyuan, M. Changlin, P. Lv and Z. Yuan, *Fuel*, 2016, **181**, 1058.
78. B. M. E. Russnueldt and W. F. Hoelderich, *Appl. Catal., A*, 2009, **362**, 47.
79. M. Kouzu, A. Nakagaito and J.-S. Hidaka, *Appl. Catal., A*, 2011, **405**, 36.
80. N. Özbay, N. Oktar and N. Alper Tapan, *Fuel*, 2008, **87**, 1789.
81. J. Fu, L. Chen, P. Lv, L. Yang and Z. Yuan, *Fuel*, 2015, **154**, 1.
82. S. Pasias, N. Barakos, C. Alexopulos and N. Papayannakos, *Chem. Eng. Technol.*, 2006, **29**, 1365.
83. E. Andrijanto, E. A. Dawson and D. R. Brown, *Appl. Catal., B*, 2012, **115–116**, 261.
84. S. Z. Abidin, K. F. Haigh and B. Saha, *Ind. Eng. Chem. Res.*, 2012, **51**, 14651.
85. S. Bhunia, B. Banerjee and A. Bhaumik, *Chem. Commun.*, 2015, **51**, 5020.
86. S. K. Kundu and A. Bhaumik, *ACS Sustainable Chem. Eng.*, 2015, **3**, 1715.
87. R. Gomes, P. Bhanja and A. Bhaumik, *J. Mol. Catal. A: Chem.*, 2016, **411**, 110.
88. F. Liu, W. Kong, C. Qi, L. Zhu and F. S. Xiao, *ACS Catal.*, 2012, **2**, 565.
89. P. Xia, F. Liu, C. Wang, S. Zuo and C. Qi, *Catal. Commun.*, 2012, **26**, 140.
90. F. Liu, X. Meng, Y. Zhang, L. Ren, F. Nawaz and F. Xiao, *J. Catal.*, 2010, **271**, 52.

91. S. M. Son, H. Kimura and K. Kusakabe, *Bioresour. Technol.*, 2011, **102**, 2130.
92. (a) R. Tesser, L. Casale, D. Verde, M. Di Serio and E. Santacesaria, *Chem. Eng. J.*, 2009, **154**, 25; (b) R. Tesser, L. Casale, D. Verde, M. Di Serio and E. Santacesaria, *Chem. Eng. J.*, 2010, **157**, 539.
93. R. Tesser, M. Di Serio, M. Guida, M. Nastasi and E. Santacesaria, *Ind. Eng. Chem. Res.*, 2005, **44**, 7978.
94. E. Santacesaria, R. Tesser, M. Di Serio, M. Guida, D. Gaetano and A. Garcia Agreda, *Ind. Eng. Chem. Res.*, 2007, **46**, 5113.
95. E. Santacesaria, R. Tesser, M. Di Serio, M. Guida, D. Gaetano, A. Garcia Agreda and A. Cammarota, *Ind. Eng. Chem. Res.*, 2007, **46**, 8355.
96. R. Tesser, M. Di Serio, L. Casale, L. Sannino, M. Ledda and E. Santacesaria, *Chem. Eng. J.*, 2010, **161**, 212.
97. K. Jeřábek, L. Hanková and L. Holub, *J. Mol. Catal. A: Chem.*, 2010, **333**, 109.
98. R. Banavali, R. T. Hanlon, K. Jerabek and A. K. Schultz, Heterogeneous catalyst and process for the production of biodiesel from high free-fatty acid containing feedstocks, in *Catalysis of Organic Reactions: Twenty-Second Conference (Chemical Industries)*, ed. M. L. Prunier, CRC, NY, 2008, pp. 279.
99. Y. Liu and L. Wang, *Chem. Eng. Process.*, 2009, **48**, 1152.
100. Y. Feng, B. He, Y. Cao, J. Li, M. Liu, F. Yan and Y. Liang, *Bioresour. Technol.*, 2010, 1518.
101. Y. Feng, Z. Aiqing, J. Li and B. He, *Bioresour. Technol.*, 2011, **102**, 3607.
102. C. Yu, Y. Feng, R. Yanbiao, L. Xhuan, G. Aoran, B. He, F. Yan and J. Li, *Bioresour. Technol.*, 2012, **113**, 66.
103. P. Centomo, I. Bonato, L. Hanková, L. Holub, K. Jeřábek and M. Zecca, *Top. Catal.*, 2013, **56**, 611.
104. T. A Werpy, J. E. Holladay and J. F. White, Top value-added chemicals from biomass. I. Results of screening for potential candidates from sugars and synthesis gas, 2004.
105. G. Shrivastav, T. S. Khan, M. Agarwal and M. A. Haider, *ACS Sustainable Chem. Eng.*, 2017, **5**, 7118.
106. A. Démolis, N. Essayem and F. Rataboul, *ACS Sustainable Chem. Eng.*, 2014, **2**, 1338.
107. *Biofuels from Lignocellulosic Biomass: Innovation beyond Bioethanol*, ed. M. Boot, Wiley-VCH, Germany, 2016.
108. D. R. Fernandes, A. S. Rocha, E. F. Mai, C. J. A. Mota and V. Teixeira da Silva, *Appl. Catal., A*, 2012, 199.
109. J. A. Melero, G. Morales, J. Iglesias, M. Paniagua, B. Hernández and S. Penedo, *Appl. Catal., A*, 2013, **466**, 116.
110. R. Maggi, N. R. Shiju, V. Santacroce, G. Maestri, F. Bigi and G. Rothenberg, *Beilstein J. Org. Chem.*, 2016, **12**, 2173.

111. M. A. Tejero, E. Ramírez, C. Fité, J. Tejero and F. Cunill, *Appl. Catal., A*, 2016, **517**, 56.
112. V. Trombettoni, L. Bianchi, A. Zupanic, A. Porciello, M. Cuomo, O. Piermatti, A. Marrocchi and L. Vaccaro, *Catalysts*, 2017, **7**, 235.
113. V. Trombettoni, D. Sciosci, M. P. Bracciale, F. Campana, M. L. Santarelli, A. Marrocchi and L. Vaccaro, *Green Chem.*, 2018, **20**, 3149.

CHAPTER 6

Renewable Starting Materials, Biocatalysis, and Multicomponent Reactions: A Powerful Trio for the Green Synthesis of Highly Valued Chemicals

L. BANFI*[a], C. LAMBRUSCHINI[a], L. MONI[a] AND R. RIVA[b]

[a]Genova University, Department of Chemistry and Industrial Chemistry, via Dodecaneso 31, 16146 Genova, Italy; [b]Genova University, Department of Pharmacy, viale Cembrano 4, 16147 Genova, Italy
*E-mail: banfi@chimica.unige.it

6.1 Introduction

In the past 50 years, the main challenge in organic synthesis has been the successful accomplishment of the total synthesis of increasingly complex natural products. However, it is plausible that, in the next 50 years, the main challenge will shift from 'if' to 'how' a molecule (or a collection of similar molecules) can be prepared. Within this 'how', there are several issues

Green Chemistry Series No. 61
Green Synthetic Processes and Procedures
Edited by Roberto Ballini

Published by the Royal Society of Chemistry, www.rsc.org

related to green chemistry principles.[1] First of all, the ideal synthesis[2] should be endowed with step economy (reduction of number of unit operations) and atom economy (reduction of waste and the energy required for recycling solvents, side-products, and so on). Instead of 'complex (long) synthetic pathways constituted by simple reactions', chemists should try to conceive 'simple (short) synthetic pathways constituted by complex reactions', that are environmentally more efficient, and also more appropriate for diversity-oriented synthesis. Multicomponent reactions (MCRs) are defined as processes where three or more substrates are combined in a single step, generating a product that contains all essential parts of the starting materials. They seem to be perfectly suited for green chemistry, being endowed with high atom and step economy, and operational simplicity.[3]

Another very useful tool for more sustainable synthesis is biocatalysis,[4,5] thanks to its typical high selectivity, which reduces waste and avoids the use of protecting groups, and to the mild and eco-friendly reaction conditions.

Nevertheless, this is not yet enough to ensure real sustainable syntheses. The production of chemicals, from commodities to specialties to highly valued products, *e.g.*, drugs, nutraceuticals, cosmetic, functional materials, and so on, still heavily depends on the conversion of fossil feedstock, contributing to the unbalance of the carbon cycle caused by human activities. A whole new chemistry, which starts from renewable biological feedstocks,[6] must be developed.[7]

If sustainable synthesis profits from these three tools applied separately, the impact of their coupling in a single integrated approach should be much more effective. Yet, the use of this 'perfect trio' has been rather limited so far, whereas more examples of the combination of just two of them can be found in the literature. In this chapter, we will highlight some of the most important results published in recent years on the implementation of this synergy, hoping to prompt more and more researchers to add it to their own methodology arsenal in the quest for the ideal sustainable synthesis of the 21st century.

6.2 Coupling Multicomponent Reactions with Biocatalysis

6.2.1 General Remarks

The main advantage of integrating MCRs with biocatalysis relies in the inherent enantio- and diastereoselectivity of enzymes, which can overcome the typical low stereocontrol achieved in most multicomponent processes. In particular, the enantioselectivity of enzymes can be exploited through two alternative strategies, shown in Figure 6.1.

In the first one (A), the biocatalyst provides suitable chiral enantiopure inputs for the MCRs. Then, when further stereogenic centers are generated during the MCR, the chiral, biocatalytically produced input may

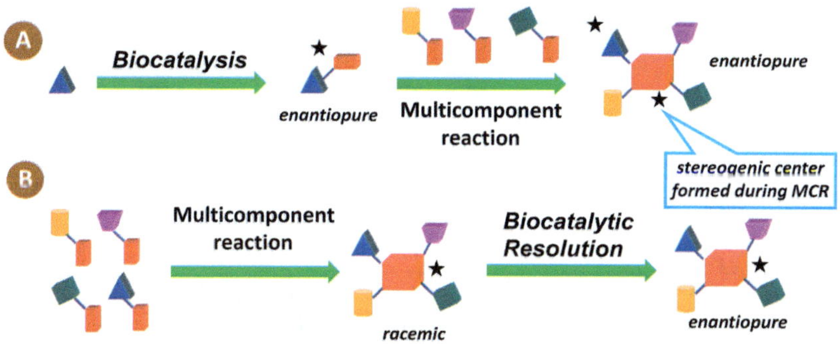

Figure 6.1 The two strategies that exploit biocatalysis for the synthesis of enantiopure MCR products.

hopefully afford good diastereoselectivity. In the second one (B), biocatalysis is employed for kinetic resolution of the racemic MCR products. In principle, it is also possible to conceive the synthesis of a *meso* or prochiral MCR adduct, followed by enzymatic desymmetrization, providing a chiral enantiopure final product. This approach, although quite attractive, has never been pursued so far to the best of our knowledge.

Finally, biocatalysis could be exploited to promote the MCR itself. In principle, this could lead to enantioselective MCRs. However, surprisingly, no useful enantioselectivity has been observed so far in MCRs that generate new stereogenic centers. Therefore, the main advantage of enzyme-catalyzed MCRs relies on the milder conditions or in the substitution of less environmentally friendly catalysts/reagents.

6.2.2 Biocatalytic Preparation of Inputs for Multicomponent Reactions

In order to exploit approach A, it is essential to consider the functionality needed for a given MCR. Aldehydes and ketones are among the most widely used functional groups in MCRs. These compounds are not easily obtained directly in enantiopure form through biocatalytic methodologies, but can be prepared by indirect methods. For example, *ene reductase* was used to convert achiral α,β-unsaturated ketone **1** into chiral enantiopure ketoacid **2**, which was used in a diastereoselective Passerini reaction to give lactone **3** (Figure 6.2).[8]

Enantiopure aldehydes have also been obtained by oxidation of chiral primary alcohols derived by enzymatic desymmetrization of *meso* or prochiral diols. Figure 6.2 shows an example of the application of a desymmetrized *meso* diol in a multicomponent approach to bicyclic pyrrolidines.[9] Diol **4** was efficiently desymmetrized using Amano PS lipase to give in high ee both monoacetate **5** (*via* acylation) and its enantiomer (*via* hydrolysis of the diacetate of **4**). After high yielding conversion into azidoaldehyde **6**, a sequence

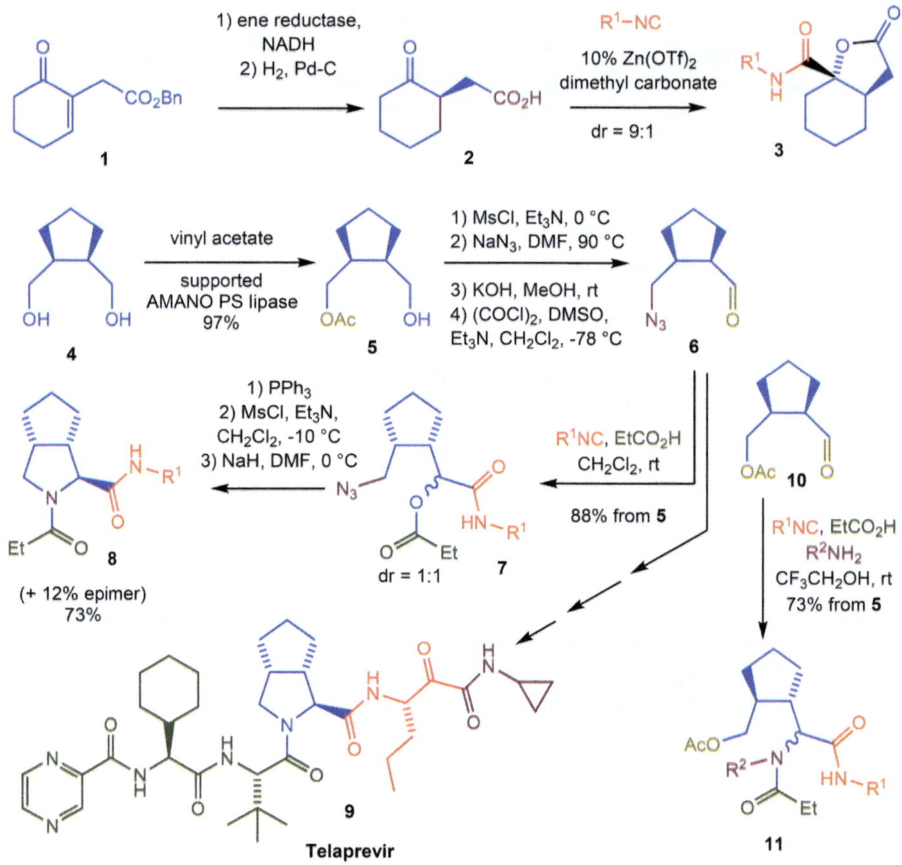

Figure 6.2 Application of ketones and aldehydes produced biocatalytically in Passerini reactions.

of Passerini reaction, followed by azide reduction, acyl migration, and cyclization, gave compound **8**. Interestingly, although the Passerini reaction proceeded with no diastereoselectivity, both stereoisomers of 7 furnished preferentially *trans*-pyrrolidine **8,** thanks to the isomerization of the less reactive isomer. This chemistry was applied to the total synthesis of antiviral Telaprevir **9**.[9] The aldehyde **10** derived from the oxidation of **5** was also used in the related Ugi reaction but, in this case, complete epimerization of the aldehyde took place, giving the two *trans* adducts **11** with no diastereoselectivity.[9] Other applications of this strategy to renewable sources will be described in Section 6.3.

Enzymes may also serve for environmentally friendly transformations of a primary alcohol into an aldehyde, followed by an *in situ* Passerini reaction. Recently, the aerobic oxidation of primary or secondary alcohols to aldehydes or ketones, performed by the synergic action of laccase and TEMPO, was combined in a one-pot procedure with the Passerini reaction.

It is noteworthy that the latter could be efficiently performed in water. Yields ranged from poor to good, depending on the structure of the alcohol. Benzyl alcohols behave much better than aliphatic ones.[10]

Several other MCRs employ a primary amine as a component. Primary chiral amines may be obtained, again through biocatalysis, by a kinetic resolution process or by derivatization of desymmetrized diols (Figure 6.3). For example, amine **12** was smartly designed in order to work as a removable chiral auxiliary in Ugi MCRs. Acylation of racemic amine **12** with methyl methoxyacetate was extremely enantioselective when using Novozym 435 lipase (enantioselectivity factor E = 750). Thus, both amide (*R*)-**13** and unreacted amine (*S*)-**12** can be obtained in high ee and nearly quantitative yield. Unfortunately, the diastereoselectivity in a series of Ugi reactions was only moderate. However, the two diastereomers of **14** could be separated, affording enantiopure Ugi adducts **15** after removal of the auxiliary.[11]

Some desymmetrized 1,3-propanediols **16** have been converted in a series of high yielding steps into azides **17**. Then, reduction to the primary amine,

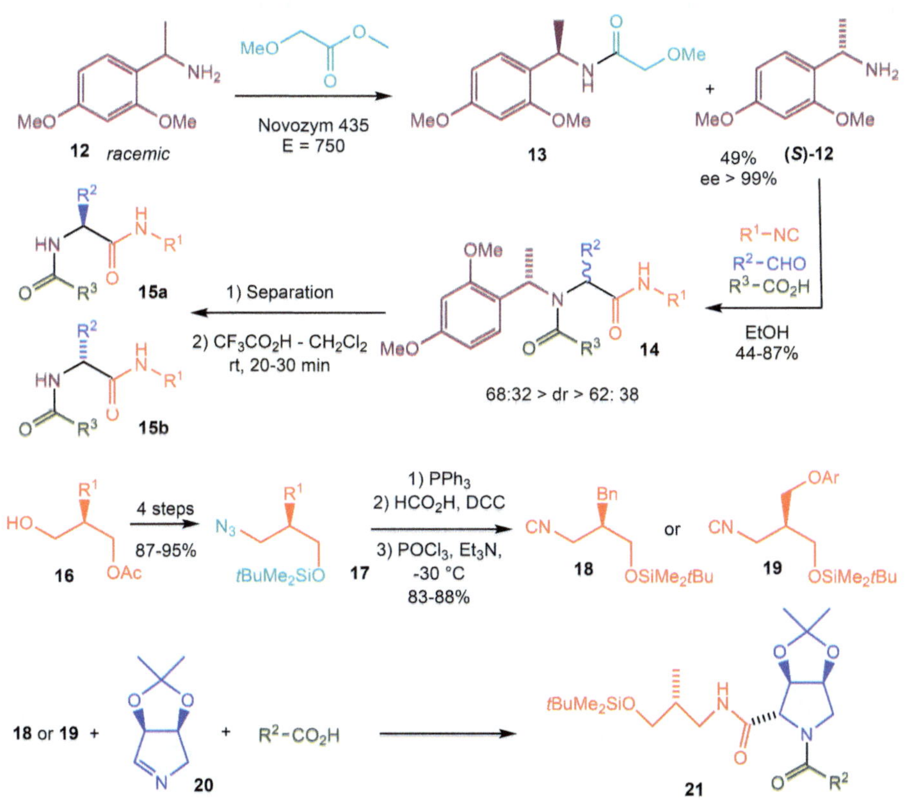

Figure 6.3 Biocatalytic synthesis of enantiopure amines and isocyanides and their use in IMCRs. DCC = dicyclohexyl carbodiimide.

formylation, and dehydration furnished chiral isocyanides **18** and **19**,[12] which have been used as chiral components in some Ugi–Joullié reactions with chiral pyrroline **20** (or its enantiomer) to give, with complete diastereoselectivity, pyrrolidines **21**.[12] The stereochemical control is in this case given by the chiral pyrroline and not by the isocyanide.

The Ugi–Joullié is a particularly useful modification of the classical Ugi reaction. It is a three-component reaction that employs cyclic imines. When they are chiral, often good levels of diastereoselectivity are observed, especially when they are fairly rigid. The main issue is to obtain them in enantiopure form, and for this task biocatalysis is an ideal tool. There are two general biocatalytic approaches to chiral pyrrolines of general formula **24** (Figure 6.4). The first one starts from *meso* diols **22** and involves enzymatic desymmetrization, followed by substitution of one of the hydroxymethyl arms with an azide, oxidation of the other arm, and Staudinger–aza-Wittig formation of the imine. For example, erythritol derived pyrroline **20** was prepared in this way starting from a desymmetrized monoester **23** that will be described later in Section 6.3.1. The other pathway involves enzymatic oxidative desymmetrization of pyrrolidine **25**, using an engineered Mono Amine Oxidase (MAO). This reaction is typically carried out on whole cells. Both approaches have pros and cons. The first one is longer, although the

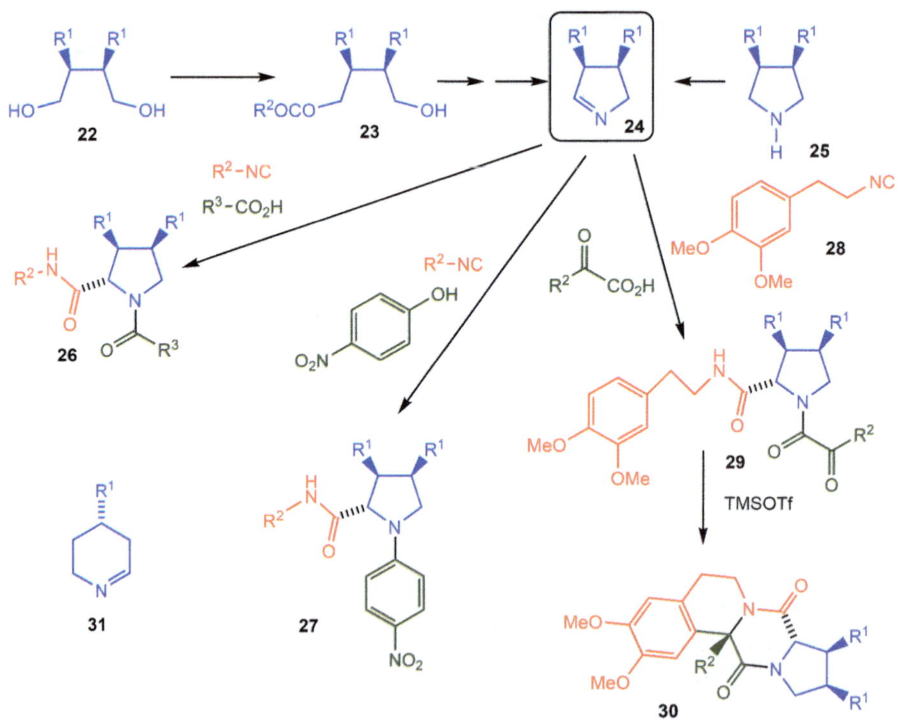

Figure 6.4 Biocatalytic synthesis and applications of chiral cyclic imines.

various steps usually proceed in excellent yield, but it can easily lead to both enantiomers thanks to the complementarity of enzymatic acylation and hydrolysis. The second one is shorter but it may be troublesome to find complementary MAOs able to afford the two possible enantiomers in comparable ee.

The group of Orru and Ruijter in Amsterdam has been very active in applying MAO desymmetrized bicyclic or tricyclic *meso* pyrrolines **24** in various isocyanide-based MCRs. For example, the already cited Ugi–Joullié reaction gives acylated pyrrolidines **26** with high levels of diastereoselectivity.[13] This chemistry was applied to the highly convergent synthesis of antiviral Telaprevir **9**.[14] Pyrrolines **24** were also used for the Ugi–Smiles variant, which employs nitrophenols instead of carboxylic acids and gives pyrrolidines **27**.[15] When the suitably indole-containing isocyanide **28** is applied in combination with α-oxoacids, the Ugi–Joullié process can be followed by a Pictet–Spengler cyclization of **29** to give polycyclic adduct **30**, similar to natural alkaloids.[16] Very recently, the same authors have studied the MAO-catalyzed desymmetrization of symmetric piperidines to give tetrahydropyridines **31** and applied them in various Ugi reactions, also in the context of the total synthesis of pharmaceuticals.[17]

6.2.3 Biocatalytic Transformation of Multicomponent Reaction Products

The typical isocyanide-based MCRs generate a new stereogenic center and, in the absence of efficient asymmetric catalysts, lead to racemic compounds. A way to overcome this problem is to subject these adducts to biocatalytic kinetic resolution. For example, the products of Passerini reactions are esters, which may be in principle hydrolyzed by various esterases. However, due to high steric biases, most lipases and esterases resulted inactive on racemic Passerini adducts **32** (Figure 6.5). Eventually, lipase from wheat germ was found to be active enough, and the hydrolysis proceeded with an enantioselectivity factor (E) of 55.[18] While the maximum achievable yield of kinetic resolution is normally 50%, in this case it was possible to achieve a higher yield by conversion *via* an S_N2 process of alcohol (*S*)-**33** into acetate (*R*)-**32**, which is identical to the unreacted ester of the kinetic resolution. In a similar way, but with a longer route, unreacted acetate (*R*)-**32** was also converted into (*S*)-**33**. This approach has been extended to Passerini adducts derived from ketones, this time employing a metagenome-derived esterase.[19] Substitution of the hydroxy group in **33** by an azide, followed by reduction, allowed the enantioselective preparation of amides **34** of unnatural α-aminoacids.[20,21]

Recently, an interesting domino process based on this chemistry was reported. An isocyanide and a vinyl ester were mixed in CH_2Cl_2 in the presence of a lipase and a small amount of water, affording a separable mixture of enantioenriched **35** and **36**. The enzyme first catalyzes the hydrolysis of the vinyl ester affording the remaining two components

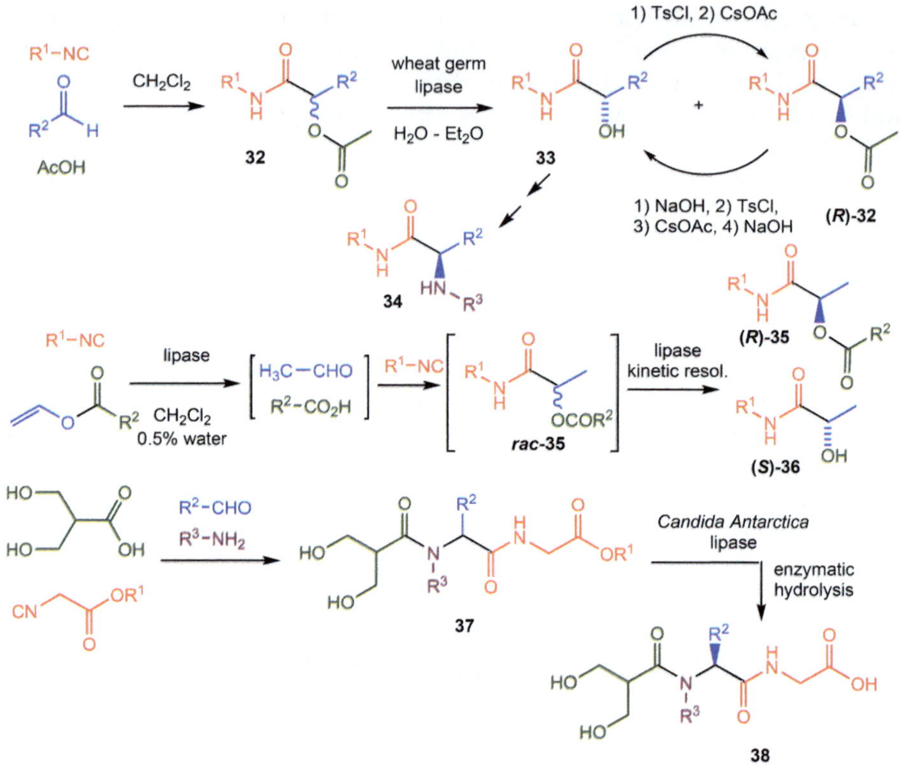

Figure 6.5 Biocatalytic kinetic resolution on Passerini or Ugi adducts.

needed for a Passerini reaction: acetaldehyde and a carboxylic acid. Then, the resulting racemic Passerini adducts **35** are selectively hydrolyzed (with kinetic resolution) by the same lipase, to give enantioenriched alcohols **36** and esters **35**.[22] The best enzymes were found to be lipases from *Pseudomonas cepacia* and *Candida antarctica* depending on the nature of the substrates.

Application of biocatalytic kinetic resolution to Ugi adducts is more difficult, since the two amide bonds generated by this reaction are resistant to hydrolysis. Thus, the application of this strategy required the introduction of a suitable additional alcohol or ester moiety in one of the components. For example, Ugi adduct **37** contains two sites that are able to react with hydrolytic enzymes. Under hydrolysis conditions, acid **38** was obtained in excellent ee using *Candida antarctica* lipase. On the contrary, monoacylation of the diol afforded poor enantio- and diastereoselectivity with all tested enzymes.[23]

Kinetic resolution of esters has been applied also to products of other famed MCRs.[24–26] As an example, Figure 6.6 shows compounds **39**, which are the products of Hantzsch reaction using Meldrum's acid as a synthetic equivalent of acetic acid.[26,27] Unfortunately, methyl esters **39** were found to be

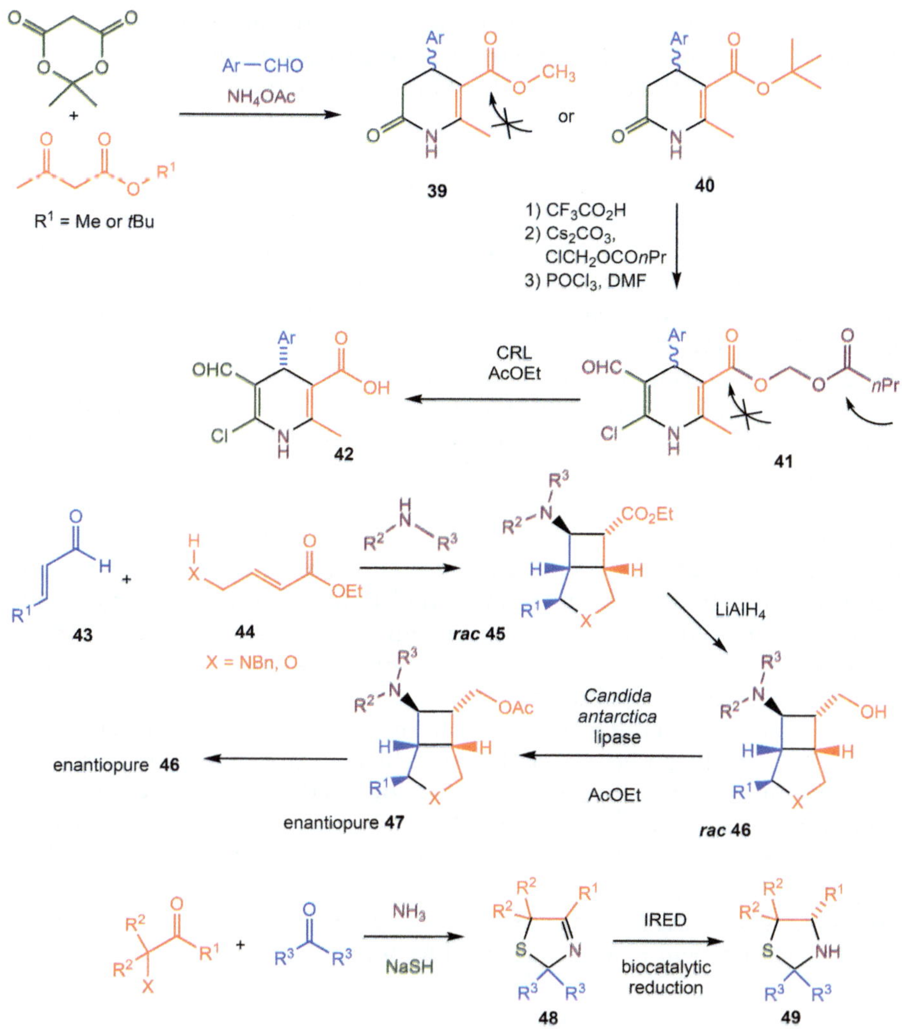

Figure 6.6 Biocatalytic transformations after non-isocyanide-based MCRs.

unreactive under the action of a large panel of esterases or lipases. This problem, likely due to the proximity of the dihydropyridine ring, was solved by the conversion of *tert*-butyl ester **40** into compound **41** (also including a formylation of the dihydropyridine ring). The lipase from *Candida rugosa* (CRL) was then able to catalyze exclusively the hydrolysis of the butyrate in **41**, giving acid **42** with enantioselectivity ranging from moderate to high depending on the Ar group.

A new and very powerful MCR that gives [3.2.0]-heterobicyclic compounds **45** is shown in Figure 6.6. This reaction involves an aza-Michael or oxa-Michael addition of amine/alcohol **44** onto the iminium ion derived from unsaturated aldehyde **43** and a secondary amine, followed by an intramolecular Stork–Michael reaction of the resulting enamine that closes the five-membered ring. Finally, the cascade reaction sequence is concluded by a final intramolecular Mannich reaction, which generates the four-membered ring.[28] This powerful MCR creates five new contiguous stereogenic centers with excellent diastereoselectivity, but leads to racemic adducts as no valid enantioselective catalyst has been found. The successful isolation of enantiopure compounds was thus solved again by kinetic resolution of alcohols **46**, obtained by reduction of esters **45**. CAL-B (lipase from *Candida antarctica*) was found to be the best enzyme for the enantioselective acetylation of **46** with ethyl acetate to give enantiopure **47** and **46**.[24]

Another very promising approach entails the generation by MCR of an achiral adduct, followed by enantioselective generation of a stereogenic center, for example by reduction of a double bond. Despite the great potential of this strategy, its successful accomplishment has been reported only very recently. An Asinger MCR gives thiazolines starting from α-haloketones, ketones, ammonia, and hydrogen sulfide. Upon suitably choosing substrates that do not generate stereogenic centers, achiral thiazolines **48** are obtained.[29] The reduction of **48** to thiazolidines **49** by chemical means is troublesome, even when racemic products are targeted. Thus, an enzymatic reduction is very appealing not only from the point of view of enantioselectivity. The authors of this work chose to employ a panel of recombinant imine reductases (IREDs). These enzymes need stoichiometric NADPH, which can be recycled *in situ* through the enzymatic oxidation of glucose by glucose dehydrogenase. The optimal catalyst allows efficient reduction even of quite encumbered thiazolines **48** with excellent ee's. Process optimization was also carried out using whole-cell catalysts[29] and/or one-pot methodologies.[30]

Interestingly, to the best of our knowledge, no enzymatic desymmetrization of *meso* or prochiral MCR adducts has been reported to date.

6.2.4 Biocatalytic Multicomponent Reactions

The commonest MCRs do not exist in nature. However, in recent years, several reports on the use of various enzymes, especially hydrolases, to catalyze 'unnatural' reactions have emerged. These unconventional biotransformations have been grouped inside the term 'enzyme promiscuity'.[31] Application of this concept to MCRs has rapidly gained importance in recent years. Since all MCRs are domino processes, enzymes can either (a) catalyze in a promiscuous way one or more individual steps of the domino process that are outside the normal competence of the given enzyme, or (b) catalyze an individual step of the domino process, which lies inside the typical aptitude of the enzyme.

In this context, the main advantage of using enzymes instead of more conventional catalysts relies on the greener nature of enzymes because, surprisingly, no enantioselectivity has been observed so far in MCRs that create new stereogenic centers. Due to space limitations, it is impossible to cite here all the reported enzyme-catalyzed MCRs. We will show only selected examples from the most recent literature, just to highlight the concept.

The first example of biocatalyzed MCRs appeared in 2005, involving the Hantzsch synthesis of dihydropyridine **51** in the presence of fermenting baker's yeast (Figure 6.7). Interestingly, only a β-ketoester **50** and ammonium acetate are used as substrates. The third substrate, acetaldehyde, is produced by the yeast during its fermenting metabolism.[32] In this paper, it was not stated whether baker's yeast was a real catalyst or it just served as the producer of acetaldehyde (derived by decarboxylation of pyruvate under anaerobic conditions). The actual catalytic capability of yeast was confirmed two years later thanks to a more diversity-oriented 'asymmetric'

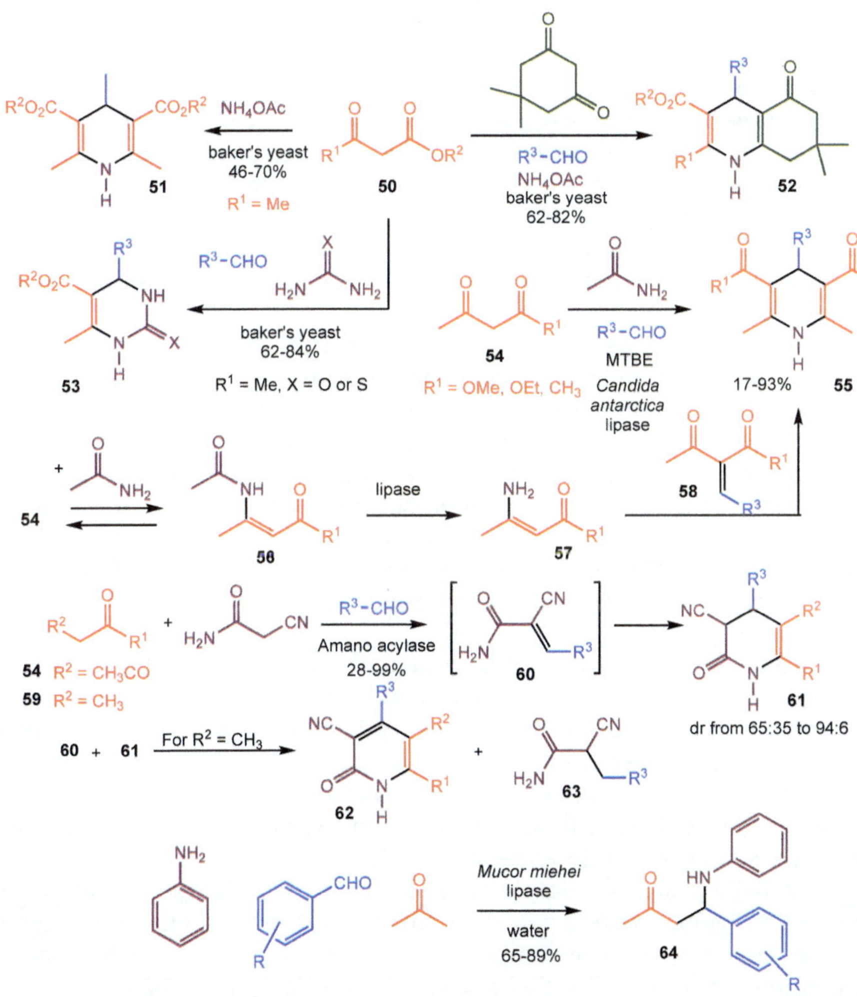

Figure 6.7 Biocatalytically promoted Hantzsch, Biginelli, and Mannich reactions. MTBE = methyl *tert*-butyl ether.

Hantzsch synthesis employing various aldehydes, β-ketoesters, and dimedone.[33] In the absence of baker's yeast, no dihydropyridine **52** was formed, proving the essential role of the microorganism. Baker's yeast was also used as a promoter for the related Biginelli reaction to give pirimidones **53**.[34] Later, other researchers have used isolated enzymes (esterases) to promote these two MCRs.[35,36] For example, Figure 6.7 shows the Hantzsch reaction using acetamide as synthetic equivalent of ammonia to give dihydropyridine **55**. In this case, the lipase plays a multiple role: it first catalyzes the Knoevenagel condensation of diketone or ketoester **54** with the aldehyde to give **58**; then, it deacetylates intermediate **56** to give enamine **57**, and finally it catalyzes the Stork–Michael reaction of enamine **57** with unsaturated ketone **58**. Thus, this reaction is an example of an enzyme involved in a 'normal' reaction (hydrolysis of **56**) as well as two 'promiscuous' processes.[36] Interestingly, only *Candida antarctica* lipase was able to promote this MCR, whereas other lipases behaved poorly. It was proven that deacetylation takes place at the stage of **56**, since acetamide is not hydrolyzed by the enzyme.

The Hantzsch-like three-component reaction between a β-ketoester or a β-diketone **54**, cyanoacetamide, and an aldehyde to give dihydropyridones **61** can also be catalyzed by a hydrolase. In this case, the best one was Amano acylase in ethylene glycol. The reaction involves a domino Knoevenagel–Michael process, followed by lactam formation, which was found to afford moderate to very good diastereoselectivity. Once again, no enantioselection was detected.[37]

Later, the reaction was extended to simple monoketones **63** ($R^2 = CH_3$). Surprisingly, this time, the formation of dihydropyridine **61** was not observed. The reaction gave instead a mixture of aromatized pyridone **62** and nitrile **63** in equimolar amounts. The latter must derive from the reduction of Knoevenagel adduct **60**. Thus, a hydrogen transfer process takes place causing at the same time aromatization of **61** and reduction of Knoevenagel adduct **60**. In order to gain insight into this reaction, the authors mixed the independently prepared compounds **60** and **61** and found that they actually undergo this unusual hydrogen transfer, but only in the presence of the enzyme.[38]

The promiscuous activity of lipases was also exploited in the context of three-component Mannich reactions to give **64**.[39] The reaction was carried out in water and proceeded only to a very low extent in the absence of a lipase. Only a few lipases were able to give good conversions. Among them, the best one was this time the lipase from *Mucor miehei*.

It was suggested that, in all the reactions depicted in Figure 6.7, lipases use their active site for multiple acid–base catalysis.

A particularly complex example of the multiple roles of a lipase is shown in Figure 6.8.[40] Again, *Candida antarctica* lipase was the most efficient enzyme for the catalysis of the five-component reaction of *p*-nitrobenzaldehyde, a cyclic ketone, a nitroolefin, and acetamide to give spiro-oxazines **65**. Also

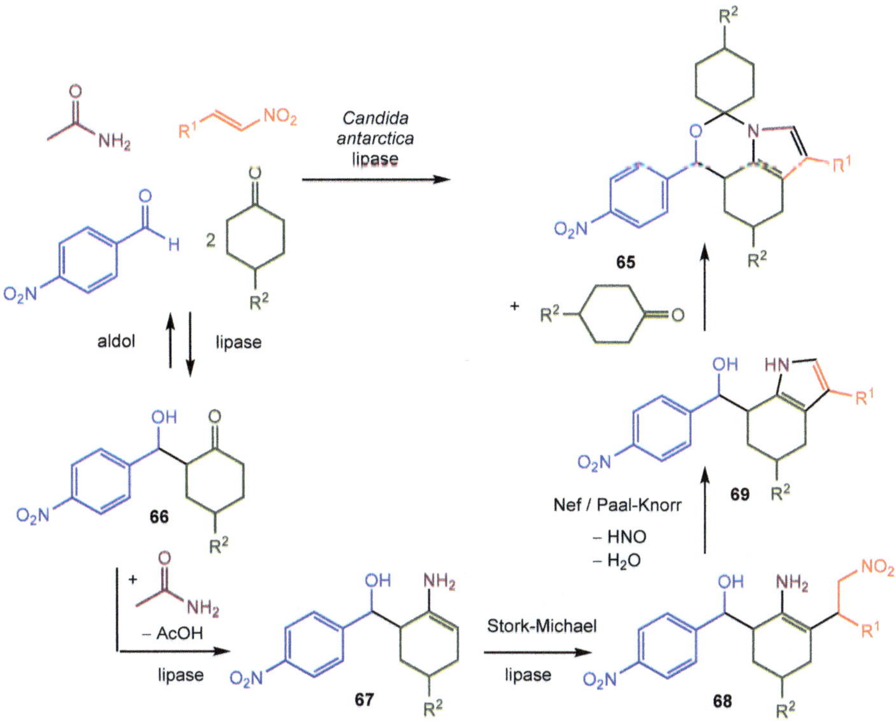

Figure 6.8 Biocatalytically promoted synthesis of spiro-oxazines.

in this case, acetamide is used as a synthetic equivalent of ammonia. Apart from the deacetylation to give **67**, the lipase probably catalyzes the aldol reaction to give **66**, and the Stork–Michael reaction of **67** with the nitroolefin to furnish **68**. Finally, after an intramolecular Nef reaction to give **69**, the second equivalent of cyclic ketone forms an aminal to end the domino process. For the full domino reaction, a benzaldehyde bearing strong electron-withdrawing substituents is necessary, but later the same authors[41] were able to expand the diversity using also other benzaldehydes and differently substituted cyclohexanones. The trick was the preformation of the aldol adduct in a separate step. The whole procedure can be carried out in one pot, and the first aldol addition is also biocatalyzed (but with a different enzyme: pig pancreatic lipase).

Spiro-oxazines **65** possess at least two stereogenic centers, whose number may increase when using substituted cyclohexanones. The reaction seems to be highly diastereoselective, leading to a single compound by nuclear magnetic resonance (NMR) spectroscopy. However, no enantioselectivity was observed. This lack of enantioselectivity may be ascribed to the reversibility of the aldol reaction. When starting from a preformed enantiopure aldol **66**, slightly racemized product **65** was indeed obtained.[41]

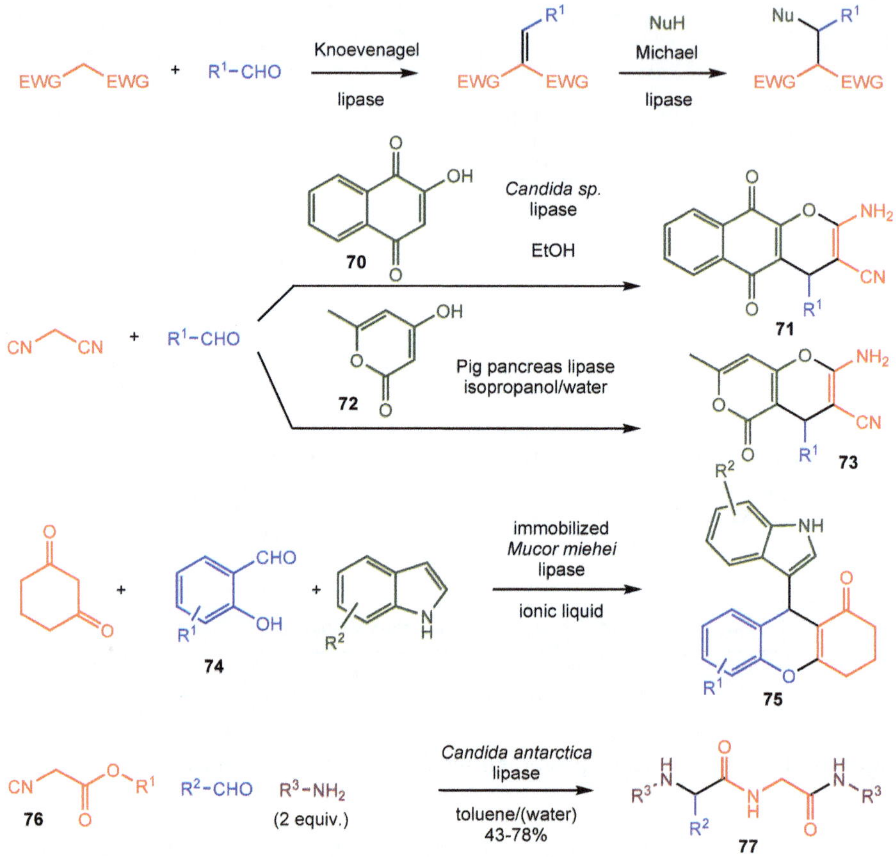

Figure 6.9 Biocatalytic synthesis of chromenes and truncated Ugi adducts.

As seen above, lipases have been shown to catalyze promiscuous processes such as Knoevenagel and Michael reactions. Figure 6.9 shows a general scheme and some examples of MCRs based on domino Knoevenagel–Michael processes. When one of the starting materials contains an additional phenol or enol, the domino process is terminated by a cyclization to afford chromene derivatives. For example, using naphthoquinone **70** or pyranone **72** as the nucleophile for the Michael reaction, benzo[*g*] chromenes **71**[42] or dihydropyrano[4,3-*b*]pyranes **73**[43] are respectively obtained, whereas using salicyladehyde **74** as the aldehyde counterpart and an indole as the nucleophile affords indolyl-4*H*-chromenes **75** as the final products.[44] For each of these reactions, careful optimization was carried out to find the best catalyst and solvent and the right water content. The catalytic activity is indeed in all cases highly dependent on the amount of water, although no general trend exists.

Biocatalysis was applied to isocyanide-based MCRs too. In the classical Ugi reaction, the intermediate imine is activated by the carboxylic acid. In the so called three-component 'truncated' Ugi reaction, where the carboxylic component is absent, some sort of imine activation is needed anyway. This can be in principle provided by the active site of a lipase. This concept was first demonstrated in 2013, for the reaction shown in Figure 6.9 leading to truncated adduct **77**.[45] Here, the enzyme not only catalyzes the MCR, but it also promotes the amidation of the ester with a second equivalent of the amine. The best enzyme was found to be *Candida antarctica* lipase and the ideal solvent toluene. The addition of a small amount of water had a significant effect, although highly dependent on the nature of the components. With some components, water leads to an increase of the yield, whereas other combinations caused a sharp yield drop. Interestingly, the ester moiety in isocyanide **76** is essential, since the reaction does not occur with isocyanides bearing amido groups or simple alkyl chains. This fact may be due to complexation of the ester in the active site. Unfortunately, no enantioselectivity was observed.

6.3 Application of Multicomponent Reactions to Renewable Feedstocks

6.3.1 Biobased Diols

Erythritol is a polyol that can be obtained by fermentation of sugar biomass. It is an achiral compound, but its protected derivative **78** has been successfully desymmetrized to both chiral enantiopure monoester **79** and its enantiomer by exploiting the complementary lipase-catalyzed acylation or ester hydrolysis (Figure 6.10).[12] These monoesters were directly oxidized to the corresponding aldehydes and subjected to a multicomponent Passerini reaction. When the oxidation was performed with catalytic TEMPO and stoichiometric BAIB, the sequence could be carried out in one pot and the acetic acid released from BAIB served as the carboxylic component. Many efforts have been made to improve the diastereoselectivity for the *anti* adduct **80**, and it was eventually found that good to excellent dr's could be achieved using $ZnBr_2$ as the catalyst in THF.[46] Similar dr's were achieved also performing the Passerini reaction on different chiral erythritol derivatives **81** and **82**, straightforwardly synthesized from **79**. Compounds **80–82** have two functionalized arms that can be exploited for cyclization processes leading to a large variety of interesting polyoxygenated heterocycles.[46] For example, **83** derives from a stereospecific Mitsunobu reaction on the diol obtained from **80** by solvolysis. **84** and **85** are synthesized by highly chemospecific oxidation of the same diol. Pyrrolidone **86** is obtained from **81** through substitution of the left hydroxymethyl group by the isocyanide-derived secondary amide. Finally, **87** is generated from azide **82** by deacetylation, reduction, acylation, and final S_N2 of the resulting amide on the right-hand secondary alcohol. These transformations show the potential of combining renewable sources, biocatalysis, and MCRs.

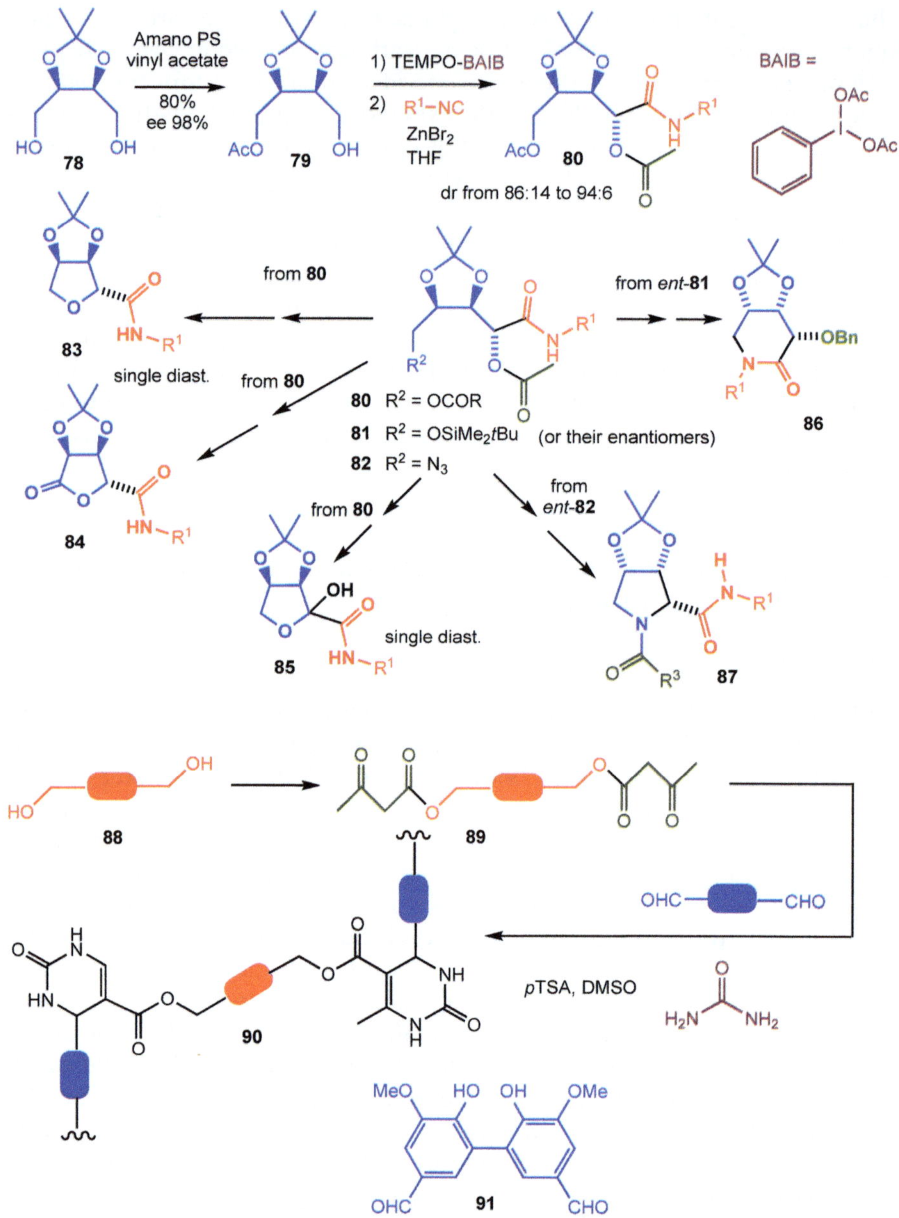

Figure 6.10 Application of MCRs to biobased diols. *p*-TSA = *p*-toluenesulphonic acid.

Achiral diols from renewable sources have been used, in combination with other biobased substances, for the synthesis of polymers *via* the already cited Biginelli reaction (Figure 6.10). The selected renewable diols **88** were propane-1,3-diol, hexane-1,6-diol, decane-1,10-diol, and icos-10-ene-1,20-diol. The first two can be obtained from glycerol and adipic acid, whereas the other two may be produced from castor oil. These diols were converted into esters **89** by reaction with β-ketoesters, providing the first bifunctionalized building block. The second bifunctionalized building block was terephthalaldehyde (which is currently accessible from renewable sources thanks to the effort to produce biobased terephthalic acid) or divanillin **91**. The third component was urea. By tuning the reaction parameters and the diversity inputs, polymers **90** with different thermal properties were obtained. In general, these new thermoplastic polymers display high glass transition temperatures.[47]

6.3.2 Furfural and Derivatives

Furfural **92** and 5-(hydroxymethylfurfural) (5-HMF) **93** are renewable chemical feedstocks obtained from the hydrolysis and dehydration of cellulosic biomass, which is available from plant waste matter. Thus, it is highly important to develop green methodologies for their transformation into a variety of compounds.

One of the big challenges in the synthesis of biobased products and intermediates is the transformation of polyoxygenated aliphatic molecules, such as monosaccharides or their derivatives, into benzene derivatives. This has been recently accomplished thanks to the ability of the furan ring to undergo Diels–Alder cycloadditions, followed by rearrangement (Figure 6.11).[48] This process is not a strict MCR, since the two reagents (dimethylhydrazine and the dienophile) are not added simultaneously. However, it is a one-pot process that takes place in water at moderate temperature (50 °C). Using a maleimide as the dienophile, benzene derivatives **94** are obtained, which have been converted into a variety of compounds. For example, products **95** and **96** are generated, respectively, from a reduction or an oxidation of the hydrazone. The reaction, which has been extended to other dienophiles (R^3, R^4 = electron withdrawing groups or H) to give **97**, proceeds through a Diels–Alder reaction to **98**, followed by rearrangement to **99** and aromatization.

A similar approach exploited an Ugi MCR using furfural **92** and α,β-unsaturated acids **100**, followed by intramolecular Diels–Alder of **101** and rearrangement. Under the conditions of the Ugi reaction, the intramolecular Diels–Alder addition takes place directly as a domino process, but the rearrangement requires a separate acid-catalyzed step. While the yields of bridged bicyclic systems **102** are generally good, the rearrangement to **103** proceeds only in moderate yield.[49]

5-Hydroxyfuran-2(5*H*)-one 1 **104** can be obtained at large scale by photooxygenation of furfural **92**. It has been employed in a sequence of two reactions involving first a Friedel–Crafts alkylation of an indole, followed by an intramolecular Ugi reaction. The Friedel–Crafts product **106** is indeed a bifunctional component containing both oxo and carboxylic functionalities. The Friedel–Crafts step affords good enantioselectivity under the organocatalysis

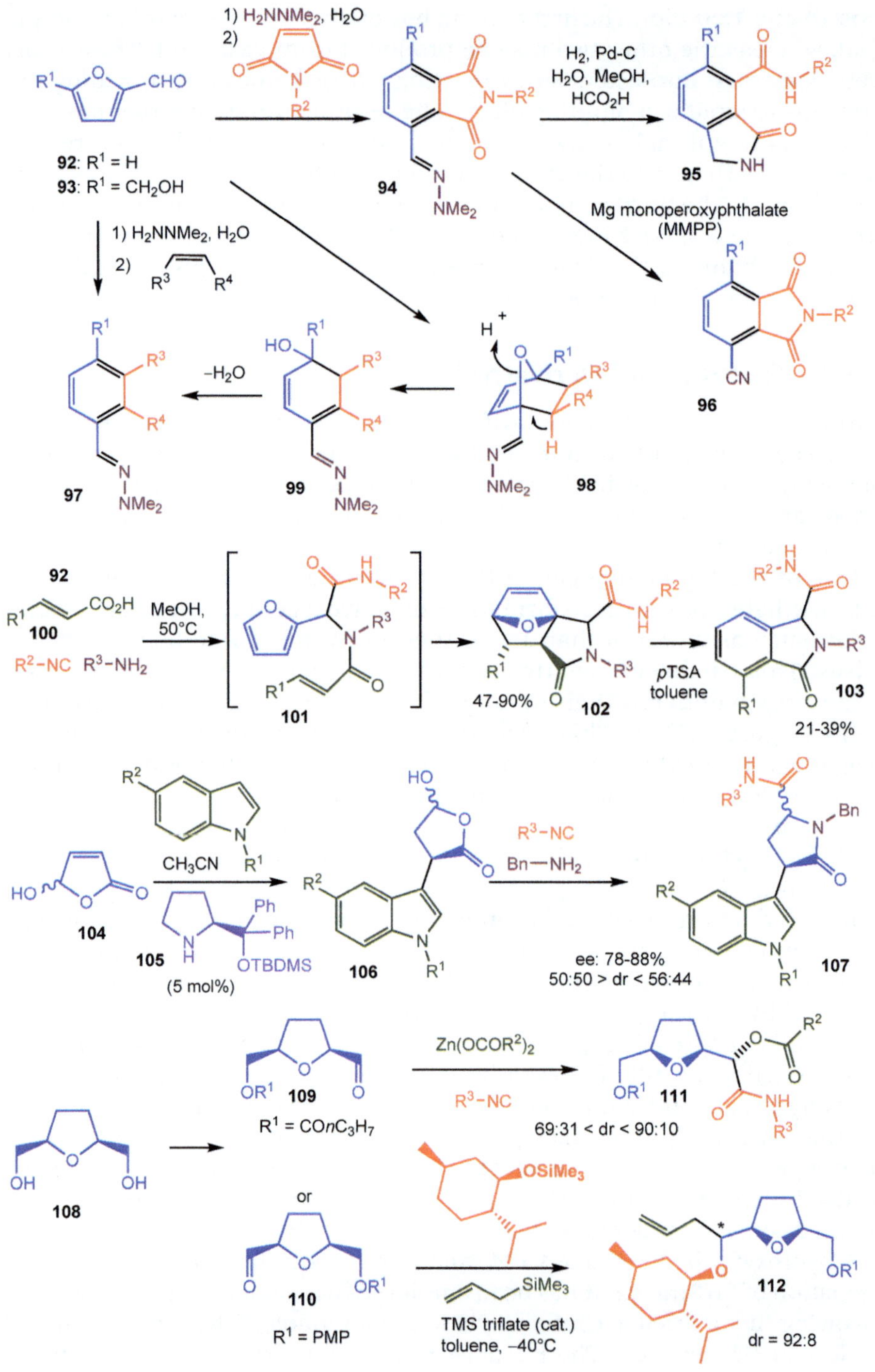

Figure 6.11 Application of MCRs to furfural, HMF, and their derivatives. TMS = trilmethylsilyl.

of prolinol derivative **105**.[50] On the other hand, the Ugi step is, as usual, very poorly diastereoselective, affording the two possible stereoisomers of **107** in nearly equal amounts. This two-step protocol can be carried out as a one-pot process.

Hydrogenation of 5-HMF **93** affords *cis* 2,5-*bis*(hydroxymethyl)tetrahydrofuran **108**,[51] a *meso* compound that can be considered a renewable building block as well. Enzymatic desymmetrization allows the efficient preparation in high ee of aldehydes such as **109** and **110**. They have been subjected to two MCRs, trying to optimize the diastereoselectivity. In particular, the diastereoselection of the Passerini reaction on **109** to give **111** was found to be only poor under classical conditions. It was remarkably improved by the use of catalytic $ZnBr_2$ (similarly to what was shown in Figure 6.10). However, the use of stoichiometric zinc carboxylates was equally efficient from the stereochemical point of view, but led to higher yields by completely suppressing the 'truncated' Passerini side reaction, where the acyl group R^2CO is not incorporated into the multicomponent product.[52]

Aldehyde **110** (R^1 = *p*-methoxyphenyl) was instead subjected to a multicomponent Sakurai reaction using the trimethylsilyl ethers of various secondary alcohols and allyl trimethylsilane. With some terpenic alcohols, such as (−)-menthol, quite high diastereoselectivities were observed for the obtained ethers **112**.[53]

6.3.3 Levulinic Acid

Levulinic acid **113** can be easily synthesized by acid treatment of hexoses. Despite the complex mechanism of this reaction, which goes through the intermediacy of 5-HMF **93**, the overall yield from crude lignocellulosic biomass is quite good and it is thus one of the most important renewable building blocks. The use of **113** in isocyanide-based MCRs is nearly 100 years old, since Mario Passerini himself employed it in his celebrated reaction.[54] Later, monocyclic pyrrolidinones have been prepared by applying intramolecular Ugi reactions to **113**,[55–58] but here we will mainly focus on its straightforward transformation into bicyclic nitrogen heterocycles by an Ugi reaction followed by post-MCR cyclization (Figure 6.12). When using α-aminoesters as the amine component, Ugi adducts **114** can be cyclized under basic conditions to bicyclic imides **115**.[58] The Ugi reaction shows no diastereoselectivity, but only one of the two diastereomers seems to be able to cyclize. The cyclization yields are only moderate and the imidic nature of the products make them not very useful in medicinal chemistry. More valuable is the synthesis of pyrrolopiperazines **118** using Boc monoprotected diamines as bifunctional synthons and the 'convertible' resin-bound isocyanide **116**. After the Ugi reaction to give **117**, acid catalysis results in Boc removal and promotes substitution of

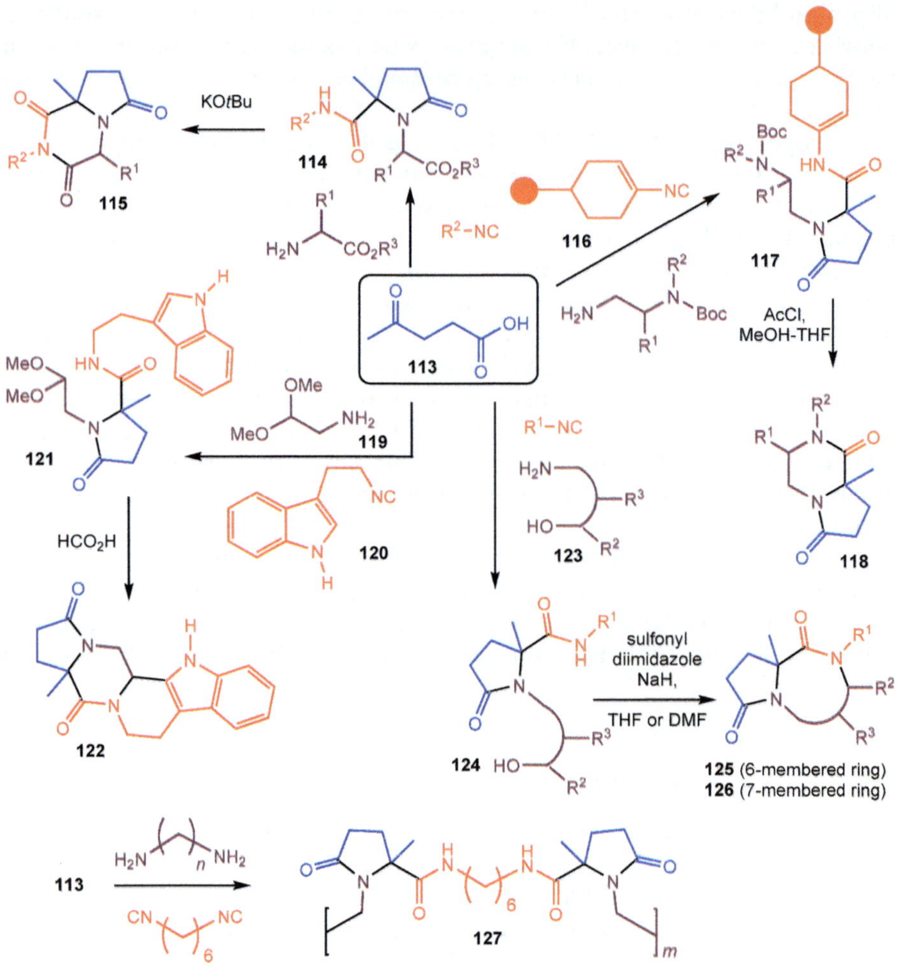

Figure 6.12 Levulinic acid in MCRs.

the convertible isocyanide by the liberated amine.[59] Unfortunately, in this methodology, the diversity of isocyanides cannot be exploited. Amine **119**, a synthetic equivalent of glycinal, has been reacted with levulinic acid and an indole containing isocyanide **120**. The resulting Ugi adduct was then cyclized under acidic catalysis through a Pictet–Spengler reaction to give alkaloid-type compound **122**.[60] This two-step procedure affords a very fast increase of structural complexity, but is limited from the point of view of diversity. Recently a two-step procedure for the efficient preparation of pyrrolopiperazines **125** and pyrrolotetrahydrodiazepines **126** starting from levulinic acid has been developed. This time, the procedure exploits unprotected aminoalcohol **123** as a bifunctional synthon.[61] After the Ugi reaction, cyclization involves an intramolecular S_N2 reaction, where the isocyanide derived secondary amide acts as a nucleophile. For this step, *in situ* activation of the alcohol in **124** is accomplished by sulfonyl diimidazole. The overall sequence is operationally simple, endowed with good atom economy, and the diversity of two inputs can be varied at will. Also, chiral aminoalcohols were used, but the diastereoselectivity in the Ugi reaction was in all cases absent.

Novel polyamides **127** were obtained by polymerization of levulinic acid, a diamine, and a diisocyanide, using the Ugi reaction.[62] The process was optimized varying several reaction parameters, and the best conditions turned out to be methanol as the solvent with μW irradiation at 100 °C. The molecular weight of the polymer can be tuned, especially by employing diamines with different chain lengths. Thus, polyamides with different glass transition and decomposition temperatures could be obtained.

6.4 Lipids

The synthesis of hybrid compounds where a central heterocyclic scaffold is bound to a lipidic chain can be relevant in medicinal chemistry in order to facilitate cell penetration. Obviously, fatty acids are lipids easily obtainable from renewable sources. Polyhydroquinolines (PHQ)-fatty acids hybrids **130** have been prepared by multicomponent Hantzsch synthesis.[63] Toward this goal, fatty acids **128** including, for example, stearic acid, oleic acid, and ricinoleic acid (the main constituent of castor oil, a non-edible feedstock) were first converted into β-ketoesters **129**. The MCR was preferably carried out with sulfamic acid, an inexpensive and not-toxic catalyst. On the other hand, hybrids dihydropyrimidinone-fatty acids have been prepared from β-ketoesters **131**, again derived from **128** through the corresponding fatty alcohols. This time, an $InCl_3$-catalyzed Biginelli MCR was employed.[64] Compounds **132** were tested against two glioma cell lines and some members of the library showed antiproliferative activity.

Ricinoleic acid **133** has attracted much interest in recent years as a useful renewable building block thanks to its unusual structure, which contains an alcohol moiety. Pyrolysis of **133** furnishes 10-undecenoic acid **134** and heptanal. Recently, **134** and its reduced aldehyde **135** have been used as inputs in multicomponent Passerini[65] or Ugi reactions[66] to affords complex monomers **136** and **138**, suitable for subsequent Acyclic Diene Metathesis (ADMET) polymerization to novel renewable polyamides **137** and **139**. The type of ruthenium catalyst influences the molecular weight of the polymer, the highest one being obtained with the Hoveyda–Grubbs catalyst. 1,4-Benzoquinone was added to inhibit the double bond positional isomerization in the products.

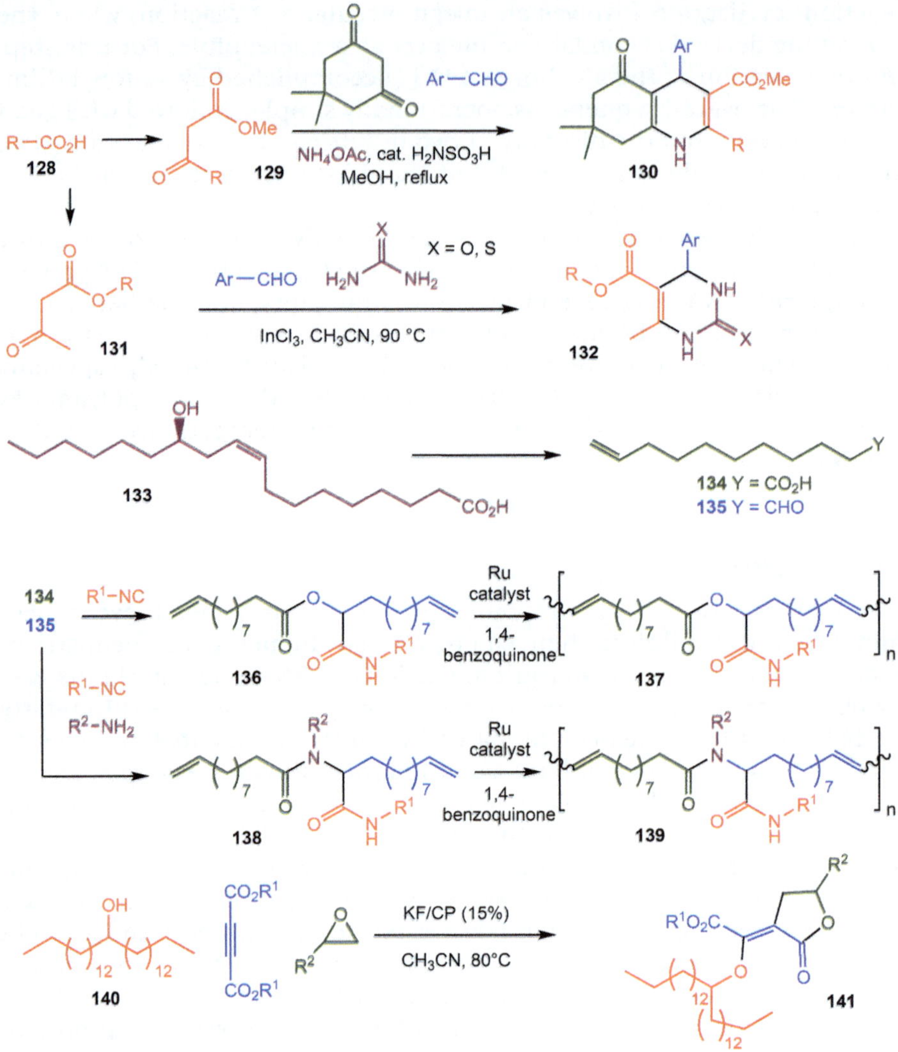

Figure 6.13 Application of renewable lipids to MCRs.

Various lipidic alcohols can be obtained from plants, and they may serve as inputs for MCRs. Unfortunately, MCRs that employs alcohols as components are not very common. An interesting example is shown in Figure 6.13, where the lipidic component is 15-nonacosanole **140**, extracted from *Fumaria officinalis*. Reaction of this secondary alcohol with dimethyl acetylenedicarboxylate and an epoxide leads to furanones **141**. The catalyst for this reaction is KF deposited on a zeolite named Clinoptilolite (KF/CP).

6.5 Concluding Remarks

The examples shown above of the utilization of renewable starting materials in MCRs are not the only ones. For example, for space limitation reasons, we have not described here the use of sugars or sugar-derived compounds in such processes.[67–69]

In any case, examples of the application of MCRs to biomass-derived feedstocks are still quite limited and even more infrequent are the combination of renewable sources, MCRs, and biocatalysis. We therefore believe that we have only seen so far the tip of the iceberg, and that combination of these three green methodologies is destined to become a very important tool for sustainable organic synthesis in the 21st century.

References

1. P. T. Anastas and J. C. Warner, *Green Chemistry: Theory and Practice*, Oxford University Press, New York, 2000.
2. R. A. Sheldon, *Green Chem.*, 2007, **9**, 1273.
3. R. C. Cioc, E. Ruijter and R. V. A. Orru, *Green Chem.*, 2014, **16**, 2958.
4. R. A. Sheldon and J. M. Woodley, *Chem. Rev.*, 2018, **118**, 801.
5. A. M. Bezborodov and N. A. Zagustina, *Appl. Biochem. Microbiol.*, 2016, **52**, 237.
6. R. A. Sheldon, *Green Chem.*, 2014, **16**, 950.
7. C. R. Bertozzi, C. J. Chang, B. G. Davis, M. Olvera de la Cruz, D. A. Tirrell and D. Zhao, *ACS Cent. Sci.*, 2016, **2**, 1.
8. R. C. Cioc, V. Estevez, D. J. Van der Niet, C. M. L. Vande Velde, N. G. Turrini, M. Hall, K. Faber, E. Ruijter and R. V. A. Orru, *Eur. J. Org. Chem.*, 2017, 1262.
9. L. Moni, L. Banfi, A. Basso, L. Carcone, M. Rasparini and R. Riva, *J. Org. Chem.*, 2015, **80**, 3411.
10. D. Paprocki, D. Koszelewski, A. Zadlo, P. Walde and R. Ostaszewski, *RSC Adv.*, 2016, **6**, 68231.
11. S. Klossowski, A. Brodzka, M. Zysk and R. Ostaszewski, *Tetrahedron: Asymmetry*, 2014, **25**, 435.
12. V. Cerulli, L. Banfi, A. Basso, V. Rocca and R. Riva, *Org. Biomol. Chem.*, 2012, **10**, 1255.
13. A. Znabet, E. Ruijter, F. J. J. de Kanter, V. Kohler, M. Helliwell, N. J. Turner and R. V. A. Orru, *Angew. Chem., Int. Ed.*, 2010, **49**, 5289.
14. A. Znabet, M. M. Polak, E. Janssen, F. J. J. de Kanter, N. J. Turner, R. V. A. Orru and E. Ruijter, *Chem. Commun.*, 2010, **46**, 7918.
15. A. Znabet, S. Blanken, E. Janssen, F. J. J. de Kanter, M. Helliwell, N. J. Turner, E. Ruijter and R. V. A. Orru, *Org. Biomol. Chem.*, 2012, **10**, 941.
16. A. Znabet, J. Zonneveld, E. Janssen, F. J. J. De Kanter, M. Helliwell, N. J. Turner, E. Ruijter and R. V. A. Orru, *Chem. Commun.*, 2010, **46**, 7706.

17. R. V. A. Orru, VII International Conference on Multicomponent Reactions and Related Chemistry, Düsseldorf, 2018.
18. W. Szymanski and R. Ostaszewski, *Tetrahedron: Asymmetry*, 2006, **17**, 2667.
19. R. Kourist, G. S. Nguyen, D. Strubing, D. Bottcher, K. Liebeton, C. Naumer, J. Eck and U. T. Bornscheuer, *Tetrahedron: Asymmetry*, 2008, **19**, 1839.
20. W. Szymanski and R. Ostaszewski, *Tetrahedron*, 2008, **64**, 3197.
21. W. Szymanski, M. Zwolinska, S. Klossowski, I. Mlynarczuk-Bialy, L. Bialy, T. Issat, J. Malejczyk and R. Ostaszewski, *Bioorg. Med. Chem.*, 2014, **22**, 1773.
22. A. Zadlo-Dobrowolska, D. Koszelewski, D. Paprocki, A. Madej, M. Wilk and R. Ostaszewski, *ChemCatChem*, 2017, **9**, 3047.
23. S. Klossowski, A. Redzej, S. Szymkuc and R. Ostaszewski, *ARKIVOC*, 2013, 134.
24. R. Reinart-Okugbeni, K. Ausmees, K. Kriis, F. Werner, A. Rinken and T. Kanger, *Eur. J. Med. Chem.*, 2012, **55**, 255.
25. S. Y. Torres, R. Brieva and F. Rebolledo, *Org. Biomol. Chem.*, 2017, **15**, 5171.
26. S. Y. Torres, Y. Verdecia and F. Rebolledo, *Tetrahedron*, 2015, **71**, 3976.
27. S. Y. Torres, E. Ochoa, Y. Verdecia and F. Rebolledo, *Tetrahedron*, 2014, **70**, 4675.
28. K. Ausmees, K. Kriis, T. Pehk, F. Werner, I. Jarving, M. Lopp and T. Kanger, *J. Org. Chem.*, 2012, **77**, 10680.
29. N. Zumbrägel, C. Merten, S. M. Huber and H. Gröger, *Nat. Commun.*, 2018, **9**, 1949.
30. N. Zumbrägel and H. Gröger, *Bioengineering*, 2018, **5**, 60.
31. M. Lopez-Iglesias and V. Gotor-Fernandez, *Chem. Rec.*, 2015, **15**, 743.
32. J. H. Lee, *Tetrahedron Lett.*, 2005, **46**, 7329.
33. A. Kumar and R. A. Maurya, *Tetrahedron Lett.*, 2007, **48**, 3887.
34. A. Kumar and R. A. Maurya, *Tetrahedron Lett.*, 2007, **48**, 4569.
35. W. Zhang, N. Wang, Z. J. Yang, Y. R. Li, Y. Yu, X. M. Pu and X. Q. Yu, *Adv. Synth. Catal.*, 2017, **359**, 3397.
36. J.-L. Wang, B.-K. Liu, C. Yin, Q. Wu and X.-F. Lin, *Tetrahedron*, 2011, **67**, 2689.
37. Z.-Q. Liu, B.-K. Liu, Q. Wu and X.-F. Lin, *Tetrahedron*, 2011, **67**, 9736.
38. Z. Q. Liu, Y. J. Hu, X. Y. Chen, Q. Wu and X. F. Lin, *Tetrahedron*, 2015, **71**, 663.
39. K. Li, T. He, C. Li, X.-W. Feng, N. Wang and X.-Q. Yu, *Green Chem.*, 2009, **11**, 777.
40. J. L. Wang, X. Y. Chen, Q. Wu and X. F. Lin, *Adv. Synth. Catal.*, 2014, **356**, 999.
41. X. Y. Chen, J. L. Wang, X. F. Lin and Q. Wu, *Tetrahedron*, 2016, **72**, 3318.
42. F. J. Yang, H. R. Wang, L. Y. Jiang, H. Yue, H. Zhang, Z. Wang and L. Wang, *RSC Adv.*, 2015, **5**, 5213.

43. X. Chen, W. A. Zhang, F. J. Yang, C. Guo, Z. Y. Zhao, D. Ji, F. Zhou, Z. Wang, R. Zhaob and L. Wang, *Green Chem. Lett. Rev.*, 2017, **10**, 54.
44. W. A. Zhang, Z. Y. Zhao, Z. Wang, C. Guo, C. Y. Wang, R. Zhao and L. Wang, *Catalysts*, 2017, **7**, 185.
45. S. Klossowski, B. Wiraszka, S. Berlozecki and R. Ostaszewski, *Org. Lett.*, 2013, **15**, 566.
46. L. Moni, L. Banfi, A. Basso, E. Martino and R. Riva, *Org. Lett.*, 2016, **18**, 1638.
47. A. C. Boukis, A. Llevot and M. A. R. Meier, *Macromol. Rapid Commun.*, 2016, **37**, 643.
48. S. Higson, F. Subrizi, T. D. Sheppard and H. C. Hailes, *Green Chem.*, 2016, **18**, 1855.
49. G. Caillot, S. Hegde and E. Gras, *New J. Chem.*, 2013, **37**, 1195.
50. E. Riguet, *J. Org. Chem.*, 2011, **76**, 8143.
51. T. Buntara, S. Noel, P. H. Phua, I. Melián-Cabrera, J. G. de Vries and H. J. Heeres, *Angew. Chem., Int. Ed.*, 2011, **50**, 7083.
52. L. Moni, L. Banfi, D. Cartagenova, A. Cavalli, C. Lambruschini, E. Martino, R. V. A. Orru, E. Ruijter, J. M. Saya, J. Sgrignani, R. Riva, submitted, 2019.
53. C. Lambruschini, A. Cherubin, L. Banfi, A. Barbero and R. Riva, VII International Conference on Multicomponent Reactions and Related Chemistry, Düsseldorf, 2018.
54. M. Passerini, *Gazz. Chim. Ital.*, 1923, **53**, 331.
55. H. Gross, J. Gloede, I. Keitel and D. Kunath, *J. Prakt. Chem.*, 1968, **37**, 192.
56. G. C. B. Harriman, *Tetrahedron Lett.*, 1997, **38**, 5591.
57. K. M. Short and A. M. M. Mjalli, *Tetrahedron Lett.*, 1997, **38**, 359.
58. C. Hanusch-Kompa and I. Ugi, *Tetrahedron Lett.*, 1998, **39**, 2725.
59. C. Hulme, L. Ma, M. P. Cherrier, J. J. Romano, G. Morton, C. Duquenne, J. Salvino and R. Labaudiniere, *Tetrahedron Lett.*, 2000, **41**, 1883.
60. W. Wang, E. Herdtweck and A. Domling, *Chem. Commun.*, 2010, **46**, 770.
61. C. Lambruschini, A. Basso, L. Moni, A. Pinna, R. Riva and L. Banfi, *Eur. J. Org. Chem.*, 2018, 5445.
62. M. Hartweg and C. R. Becer, *Green Chem.*, 2016, **18**, 3272.
63. H. D. Fontecha-Tarazona, R. C. Brinkerhoff, P. M. de Oliveira, S. B. Rosa, D. C. Flores, C. D. R. Montes D'Oca, D. Russowsky and M. G. Montes D'Oca, *RSC Adv.*, 2015, **5**, 59638.
64. T. G. M. Treptow, F. Figueiró, E. H. F. Jandrey, A. M. O. Battastini, C. G. Salbego, J. B. Hoppe, P. S. Taborda, S. B. Rosa, L. A. Piovesan, C. D. R. Montes D'Oca, D. Russowsky and M. G. Montes D'Oca, *Eur. J. Med. Chem.*, 2015, **95**, 552.
65. O. Kreye, T. Toth and M. A. R. Meier, *J. Am. Chem. Soc.*, 2011, **133**, 1790.
66. O. Kreye, O. Tueruenc, A. Sehlinger, J. Rackwitz and M. A. R. Meier, *Chem. - Eur. J.*, 2012, **18**, 5767.

67. Z. Söyler, K. N. Onwukamike, S. Grelier, E. Grau, H. Cramail and M. A. R. Meier, *Green Chem.*, 2018, **20**, 214.
68. V. K. Rai, S. Singh, P. Singh and L. D. S. Yadav, *Synthesis*, 2010, 4051.
69. L. D. S. Yadav and A. Rai, *Carbohydr. Res.*, 2009, **344**, 2329.

CHAPTER 7

Green Synthetic Procedures under Hydrodynamic and Acoustic Cavitation

GIANCARLO CRAVOTTO*, SILVIA TAGLIAPIETRA AND ZHILIN WU

Dipartimento di Scienza e Tecnologia del Farmaco, University of Turin, Via Pietro Giuria 9, 10125, Turin, Italy
*E-mail: giancarlo.cravotto@unito.it

7.1 Introduction

The intensification of organic reactions using high-intensity ultrasound has been extensively studied since the chemical effects of ultrasonic cavitation (UC) were first discovered.[1–4] However, hydrodynamic cavitation (HC), which is the originator of cavitation phenomena, has long been thought of as simply the destroyer of screw propellers, pumps, and hydraulic equipment.[5,6] HC was first observed as the initiator of screw-blade erosion and the cause of less efficient screw-propeller propulsion at high speeds.[7,8] Since the 1990s, HC technology has gradually become more frequently used to intensify chemical processes, such as fatty-oil hydrolysis,[9] nanomaterial preparation[10,11] and hydroxyl-radical formation.[12]

Green Chemistry Series No. 61
Green Synthetic Processes and Procedures
Edited by Roberto Ballini

Published by the Royal Society of Chemistry, www.rsc.org

7.2 Hydrodynamic Cavitation for Green Organic Synthesis

HC is defined as the formation, growth and implosive collapse of gas- or vapour-filled bubbles in liquids that is caused by a decrease and subsequent increase in local pressure up to the vapour pressure of the media.[12] It typically occurs as the result of an increase in the kinetic energy through area constriction, or the high-speed rotation of a rotor through a fluid.[12–14] Ways to achieve this include throttling valves, orifice plates, venturi, cavitation jet mixers, two-stage orifice valves, high-pressure liquid whistles, high-pressure jet fluidisers and high-speed rotor–stator homogenisers, among others.[15] A typical HC loop device generally consists of a circulating pump, a storage tank with a cooling jacket, main and bypass pipelines, valves, manometers, flowmeters and a cavitation generating unit (restrictor), as shown in Figure 7.1.

The processing volumes of low-pressure HC devices (0.015–1.5 MPa) are usually designed to be about 1 to 200 L, which are suited to bench-scale investigations and pilot-scale production,[9,17–19] respectively. However, HC can also be produced in a smaller volume (60 mL) in a high-pressure jet fluidiser (15–124 MPa) and in volumes of 1–800 mL in high-speed rotor-stator

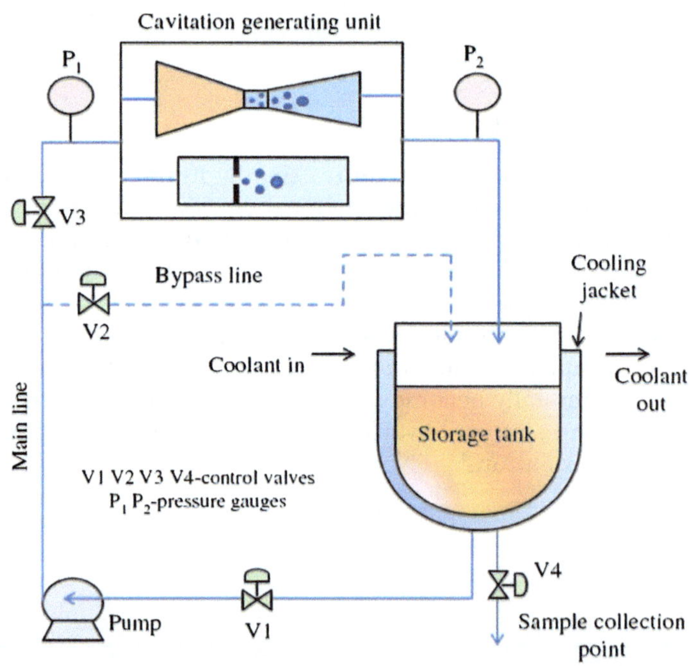

Figure 7.1 Schematic of a loop HC reaction system. Adapted from ref. 16 with permission from De Gruyter.

homogenisers (3000–60 000 rpm).[12,20] Low-pressure loop HC devices have received more attention because of their convenience, flexibility and operation safety.

Similarly to UC, the HC process of generation, growth and collapse results in extremely high energy densities at the surface of bubbles – namely, the localised hot spots with enormous impact temperatures of several thousand degrees and pressures in excess of 1000 atm – while the bulk liquid remains at ambient conditions.[2,21–24] Cavitation collapse induces the dissociation of the vapours trapped in the cavitating bubbles and the formation of chemical species that are not easily attained under conventional conditions, which can drive particular radical reactivity.[3,12,14,25] Meanwhile, the mechanical effects, such as shock-waves, high-speed jets, turbulent flow and liquid whirlpools, that are generated during cavitational collapse produce high-velocity interparticle and substrate collisions.[14,21,23,26,27] The collapse of cavitation bubbles at or near the interface of immiscible liquids can lead to powerful disruption and mixing, resulting in the formation of very fine emulsions.[2] It can therefore be stated that HC can efficiently enhance mass and heat transfer.[26–29]

The common advantages of using UC and HC processes is that they are environmentally friendly, as they use milder reactions and give improved yields and selectivities, require shorter reaction times, can easily generate reactive species and catalysts, and can replace hazardous reagents.[3,30] Higher reactant volumes can generally be processed by HC than by UC, resulting in lower energy requirements per unit of reactant volume and more energy-efficient protocols.[31,32] HC reactors are less sensitive to the reactor geometry,[32] meaning that they have immediate and realistic potential for industrial scale applications.[29] The scale up of these reactors is comparatively easy as vast amounts of information on the fluid dynamics downstream of the constriction are readily available.[29] In addition, the equipment for generating HC is simple and maintenance costs are low.[33]

HC has closely adhered to the concept of green chemistry ever since it was first used to intensify chemical processes. Firstly, organic reactions at extreme temperatures and pressures can be carried out at room temperature and atmospheric pressure using HC processes, while higher energy efficiency is achieved. For example, the hydrolysis of fatty oils with water/steam generally requires high operating temperatures of 250–350 °C and pressures of 3–15 atm. HC and UC have been used to enhance the hydrolysis of castor oil and kerdi oil at 40 °C under atmospheric pressure, while the inlet pressure of the HC process was set at 0.145–0.310 MPa (21–45 psi).[9] The volumes of the oil–water emulsion were 220 mL and 200 L for the UC and HC processes, respectively, resulting in reaction times of 270 min with UC and 20 h with HC, and an increase in the acid value from 20 to 150. Hence, the total energy dissipation of the UC and HC processes were 1384 and 1080 J mL^{-1} of emulsion, respectively, during hydrolysis.[9]

Obviously, the cavitation-assisted hydrolysis of oil can be performed under mild conditions. The energy consumption needed to achieve the same conversion value was lower using HC than using UC, while the HC process was carried out on a much larger scale than UC.

Secondly, reaction rates increase and byproduct formation is hampered when using HC processes, leading to shorter reaction times and higher product yields. For example, the oxidation of *p*-xylene with aqueous potassium permanganate was considerably accelerated at ambient temperature, and seven-times more product was obtained using an HC process than a UC process.[34] Moreover, reaction pathways can also be changed using cavitation processes. For example, while the use of an UC process in the base-catalysed autoxidation of alkylnitrobenzenes in polyethylene glycol phase-transfer systems changed the reaction selectivity from the usual dimeric products to the carboxylic acid, the corresponding nitrocarboxylic acids were obtained in low yield along with the dimers.[35] Interestingly, no dimer was formed during the oxidation of nitrotoluene using an HC process.[34]

Over the past decades, HC processes have been extensively investigated for use in the preparation of metal oxide and salts nanoparticles,[36–38] the degradation of bio-refractory contaminants in wastewater,[18,39–43] the disinfection and inactivation of microalgae,[19,44–48] the delignification and valorisation of biomass,[49–55] the disintegration of sludge and production of biogas,[56–60] the extraction of valuable compounds and food processing,[61–65] as well as the emulsification of oil and water,[66–69] among others. In particular, HC processes have also been used in the intensification of classical organic reactions, such as the preparation of biodiesel *via* transesterification of vegetable oil,[70,71] the removal of free fatty acids *via* esterification, the hydrolysis of oils,[9,72] the oxidation of alkylarenes to acids,[33,34] the oxidation and epoxidation of oils and esters,[29,73,74] polymerisation and depolymerisation,[75–77] deep desulphurisation of liquid fuels[78–80] and heavy oil upgrading,[81,82] among others.

This review will concentrate on the intensification of organic reactions, such as transesterification, esterification, glycerolysis and epoxidation of vegetable oil, the deep desulphurisation of liquid fuel, cavitational switching, the oxidation of hydrocarbons, reduction and hydrogenation, substitution, cyclopropanation and the hydrolysis of cellulose, to highlight the main contributions of HC processes to green chemistry, which include milder reaction conditions, enhanced mass and heat transfer in heterogeneous processes, increased reaction rates and reduced reaction times, improved conversions, yields, selectivities and energy efficiency, easy generation of reactive species and catalysts, and the ability to adapt to renewable and waste resources and replace hazardous reagents.[1,3,4,25] Moreover, the main limitations of the use of HC processes in organic synthesis will also be outlined.

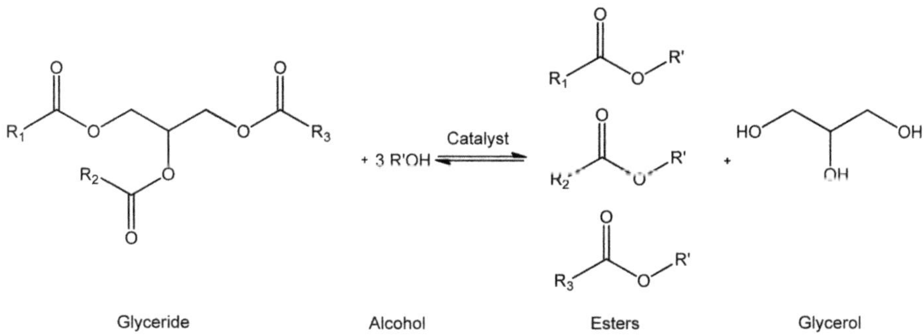

Scheme 7.1 Transesterification reaction of triglycerides with alcohols.[89]

7.2.1 Synthesis of Biodiesel (Transesterification and Esterification)

Transesterification. Biodiesel has been regarded as a sustainable and eco-friendly fuel for the last 20 years because of its unique properties, such as biodegradability, renewability and non-toxicity. Its use leads to the reduction of sulphur oxide and greenhouse gas emissions as it produces considerably lower amounts of carbon- and sulphur-based gases than conventional fossil fuels.[83] Ideally, non-edible and waste cooking oil (WCO), which are considered to be sustainable feedstocks, are used as raw materials for biodiesel synthesis to avoid the use of edible oils.[84–88] Mechanical stirring (MS) for transesterification is the most common and extensively utilised method for biodiesel production and involves the conversion of oil to glycerol and acid alkyl esters using methanol and a catalyst (Scheme 7.1). However, the poor mass and heat transfer efficiency of MS reduce the reaction rate and increase the energy consumption and production costs.[83]

Besides degrading organic pollutants in wastewater, biodiesel synthesis is the most widely developed use of HC technology. Cavitation results in turbulence and liquid circulation in the reactor, which can aid in the elimination of mass transfer resistance in heterogeneous reactions.[70,90] The UC process for the methyl transesterification of triglycerides is an order of magnitude more efficient in terms of mass of product per energy consumed than traditional MS, while the HC process is more efficient than conventional MS and heating protocols by a factor of 1000.[25,91] Cavitation reactors can therefore considerably intensify the biodiesel synthesis process with substantial reductions in processing time and raw material needs, coupled with greener and safer operation.[85] Energy efficiency, or cavitational yield, which indicates the overall performance of a reactor, is defined as the product yield per unit energy supplied to the system.[90] A comparison of the energy efficiencies and conversions of HC and UC or/and MS is provided in Table 7.1.

Table 7.1 Comparison of the energy efficiency and conversion of HC and UC and/or MS.[a]

		HC					UC or MS					EE (mg J^{-1})				
Feedstock	Catalyst	HC type	P (kW)	V (L)	t (min)	X (%)	F (kHz)	P (kW)	V (L)	t (min)	X (%)	MS	UC	HC	HC/UC, HC/MS	Reference
FFA cut (C8–C10)	H_2SO_4 for esterification	OP	1.5	10	90	92	20	0.12	0.1	120	98	—	0.005–0.020	0.1–0.2	10–20	70
Nagchampa oil	H_2SO_4 for esterification	OP	1.5	10	60	82	20	0.12	0.05	15	94	0.003	0.006	0.48	80	88, 92
Nagchampa oil	KOH	OP	1.5	10	20	92	20	0.12	0.05	40	93	0.005	0.010	0.87	87	88, 92
Used frying oil	KOH	OP, 25 × 2 mm	7.5	10	10	95	20	0.2	0.11	40	90 58[c]	—	—	1.28	—	90, 93
WCO[d]	$KOCH_3$	Slit venturi 3.7 × 0.92 mm	1.5	<15	30	89[b]	22	0.75	<0.1	30	90[b]	0.032	0.047	1.22	26	84, 94
WCO (palm olein)	KOH	OP, 21 × 1 mm	4	50	15	98	600 rpm		50	15[c]	19	0.15	—	1.25	8	95
Palm and rubber seed oil (1:1)	KOH	OP, 1 mm		50	50	97	MS		50	50[c]	87	0.08	—	1.35	17	96

[a]OP: Orifice plate, *P*: electric power; *V*: volume of oils; *t*: reaction time; *X*: conversion; *F*: acoustic frequency; EE: energy efficiency. FFA: free fatty acid.
[b]Biodiesel yield.
[c]MS.
[d]With methyl acetate.

As shown in Table 7.1, HC processes are an order of magnitude more energy efficient than UC and MS processes.[70] Furthermore, HC-based approaches afford higher biodiesel yields at significantly lower reactant molar ratios, while also providing lower reaction times.[84] Moreover, the UC technique is not yet completely feasible for biodiesel production at industrial scale, while HC offers a number of advantages over other intensification technologies.[83] More importantly, HC technology significantly improves the energy efficiency and product quality of biodiesel synthesis.[83,97] Biodiesel production using HC processes generally seems to be relatively easy to scale up to higher production values, which results in biodiesel being a viable fuel for industrial production and one that is available for use without engine modification being necessary.[90]

The performance and emission characteristics of HC-process-sourced biodiesel blends of 10%, 30% and 50% from WCO are acceptable according to the EN 14214 and ASTM D 6751 standards.[95,98] During engine performance tests, biodiesel blends showed higher brake specific fuel consumption (2.1–9.0%) and exhaust gas temperature (1.0–6.8%), and lower brake power (1.6–6.7%), torque (0.6–5.2%) and brake thermal efficiency (1.9–8.4%) than diesel fuel. Biodiesel-blend engine emissions showed higher carbon dioxide (8.7–38.5%) and nitrogen oxide (4.7–19.0%) release but, surprisingly, lower amounts of carbon monoxide (3.3–26.3%) than diesel fuel.[98]

A range of different HC devices, such as orifice plates and venturi, have typically been used to prepare biodiesel under varying operating parameters, such as the inlet pressure (2–5 bar).[84] High speed homogenisers[99,100] and rotor-stator HC systems[101] have also been used to intensify biodiesel production. For soybean oil and WCO, maximum biodiesel yields of 84% and 88% were obtained with methanol and CaO at 50 °C using a T25-UltraTurax at 12 000 rpm of rotation speed in 30 min, while it takes 2–3 h to obtain a similar biodiesel yield by conventional MS.[99] However, UC reactors consume only about one third of the energy required by a specialised mixer (UltraTurrax T25 basic operating at 11 000 rpm) to provide similar levels of conversion (around 85%).[102] On a pilot scale, the electrical energy consumption of a rotor-stator type HC reactor is 0.030 kW h L^{-1} of crude biodiesel produced *via* base-catalysed transesterification of refined and bleached palm oil and WCO, making this innovative technology rather competitive.[101] The reactor can be readily scaled-up, from producing a few hundred to thousands of litres of biodiesel per hour, and avoids the risk of orifices becoming clogged with oil impurities, which may occur in conventional HC reactors. Overall, energy efficiency calculations and synthesised biodiesel properties demonstrate that HC reactors can efficiently prepare biodiesel from non-edible oils and WCO.

Esterification. The HC process has frequently been used to treat sustainable feedstocks, such as WCO and non-edible oil (rubber seed oil).[84–88] The higher free fatty acid (FFA) contents present in these feedstocks should be reduced by an acid esterification process to avoid saponification reactions and hydrolysis. Generally, a FFA content below 1% gives favourable transesterification reactions in the presence of a base catalyst.[96,103] Therefore, the synthesis of biodiesel from high-acid-value oil entails two stages: acid esterification (catalysed by H_2SO_4) and transesterification with a base catalyst.[104]

However, this additional esterification process also requires considerable amounts of energy and time.[85] Just like the UC process,[104] the HC process can decrease the FFA content from 25% to 0.2% in shorter reaction times of *ca.* 50 min. By contrast, the conventional process needs more time to reduce the FFA content to the desired level. Furthermore, the conventional approach requires higher temperatures, whereas HC reactors successfully produce results at ambient temperature.[96] The HC and UC energy efficiencies for FFA (C8–C10) esterification using methanol in the presence of concentrated H_2SO_4 have been compared, and are listed in Table 7.1.[70]

Ambient operating conditions and <3 h reaction times have been found to be sufficient to give >90% conversion (mol%) for several combinations of acid (lower and higher)/methanol studied.[70] The maximum reduction in the acid value was obtained in Karanja oil with an initial acid value of *ca.* 14 mg of KOH g^{-1} oil. After cavitational treatment, the acid value reached 2.7 mg of KOH g^{-1} of oil using UC and 4.2 mg of KOH g^{-1} of oil using HC, at optimum oil-to-methanol molar ratios of 1:5 and 2% sulphuric acid loading at ambient temperature.[103]

The equilibration time in the esterification of Chinese vegetable tallow with methanol to remove FFA was reduced to a greater extent by HC than by MS. Moreover, three operating modes, namely methanol one time-fed (Mode 1), batch-fed (Mode 2) and recycled (Mode 3), were compared. Mode 3 reduced the methanol consumption by 5/6 compared to Mode 1, indicating that recycling of the methanol catalyst can significantly reduce the methanol consumption in the esterification process.[105]

Glycerolysis of oil. The glycerolysis of triacylglycerides (TAGs) in vegetable oils is another transesterification method commonly used to produce monoacylglycerides (MAGs) and diacylglycerides (DAGs), which are used as surfactants and emulsifiers in food, cosmetics and pharmaceutical products. Glycerolysis can also be a useful alternative for the removal of the free fatty acid content present in sustainable feedstocks, but presents limitations as its requires higher temperatures and longer processing times.[106] Glycerolysis is a two-step reversible reaction and DAG is the intermediate product. Firstly, the TAG reacts with glycerol to produce a DAG and MAG. Secondly, the DAG reacts with glycerol to produce a MAG.[107]

While enzymatic catalysis is green and proceeds under mild conditions, industrial scale applications are expensive. MAGs and DAGs are chemically produced by glycerolysis of fats and oils at high temperatures of 220–260 °C in the presence of inorganic alkalis, which are used as catalysts.[108] Microwave irradiation can significantly reduce the glycerolysis time and temperature. Reaction times have been reduced from 240 to 25 min, while the optimum temperature for maximum benefit was reduced from 200 to 105 °C with the microwave approach. Energy-consumption analysis also revealed the superiority of microwave-assisted synthesis as it had a much lower energy requirement (35.3 kJ g^{-1}) than the conventional method (203.6 kJ g^{-1}).[106] An orifice-HC system was used to enhance the miscibility of the oil and glycerol phases at 40 °C. NaOH and acetone were used as catalyst and co-solvent, respectively. Under the optimum conditions of 1.5% catalyst and 300% co-solvent, only trace amounts of MAGs (0.16 ± 0.13%) and DAGs (*ca.* 5.18%) were formed in the MS system. By comparison, substantial amounts of *ca.* 39% MAGs and 31% DAGs were formed in the HC system.[107] Obviously, cavitation affects the reaction medium and cavitation bubble collapse improves the mass transfer and overall reaction rate.

7.2.2 Epoxidation of Oil and Fatty Acid Methyl Esters

The epoxidation of vegetable oils and fatty acid methyl esters (FAMEs) refers to the reaction of the double bonds of oils with a percarboxylic acid (generally peracetic acid or performic acid). It involves two immiscible phases and is extremely exothermic, with enthalpies of up to −196 kJ mol^{-1}.[29,109] In aqueous phase, formic acid reacts with hydrogen peroxide to produce performic acid, which is then transferred to the organic (oily) phase to react with the double bonds of triglyceride molecules (Scheme 7.2):[109]

Mass and heat transfer are the two limiting factors in a conventional MS reactor. In order to prevent excessive temperature increases in conventional industrial reactors, H_2O_2 is added to the oil slowly (typically over 2 h) at moderate temperature. The reaction therefore takes about 10–12 h to complete.[29] Furthermore, the epoxidation of vegetable oil requires the assistance of a phase transfer catalyst (PTC), such as tetra-*n*-butyl ammonium bromide.[110] The iodine value of soybean oil was reduced from 132 to almost 19 at 80 °C using an UC horn under cavitation conditions. By comparison, the relative percentage conversion to oxirane using conventional MS was about 87% in 10 h, while almost 83% conversion was obtained in 4 h using the UC horn.[110]

HC technology has also been used for the epoxidation of FAMEs and comparisons with MS were carried out. HC provided a better mass transfer rate in a liquid–liquid system, and the reaction time was reduced more significantly without sulphuric acid. Furthermore, higher circulation rates resulted in much higher cavitation times in the system, which favoured the epoxy reaction. Higher upstream pressures also led to better emulsification and higher FAME conversions. However, excessively high upstream pressures resulted in the degradation of epoxy FAMEs.[73]

A novel reactor for the epoxidation of refined FAMEs (iodine value, 125 g I_2 per 100 g oil), which were produced from soybean oil, coupled HC with a static mixer, leading to both mass and heat transfer being effectively improved. After FAMEs and formic acid were mixed, a calculated amount of

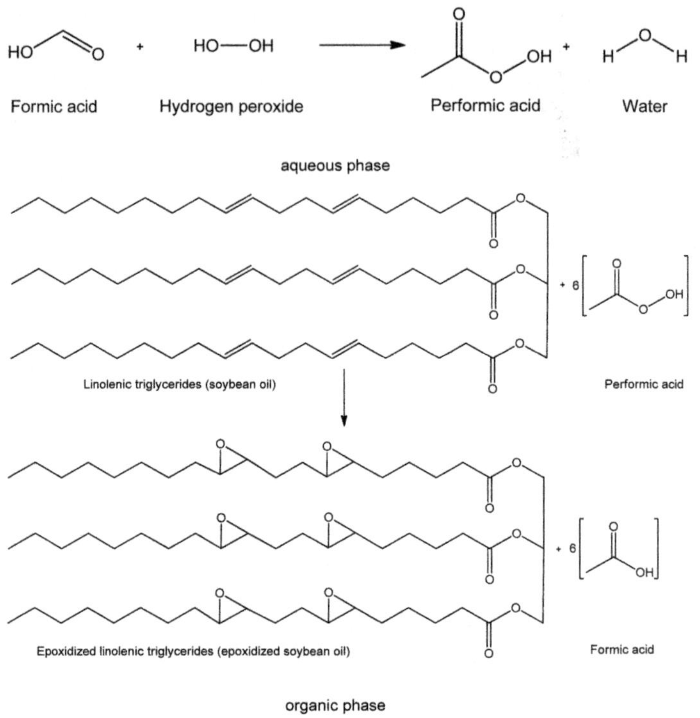

Scheme 7.2 Epoxidation of oil.[109]

H_2O_2 was added over less than 1 min to the system at 50–70 °C. As a result, a Sauter mean diameter of *ca.* 9 μm was observed in the oil/water dispersion formed using HC at a low inlet pressure of 0.25 MPa (2.5 bar). The droplets were very fine compared to the *ca.* 72 μm attained using conventional MS at 400 rpm. Moreover, the heat-transfer coefficient in the HC system was estimated to be 288 W m^{-2} K^{-1}, which is much higher than the 186 W m^{-2} K^{-1} value observed with laboratory-scale MS and the 100 W m^{-2} K^{-1} value obtained in an industrial-scale MS reactor. The HC process provides the same iodine value of 3.82 g I_2 per 100 g oil and epoxy value of 6.01 in a significantly shorter reaction time of 480 min compared to the 600 min required with MS. Meanwhile, the HC process provided a stable reaction temperature, simpler and safer operation and shorter H_2O_2 addition time (from 30–90 min to less than 1 min).[29]

HC has been used to intensify the epoxidation of soybean oil with H_2O_2 in a formic acid autocatalytic system and its reaction efficiency was compared to that of MS. As a result, the HC process was found to provide better mass transfer rate in the heterogeneous liquid–liquid system than MS. Under optimum conditions (170 g of soybean oil, 25 g of formic acid, 50 mL min^{-1} flow rate, 117 g of H_2O_2, 70 °C and 0.5 MPa of upstream pressure), the reaction time and energy consumption required to reach the same iodine value of 2.7

g I_2 per 100 g oil and epoxy value of 6.24 were significantly lower.[74] The physical effects of cavitation enhanced the turbulence and interactions between the oil and water phases, which enabled the average droplet size distribution to reach 60–90 nm. The reaction rate of the HC process was much higher than that of MS when agitation speeds below 85 rpm were used (common industrial case). Moreover, the conversion and selectivity of the HC process were almost the same as those of the MS system at agitation speeds of over 135 rpm.[74]

7.2.3 Deep Desulphurisation of Liquid Fuel

Current strict environmental regulations require very deep desulphurisation to be carried out to meet ultra-low sulphur diesel specifications (15 ppm sulphur). The catalytic hydrodesulphurisation (HDS) process commonly used in refineries is considered a satisfactory method for reducing the sulphur content down to 350 ppm, which thereafter pose limitations due to low reactivity of the remaining refractory compounds and increased cost of operation.[79,111,112] The HDS process requires high temperatures of 450 °C and high pressures of 20–40 atm, as well as Co–Mo or Ni–Mo catalysts. The new standards mean that a three-fold increase in the catalyst volume/reactor used in the HDS process is required.[79,113] The disadvantages of this hydrotreating technology, including production conditions, costs, safety and environmental issues, have led to the development of ultrasound-assisted oxidative desulphurisation (UAOD) as an alternative technology. The UAOD process selectively oxidises with H_2O_2 and carboxylic acids the sulphur in common thiophenes in diesel to sulphoxides and sulphones (Scheme 7.3), which can be removed *via* selective adsorption or be extracted.[114]

The heavier cuts (kerosene, diesel fuel cuts, vacuum gasoil) contain cyclic and polycyclic high-molecular-weight sulphur compounds whose removal entails exhaustive and complicated treatment, while the light fractions primarily contain low-molecular-weight sulphur compounds, including hydrogen sulphide and light mercaptans. The catalytic desulphurisation of diesel fuel using a cavitation mixer on a pilot scale (20 L of fuel) was reported in 2010. HC technology was developed to provide a 2.0–2.2-fold decrease of the residual content of mercaptan sulphur in diesel fuels, while also providing a two-fold reduction in the consumption of the catalytic complex (an alkaline solution of cobalt disulphophthalocyanine sodium salt in caustic soda) after 40–280 s of treatment.[115]

A non-catalytic deep desulphurisation process that made use of HC technology was reported in 2016.[79] The organic and aqueous phases (12–20 L) in this process are mixed in a predefined manner under ambient conditions and passed through a cavitating device (vortex diode with a pump at 2.2 kW and 2900 rpm). The sulphur compounds (500 ppm thiophene) are oxidised by an oxidising species generated *in situ* during HC processing, resulting in the removal of sulphur from three organic solvents (*n*-octane, toluene and

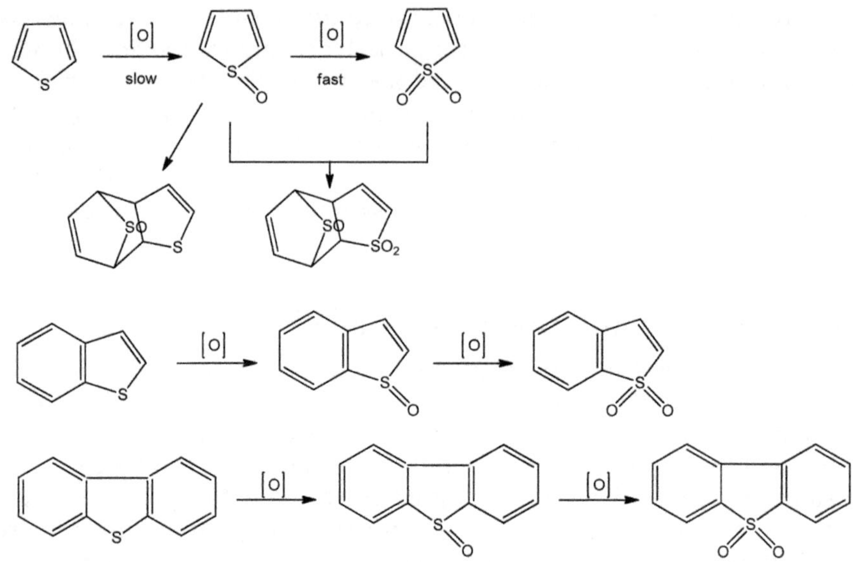

Scheme 7.3 Oxidation of thiophenes. Adapted from ref. 114 with permission from Elsevier, Copyright 2010.

n-octanol), separately. Very high sulphur content removals of up to 100% were demonstrated and also verified and substantiated using commercial diesel as the solvent.

In 2017, the cavitation yields for the desulphurisation (removal of 100–390 ppm thiophene) of 12–20 L of a mixture of water and diesel, and three organic solvents (*n*-octane, toluene, *n*-octanol, separately, 2.5–10 vol% of organic phase), as performed by two types of cavitation devices – linear flow for the generation of cavities (orifice) and vortex flow – were compared. The cavitation yield of the vortex diode was much higher than that of orifice HC for 2 h at 0.2–1.0 MPa (2–10 bar) of inlet pressure. This resulted in the vortex flow-type HC process presenting particularly low energy consumption and costs.[80] The HC process essentially exploits the *in situ* generation of oxidising agents, such as hydroxyl radicals, for the oxidative removal of sulphur, and provides a green approach for the desulphurisation of fuels and organics without the use of a catalyst or external chemicals/reagents. In addition, novel engineering configurations for the effective implementation of the HC process in industrial practice were established and the protocol appears to be economically sustainable.[80]

7.2.4 Cavitational Switching

The competition between two nucleophiles, one charged (KCN) and one neutral (toluene), for the same substrate, benzyl bromide, in the presence of alumina and under agitation by a vibration device for 3 h at 50 °C,

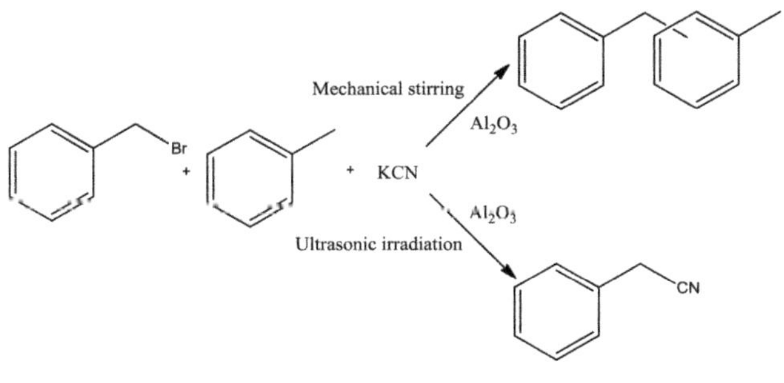

Scheme 7.4 Heterogeneous sonochemical switching.[116,117]

gave 83% substitution by the neutral aromatic derivative (Friedel–Crafts type product) and no benzyl cyanide. However, after 24 h sonication at the same temperature, benzyl cyanide was obtained at 76% yield and substitution by toluene was not observed. This effect is called 'sonochemical switching' (Scheme 7.4); sonication changes the main reaction pathway from aromatic electrophilic to aliphatic nucleophilic substitution.[116,117]

This same reaction was performed at the same temperature under MS, sonication and high-speed agitation with an UltraTurax system (UT, 8000–24 000 rpm). Traditional MS gave a Friedel–Crafts product yield of ~82% and no $PhCH_2CN$ was observed. By contrast, UC and HC (UT) at 24 000 rpm (a type of HC process) gave $PhCH_2CN$ yields of ~16% and ~28%, respectively, and no Friedel–Crafts product was observed. Obviously, this switching not only occurs when using UC, but also with HC. Interestingly, a rough linear relationship exists between the benzyl cyanide yield and the rotation rate of the stirrer, while sonication was equivalent to stirring at 14 000 rpm and was certainly less efficient than stirring at higher rates.[116]

The autoxidation of alkylnitrobenzenes under base-catalysed phase-transfer catalysis by polyethylene glycol in a UC process afforded the corresponding nitrocarboxylic acids in low yield along with the dimer. However, MS was only able to produce the dimer, as shown in Scheme 7.5.[35]

The UC process changes the reaction selectivity from the usual dimeric products to the carboxylic acid. Similarly, only carboxylic acid and no byproducts, such as the dimer, were formed during the oxidation of nitrotoluene with $KMnO_4$ using an HC process in aqueous solution.[34]

Mineral and vegetable oils, such as oil/water emulsions, are used in the leather industry during the production of fatliquors, which are then used to lubricate tanned leather fibres to achieve softness and improve the robustness of the final product. The preparation of fatliquor includes either a sulphation or sulphonation process that facilitates the emulsion of the fatliquor oils in water.[118] A common means of producing sulphonated fatliquors is the reaction of oils with air at high temperatures to promote the formation of high levels of hydroperoxides and epoxides, which are subsequently reacted with sodium bisulphite (Scheme 7.6).[76]

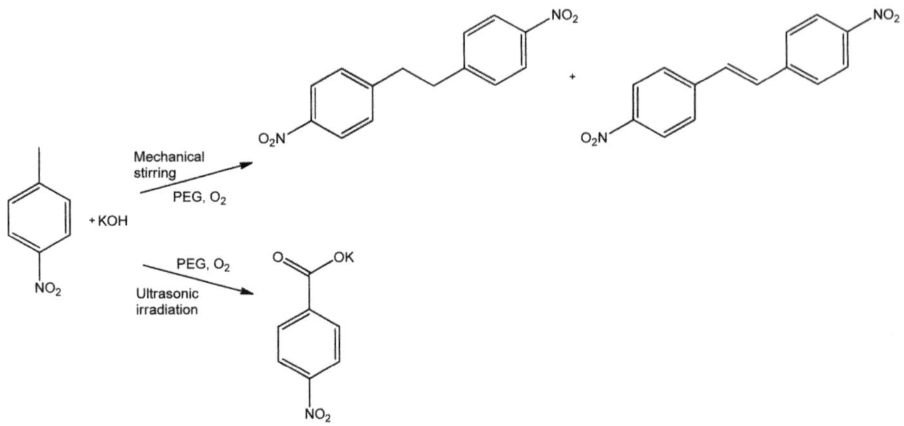

Scheme 7.5 Sonochemical switching during the autoxidation of alkylnitrobenzenes under base-catalysed phase-transfer catalysis by polyethylene glycol (PEG). Adapted from ref. 35 with permission from the Royal Society of Chemistry.

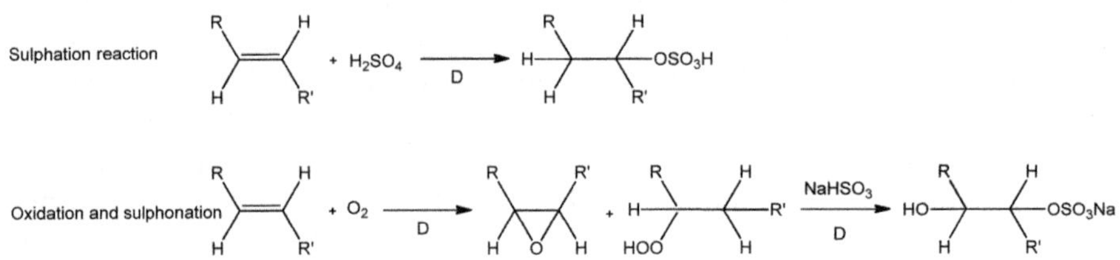

Scheme 7.6 Sulphation and sulphonation processes of oils to facilitate fatliquor production.[76]

The challenge here derives from the fact that mass and heat transfer limitations make the oxidation and sulphonation reactions time-consuming processes. Aside from these oxidised intermediates, polymerisation takes place during the oxidation process to form dimers, oligomers and polymers, which are high molecular-weight structures formed by cross-linking bridges of –C–C–, –C–O–C– and –C–O–O–C– bonds, leading to a progressive increase in viscosity.[76] The use of a lab-scale suction-HC device with 0.8 L oil volume gave a product with the desired viscosity in a much shorter time (7 h). In addition, 10 L of waste cooking oil reached the target viscosity under the action of high-speed rotor HC in 10 h at 90 °C, while the conventional MS procedure required 37 h at 90 °C to obtain a viscosity of about 70 cSt. Air consumption was thus reduced from 500 NL kg^{-1} oil under conventional heating to 80 NL kg^{-1} oil under suction HC, and to 10.3 NL kg^{-1} oil with rotor-stator HC. The unit energy consumption decreased from 15 kW h kg^{-1} oil under conventional heating to 7.4 kW h kg^{-1} oil under suction HC, and to 4.6 kW h kg^{-1} oil with rotor-stator HC.[76] However, the peroxide value (PV) was higher in the conventional heating process than in the two HC systems.[76] Cavitational events may promote peroxide scission and the formation of new –O–O– cross-linking bridges.[119] In HC systems, the formation of dimers and trimers is favoured at the expense of peroxide species. HC accelerates the radical formation and, consequently, promotes the growth of polymeric chains.

7.2.5 Oxidation of Hydrocarbons

The liquid-phase oxidation of hydrocarbons, such as *p*-xylene, cumene, ethylbenzene/isobutane, cyclohexane and *n*-butane, is of great scientific, technological and commercial importance. Mass transfer significantly influences the reaction rate and selectivity towards the desired product.[120] Such heterogeneous reactions are particularly accelerated by sonication thanks to the well-known effect of cavitation. The oxidation of alkylarenes to the corresponding aryl carboxylic acids is also an industrially important process. In industry, these oxidations are typically carried out with either dilute HNO_3 or air under high-temperature and high-pressure conditions.[33]

Considerable acceleration of the oxidation of toluene with aqueous potassium permanganate has been observed at ambient temperature using both UC and HC, as shown in Scheme 7.7. Cavitational effects can generate sufficiently high interfacial area to promote the reaction and avoid the need for high temperatures, high speed agitation and the use of a phase-transfer catalyst.[33]

Two mol $KMnO_4$ (316 g) and 5 L water were added to an HC reactor, followed by addition of 1 mol toluene (106.4 mL), and the mixture was circulated in an orifice HC system at 0.29 MPa (3 kg cm^{-2}) for 3 h. By comparison, 20 mmol $KMnO_4$ and 50 mL water and 10 mmol toluene were sonicated and stirred for 3 h. As a result, the yields of benzoic acid were 60% and 44% using UC and HC for 3 h, respectively. Considering the variation in input density of electrical energy (2.4 kW L^{-1} for UC and 0.3 kW L^{-1} for HC), the energy efficiencies for the oxidative product were 4.63×10^{-6} and 2.70×10^{-5} mol kJ^{-1} for UC and HC, respectively. It can be seen that the HC process is about six times more energy efficient than the UC process.[33]

Further studies indicated that the oxidation of xylenes, mesitylene, nitrotoluenes and chlorotoluenes was also accelerated at ambient temperature using UC and HC.[34] The yields of terephthalic acid from the

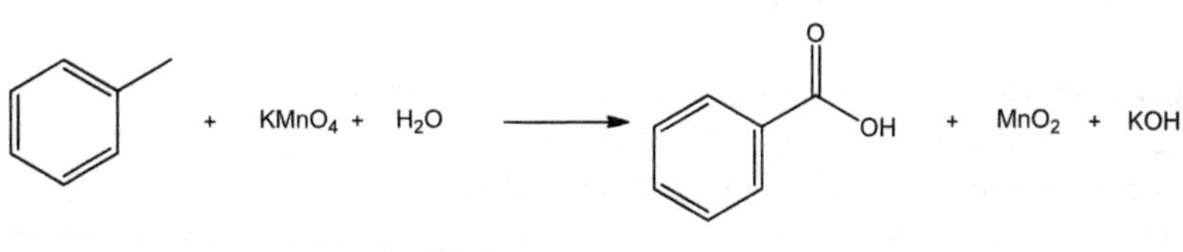

Scheme 7.7 Oxidation of toluene with aqueous potassium permanganate. Adapted from ref. 33 with permission from Elsevier, Copyright 2004.

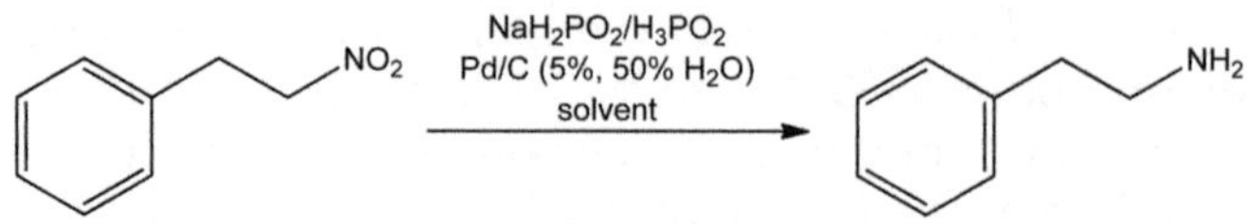

Scheme 7.8 Reduction of nitro compounds to the corresponding amines.[123]

oxidation of *p*-xylene with aqueous $KMnO_4$ after 5 h were 86% and 60% with UC and HC, respectively. The different levels of energy dissipation through UC and HC provided energy efficiencies for the oxidative product of 1.98×10^{-6} and 1.10×10^{-5} mol kJ^{-1} for UC and HC, respectively. Hence, the energy efficiency of the HC process is about seven times higher than that of the UC process. Similarly, the energy efficiencies for the oxidation of *o*-xylene, *m*-xylene and mesitylene are also higher for HC than UC.[34]

7.2.6 Reduction and Hydrogenation

HC was applied for the preparation of tetralin *via* the hydrogenation of naphthalene by throttling the reactant stream through the gaps between orifice plates which act as orifices. Slurry pumps can be used in slurry reactions, such as the reduction of nitrobenzene to aniline by catalytic hydrogenation. However, in such cases, the discharge pressures should be very high compared to gas–liquid systems. Moreover, cyclohexane, 4-methyl pentane-2-ol and sorbitol have also been prepared *via* the hydrogenation of benzene, mesityl oxide and D-glucose, respectively, using HC processes. Energy input requirements per unit of reactant volume arc low because of the higher reactant volumes.[17]

Amines are classically prepared *via* reduction of a wide variety of nitro precursors in either heterogeneous or homogeneous hydrogenation reactions that are catalysed by metals, such as Pd/C and RANEY®, Ni, *etc.*[121] However, these protocols sometimes lack functional group compatibility, often require high-pressure equipment and can suffer from the use of hazardous reagents (*e.g.*, hydrazine) and the presence of potentially toxic transition metals.[122] In order to develop an efficient, eco-friendly, economical and safe general method for nitro-to-amine conversion, the reduction of nitro compounds to the corresponding amines was performed with Pd/C (0.6 mol%) in H_2O/2-MeTHF at 60 °C in the presence of sodium hypophosphite and hypophosphorous acid (Scheme 7.8), which are well-known mild, cheap and powerful reducing agents.[123]

Meanwhile, the influence of UC and a vibromixer (HC) on this reaction was investigated, both in terms of efficiency and kinetics. The reaction was carried out for 15 min in two solvents, H_2O and H_2O/2-MeTHF (2/1, v/v), respectively. The amine yields observed under UC activation and HC were similar in both water (Figure 7.2a) and the biphasic solvent system H_2O/2-MeTHF (Figure 7.2b).

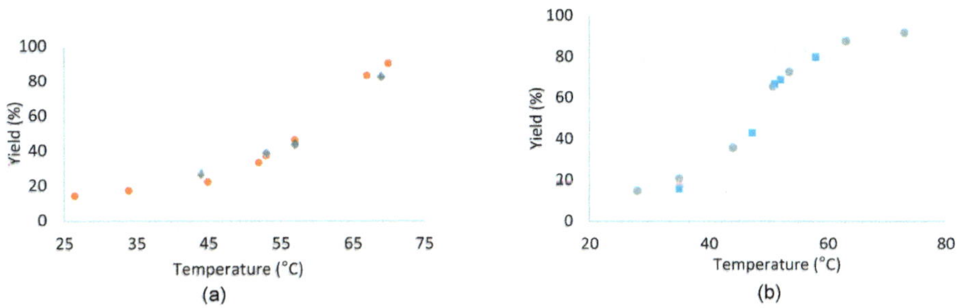

Figure 7.2 Comparison of UC and HC activation for 15 min of reaction with NaH_2PO_2/H_3PO_2 (4 : 1), 0.6 mol% Pd/C (5%, 50% water) in H_2O/2-MeTHF (2 : 1): (a) in water: UC (●), vibromixer UltraTurax (HC, ◆) and (b) in water/2-MeTHF: UC (●), vibromixer UltraTurax (HC, ■). Adapted from ref. 123 with permission from the Royal Society of Chemistry.

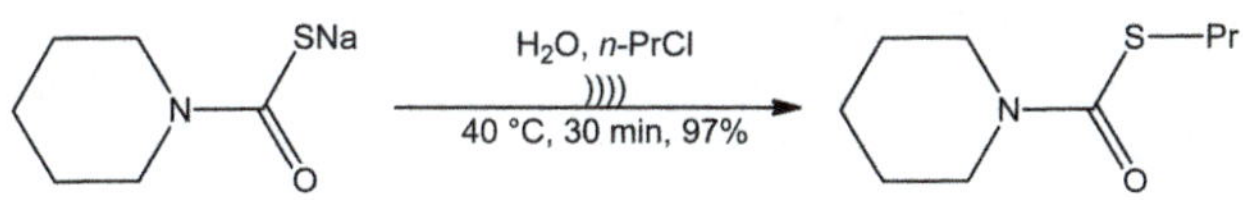

Scheme 7.9 Preparation of thiocarbamates. Adapted from ref. 3 with permission from the Royal Society of Chemistry.

As a result, the reaction was shown to be efficient in water at 70 °C, giving quantitative conversion and a maximal yield after only 15 min thanks to the cavitation activation of UC and HC, which act as the physical agents for phase transfer.[123]

7.2.7 Other Reactions

Substitution: Scheme 7.9 illustrates the preparation of thiocarbamates by UC, HC and conventional MS.

Upon comparison, the rate constants were in relative ratios of 10, 12, and 1 for UC, HC and MS, respectively. Obviously, UC and HC activation improves the yield, probably because the rate increase that occurs even at ambient temperature avoids the decomposition of thermally sensitive compounds.[3,124]

An efficient zinc/copper couple has been observed to induce the addition of carbohydrates or amino acid residues, which were derived from their corresponding iodides, to various activated olefins, *via* simple vibromixing of the heterogenous reaction mixture, to prepare sinefungin analogues (HC process).[125] Simple reaction conditions were used for the addition of the species, which were generated by the Zn/Cu couple from the carbohydrate and amino acid series, to electron deficient olefins. Sonication provided the desired adduct in modest yield. Improved HC-process stirring of the heterogeneous medium was found to be more successful.[125,126] The reaction is shown in Scheme 7.10.

Cyclopropanation. Higher yields of 7,7-dichlorobicyclo[4.1.0]heptane have been obtained using HC than UC (70% using UT at 24 000 rpm of rotation speed *vs.* 20% using UC) in the cyclopropanation of cyclohexene by dichlorocarbene, which was formed *in situ* from chloroform and sodium hydroxide (Scheme 7.11).[116]

Hydrolysis of cellulose. As much as 5% of starch remains unconverted after fermentation during dry mill corn ethanol production. Furthermore, the cellulose present in the grain is not converted by the enzymes used for starch hydrolysis. Controlled-flow cavitation, with the addition of cellulase, has therefore been used to enhance ethanol yields in a commercial scale dry mill ethanol plant (100 million gal year^{-1}).[127] As a result, cavitation significantly increased the DP4+, DP3 and total sugar concentrations by 2%, 6% and 3% (CI > 98%), respectively, during liquefaction. The greatest yield improvement of 4.1% (CI = 99.9%) was observed in samples that were both cavitated and amended with cellulase enzymes. These results indicate that controlled-flow cavitation and cellulase amendment can significantly improve the economics and resource-use efficiency of bioethanol production in dry mill ethanol plants.

7.2.8 HC-process Limitations

Although HC processes present many advantages over conventional MS and UC processes, they also have some drawbacks. For example, the earliest use of HC technology in chemical processes is the intensification of fatty-oil hydrolysis.[9] The effect of UC and HC (UT) on the alkaline

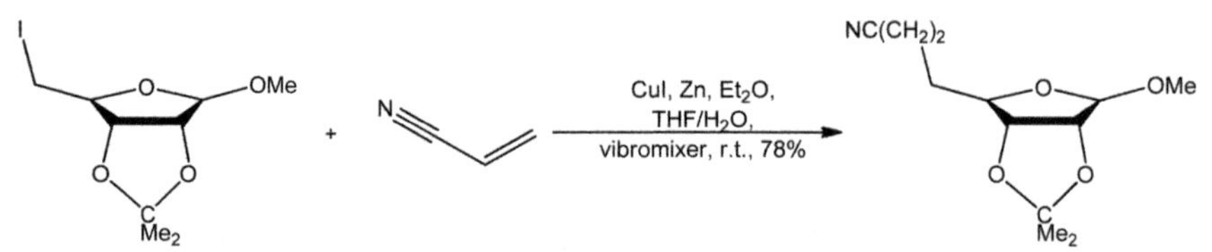

Scheme 7.10 Preparation of Sinefungin analogues. Adapted from ref. 126 with permission from Elsevier, Copyright 2001.

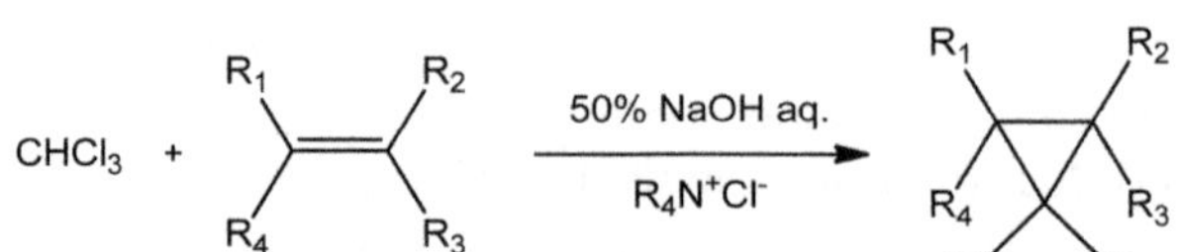

Scheme 7.11 Cyclopropanation of cyclohexene by dichlorocarbene.[126]

hydrolysis of benzoate esters has been compared to that of traditional MS. No reaction was observed for the alkaline hydrolysis of benzoate esters using MS, while both UC and HC were able to induce hydrolysis. However, higher benzoate yields were achieved using UC than HC (44% using UC *vs.* 33% using UT at 24 000 rpm rotation speed).[116] Moreover, HC has been found to inhibit the enzymatic hydrolysis of canola oil using lipase from *Candida rugose*.[72] HC generally leads to the production of fine oil-in-water and water-in-oil emulsions with the enzyme in the water phase.[116] Unfortunately, fatty acid yields were only about 60% using a venturi at 0.8 MPa (8 bar) inlet pressure. By contrast, a simple MS reactor produced a maximum yield of over 90%. Further HC-process limitations are denoted below.

Firstly, pressure recovery downstream of the orifice is very poor; pressure losses are typically about 73% of the pipe differential for an orifice-to-pipe diameter ratio of 0.5. Pumping costs for this reactor can therefore run very high, especially when a solid and liquid mixture is handled. Specifically, only reactions that occur at up to 10–15 atm and temperatures of up to 300 °C can be conducted efficiently in such reactors. Those requiring high temperatures and pressures are difficult to process. Furthermore, the performance of HC reactors when viscous reacting mixtures are used is quite low.[17]

Secondarily, UC displays higher energy efficiency than HC for the direct pyrolysis of volatile organic compounds and water vapour inside or near the hot-spots. However, the suction-HC process can overcome the shortcoming in pressure recovery of the extrusion HC process to some extent.[18] The energy efficiencies for the degradation of halogen compounds (a-chlorotoluene, a-bromotoluene, chlorobenzene, bromobenzene, chloroform and carbon tetrachloride) in aqueous solution by HC and UC (850 kHz) have been compared, under their respective optimum conditions, but with the same processing volumes and initial substrate concentrations. The energy efficiency of the degradation by HC was still lower than that of 850 kHz UC treatment.[18] UC processes are therefore more effective in initiating and promoting free radical reactions than HC processes.

Thirdly, organic solvents slowly decompose under sonication at ambient conditions. However, solvent decomposition is normally only a minor contribution to the sonochemical reactions that take place in the medium.[2] Similarly, the dynamics of cavitation bubbles in organic solvents are relatively insensitive to the operating parameters because of the physicochemical properties of organic media.[128,129] For example, the intensity of cavitation collapse in toluene is higher than in hexane. The temperature peaks attained at transient bubble collapse in both toluene and hexane, under air or argon atmosphere, within HC reactors are too low to drive the formation of radical species in appreciable quantities. Thus, HC processes in organic solvents with higher vapour pressures are unsuitable for organic reactions, although some physical processes and chemical reactions with external radical/ion sources may well be carried out.[130] Therefore, the use of UC and HC treatments in the cracking of organic compounds, including linear alkanes (octane, undecane and hexadecane), carboxylic acids (decanoic and oleic acids), alcohols (decanol) and esters (ethyl caprate) has resulted in transformations that are no more than 1.1%. The yield of fatty acid esters in the transesterification of rapeseed oil, methanol and ethanol without a catalyst and under the action of cavitation did not exceed 2.1 mol%. The energy efficiency of the cavitation apparatus was estimated to be 0.002% for the endothermic cracking of alkanes. It can therefore be stated that the use of cavitation directly for the cracking of various organic compounds and the non-catalytic transesterification of vegetable oils is ineffective.[131]

7.3 Ultrasound Methods for Green Organic Synthesis under Heterogeneous Catalysis

The past few years have seen nanotechnology-based processes promoting the development of economical and environmentally benign synthetic pathways to produce high-performance catalytic nanomaterials. The scientific community has recently paid considerable attention to the synthesis of superparamagnetic nanoparticles for the design of magnetically retrievable nanocatalytic systems, as their exceptional physico-chemical properties and rapid response to applied magnetic fields makes them particularly interesting. Furthermore, these magnetite-based nanocatalysts are robust, inexpensive, easy to prepare, non-toxic, magnetically recoverable and can be reused over several reaction cycles. Consequently, they have emerged as viable alternatives to existing solid-supported heterogeneous catalysts. A number of sonochemical procedures for the fabrication of MNP-based core–shell nanostructured catalysts are herein discussed with special emphasis being placed on their use in coupling reactions, such as Suzuki, Heck, Sonogashira, Stille, Hiyama, C–O, C–N, C–S and multicomponent reactions.[132,133] Performing multicomponent reactions with heterogeneous catalysts under ultrasonic irradiation may improve the effectiveness in terms of the costs and environmental impact. Javad *et al.* have reported a novel protocol for the synthesis of benzo[*g*]chromenes *via* one-pot condensation of aldehydes, malononitrile and 2-hydroxy-1,4-naphthoquinone. This protocol is catalysed by core–shell Fe_3O_4/PEG nanoparticles, which are used as a magnetic catalyst under ultrasonic irradiation[134] (Scheme 7.12).

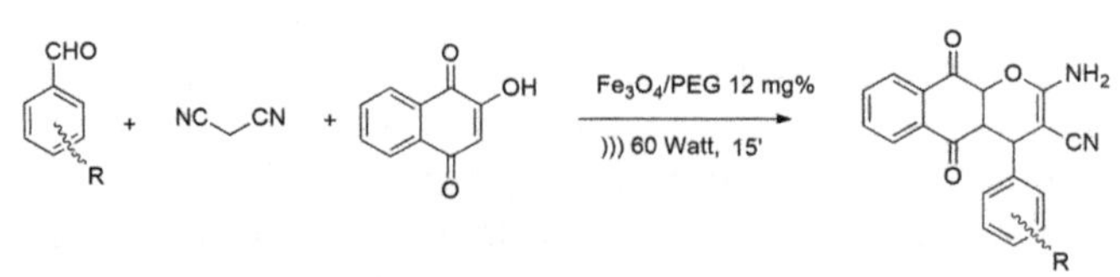

Scheme 7.12 Synthesis of benzo[*g*]chromenes with nano-Fe_3O_4/PEG. Adapted from ref. 134 with permission from Elsevier, Copyright 2016.

Catalyst nanoparticles were prepared by precursor sonication (power 60 W, 15 min) to generate Fe_3O_4/PEG in good yield (95%) (Scheme 7.13).

The FE-SEM analysis showed particles with diameters in the range of 8–12 nm. Ethanol was the preferred solvent and Fe_3O_4/PEG (12 mg%) was selected from a range of oxides (ZnO, SiO_2, NiO, SnO, *etc.*). The reaction was also carried out under reflux but afforded lower yields (18–35%) and required longer reaction times (300 min). The proposed reaction mechanism entails the activation of C=O and C–N groups by nucleophilic attack. The catalyst was recovered by an external magnetic field and reused five times with very low activity losses.

An efficient Heck cross-coupling of aryl halides with olefins was performed in the presence of poly(3,4-ethylenedioxythiophene) (PEDOT)-nanofibre (NF)/Pd(0) composites, which were used as a green and recyclable catalyst in water under ultrasonic irradiation. This protocol was reported by Rezaei[135] (Scheme 7.14).

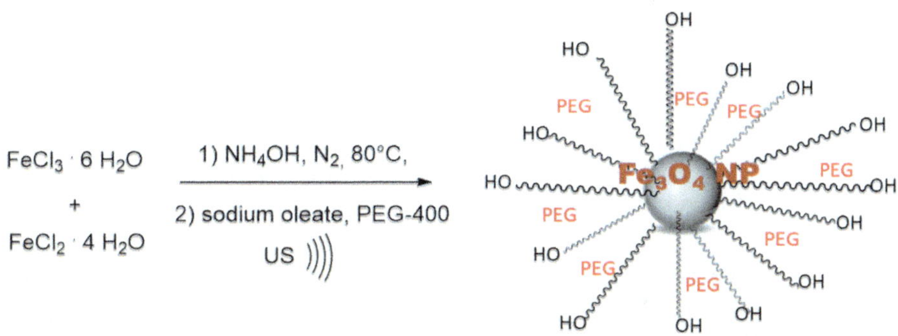

Scheme 7.13 Synthesis of nano-Fe_3O_4/PEG. Adapted from ref. 134 with permission from Elsevier, Copyright 2016.

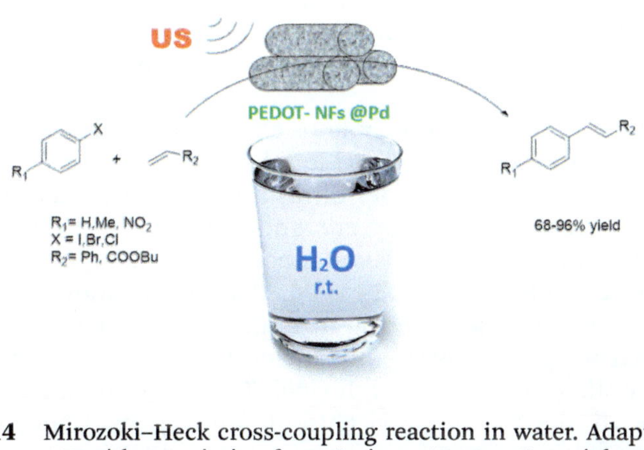

Scheme 7.14 Mirozoki–Heck cross-coupling reaction in water. Adapted from ref. 135 with permission from Springer Nature, Copyright 2016.

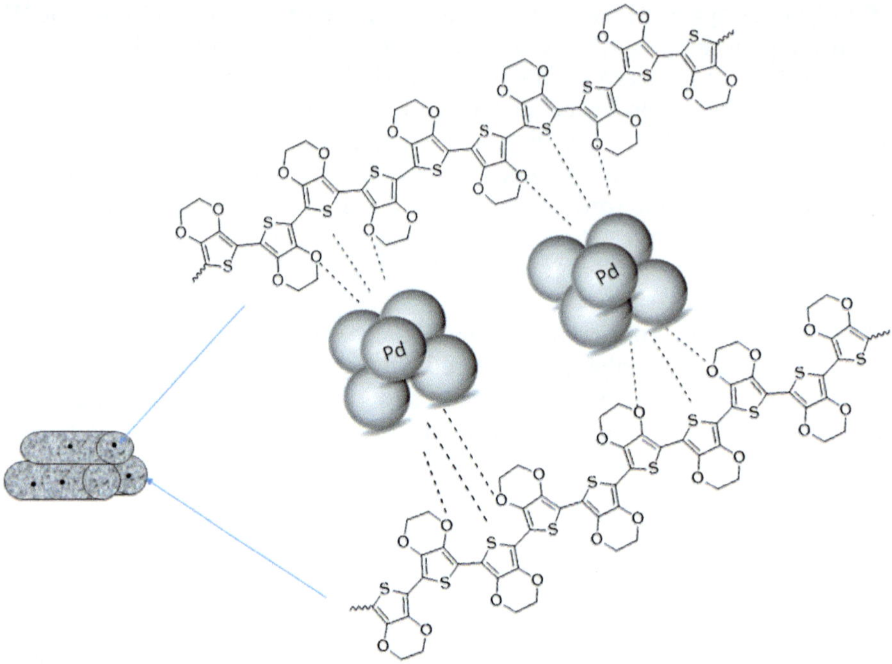

Figure 7.3 Molecular structure of PEDOT-NF@Pd. Adapted from ref. 135 with permission from Springer Nature, Copyright 2016.

The PEDOT-NF/Pd composite was prepared by first synthesising PEDOT-NFs in a rapidly initiated polymerisation reaction from 3,4-ethylenedioxythiophene (EDOT) in HCl and $FeCl_3$ under ultrasonication. The Pd nanoparticles were grown by combining an aqueous solution of PEDOT-NFs with a solution of palladium(II) nitrate, which were incubated for one day. The resulting solution was then dialysed to remove any remaining palladium salts in fresh deionised water. SEM images of PEDOT-NFs showed a 10 μm length and 250–500 nm diameter (Figure 7.3). TEM experiments demonstrated that palladium nanoparticles were uniformly dispersed on the surface of the PEDOT-NFs.

PEDOT-NF/Pd 0.05 mol% was able to efficiently catalyse a model Heck coupling reaction between 4-methyl-iodobenzene and *n*-butyl acrylate in water and K_2CO_3 as the base under ultrasonic irradiation (170 W) (conversion 99%). The reaction with styrene and *n*-butyl acrylate was extended to a range of aryl halides under similar conditions in an effort to confirm the catalytic versatility of PEDOT-NF/Pd. The results showed that several aryl bromides and iodides bearing a range of electron-withdrawing and electron-donating groups were all efficiently converted into the corresponding coupling products in high yield. The reaction also proceeded in good yield with less reactive aryl chloride derivatives. Catalyst recycling was fundamental (good activity after six cycles) to the scalability of this sonochemical protocol.

The well-known catalytic effect of UC has recently been reviewed by Banerjee,[136] who summarised the latest developments in catalyst-free organic synthesis.

7.3.1 Enzyme Catalysis under Ultrasound

The effect of UC in enzymatic reactions can be divided in two aspects: (a) the use of UC during pre-treatment to reduce the size of particles and consequently enhance the surface area between enzyme and substrate, thus contributing to reductions in mass-transport limitations; and (b) the use of UC during the reaction. While cavitation energy is believed to accelerate reaction rates, the mechanism by which this occurs is unclear. It may be that increasing the movement of liquid molecules leads to the substrate having improved access to the active sites. Of the various enzymes with industrial applications, lipases (triglyceride ester hydrolases) have had a significant impact on biotechnological applications. Microbial lipases occupy a prominent place among biocatalysts due to their ability to catalyse a wide variety of reactions in aqueous and non-aqueous media. The specific characteristics of these enzymes have generated interest among researchers as they can catalyse reactions with reduced by-product formation, reduced waste-treatment costs and mild temperature and pressure conditions. Ultrasound has also been used to extract and release intracellular enzymes, whose activity can subsequently be further enhanced by the technology. Naturally, enzyme activity and stability strictly depend on the sonication parameters and specific enzyme preparation as cavitation can damage the enzyme structure. The application of this technology to lipase-catalysed reactions is mainly for hydrolysis and alcoholysis, for the production of free-fatty acids, esterification and transesterification, biodiesel production, and glycerolysis for the production of mono and diglycerides and polymerisation.[137,138]

Ultrasound-assisted extraction gave excellent results as one of the range of technologies tested.[139] Its mild operating conditions have led to several ultrasound-assisted extraction applications being reported for use in pharmaceutical, cosmetics and water treatment fields. Unfortunately, the generally high cost of enzymes and slow reaction rates pose several limitations. Ultrasound-assisted enzymatic hydrolysis is a promising solution to resolve these shortcomings. Besides the environmental benefits, the extraction of phenolic compounds from olive pomace may generate added value from the agroindustry waste. Many factors have therefore been considered to improve the extraction efficacy. These include time, temperature, solvent type and treatment method, while combinations of these factors and their interactions can lead to the desired response. Response surface methodology (RSM) is an effective tool with which to optimise the extraction process. RSM is a multivariate statistical tool capable of deriving a polynomial equation that describes an experimental process, can provide statistical predictions and find the combination of factors that will yield the optimum response. RSM was employed by Wang to determine the optimised conditions for the extraction of phenolics using UAEH (ultrasound-assisted enzymatic hydrolysis).[140] The optimised extraction conditions gave a 4.0% yield (pH 5.75, for 40 min at 55 °C) (Figure 7.4).

The SEM images revealed that the effect of UAEH on the tissue and microstructure of olive pomace was more effective than that of other methods. UAEH has huge potential for phenolic extraction from OW as it can provide higher phenolic yields in shorter extraction times. Compared to those of petroleum ether and *n*-butyl alcohol, the ethyl acetate extract fraction displayed higher phenolic content, especially in hydroxytyrosol and tyrosol, and exhibited higher antioxidant activities (Figure 7.5).

7.3.2 Sonochemical Biomass Conversion

Lignocellulosic biomass is a natural renewable chemical feedstock that can be used to produce high value-added chemicals and platform molecules. A range of extensive studies on a variety of aspects of the valorisation of lignocellulosic biomass into desirable products are currently available. Some of the current technologies for biomass conversion entail extreme conditions, with high temperatures and pressures. The conversion of lignocellulosic feedstocks leads to the production of a variety of added-value platform chemicals, including phenolic compounds (*p*-coumaryl alcohol, coniferyl alcohol, and sinapyl alcohol), aliphatic acids (formic acid, acetic acid, levulinic acid), and furan aldehydes. The depolymerisation of lignocellulosic biomass includes physical, chemical and biological treatment processes. However, the energy consumption required for physical treatment is higher than the theoretical energy content available in lignocellulose, which makes it

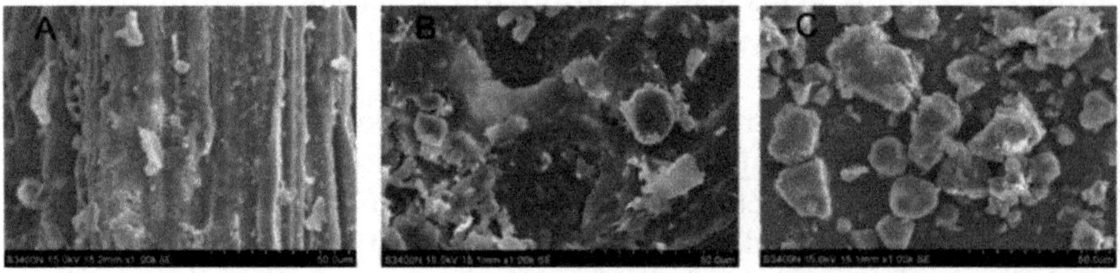

Figure 7.4 SEM images of olive pomace: (A) untreated, (B) UC treatment alone, and (C) UC-assisted enzyme hydrolysis treatment. Reproduced from ref. 140 with permission from Springer Nature, Copyright 2017.

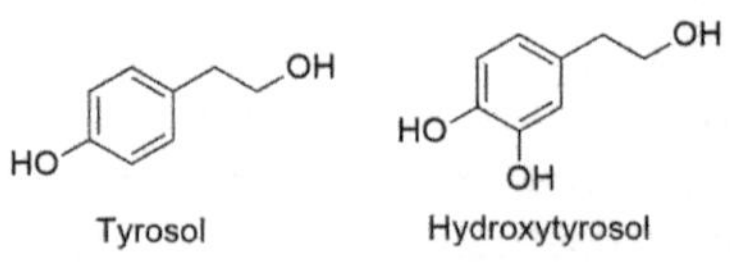

Figure 7.5 Tyrosol and hydroxytyrosol main phenolic compounds in the ethyl acetate extract.

prohibitively expensive for large-scale use. Chemical pre-treatment is a viable alternative with which to deconstruct lignocellulosic biomass. This method is based on catalytic processes, such as acid/alkaline hydrolysis, oxidative delignification, cracking and reduction reactions. Ultrasound-assisted catalysis is an alternative method, in this context, and offers new strategies for the treatment of biomass and the valorisation of lignocellulosic compounds into value-added chemical feedstocks.

A recent review by Kuna *et al.* has demonstrated that sonocatalysis avoids harsh chemical conditions, as well as reduced reaction times and improved heat and mass transfer. It therefore increases the chemical rate constants, yields and selectivities.[141] Hence, ultrasound-assisted catalysis can be successfully used for the pre-treatment and chemical conversion of lignocellulosic biomass and its derivatives in a variety of processes, such as hydrolysis, oxidation and hydrogenation reactions. The use of ultrasound in homogeneous and heterogeneous catalytic reactions for lignocellulose valorisation is growing. UC and the nature of the solvents used both play a crucial role in lignocellulose depolymerisation, as documented by the use of aqueous ionic liquids (*i.e.*, 1-butyl-3-methylimidazolium acetate). Recent studies on the use of ultrasound to assist catalytic reactions have clearly shown the great advantages and technological potential that this tool can offer to the chemical industry, especially in flow processes. More significant scientific breakthroughs in biomass valorisation are expected to occur in this innovative field in the near future.

7.3.3 Sonochemical Photocatalysis

A number of methods for the preparation of photoactive materials for application in photocatalysis and solar cells have become available in recent years. The reviews by Colmenares *et al.* in 2014 and 2016 highlighted the role that ultrasound can play as the synthetic procedure of choice and its ability to control the size, morphology, nanostructure and the tuning of catalytic properties.[142,143] Ultrasound-based procedures offer a facile and versatile synthetic tool for the preparation of light-activated materials that would often be inaccessible using conventional methods. Unlike traditional sources of energy, ultrasound can produce cavitation phenomena that provide unusual reaction conditions in liquid-phase reactions (extremely high temperatures and pressures are formed in very short times in liquids). Furthermore, the application of ultrasonic irradiation to a photocatalytic system may also enhance both bulk and localised mass transport and consequently expedite the molecular contact and photocatalytic activity. A considerable number of novel photocatalysts with novel nano/microstructures (*e.g.*, $ZnWO_4$, $CdMoO_4$, V_2O_5, $CuMoO_4$, TiO_2/Bi_2O_3) have been prepared by means of UC-assisted protocols and potential applications for these materials for the photocatalytic conversion of selected organic waste materials *via* oxidation processes have been reported.

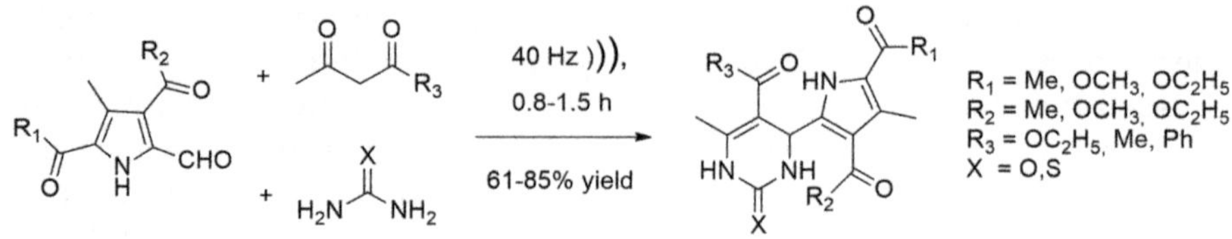

Scheme 7.15 Synthetic route for dihydropyrimidinones in lactic acid as the solvent. Adapted from ref. 145 with permission from John Wiley and Sons, © 2016 Wiley-VCH Verlag GmbH & Co. KGaA, Weinheim.

Titanium dioxide (TiO_2) is the most widely used oxide semiconductor and has attracted considerable interest thanks to its various applications in solar-driven hydrogen production, photocatalytic decomposition of pollutants, solar cells and so forth. Titania-based photocatalysts have become one of the most actively studied materials for waste water remediation. A recent review has dealt with the recent advances, challenges and future prospects of ultrasound use in the development and modification of titania-based photocatalysts for pollutant degradation.[144]

7.3.4 UC-assisted Heterocycle Synthesis

Dihydropyrimidinone derivatives have recently been attracting attention due to their potential anti-inflammatory, antitumour, calcium channel inhibition, antimicrobial and antioxidant properties. Dihydropyrimidinone derivatives are typically prepared *via* the Biginelli reaction, a one-pot condensation of 1,3-dicarbonyl compounds, aldehydes and urea/thiourea. However, the organic solvents and catalysts used have a considerable environmental impact. An eco-friendly, efficient and facile protocol for the sonochemical synthesis of novel dihydropyrimidinones in lactic acid has been recently reported (Scheme 7.15).[145]

It is well known that lactic acid (2-hydroxy propionic acid) is a bio-based eco-friendly weak acid that plays the dual roles of solvent and catalyst in some sonochemical reactions, such as the Biginelli reaction (Scheme 7.16).

High yields were obtained at a ratio of pyrrole aldehyde, acetoacetate and urea/thiourea of 1:1:1.1. Moreover, subsequent product purification was found to be rather time-consuming when more than 1.1 equiv. of urea/thiourea were used. Further research indicated that, while the reactants are soluble in lactic acid, the products generated are poorly soluble in lactic acid aqueous solution, meaning that they can be easily isolated by means of suction filtration. The reaction temperature is lower (80 °C) than that of the conventional Biginelli reaction, while the yield increased to 70–80%. This sonochemical method can be extended to Biginelli condensations in which the aldehydes are sensitive to strong acids and high temperatures (*i.e.* pyrrole aldehydes).

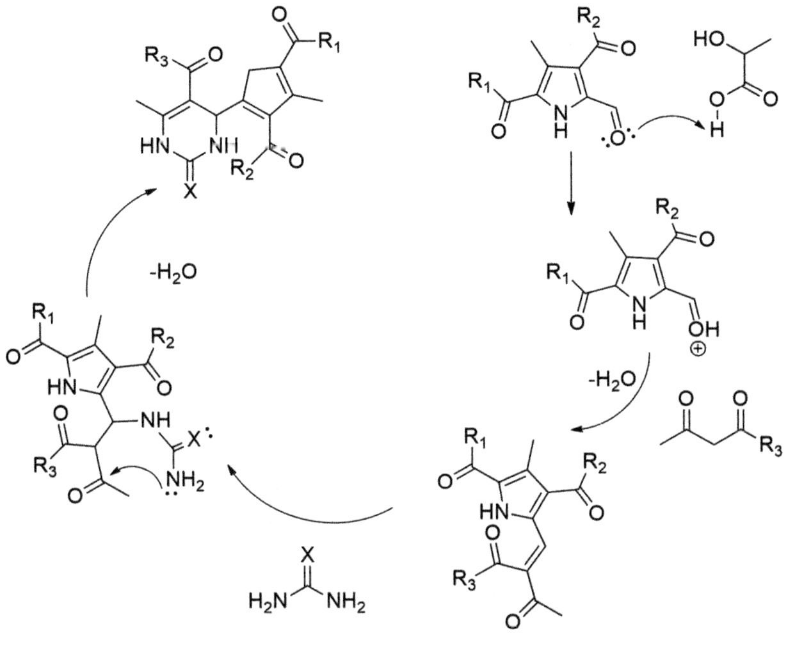

Scheme 7.16 Biginelli reaction mechanism in lactic acid as the solvent. Adapted from ref. 145 with permission from John Wiley and Sons, © 2016 Wiley-VCH Verlag GmbH & Co. KGaA, Weinheim.

A recent review on MCRs has described recent developments in the synthesis of biologically active heterocycles using one-pot multicomponent protocols under sonochemical conditions.[146] It is a useful and green approach that will certainly undergo further developments and applications.

7.3.5 Organic Reactions in Water

Catalytic procedures in water assisted by UC and/or HC are environmentally friendly and provide milder conditions, shorter reaction times and often give higher yields. Sonochemical processes can reduce the formation of hazardous by-products, the generation of waste and also produce energy savings. The majority of synthetic protocols have focused on heterogeneous reactions performed in water or aqueous media. Besides being inexpensive, non-toxic and environmentally friendly, water can also be used as a dispersing medium under ultrasound. It is often the solvent of choice in sonochemistry as it furnishes excellent cavitation from room temperature up to 50–60 °C. Water also causes special effects in reactions that arise from intra- and inter-molecular non-covalent interactions. Sonochemical reactions in water can often lead to peculiar reactivities and selectivities that are not observed under anhydrous organic solvents. Cavitational implosion generates mechanical and chemical effects, such as catalyst surface cleaning and the formation of free radicals *via* the sonolysis of water. Ultrasound is especially efficient in activating hard metal surfaces *via* the well-known cleaning effect of ultrasonic waves that can cause metal surface depassivation. UC promotes reagent diffusion from solution to the metal, electron transfer from the active surface to the reducible point of the organic substrate and, finally, the extraction of ions from the surface to generate a soluble product. Organic reactions in water are often difficult as most organic reactants, including catalysts, are insoluble. The dual hydrophobic and hydrophilic nature of surfactants means that they can help lessen the impact of this issue by forming micelles and promoting reactions in water. An example of this is *p*-dodecyl-benzenesulphonic acid (DBSA), a Brønsted acid-surfactant that combines catalytic action and the ability to form stable colloidal dispersions with water-insoluble substrates that facilitate reaction outcomes. A comprehensive overview of sonochemical reactions in water was published by Cravotto *et al.* in 2015, and includes oxidations, bromination, aza-Michael, C–C couplings, MCRs and aldol reactions as useful models with which to further the progress of organic synthesis using harmless and green sound energy.[30]

7.4 Conclusion

Both HC and UC are mature techniques that can boast green organic syntheses. Their real potential lies in their 'true sonochemical effects' that can provide reaction pathway switching and lead to unexpected products and selectivities. In other words, they can open the way for new mechanistic routes while working under greener conditions. Cavitational chemistry will undoubtedly continue to help chemists to overcome synthetic limitations and broaden the range of green chemistry applications, while also allowing for new targets to be found. It is hoped that these pursuits will be a valuable research domain for years to come.

References

1. G. Cravotto and P. Cintas, *Chem. Soc. Rev.*, 2006, **35**, 180.
2. T. J. Mason, *Chem. Soc. Rev.*, 1997, **26**, 443.
3. P. Cintas and J.-L. Luche, *Green Chem.*, 1999, **1**, 115.
4. G. Chatel, *Ultrason. Sonochem.*, 2018, **40**, 117–122.
5. R. Singh, S. K. Tiwari and S. K. Mishra, *J. Mater. Eng. Perform.*, 2012, **21**, 1539.
6. F. G. Hammitt, *Appl. Mech. Rev.*, 1979, **32**, 665.
7. J. I. Thornycroft and S. W. Barnaby, *Minutes Proc. Inst. Civ. Eng.*, 1895, **122**, 51.

8. J. I. Thornycroft, *J. Am. Soc. Nav. Eng.*, 1895, **7**, 711.
9. A. B. Pandit and J. B. Joshi, *Chem. Eng. Sci.*, 1993, **48**, 3440.
10. W. R. Moser, B. J. Marshik, J. Kingsley, M. Lemberger, R. Willette, A. Chan, J. E. Sunstrom and A. Boye, *J. Mater. Res.*, 1995, **10**, 2322.
11. J. E. Sunstrom, W. R. Moser and B. Marshik-Guerts, *Chem. Mater.*, 1996, **8**, 2061.
12. K. S. Suslick, M. M. Mdleleni and J. T. Ries, *J. Am. Chem. Soc.*, 1997, **119**, 9303.
13. P. S. Kumar and A. B. Pandit, *Chem. Eng. Technol.*, 1999, **22**, 1017.
14. V. S. Moholkar and A. B. Pandit, *AIChE J.*, 1997, **43**, 1641.
15. Z. Wu, S. Tagliapietra, A. Giraudo, K. Martina and G. Cravotto, *Ultrason. Sonochem.*, 2019, **52**, 530.
16. S. Rajoriya, J. Carpenter, V. K. Saharan and A. B. Pandit, *Rev. Chem. Eng.*, 2016, **32**, 379.
17. Y. T. Shah, A. B. Pandit and V. S. Moholkar, Gas-Liquid-Solid Cavitation Chemistry, in *Cavitation Reaction Engineering*, Springer, Boston, MA, 1999, pp. 155–192.
18. Z. Wu, B. Ondruschka, Y. Zhang, D. H. Bremner, H. Shen and M. Franke, *Green Chem.*, 2009, **11**, 1026.
19. Z. Wu, H. Shen, B. Ondruschka, Y. Zhang, W. Wang and D. H. Bremner, *J. Hazard. Mater.*, 2012, **235–236**, 152.
20. A. K. Lee, D. M. Lewis and P. J. Ashman, *Biomass Bioenergy*, 2012, **46**, 89.
21. K. S. Suslick, *Science*, 1990, **247**, 1439.
22. Y. L. Chen and J. Israelachvili, *Science*, 1991, **252**, 1157.
23. C. F. Naudé and A. T. Ellis, *J. Fluids Eng. Trans. ASME*, 1961, **83**, 648.
24. E. Pattantyús-H, *Int. J. Heat Mass Transfer*, 1972, **15**, 2419.
25. E. S. Beach, Z. Cui and P. T. Anastas, *Energy Environ. Sci.*, 2009, **2**, 1038.
26. R. M. Martín-Aranda and V. Calvino-Casilda, *Recent Pat. Chem. Eng.*, 2010, **3**, 82.
27. R. B. N. Baig and R. S. Varma, *Chem. Soc. Rev.*, 2012, **41**, 1559.
28. Z. Qiu, L. Zhao and L. Weatherley, *Chem. Eng. Process. Process Intensif.*, 2010, **49**, 323.
29. Z. Wu, T. Zheng, L. Wu, H. Lou, Q. Xie, M. Lu, L. Zhang, Y. Nie and J. Ji, *Ind. Eng. Chem. Res.*, 2017, **56**, 5231.
30. G. Cravotto, E. Borretto, M. Oliverio, A. Procopio and A. Penoni, *Catal. Commun.*, 2015, **63**, 2.
31. K. Nakashima, Y. Ebi, N. Shibasaki-Kitakawa, H. Soyama and T. Yonemoto, *Ind. Eng. Chem. Res.*, 2016, **55**, 1866.
32. V. S. Moholkar, P. S. Kumar and A. B. Pandit, *Ultrason. Sonochem.*, 1999, **6**, 53.
33. G. V. Ambulgekar, S. D. Samant and A. B. Pandit, *Ultrason. Sonochem.*, 2004, **11**, 191.

34. G. V. Ambulgekar, S. D. Samant and A. B. Pandit, *Ultrason. Sonochem.*, 2005, **12**, 85.
35. R. Neumann and Y. Sasson, *J. Chem. Soc., Chem. Commun.*, 1985, 616.
36. J. Find, S. C. Emerson, I. M. Krausz and W. R. Moser, *J. Mater. Res.*, 2001, **16**, 3503.
37. S. H. Sonawane, S. P. Gumfekar, S. H. Kate, S. P. Meshram, K. J. Kunte, L. Ramjee, C. M. Mahajan, M. G. Parande and M. Ashokkumar, *Int. J. Chem. Eng.*, 2010, **10**, 1.
38. W. R. Moser, J. E. Sunstrom IV and B. Marshik–Guerts, The synthesis of nanostructured pure–phase catalysts by hydrodynamic cavitation, in *Advanced Catalysts and Nanostructured Materials - Modern Synthetic Methods*, ed. M. R. Moser, Academic Press, San Diego, CA, USA, 1996, pp. 285–306.
39. Z. Wu, M. Franke, B. Ondruschka, Y. Zhang, Y. Ren, P. Braeutigam and W. Wang, *J. Hazard. Mater.*, 2011, **190**, 375.
40. A. G. Chakinala, P. R. Gogate, A. E. Burgess and D. H. Bremner, *Chem. Eng. J.*, 2009, **152**, 498.
41. M. Zupanc, T. Kosjek, M. Petkovšek, M. Dular, B. Kompare, B. Širok, Ž. Blažeka and E. Heath, *Ultrason. Sonochem.*, 2013, **20**, 1104.
42. Z. Wu, B. Ondruschka and P. Bräutigam, *Chem. Eng. Technol.*, 2007, **30**, 642.
43. X. Wang and Y. Zhang, *J. Hazard. Mater.*, 2009, **161**, 202.
44. S. Arrojo, Y. Benito and A. Martínez Tarifa, *Ultrason. Sonochem.*, 2008, **15**, 903.
45. P. R. Gogate, *J. Environ. Manage.*, 2007, **85**, 801.
46. K. K. Jyoti and A. B. Pandit, *Water Res.*, 2004, **38**, 2248.
47. S. S. Sawant, A. C. Anil, V. Krishnamurthy, C. Gaonkar, J. Kolwalkar, L. Khandeparker, D. Desai, A. V. Mahulkar, V. V. Ranade and A. B. Pandit, *Chem. Eng. J.*, 2008, **42**, 320.
48. K. K. Jyoti and A. B. Pandit, *Biochem. Eng. J.*, 2001, **7**, 201.
49. A. Iskalieva, B. M. Yimmou, P. R. Gogate, M. Horvath, P. G. Horvath and L. Csoka, *Ultrason. Sonochem.*, 2012, **19**, 984.
50. K. Thangavelu, R. Desikan, O. P. Taran and S. Uthandi, *Biotechnol. Biofuels*, 2018, **11**, 203.
51. R. Terán Hilares, J. C. dos Santos, M. A. Ahmed, S. H. Jeon, S. S. da Silva and J.-I. Han, *Bioresour. Technol.*, 2016, **214**, 609.
52. M. P. Badve, P. R. Gogate, A. B. Pandit and L. Csoka, *Ultrason. Sonochem.*, 2014, **21**, 162.
53. A. K. Lee, D. M. Lewis and P. J. Ashman, *J. Appl. Phycol.*, 2015, **27**, 1881.
54. P. B. Baxi and A. B. Pandit, *Bioresour. Technol.*, 2012, **110**, 697.
55. R. T. Hilares, L. Ramos, S. S. da Silva, G. Dragone, S. I. Mussatto and J. C. dos Santos, *Crit. Rev. Biotechnol.*, 2018, **38**, 483.
56. N. Habashi, N. Mehrdadi, A. Mennerich, A. Alighardashi and A. Torabian, *Ultrason. Sonochem.*, 2016, **31**, 362.

57. K. Grűbel and A. Machnicka, *Water Environ. Res.*, 2009, **81**, 2420.
58. P. N. Patil, P. R. Gogate, L. Csoka, A. Dregelyi-Kiss and M. Horvath, *Ultrason. Sonochem.*, 2016, **30**, 79.
59. G. Mancuso, M. Langone and G. Andreottola, *Ultrason. Sonochem.*, 2017, **35**, 489.
60. M. Zieliński, M. Dębowski, M. Kisielewska, A. Nowicka, M. Rokicka and K. Szwarc, *Waste Biomass Valor.*, 2019, **10**, 747.
61. L. Albanese, R. Ciriminna, F. Meneguzzo and M. Pagliaro, *J. Cleaner Prod.*, 2017, **142**, 1457.
62. D. Crudo, V. Bosco, G. Cavaglià, S. Mantegna, L. Battaglia and G. Cravotto, *Agro Food Ind. Hi-Tech*, 2014, **25**, 55.
63. K. E. Preece, N. Hooshyar, A. J. Krijgsman, P. J. Fryer and N. J. Zuidam, *Innovative Food Sci. Emerging Technol.*, 2017, **41**, 47.
64. I. Lee and J.-I. Han, *Bioresour. Technol.*, 2015, **186**, 246.
65. P. R. Gogate, *Food Bioprocess Technol.*, 2011, **4**, 996.
66. S. Y. Tang and M. Sivakumar, *AIChE J.*, 2013, **59**, 155.
67. Z. Zhang, G. Wang, Y. Nie and J. Ji, *Chin. J. Chem. Eng.*, 2016, **24**, 1477.
68. M. Sivakumar, S. Y. Tang and K. W. Tan, *Ultrason. Sonochem.*, 2014, **21**, 2069.
69. J. Carpenter, M. Badve, S. Rajoriya, S. George, V. K. Saharan and A. B. Pandit, *Rev. Chem. Eng.*, 2017, **33**, 433.
70. M. A. Kelkar, P. R. Gogate and A. B. Pandit, *Ultrason. Sonochem.*, 2008, **15**, 188.
71. J. Ji, J. Wang, Y. Li, Y. Yu and Z. Xu, *Ultrasonics*, 2006, **44**, e411.
72. R. S. Beuve and K. R. Morison, *Chem. Eng. Process. Process Intensif.*, 2010, **49**, 1101.
73. Y. Nie, W. Chen, M. Z. Lu, L. Yu, Z. Y. Wu, J. B. Ji, G. Wei, M. Min and Y. Li, in *Advances in Energy Science and Equipment Engineering – Proceedings of International Conference on Energy Equipment Science and Engineering, ICEESE 2015*, 2015, vol. 1, p. 295.
74. W. Chen, Y. Nie, Z.-Y. Wu, M.-Z. Lu, G. Wei and J.-B. Ji, *Chin. Modern Chem. Ind. (Xiandai Huagong)*, 2015, **35**, 66.
75. A. L. Prajapat and P. R. Gogate, *Chem. Eng. Process. Process Intensif.*, 2015, **93**, 1.
76. L. Rinaldi, Z. Wu, S. Giovando, M. Bracco, D. Crudo, V. Bosco and G. Cravotto, *Green Process. Synth.*, 2017, **6**, 425.
77. A. L. Prajapat and P. R. Gogate, *Ultrason. Sonochem.*, 2016, **31**, 371.
78. B. Avvaru, N. Venkateswaran, P. Uppara, S. B. Iyengar and S. S. Katti, *Ultrason. Sonochem.*, 2018, **42**, 493.
79. N. B. Suryawanshi, V. M. Bhandari, L. G. Sorokhaibam and V. V. Ranade, *Sci. Rep.*, 2016, **6**, 33021.
80. N. B. Suryawanshi, V. M. Bhandari, L. G. Sorokhaibam and V. V. Ranade, *Fuel*, 2017, **210**, 482.
81. M. Askarian, A. Vatani and M. Edalat, *J. Pet. Sci. Eng.*, 2017, **151**, 55.
82. K. B. Ansari, N. H. Loke, A. B. Pandit, V. G. Gaikar, R. Sivakumar, R. Kumar and S. Das, *Indian Chem. Eng.*, 2015, **57**, 256.

83. L. F. Chuah, J. J. Klemeš, S. Yusup, A. Bokhari and M. M. Akbar, *J. Cleaner Prod.*, 2017, **146**, 181.
84. G. L. Maddikeri, P. R. Gogate and A. B. Pandit, *Fuel*, 2014, **137**, 285.
85. V. L. Gole and P. R. Gogate, *Chem. Eng. Process. Process Intensif.*, 2012, **53**, 1.
86. L. F. Chuah, S. Yusup, A. R. A. Aziz, A. Bokhari and M. Z. Abdullah, *J. Cleaner Prod.*, 2016, **112**, 4505.
87. A. Bokhari, L. F. Chuah, S. Yusup, J. J. Klemeš, M. M. Akbar and R. N. M. Kamil, *J. Cleaner Prod.*, 2016, **136**, 31.
88. V. L. Gole, K. R. Naveen and P. R. Gogate, *Chem. Eng. Process. Process Intensif.*, 2013, **71**, 70.
89. F. Ma and M. A. Hanna, *Bioresour. Technol.*, 1999, **70**, 1.
90. D. Ghayal, A. B. Pandit and V. K. Rathod, *Ultrason. Sonochem.*, 2013, **20**, 322.
91. P. R. Gogate, *Chem. Eng. Process. Process Intensif.*, 2008, **47**, 515.
92. V. L. Gole and P. R. Gogate, *Ind. Eng. Chem. Res.*, 2012, **51**, 11866.
93. S. M. Hingu, P. R. Gogate and V. K. Rathod, *Ultrason. Sonochem.*, 2010, **17**, 827.
94. G. L. Maddikeri, A. B. Pandit and P. R. Gogate, *Fuel Process. Technol.*, 2013, **116**, 241.
95. L. F. Chuah, S. Yusup, A. R. Abd Aziz, A. Bokhari, J. J. Klemeš and M. Z. Abdullah, *Chem. Eng. Process. Process Intensif.*, 2015, **95**, 235.
96. S. Yusup, A. Bokhari, L. F. Chuah and J. Ahmad, *Chem. Eng. Trans.*, 2015, **43**, 517.
97. J. A. Ladino, J. Herreraa, D. H. Malagónb, M. Prisciandarod, V. Piemonte and M. Capocellic, *Chem. Eng. Trans.*, 2016, **50**, 319.
98. L. F. Chuah, A. R. A. Aziz, S. Yusup, A. Bokhari, J. J. Klemeš and M. Z. Abdullah, *Clean Technol. Environ. Policy*, 2015, **17**, 2229.
99. S. Joshi, P. R. Gogate, P. F. Moreira and R. Giudici, *Ultrason. Sonochem.*, 2017, **39**, 645.
100. A. V. Mohod, P. R. Gogate, G. Viel, P. Firmino and R. Giudici, *Chem. Eng. J.*, 2017, **316**, 751.
101. D. Crudo, V. Bosco, G. Cavaglià, G. Grillo, S. Mantegna and G. Cravotto, *Ultrason. Sonochem.*, 2016, **33**, 220.
102. J. Lifka and B. Ondruschka, *Chem. Eng. Technol.*, 2004, **27**, 1156.
103. A. V. Mohod, A. S. Subudhi and P. R. Gogate, *Ultrason. Sonochem.*, 2017, **36**, 309.
104. S. M. Dubey, V. L. Gole and P. R. Gogate, *Ultrason. Sonochem.*, 2015, **23**, 165.
105. Y. Zhou, X. Lu, Y. Yu and J. Ji, *Acta Energ. Sol. Sin.*, 2010, **31**, 1532.
106. V. L. Gole and P. R. Gogate, *Fuel Process. Technol.*, 2014, **118**, 110.
107. Satriana, N. Arpi, M. D. Supardan, R. T. Gustina and W. A. W. Mustapha, in *AIP Conference Proceedings*, 2018, vol. 1940.
108. N. Zhong, L. Li, X. Xu, L.-Z. Cheong, X. Zhao and B. Li, *Food Chem.*, 2010, **122**, 228.
109. J. V. de Quadros Jr and R. Giudici, *Chem. Eng. Process. Process Intensif.*, 2016, **100**, 87.

110. V. P. Chavan, A. V. Patwardhan and P. R. Gogate, *Chem. Eng. Process. Process Intensif.*, 2012, **54**, 22.
111. Y. Huang, Z. Zhou, Y. Qi, X. Li, Z. Cheng and W. Yuan, *Chem. Eng. J.*, 2011, **172**, 444.
112. V. Chandra Srivastava, *RSC Adv.*, 2012, **2**, 759.
113. C. Song and X. Ma, *Appl. Catal., B*, 2003, **41**, 207.
114. Z. Wu and B. Ondruschka, *Ultrason. Sonochem.*, 2010, **17**, 1027.
115. S. I. Kolesnikov, M. Y. Kil'Yanov and O. M. Chekhovskaya, *Chem. Technol. Fuels Oils*, 2010, **46**, 70.
116. Y. Kegelaers, O. Eulaerts, J. Reisse and N. Segebarth, *Eur. J. Org. Chem.*, 2001, 3683.
117. T. Ando, S. Sumi, T. Kawate, J. I. Née Yamawaki and T. Hanafusa, *J. Chem. Soc., Chem. Commun.*, 1984, 439.
118. Y. Zhang and L. Wang, *Polym.-Plast. Technol. Eng.*, 2009, **48**, 285.
119. M. V. Encina, E. Lissi, M. Sarasúa, L. Gargallo and D. Radic, *J. Polym. Sci., Polym. Lett. Ed.*, 1980, **18**, 757.
120. A. K. Suresh, M. M. Sharma and T. Sridhar, *Ind. Eng. Chem. Res.*, 2000, **39**, 3958.
121. H. J. Janssen, A. J. Kruithof, G. J. Steghuis and K. R. Westerterp, *Ind. Eng. Chem. Res.*, 1990, **29**, 754.
122. M. Orlandi, F. Tosi, M. Bonsignore and M. Benaglia, *Org. Lett.*, 2015, **17**, 3941.
123. S. Letort, M. Lejeune, N. Kardos, E. Métay, F. Popowycz, M. Lemaire and M. Draye, *Green Chem.*, 2017, **19**, 4583.
124. R. B. Valitov, R. N. Galiakhmetov, A. K. Kurochkin and M. A. Margulis, *Russ. J. Phys. Chem.*, 1985, **59**, 2973.
125. P. Blanchard, M. S. El Kortbi, J.-L. Fourrey and M. Robert-Gero, *Tetrahedron Lett.*, 1992, **33**, 3319.
126. N. Kardos and J.-L. Luche, *Carbohydr. Res.*, 2001, **332**, 115.
127. D. A. Ramirez, O. Kozyuk and F. C. Michel, *2012 Dallas, Texas, July 29–August 1, 2012*, American Society of Agricultural and Biological Engineers, 2012, 121337884.
128. Z. Wu, B. Ondruschka and A. Stark, *J. Phys. Chem. A*, 2005, **109**, 3762.
129. Z. Wu, J. Lifka and B. Ondruschka, *Ultrason. Sonochem.*, 2005, **12**, 127.
130. P. Kumar and V. S. Moholkar, *Ind. Eng. Chem. Res.*, 2011, **50**, 4769.
131. V. A. Yakovlev, S. G. Zavarukhin, V. T. Kuzavov, S. V. Stebnovskii, N. V. Malykh, L. I. Mal'tsev and V. N. Parmon, *Russ. J. Phys. Chem.*, 2010, **4**, 227.
132. R. K. Sharma, S. Dutta, S. Sharma, R. Zboril, R. S. Varma and M. B. Gawande, *Green Chem.*, 2016, **18**, 3184.
133. C. E. Domini, M. B. Alvarez, G. F. Silbestri, G. Cravotto and P. Cintas, *Catalysts*, 2017, **7**, 121.
134. J. Safaei-Ghomi, F. Eshteghal and H. Shahbazi-Alavi, *Ultrason. Sonochem.*, 2016, **33**, 99.
135. S. J. Tabatabaei Rezaei, *J. Iran. Chem. Soc.*, 2017, **14**, 585.
136. B. Banerjee, *Ultrason. Sonochem.*, 2017, **35**, 1.
137. S. R. Bansode and V. K. Rathod, *Ultrason. Sonochem.*, 2017, **38**, 503.

138. L. A. Lerin, R. A. Loss, D. Remonatto, M. C. Zenevicz, M. Balen, V. O. Netto, J. L. Ninow, C. M. Trentin, J. V. Oliveira and D. de Oliveira, *Bioprocess Biosyst. Eng.*, 2014, **37**, 2381.
139. D. Frascari, A. E. M. Bacca, F. Zama, L. Bertin, F. Fava and D. Pinelli, *Chem. Eng. J.*, 2016, **283**, 293.
140. Z. Wang, C. Wang, C. Zhang and W. Li, *Innovative Food Sci. Emerging Technol.*, 2017, **44**, 224.
141. E. Kuna, R. Behling, S. Valange, G. Chatel and J. C. Colmenares, *Top. Curr. Chem.*, 2017, **375**, 411.
142. J. C. Colmenares, *ChemSusChem*, 2014, **7**, 1512.
143. J. C. Colmenares, E. Kuna and P. Lisowski, *Top. Curr. Chem.*, 2016, **374**, 59.
144. C. Y. Teh, T. Y. Wu and J. C. Juan, *Chem. Eng. J.*, 2017, **317**, 586.
145. Y.-H. Li, L. Wang, Z. Wang, S. Yuan, S. Wu and S.-F. Wang, *ChemistrySelect*, 2016, **1**, 6855.
146. B. Banerjee, *Ultrason. Sonochem.*, 2017, **35**, 15.

CHAPTER 8

Mechanochemical Synthesis of Biologically Relevant Heterocycles

MARCO LEONARDI, MERCEDES VILLACAMPA AND
J. CARLOS MENÉNDEZ*

Unidad de Química Orgánica y Farmacéutica, Departamento de Química en Ciencias Farmacéuticas, Facultad de Farmacia, Universidad Complutense, Plaza de Ramón y Cajal s.n., 28040 Madrid, Spain
*E-mail: josecm@ucm.es

8.1 Introduction

Volatile organic solvents constitute the vast majority of chemical waste. Furthermore, solvent cooling or heating consumes most of the energy required to perform chemical reactions, both at laboratory and industrial scales. Thus, one of the goals of green chemistry is the development of new synthetic methods that require lower amounts of solvents. Mechanochemistry, which is characterized by the application of mechanical energy to achieve chemical transformations, can contribute to this goal by enabling solid-state reactions, also leading to lower energy requirements by operating at close to ambient temperature. Furthermore, by working under solvent-free conditions, reagents are highly concentrated and solvation phenomena are not relevant, and the combination of these two factors often leads to accelerated reactions. Importantly, the absence of solvents implies the absence of

Green Chemistry Series No. 61
Green Synthetic Processes and Procedures
Edited by Roberto Ballini

Published by the Royal Society of Chemistry, www.rsc.org

dissolved water, meaning that the environment inside the milling jar is often sufficiently dry to allow reactions that under conventional conditions would require dry solvents and inert atmosphere.

In this context, the present chapter is devoted to provide an overview of the application of mechanochemical conditions to the synthesis of heterocycles, the class of compounds with the highest relevance for the pharmaceutical and agrochemical industries. For further in-depth discussion of various aspects of mechanochemistry and its synthetic applications, the reader is directed to several recent reviews[1–7] and monographs.[8,9]

8.2 Mechanochemical Synthesis of Five-membered Heterocycles and Their Benzo-fused Derivatives

8.2.1 Furans

Chuang and Chem have developed a one-pot three-component reaction leading to *trans*-2,3-disubstituted-2,3-dihydrofurans from 1,3-dicarbonyl compounds, aromatic aldehydes, and *N*-phenacylpyridinium bromides **1**. Among other experimental protocols, they performed the reaction under grinding conditions with a mortar and pestle in the presence of piperidine.[10] The mechanism proposed to explain this transformation involves the initial formation of the Knoevenagel product from the β-dicarbonyl and aldehyde starting materials (**2**) and that of pyridinium ylide **3** arising from the α-deprotonation of compound **1**. Michael addition of **3** to **2** then furnishes intermediate **4**, which finally provides the final product **5** *via* intramolecular nucleophilic displacement (Figure 8.1).

8.2.2 Thiophenes

The Gewald reaction gives access to 2-aminothiophenes **8** from carbonyl compounds bearing an active α-methylene (**6**), α-cyanoesters (**7**), and sulfur. Mack and co-workers performed this reaction under high-speed vibration milling (HSVM) aerobic conditions at 130 °C using a mixer mill equipped with a heat gun. The yields were only moderate, but better that under conventional conditions (Figure 8.2).[11]

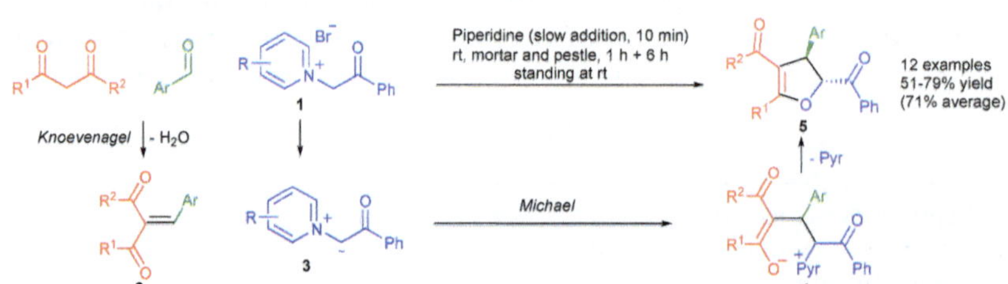

Figure 8.1 Diastereoselective synthesis of *trans*-2,3-disubstituted-2,3-dihydrofurans under mechanochemical conditions.

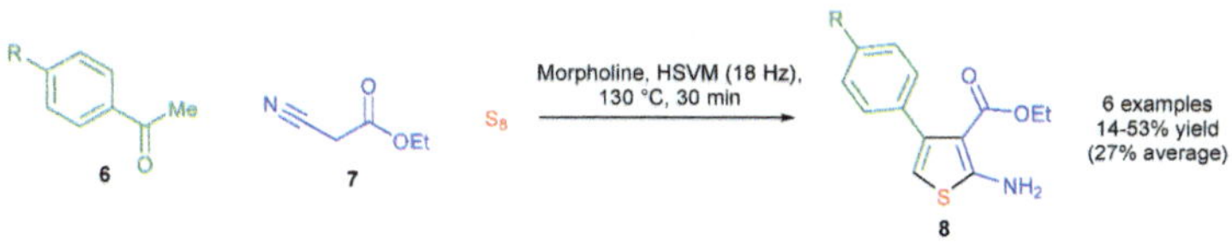

Figure 8.2 Mechanochemical Gewald thiophene synthesis.

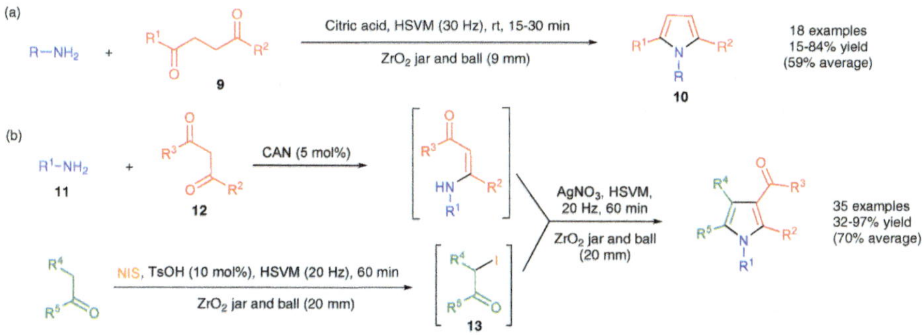

Figure 8.3 Traditional pyrrole syntheses performed under HSVM conditions: (a) Paal–Knorr synthesis and (b) Hantzsch synthesis, telescoped with the *in situ* synthesis of an α-iodoketone; NIS, *N*-iodosuccinimide.

8.2.3 Pyrroles

The Paal–Knorr reaction, starting from 1,4-dicarbonyl compounds **9** and primary amines or ammonia, is one of the classical approaches to pyrrole synthesis. Rousseau and co-workers have described the use of HSVM to promote the Paal–Knorr reaction in the presence of citric acid as the catalyst (Figure 8.3a).[12] In another mechanochemical approach to pyrroles developed by Xu, Zhang, and co-workers, a 1,4-dicarbonyl component for the Paal–Knorr reaction was generated *in situ* from two molecules of a β-ketoester.[13]

The Hantzsch reaction from primary amines **11**, β-dicarbonyl compounds **12**, and α-haloketones **13** is another classical pyrrole synthesis that can be promoted by HSVM, as we reported in 2013.[14] In this case, the use of ball milling allowed telescoping the synthesis of the α-haloketone with the Hantzsch reaction itself. This transformation was catalyzed by Ce(IV) ammonium nitrate (CAN) and required the addition of an equimolecular amount of silver nitrate in order to prevent the reductive dehalogenation of the intermediate α-iodoketone by the HI liberated during the enaminone alkylation step (Figure 8.3b). This method was far more general than previous versions of the Hantzsch reaction and was extended to the synthesis of symmetrical systems containing two pyrrole units connected by a spacer[15] and fused pyrrole derivatives derived from the indole, homoindole, benzo[*g*]indole, and indeno[1,2-*b*]pyrrole frameworks.[16] By using 2-aminoacetaldehyde dimethylacetal as the starting material, it also allowed the preparation of a number of polyheterocyclic ring systems *via* a Pommeranz–Fritsch-type cyclization step.[17]

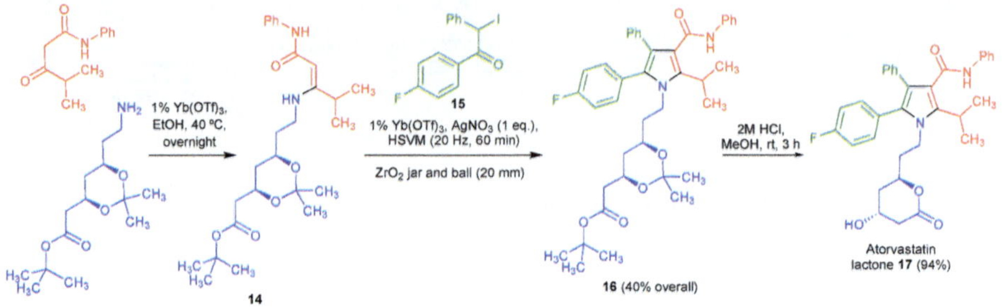

Figure 8.4 Synthesis of atorvastatin lactone based on a mechanochemical Hantzsch reaction.

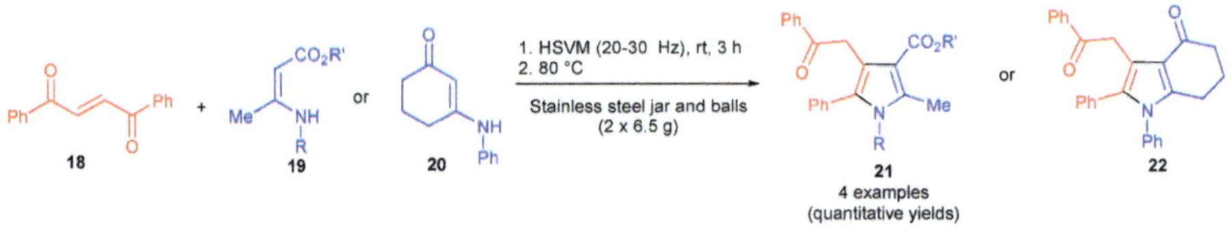

Figure 8.5 Mechanochemical synthesis of pyrroles from chalcones and β-enaminoesters or β-enaminones.

Such mechanochemical Hantzsch pyrrole synthesis was applied to the synthesis of the antihyperlipidemic agent atorvastatin, which was the top-selling drug for more than a decade and can be argued to be the most important pyrrole-based non-natural molecule ever made. The formation of the required enaminone **14** failed under the previously developed conditions; fortunately, ytterbium triflate was a suitable catalyst, although the reaction had to be performed at 40 °C overnight. The crude enaminone thus generated and the suitable α-iodoketone **15** (prepared in a separate step) was subjected to HSVM at 20 Hz for 1 h with prior addition of silver nitrate, leading to pyrrole derivative **16** in 40% overall yield. A final hydrolytic *O*-deprotection, with subsequent *in situ* acid-promoted lactonization, afforded atorvastatin lactone **17** in 94% yield (Figure 8.4). This route constitutes the shortest synthesis of atorvastatin lactone published to date.[18]

Kaupp, Atfeh, and co-workers reported an early example of a ball milling-promoted multistep mechanochemical transformation in the solid state, which in some cases was followed by heating to eliminate water. These reactions afforded quantitative yields of pyrroles **21** or tetrahydroindole **22** from diketones **18** and β-enaminoesters **19** or cyclic β-enaminones **20**, respectively (Figure 8.5).[19]

8.2.4 Indoles

Stolle and co-workers reported a mechanochemical method for the intermolecular hydroamination of 2-alkynylanilines **23** using a planetary ball mill in the presence of zinc bromide, although the starting materials had to be

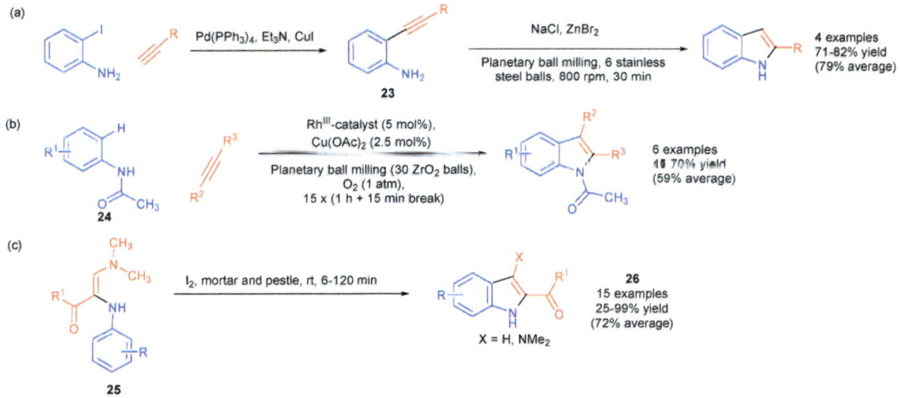

Figure 8.6 Mechanochemical synthesis of indoles.

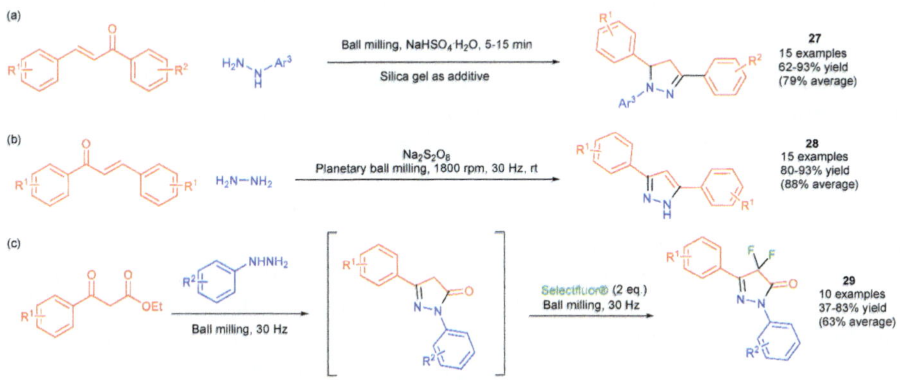

Figure 8.7 Mechanochemical synthesis of pyrazoles.

prepared in a separate step (Figure 8.6a).[20] In subsequent work, Bolm and co-workers developed a mechanochemical method that allowed the C–H activation and hydroamination steps to be performed in one pot from acetanilides **24** (Figure 8.6b).[21] In another mechanochemical approach to indoles, Tamariz and co-workers carried out the iodine-promoted cyclization of β-arylamino-β-enaminones **25** under grinding conditions, leading to 2-acylindoles **26** in good yield (Figure 8.6c).[22]

8.2.5 Five-membered Rings with Two Heteroatoms

Su and co-workers reported the solvent-free synthesis of 1,3,5-triaryl-2-pyrazolines **27** under HSVM conditions from chalcones and phenylhydrazines (Figure 8.7a).[23] Later, Zhang and co-workers studied a similar reaction starting from hydrazine and chalcones and proved that the addition of solid sodium persulfate to the reaction mixture as an oxidant allowed the direct preparation of 3,5-diarylpyrazoles **28** (Figure 8.7b).[24] Browne and

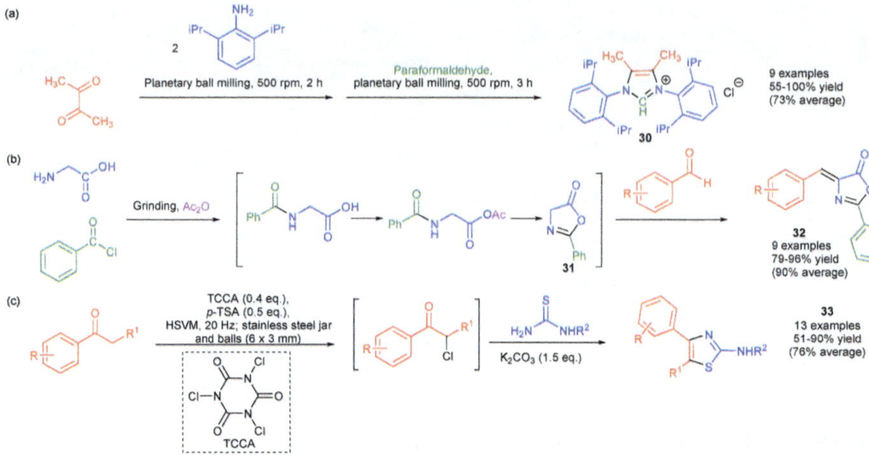

Figure 8.8 Mechanochemical synthesis of imidazoles, oxazoles, and thiazoles.

co-workers showed that pyrazolones were also accessible by a similar mechanochemical process, in this case starting from phenylhydrazine derivatives and β-ketoesters, which could be coupled to the difluorination of the heterocyclic ring to furnish compounds **29** by incorporating 1-chloromethyl-4-fluoro-1,4-diazoniabicyclo[2.2.2]octane bis-tetrafluoroborate (Selectfluor®) to the reaction mixture (Figure 8.7c).[25]

1,3-Diarylimidazolium derivatives are useful precursors to *N*-heterocyclic carbenes (NHCs). Compounds **30** have been recently obtained *via* a mechanochemical three-component reaction from anilines, diacetyl, and paraformaldehyde using the planetary ball milling technique (Figure 8.8a).[26] Regarding oxazoles, Fahmy *et al.* have described a multicomponent one-pot synthesis of 4-arylidene-2-phenyl-5(4*H*)-oxazolones **32** under solvent-free grinding conditions. This transformation comprises the synthesis of the azlactone derived from glycine and benzoyl chloride (**31**) and its Knoevenagel condensation with aldehydes (Figure 8.8b).[27] Thiazoles are also accessible by mechanochemical routes, as shown by the Hantzsch-type synthesis of 2-aminothiazoles **33** shown in Figure 8.8c[28] and related literature examples.[29]

8.2.6 Benzo-fused Five-membered Rings with Two Heteroatoms

Access to benzimidazoles by mechanochemical reactions has been achieved by several strategies. Indeed, Hagar and co-workers have disclosed the use of ball-milling conditions for the solvent and catalyst-free reaction between 1,2-phenylenediamines and carboxylic acids or aldehydes (Figure 8.9a).[30] Jang and co-workers studied the reaction with aldehydes under planetary ball milling in the presence of ZnO nanoparticles (NPs), finding that it

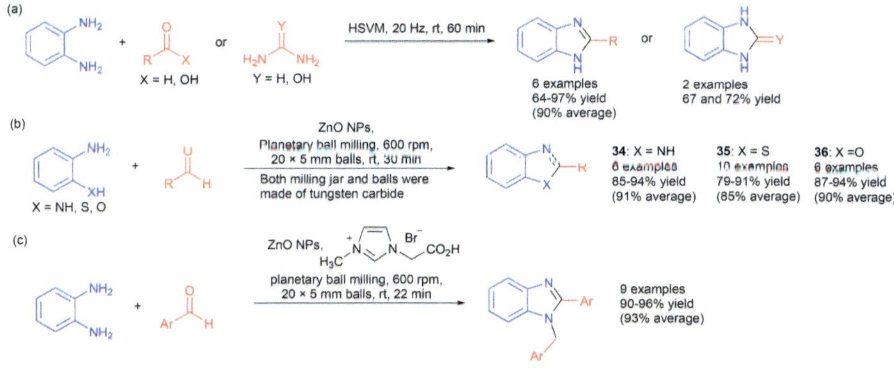

Figure 8.9 Mechanochemical synthesis of benzimidazoles.

constituted a general method for the synthesis of benzimidazoles **34**, applicable also to the preparation of benzothiazoles **35** and benzoxazoles **36** (Figure 8.9b).[31] On the other hand, in the presence of ionic liquid-coated ZnO NPs, the reaction with aldehydes afforded 1,2-disubstituted benzimidazoles under planetary ball milling conditions, with participation of two molecules of the aldehyde, as shown by the same group (Figure 8.9c).[32] Another mechanochemical synthesis of benzoxazoles and benzothiazoles from *o*-hydroxy- and *o*-mercaptoanilines, respectively, involved the use of isothiocyanates as electrophiles and afforded good to excellent yields of 2-anilino derivatives of the target heterocycles.[33]

8.2.7 Triazoles

Solvent-free copper-catalyzed azide-alkyne cycloaddition (CuAAC) reactions have been performed by several authors by milling together alkynes and azides in the presence of suitable copper catalysts.[34,35] It has also been shown that the use of copper reaction vials and balls, which provide the catalyst by shedding copper particles, makes unnecessary the use of additional catalysts. This study mentions explicitly that, despite their initial concerns, the authors did not "observe any explosion or increased exothermicity due to milling azides".[36]

8.3 Mechanochemical Synthesis of Six-membered Heterocycles and Their Benzo-fused Derivatives

8.3.1 1,4-Dihydropyridines

Hundal *et al.* described a solvent-free version of the well-known multicomponent Hantzsch dihydropyridine synthesis performed by grinding together aldehydes, ammonium acetate, 1,3-cyclohexanediones, and an acyclic active methylene compound to yield fused 1,4-dihydropyridines **37** in good

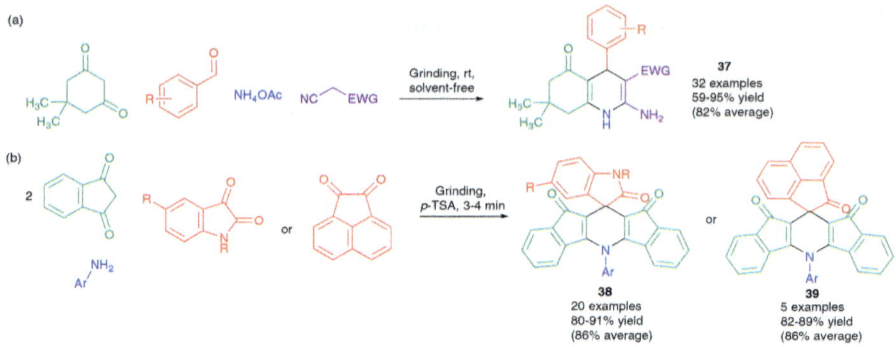

Figure 8.10 Examples of mechanochemical Hantzsch dihydropyridine syntheses from aromatic aldehydes (a) or from isatins (b).

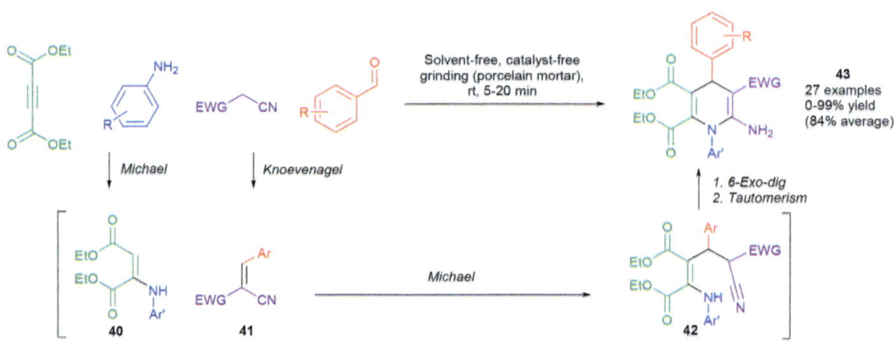

Figure 8.11 A mechanochemical Hantzsch-like dihydropyridine synthesis.

to excellent yield (Figure 8.10a).[37] Bazgir and co-workers later replaced the aldehyde component by sufficiently reactive cyclic α-diketones such as isatin derivatives and were thus able to prepare spiro derivatives **38** or **39** by Hantzsch chemistry in the presence of *p*-toluenesulfonic acid (*p*-TSA) as the catalyst (Figure 8.10b).[38]

Sharma *et al.* reported a Hantzsch-like synthesis of dihydropyridines **43** in excellent yield from diethyl acetylenedicarboxylate, amines, malononitrile or ethyl cyanoacetate, and aromatic aldehydes. The mechanism proposed starts with an initial Michael reaction between the amine and alkyne components, concomitant with a Knoevenagel reaction between the active methylene compound and the aldehyde. Michael addition of **40** to **41** would give **42**, and a 6-*exo-dig* cyclization, followed by a final imine-enamine tautomerism, would explain the formation of the final products (Figure 8.11).[39]

8.3.2 Pyrimidines

The combination of ethyl acetoacetate, aromatic aldehydes, and urea to generate 3,4-dihydropyrimidin-2(1*H*)-ones is known as the Biginelli reaction and is one of the oldest known multicomponent reactions. The Biginelli reaction

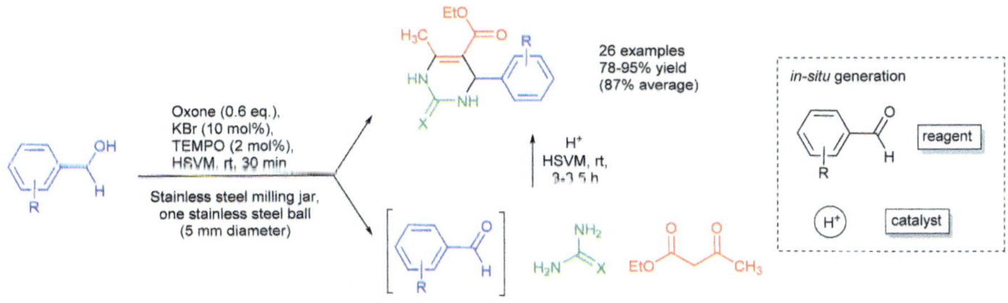

Figure 8.12 Mechanochemical variation of the Biginelli dihydropyridine synthesis.

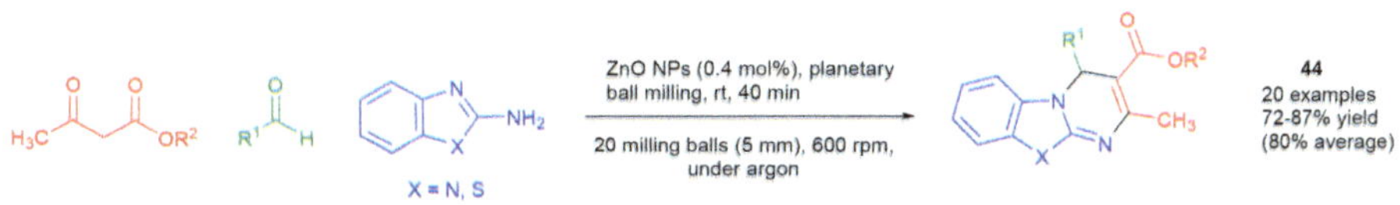

Figure 8.13 Mechanochemical synthesis of fused pyrimidines using the Biginelli reaction.

has been widely investigated under mechanochemical conditions, specially using the planetary ball milling approach.[40]

Mal *et al.* studied a one-pot solvent-free procedure, which is an example of a cascade reaction where one of the components and the catalyst are generated *in situ*. Thus, a bromonium-catalyzed oxidation of a carbinol, followed by addition to the same pot of ethyl acetoacetate and urea or thiourea yielded pyrimidines in good to excellent yield (Figure 8.12). Interestingly, the reaction did not work at all in solution.[41]

The mechanochemical Biginelli reaction was extended to the generation of fused pyrimidines **44** by Jang and co-workers using as starting materials β-dicarbonyl compounds, aldehydes, and benzo[*d*]imidazol-2-amine or benzo[*d*]thiazol-2-amine in the presence of ZnO NPs (Figure 8.13).[42]

8.3.3 Quinolines: The Mechanochemical Povarov Reaction

The Povarov reaction, *i.e.*, the formal [4 + 2] cycloaddition of imines and electron-rich olefins, is one of the most widely employed methods for the synthesis of 1,2,3,4-tetrahydroquinolines. Zang *et al.* described a $FeCl_3$-catalyzed mechanochemical version of the Povarov reaction that yielded *cis*-2,4-diphenyl-1,2,3,4-tetrahydroquinolines **45** with complete diasteroselectivity, which is normally not achieved in solution. When an alkyne was used as the dienophile, fully aromatic quinolines **46** were obtained (Figure 8.14).[43,44] Similarly, Kouznetsov and co-workers reported the application of the HSVM technique to the Povarov reaction in the presence of phosphomolybdic acid (PMA).[45]

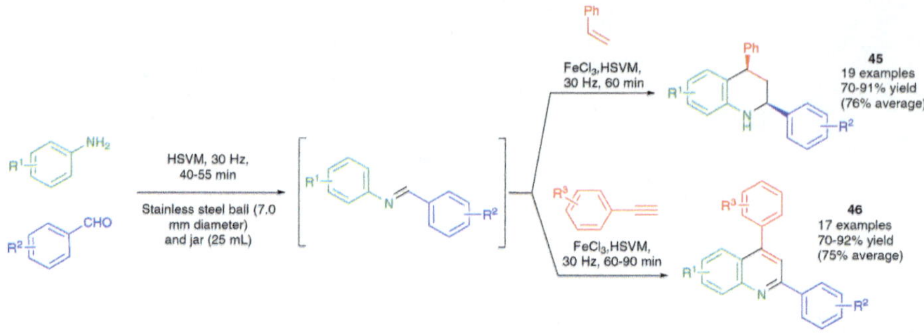

Figure 8.14 Mechanochemical Povarov reactions.

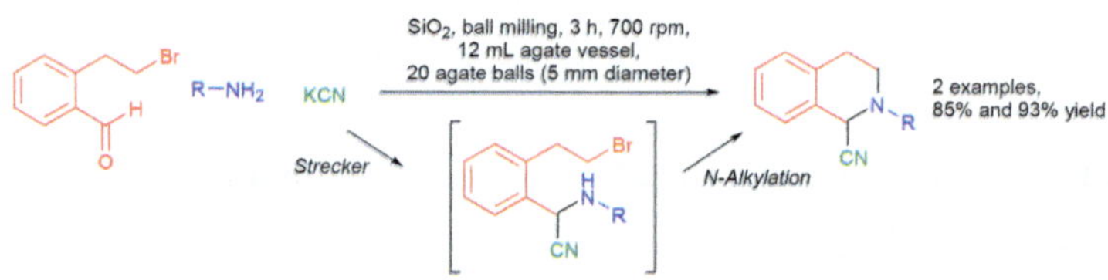

Figure 8.15 Strecker-initiated mechanochemical synthesis of isoquinolines.

8.3.4 Tetrahydroisoquinolines

Bolm *et al.* reported an interesting mechanochemical three-component synthesis of tetrahydroisoquinolines catalyzed by SiO_2 *via* a one-pot domino process initiated by a classical Strecker reaction, which was followed by intramolecular *N*-alkylation (Figure 8.15).[46]

8.3.5 Pyran and Fused Pyran Derivatives

Eslami *et al.* reported the three-component synthesis in excellent yield of 2-amino-3-cyano-4*H*-pyrans **47** and some fused derivatives of this ring system (**48–51**) from aldehydes, malononitrile, and active methylene compounds (also phenols). These reactions employed only mechanochemical energy at room temperature and were catalyzed by potassium phthalimide (POPI), a mild base (Figure 8.16).[47]

Interestingly, the reaction involving dimedone as the dicarbonyl component and leading to **51** has also been performed by Crawford and James using the twin screw extrusion (TSE) technique, a methodology that allows performing mechanochemical syntheses under flow conditions. In their setup, each of the twin screw extruders contains six heating zones that can be heated up to 160 °C with a screw speed of 55 rpm.[48]

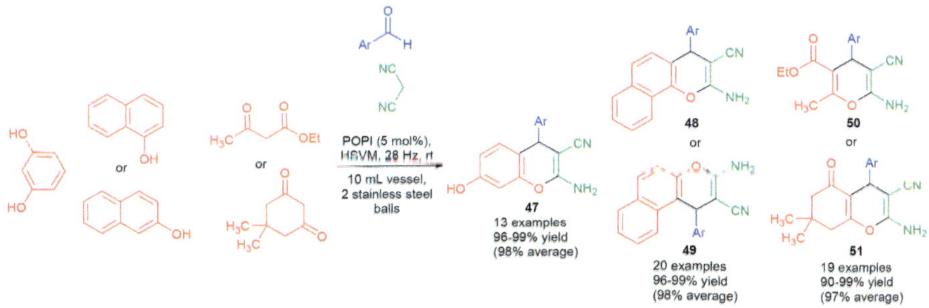

Figure 8.16 Mechanochemical synthesis of pyran and fused pyran derivatives.

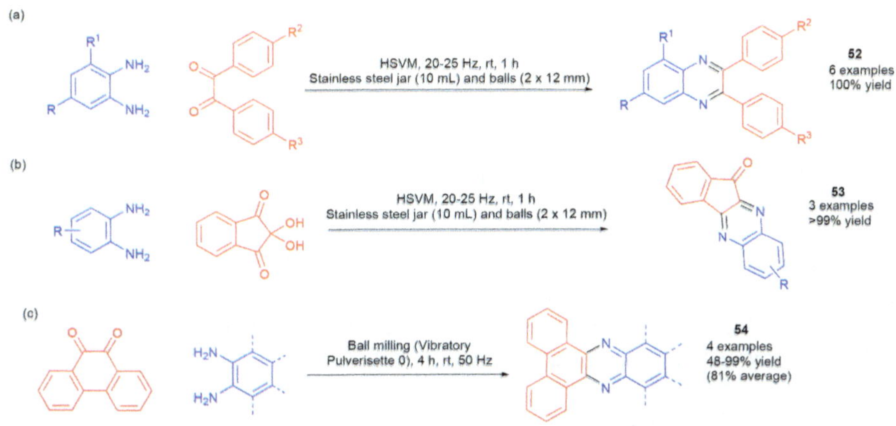

Figure 8.17 Mechanochemical synthesis of quinoxalines and their fused derivatives from α-diketones, using as starting materials benzils (a), ninhydrin (b) and 9,10-phenanthrenediones (c).

8.3.6 Quinoxaline and Its Fused Derivatives

Several authors have studied the reaction of *o*-phenylenediamines with various 1,2-dicarbonyl compounds under ball-milling conditions. For instance, Kaupp reported in 2002 the efficient synthesis of quinoxaline derivatives **52** in 1 h under solvent-free ball-milling conditions (Figure 8.17a),[49] and the same group[50] and that of Mellaway[51] applied the reaction to ninhydrin, leading to the very efficient preparation of indeno[1,2-*b*]quinoxalines **53** (Figure 8.17b). Similarly, Carlier and co-workers reported a mechanochemical method for the synthesis of dibenzophenazines and their aza-derivatives (**54**) based on the same reaction (Figure 8.17c).[52] Baron and co-workers studied the solid-state synthesis of dibenzo[*a*,*c*]phenazine under ball milling by ^{13}C CP-MAS NMR spectroscopy and calorimetry, finding evidence of the formation of a hemiaminal intermediate that was not detected in the course of a similar reaction leading to 2,3-diphenylquinoxaline.[53]

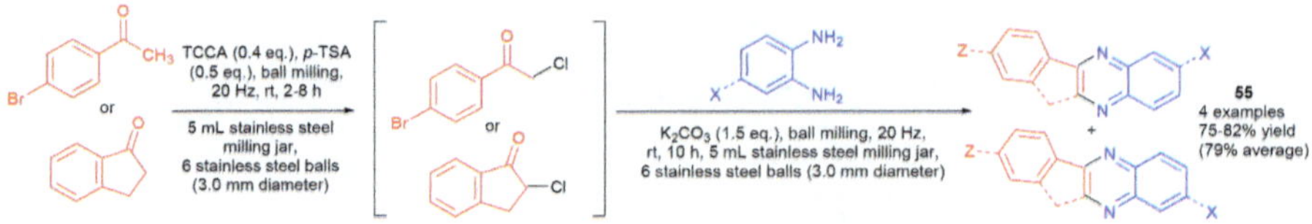

Figure 8.18 Mechanochemical synthesis of quinoxalines and their fused derivatives from *in situ*-generated α-haloketones.

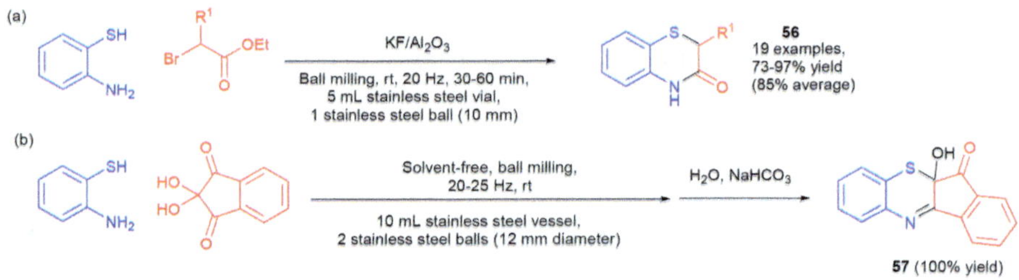

Figure 8.19 Mechanochemical synthesis of benzothiazines.

In a related transformation, Nagarajaiah *et al.* studied the efficient ball-milling synthesis of quinoxalines and their fused-derivatives **55** starting from *o*-phenylenediamine and ketones, which were α-chlorinated *in situ* with trichloroisocyanuric acid (TCCA, Figure 8.18).[54]

8.3.7 Benzothiazines

The reaction of 2-aminothiophenols with 2-bromoalkanoate esters in the presence of potassium fluoride on alumina (KF-Al_2O_3) in a ball-mill was used to generate thiazine derivatives **56** in excellent yield, as reported by Sharifi *et al.* (Figure 8.19a). A comparison with conventional methods highlighted the advantages of the mechanochemical protocol in terms of the yield, reaction time, and mildness of conditions.[55] The use of ninhydrin as the bis-electrophile allowed the preparation of hydroxyindeno[2,1-*b*]benzo[1,4]thiazine **57** (Figure 8.19b).[50]

8.4 Mechanochemical Synthesis of Complex Fused or Bridged Heterocycles

8.4.1 Bridged Diazocines

Diazocine derivatives with two bridgehead stereogenic nitrogen atoms **58**, belonging to the Tröger's base family of compounds, were synthesized from *N*-arylmethyleneiminium salts *via* a cascade of three reactions promoted by water, added as $MgSO_4{\cdot}7H_2O$, as reported by Kaupp *et al.* (Figure 8.20).[56]

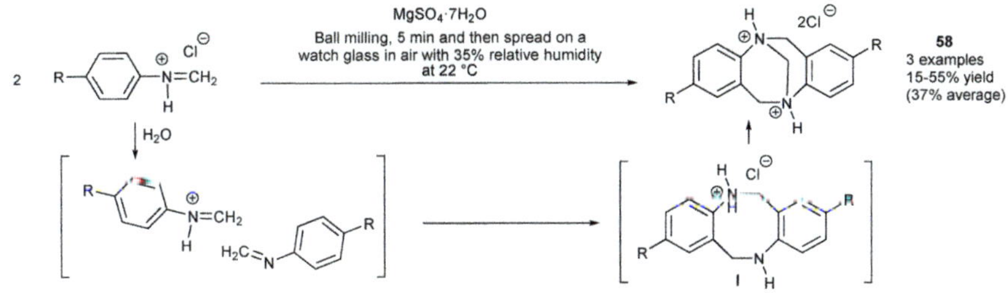

Figure 8.20 Mechanochemical synthesis of Tröger's bases.

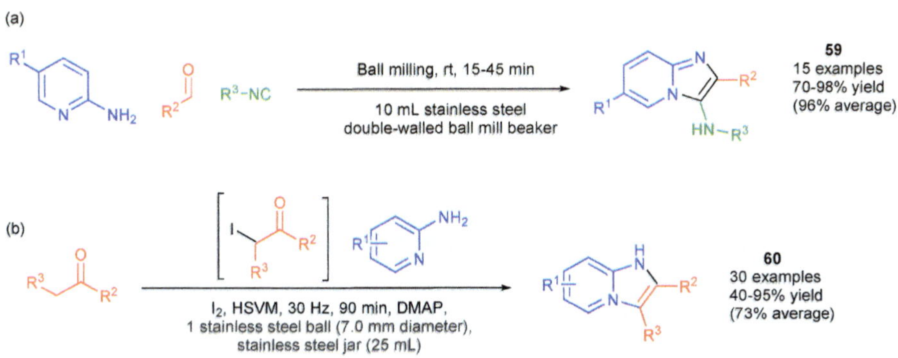

Figure 8.21 Mechanochemical synthesis of imidazo[1,2-*a*]pyridines from 2-aminopyridines using isonitrile-based three-component reactions (a) or a cyclocondensation with an *in situ*-generated α-iodoketone (b).

8.4.2 Imidazo[1,2-*a*]pyridines

3-Aminoimidazo[1,2-*a*]pyridines **59** were obtained in good to excellent yield by mechanochemical one-pot Ugi-type multicomponent reactions of 2-aminoazines, aldehydes, and isonitriles, as described by Naimabadi *et al.* (Figure 8.21a).[57] One year later, Zhang *et al.* reported the synthesis of another family of imidazopyridine derivatives **60** by reaction of 2-aminopyridines and methyl ketones or 1,3-dicarbonyl compounds promoted by I_2 *via* two steps of condensation and cyclization (Figure 8.21b).[58]

8.4.3 Pyrano[2,3-*c*]pyrazoles

The solvent- and catalyst-free mechanochemical reaction between hydrazine hydrate, ethyl acetoacetate, aromatic aldehydes, and malononitrile furnished 2,4-dihydropyrano[2,3-*c*]pyrazoles **61**, as described by Dekamin and co-workers (Figure 8.22). The proposed reaction mechanism involved the initial *in situ* generation of 4-pyrazolin-2-one, a Knoevenagel condensation between ethyl acetoacetate, and the final combination of both halves of the molecule *via* a Michael addition and final cyclocondensation step.[59]

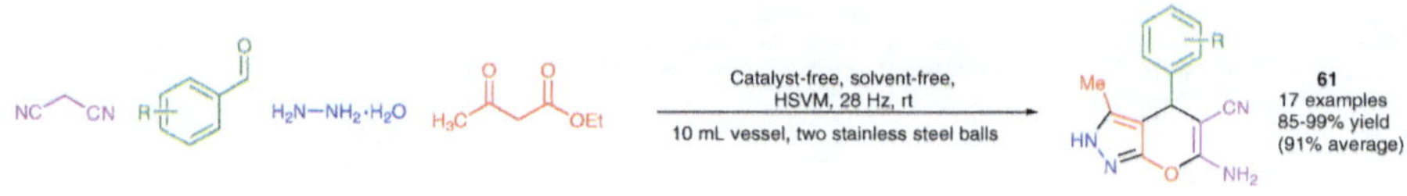

Figure 8.22 Mechanochemical synthesis of pyrano[2,3-*c*]pyrazoles.

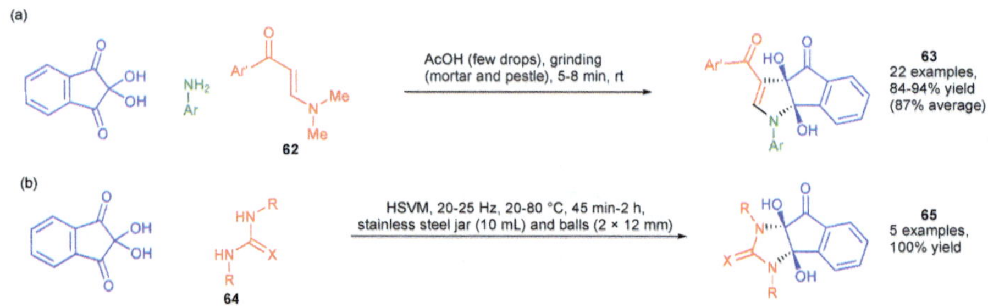

Figure 8.23 Mechanochemical synthesis of indeno[1,2-*b*]pyrroles (a) and indeno[1,2-*d*]imidazoles (b), starting from ninhydrin.

8.4.4 Indeno[1,2-*b*]pyrroles and Indeno[1,2-*d*]imidazoles

Perumal *et al.* synthesized fused pyrrole derivatives **63** in fully diastereoselective fashion starting from (*E*)-3-(dimethylamino)-1-arylprop-2-en-1-ones **62**, anilines and ninhydrin by grinding the starting materials in a mortar in the presence of a few drops of acetic acid as the catalyst (Figure 8.23a). One advantage of this reaction is the high purity of the products, which avoided the need for chromatography.[60] One example of a similar framework bearing an ester group at the C-3 position had been previously obtained by Kaupp and coworkers under ball milling conditions.[50] The Kaupp group also described the ball milling-assisted synthesis of the related tricyclic compounds **65** from ninhydrin and urea or thiourea derivatives **64**, in quantitative yield (Figure 8.23b).[50]

8.5 Conclusion

Mechanochemical synthesis, *i.e.*, the induction of chemical transformations by direct absorption of mechanical energy, is set to become a standard tool in synthetic laboratories in the near future. Mechanochemical processes normally involve unique reaction conditions, with very high reagent concentrations and in the absence of solvation effects. The application of mechanochemistry to the synthesis of heterocycles is a highly promising tool for the generation of new bioactive chemical entities.

Acknowledgements

Financial support from the Spanish Ministerio de Economía y Competitividad (MINECO) through grant CTQ2015-68380-R is gratefully acknowledged.

References

1. J. G. Hernández and T. Friščić, *Tetrahedron Lett.*, 2015, **56**, 4253.
2. G. W. Wang, *Chem. Soc. Rev.*, 2013, **42**, 7668.
3. J. G. Hernández, C. G. Vila-Ortiz and E. Juaristi, in *Comprehensive Organic Synthesis*, ed. G. A. Molander and P. Knochel, Elsevier, Oxford, 2nd edn, 2014, vol. 9, p. 287.
4. R. M. Claramunt, C. López, D. Sanz and J. Elguero, *Adv. Heterocycl. Chem.*, 2014, **112**, 117.
5. D. Tan and T. Friščić, *Eur. J. Org. Chem.*, 2018, 18.
6. M. Leonardi, M. Villacampa and J. C. Menéndez, *Chem. Sci.*, 2018, **9**, 2042.
7. J. Anderson and J. Mack, *Green Chem.*, 2018, **20**, 1435.
8. *Ball Milling Towards Green Synthesis. Applications, Projects, Challenges*, ed. B. Ranu and A. Stolle, Royal Society of Chemistry, 2015.
9. *Mechanochemical Organic Synthesis*, ed. D. Margetić and V. Štrukil, Elsevier, 2016.
10. C. P. Chuang and K.-P. Chen, *Tetrahedron*, 2012, **68**, 1401.
11. W. C. Shearouse, M. Z. Shumba and J. Mack, *Appl. Sci.*, 2014, **4**, 171.
12. L. Akelis, J. Rousseau, R. Juskenas, J. Dodonova, C. Rousseau, S. Menuel, D. Prevost, S. Tumkevičius, E. Monflier and F. Hapiot, *Eur. J. Org. Chem.*, 2016, 31.
13. J.-C. Zeng, H. Xu, F. Yu and Z. Zhang, *Tetrahedron Lett.*, 2017, **58**, 674.
14. V. Estévez, M. Villacampa and J. C. Menéndez, *Chem. Commun.*, 2013, **49**, 591.
15. M. Leonardi, M. Villacampa and J. C. Menéndez, *Beilstein J. Org. Chem.*, 2017, **13**, 1957.
16. V. Estévez, V. Sridharan, S. Sabaté, M. Villacampa and J. C. Menéndez, *Asian J. Org. Chem.*, 2016, **5**, 65.
17. M. Leonardi, M. Villacampa and J. C. Menéndez, *J. Org. Chem.*, 2017, **82**, 2570.
18. V. Estévez, M. Villacampa and J. C. Menéndez, *Org. Chem. Front.*, 2014, **1**, 458.
19. G. Kaupp, J. Schmeyers, A. Kuse and A. Atfeh, *Angew. Chem., Int. Ed.*, 1999, **38**, 2896.
20. M. Zille, A. Stolle, A. Wild and U. S. Schubert, *RSC Adv.*, 2014, **4**, 13126.
21. G. N. Hermann, C. L. Jung and C. Bolm, *Green Chem.*, 2017, **19**, 2520.
22. A. V. Jerezano, E. M. Labarrios, F. E. Jiménez, M. C. Cruz, D. C. Pazos, R. U. Gutiérrez, F. Delgado and J. Tamariz, *ARKIVOC*, 2014, (iii), 18.
23. X. Zhu, Z. Li, C. Jin, L. Xu, Q. Wu and W. Su, *Green Chem.*, 2009, **11**, 163.
24. Z. Zhang, Y.-J. Tan, C.-S. Wang and H.-H. Wu, *Heterocycles*, 2014, **89**, 103.
25. J. L. Howard, W. Nicholson, Y. Sagatov and D. L. Browne, *Beilstein J. Org. Chem.*, 2017, **13**, 1950.
26. A. Beillard, X. Bantreil, T.-X. Métro, J. Martinez and F. Lamaty, *Green Chem.*, 2018, **20**, 964.
27. A. A. El-Sayed, M. M. Hemdan and A. F. M. Fahmy, *Chem. Cent. J.*, 2016, **10**, 59.

28. H. Nagarajaiah, A. K. Mishra and J. N. Moorthy, *Org. Biomol. Chem.*, 2016, **14**, 4129.
29. S. Bondock, H. El-Azap, E.-E. M. Kandeel and M. A. Metwally, *Monatsh. Chem.*, 2008, **139**, 1329.
30. T. H. El-Sayeed, A. Aboelnaga and M. Hagar, *Molecules*, 2016, **21**, 1111.
31. H. Sharma, N. Kaur and D. O. Jang, *Green Chem.*, 2014, **16**, 4922.
32. H. Sharma, N. Kaur, N. Singh and D. O. Jang, *Green Chem.*, 2015, **17**, 4263.
33. Z. Zhang, F.-J. Wang, H.-H. Wu and Y.-J. Tan, *Chem. Lett.*, 2015, **44**, 440.
34. R. Thorwirth, A. Stolle, B. Ondruschka, A. Wild and U. S. Schubert, *Chem. Commun.*, 2011, **47**, 4370.
35. M. Tireli, S. Maračić, S. Lukin, M. Juribašić Kulcsár, D. Žilić, M. Cetina, I. Halasz, S. Raić-Malić and K. Užarević, *Beilstein J. Org. Chem.*, 2017, **13**, 2352.
36. T. L. Cook, J. A. Walker and J. Mack, *Green Chem.*, 2013, **15**, 617.
37. S. Kamur, P. Sharma, K. K. Kapoor and M. S. Hundal, *Tetrahedron*, 2008, **64**, 536.
38. R. Ghahremanzadeh, S. Ahadi, G. I. Shakibaei and A. Bazgir, *Tetrahedron*, 2010, **51**, 499.
39. A. Kamur and S. Sharma, *Green Chem.*, 2011, **13**, 2017.
40. A. K. Bose, S. Pednekar, S. N. Ganduly, G. Chakraborty and M. S. Manhas, *Tetrahedron Lett.*, 2004, **45**, 8351.
41. P. K. Sahoo, A. Bose and P. Mal, *Eur. J. Org. Chem.*, 2015, **32**, 6994.
42. T. Raj, H. Sharma, Mayank, A. Singh, T. Aree, N. Kaur, N. Singh and D. O. Jang, *ACS Sustainable Chem. Eng.*, 2017, **5**, 1468.
43. Y.-J. Tan, Z. Zhang, F.-J. Wang, H.-H. Wu and Q.-H. Li, *RSC Adv.*, 2014, **4**, 35635.
44. Y.-J. Tan, F.-J. Wang, A. A. Asirib, H. D. Marwanib and Z. Zhang, *J. Chin. Chem. Soc.*, 2018, **65**, 65.
45. V. V. Kouznetsov, D. R. Merchán-Arenas, C. A. Martínez-Bonilla, M. A. Marcías, P. Roussel and G. H. Gauthier, *J. Braz. Chem. Soc.*, 2016, **27**, 2246.
46. J. G. Hernández, M. Turberg, I. Schiffers and C. Bolm, *Chem. - Eur. J.*, 2016, **22**, 1451.
47. M. G. Dekamin and M. Eslami, *Green Chem.*, 2014, **16**, 4914.
48. D. E. Crawford, C. K. Miskimmin, J. Cahir and S. L. James, *Chem. Commun.*, 2017, **53**, 13067.
49. G. Kaupp and M. R. Naimi-Jamal, *Eur. J. Org. Chem.*, 2002, **2002**, 1368.
50. G. Kaupp, M. R. Naimi-Jamal and J. Schmeyers, *Chem. - Eur. J.*, 2002, **8**, 594.
51. H. A. Etman, H. M. Metwally, M. M. Elkasaby, A. M. Khalil and M. A. Metwally, *Am. J. Org. Chem.*, 2011, **1**, 10.
52. L. Carlier, M. Baron, A. Chamayou and G. Couarraze, *Tetrahedron Lett.*, 2011, **52**, 4686.
53. P. F. M. Oliveira, N. Haruta, A. Chamayou, B. Guidetti, M. Baltas, K. Tanaka, T. Sato and M. Baron, *Tetrahedron*, 2017, **73**, 2305.
54. H. Nagarajaiah, A. K. Mishra and J. N. Moorthy, *Org. Biomol. Chem.*, 2016, **14**, 4129.

55. A. Sharifi, M. Ansari, H. R. Darabi and M. S. Abaee, *Tetrahedron Lett.*, 2016, **57**, 529.
56. G. Kaupp, J. Schmeyers and J. Boy, *J. Prakt. Chem.*, 2000, **342**, 269.
57. A. Maleki, S. Javanshir and M. Naimabadi, *RSC Adv.*, 2014, **4**, 30229.
58. F. J. Wang, H. Xu, M. Xin and Z. Zhang, *Mol. Diversity*, 2016, **20**, 659.
59. M. G. Dekamin, M. Alikhani, A. Emami, H. Ghafuri and S. Javanshir, *J. Iran. Chem. Soc.*, 2016, **13**, 591.
60. S. Muthusaravanan, C. Sasikumar, B. Devi Bala and S. Perumal, *Green Chem.*, 2014, **16**, 1297.

CHAPTER 9

New and Up-and-coming Perspectives for Unconventional Chemistry: From Molecular Synthesis to Hybrid Materials by Mechanochemistry†

CLARENCE CHARNAY[a], ANDREA PORCHEDDU[b], FRANCESCO DELOGU[c] AND E. COLACINO*[d]

[a]Institut Charles Gerhardt de Montpellier (ICGM), UMR-5253 CNRS-UM, Université de Montpellier, Campus Triolet cc 1701, Place Eugène Bataillon, 34095 Montpellier Cedex 05, France; [b]Università degli Studi di Cagliari, Dipartimento di Scienze Chimiche e Geologiche, Cittadella Universitaria, SS 554 bivio per Sestu, 09042 Monserrato (CA), Italy; [c]Università degli Studi di Cagliari, Dipartimento di Ingegneria Meccanica, Chimica, e dei Materiali, via Marengo 2, 09123 Cagliari, Italy; [d]Institut Charles Gerhardt de Montpellier (ICGM), UMR-5253 CNRS-UM, Ecole Nationale Supérieure de Chimie de Montpellier, 8 Rue de l'Ecole Normale, 34296 Montpellier Cedex 05, France
*E-mail: evelina.colacino@umontpellier.fr

†The contribution of Clarence Charnay and Andrea Porcheddu to this work is equivalent.

Green Chemistry Series No. 61
Green Synthetic Processes and Procedures
Edited by Roberto Ballini

Published by the Royal Society of Chemistry, www.rsc.org

9.1 Introduction

The demand for clean processes and environmentally friendly reaction conditions has stimulated the interest on mechanochemistry,[1,2] as witnessed by excellent reviews addressing the benefits of this powerful solvent-free methodology as a valid alternative to many solution-based procedures. This methodology is becoming increasingly competitive, finding more and more space in organic[3–5] and organometallic[6] syntheses, supramolecular chemistry,[7] enzymatic catalysis,[8] preparation of pharmaceutical materials,[9–12] co-crystals,[13] metal–organic[14] and nano-materials,[15,16] with reduced effort and time and with reagent and waste economy. Mechanochemistry deals with "chemical reactions involving reagents in any aggregate state that are induced by the input of mechanical energy",[17] promptly promoting quantitative reactions between solids by neat grinding with very small amounts of solvents (liquid-assisted grinding, LAG), by ion and liquid-assisted grinding (ILAG),[18] or by adding a polymer (POLAG).[19] It has also been used as a green technology for more specific applications in the pharmaceutical industry to reduce the size of particles, discover additional crystalline forms[10] or new drugs,[20] and improve the properties of existing Active Pharmaceutical Ingredients (APIs)[21] (*e.g.*, dissolution rate, solubility, thermal and hydration stability or compressibility)[22] upon modification of their solid-state arrangements, with evident advantages during industrial formulation. Mechanochemistry avoids the use of solvents and high temperatures, sometimes responsible for API degradation, providing sustainable and low-energy chemical processing, also at manufacturing scale through twin screw extrusion processes.[23–25] The aims of this chapter are to highlight cases of the successful symbiosis of mechanochemistry and APIs and to illustrate the unique opportunities offered by mechanochemistry in the preparation of advanced materials for future biomedical applications.

9.2 Influence of Milling on Polymorphism and Amorphization of Pharmaceutical Materials: The Case of Indomethacin

In recent years, a plethora of beneficial drug products have been placed on the market. Most drug candidates in the pharmaceutical industry show poor water-solubility and the percentage of these is increasing continuously.[26] Furthermore, about 40% of oral drug products that have reached the marketplace appear to be highly insoluble in water,[27] while over 70% of new chemical entities (NCEs) discovered by a combinatorial screening approach can be classified (Biopharmaceutics Classification System, BCS Class II compounds) as poorly water-soluble[28] with consequent poor bioavailability. In the pharmaceutical industry, many of the issues that typically hamper the process of bringing new drugs to the market are due more to their poor biopharmaceutical properties rather than toxicity or lack of efficacy.[29] Among these physicochemical properties, solubility plays a key role since it affects the profile of drug bioavailability and any other biological activity of new drug compounds.[30] Within this context, the aqueous solubility of an API is a crucial factor, responsible for various aspects concerning the formulation and delivery of a drug.[31] Mechanical stress during milling or grinding, commonly used during manufacturing processes in the pharmaceutical industry, are responsible for structural and thermodynamic changes in the organic molecular crystals of APIs, inducing particle size reduction, solid-state polymorphic transformations, altering the stability of solid forms of drug candidates, or leading to amorphous drugs. Thermodynamically stable crystalline forms combining water solubility and acceptable physicochemical properties have been proposed; at the same time, amorphous drugs are becoming more and more popular for pharmaceutical formulation, displaying a faster dissolution rate and higher apparent solubility compared to their respective crystalline forms.[32,33] As such, amorphization processes, often considered as a side effect of milling procedures, are being revaluated. However, amorphous drugs are thermodynamically unstable, with a tendency to revert to the original form or convert into a more stable polymorphic form,[34,35] limiting their use in drug development. Different approaches are used to inhibit crystallization processes and obtain sufficient kinetic stability of the formulation through careful choice of stabilizing excipients. Amorphous solid dispersions of the drug in hydrophilic polymers are already on the market and co-amorphous systems have also been described; stabilizing effects involving low molecular weight molecule excipients (*e.g.*, amino acids[36] or by combining two drugs[37–39]) also have the advantage of higher drug loading and reduced hygroscopicity compared to those of polymer-based dispersions.

Indomethacin **1** (IND, Figure 9.1) is one of the most studied API models to investigate polymorphic transformations and amorphization behavior during milling processes. This research has highlighted important correlations between the nature of the transformation and the milling conditions,[40,41] with special attention being paid to the temperature of milling with respect to the glass transition temperature (T_g)[42] or the intensity of

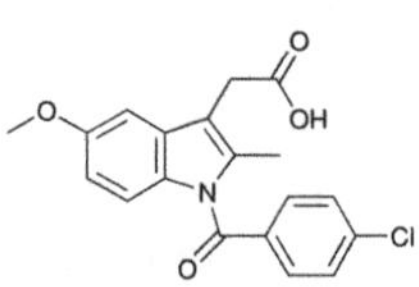

1

Indomethacin (IND)

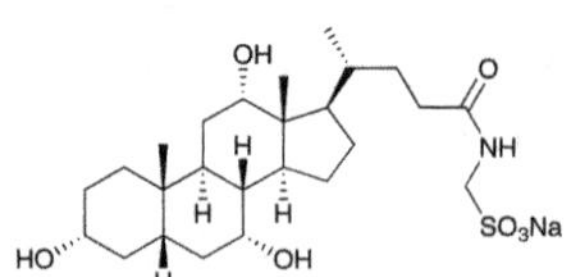

2

Sodium taurocholate (NaTC)

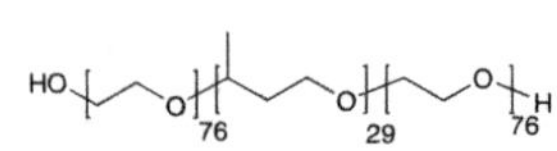

3

Pluronics F68

Figure 9.1 IND and some additives used for amorphization studies.

milling.[43] It was demonstrated that ball milling the polymorphic forms γ-indomethacin (stable) and α-indomethacin (metastable) at 30 °C (near its T_g ~42 °C) led to a 1 : 1 mixture of metastable α-indomethacin and amorphous phase in less than 10 h,[42] while cryogenic grinding at 4 °C (far below its T_g) led to complete amorphization within 2 to 4 h, independently on the starting crystalline state, with a faster amorphization rate (20 min) when using liquid nitrogen.[40,41] If these results reveal to some extent the way in which molecular compounds transform, this trend is far from general: the inverse situation was observed for caffeine, switching from a metastable form I to the stable form II when milled at room temperature.[44] Milling amorphous IND **1** at room temperature at low intensities (150 rpm) led to crystallization of stable γ-indomethacin; at intermediate intensity (200 rpm), metastable α-indomethacin was formed after a transient conversion to γ-indomethacin; while, at higher intensities (>300 rpm), amorphization was mantained.[43]

Mechanical milling also served to prepare a gram-scale co-amorphous system (at 650 rpm for 6 h) from a 1 : 1 molar ratio of crystalline γ-indomethacin and tryptophan (TRP). Cohesive intermolecular interactions (*e.g.*, hydrogen bonding and π–π interactions) within the IND-TRP blend might be responsible for its amorphization by ball milling, which occurred faster than through individual milling of each component (IND remained crystalline when milled alone for 6 h).[45,46] Stable amorphous ternary solid dispersions of IND **1** with $Mg(OH)_2$–SiO_2 (1 : 3 : 4 molar ratio) were also prepared by manual grinding: the prior mechanochemical reaction between $Mg(OH)_2$ and SiO_2 generated strong acidic sites immobilizing IND molecules and suppressing the crystallization.[47] Enhanced dissolution rates compared to crystalline IND **1** were demonstrated for a co-amorphous system prepared by ball milling a 1 : 1 molar ratio (at 25 Hz for 180 min) of IND and natural bile acid surfactant sodium taurocholate (**2**, NaTC, Figure 9.1), remaining X-ray amorphous up to eleven months.[48]

Water-soluble IND nanosuspensions were also prepared by milling (600 rpm, during 12 cycles of 15 min each) using Pluronics F68 as a stabilizer (**3**, Figure 9.1).[49] Different poloxamers were investigated for their ability to stabilize the nanosuspensions: amphiphilic non-ionic copolymers with similar average molecular weights performed differently depending on the hydrophilic–lipophilic balance (HLB) between the propylene oxide and ethylene oxide monomers. The best results were obtained for Pluronics F68 **3** (HLB > 24), with a coating thickness of 7.6 nm, rendering sufficiently hydrophilic the hydrophobic drug surface. Moreover, particle size reduction by milling resulted in accelerated dissolution as a result of increased saturation solubility due to the large surface of nanoparticles.[50] The IND crystallinity was preserved for nanoparticle agglomerates with or without a matrix by wet bead milling, with the advantage of long-term physical stability compared to other size reduction methods resulting in amorphization.

9.3 Pharmaceutical Co-crystals and Inclusion Complexes

Recently, the mechanical processing of solids has attracted growing interest as an alternative route to classical pathways for the preparation of organic compounds and phases.[51] Among others, this technology shows significant promise for the fine preparation of inclusion complexes and molecular co-crystals. Susceptible of application in different areas of investigation ranging from pharmaceuticals to catalysis and ferromagnetics,[52–58] co-crystals exhibit a definite crystalline structure that stems from the complementary combination of two or more supramolecular synthones.[59] The structure ultimately depends on the intra- and inter-molecular interactions between functional groups, as well as on the chemical composition and processing conditions.[59,60] The literature extensively demonstrates the capability of grinding, activating, and sustaining the intimate mixing of molecular species and the consequent formation of co-crystals. Within this framework, mixing represents the crucial stage that can negatively affect the whole system behavior if not effective. In this regard, early studies in the field found that the use of very small amounts of solvent somehow promoted the co-crystal formation.[61]

Ball milling technology has been proven to be highly suitable for the preparation of complex supramolecular structures. For instance, mechanical treatment of 1,2-bis(4-pyridyl)ethylene (bpe) and benzene-1,2,4,5-tetracarboxylic acid (bta) resulted in the formation of co-crystals susceptible of further chemical and structural evolution under UV irradiation conditions. Indeed, the incident radiation was able to induce the photodimerization of bpe with formation of the *syn*-dimer of *trans*-bpe, *i.e.*, *rctt*-tetrakis(4-pyridyl)cyclobutane, allowing the formation of a large channel structure by further grinding in the presence of dimethylsulfoxide.[62]

Accurate kinetic studies based on the use of thiomorpholine (tmo), tetrafluoro-1,4-diiodobenzene (1,4-tfib), and tetrafluoro-1,2-diiodobenzene (1,2-tfib) highlighted the fundamental role played by molecular recognition between compounds, providing evidence of the occurrence of competition between functional groups.[63] Competition proceeds through the possible formation of intermediates and their subsequent evolution toward the final products upon further milling in the presence of reactants. These results can be interpreted invoking a kinetic scheme involving kinetically and thermodynamically controlled stages.[63]

The mechanical treatment of piracetam (**4**, Figure 9.2) in the presence of LiCl and LiBr has been shown to result in the formation of ionic co-crystals characterized by thermal properties quite different from those of the pure components.[64] Molecular salts have also been obtained by grinding isoniazid (**5**, Figure 9.2), a supramolecular reactant, with different carboxylic acids.[65] Depending on the stoichiometry of the initial mixture, isoniazid **5** can form neutral co-crystals or it can give rise to suitably selected carboxylic acids, which in turn can allow the formation of two molecular salts.[65]

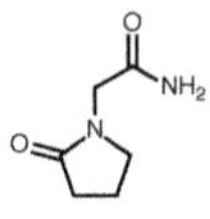

4 Piracetam

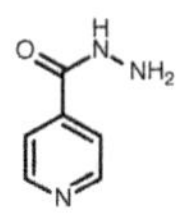

5 Isoniazid

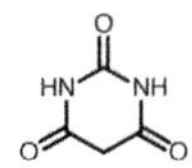

6 Barbituric acid (HBA)

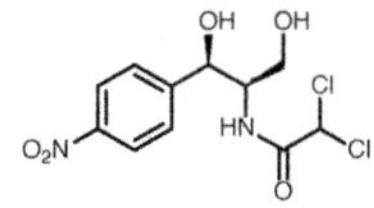

7 Chloramphenicol

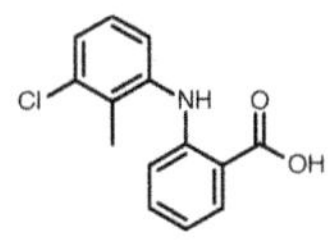

8 Tolfenamic acid (HTA)

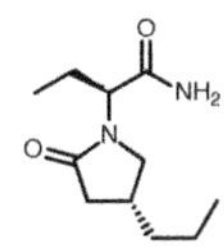

9 Brivaracetam (BRV)

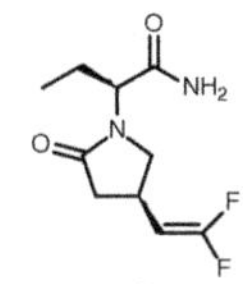

10 Seletracetam (SEL)

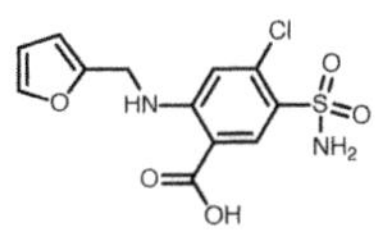

11 Furosemide

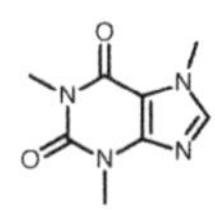

12 Caffeine (CF)

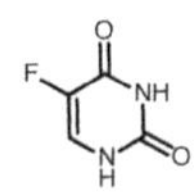

13 5-Fluorouracil

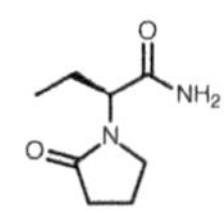

14 Levetiracetam (LEV)

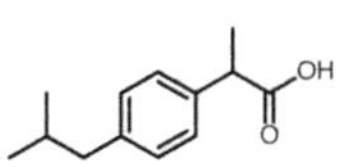

15 Ibuprofen (IBU)

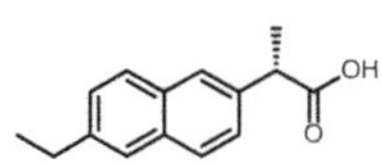

16 Naproxen (NPX)

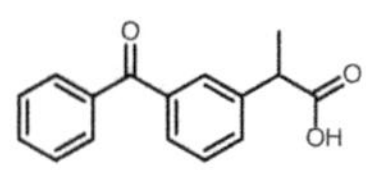

17 Ketoprofen (KTP)

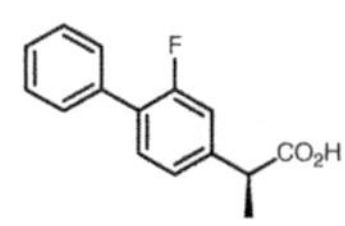

18 Flurbiprofen (FBP)

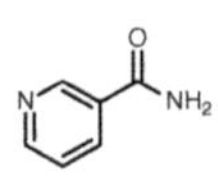

19 Nicotinamide

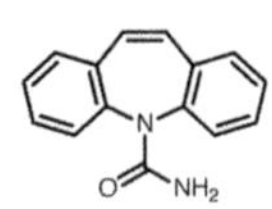

20 Carbamazepine

Figure 9.2 APIs used for co-crystal and inclusion complex preparation by mechanochemistry.

The ball milling of mixtures of barbituric acid **6** (HBA, Figure 9.2) and alkaline metal (M = Na or K) carbonates (M_2CO_3) or bicarbonates ($MHCO_3$) has been used to prepare co-crystals comprising neutral HBA molecules and alkaline barbiturate salts.[66] The mechanical processing induced the release of CO_2 and H_2O from the carbonate salts with consequent formation of the corresponding barbiturate salts. The direct grinding of 4 : 1 HBA/M_2CO_3 mixtures or 2 : 1 HBA/$MHCO_3$ mixtures finally results in the formation of salt co-crystals with general formula $M^{x+}(BA^-)^x{\cdot}HBA{\cdot}nH_2O$. Alternatively, the same compounds can be obtained using a two-step processing that involves the preparation of the BA salt first and then further treatment in the presence of a stoichiometric amount of neutral HBA.[66]

Inclusion complexes of chloramphenicol (7, Figure 9.2) in β-cyclodextrin (β–CD) have been prepared by grinding a 1:1 mixture in the absence of solvent. Co-grinding did not succeed in inducing complete inclusion and resulted in an amorphous phase. In contrast, co-crystallization from an aqueous solution allowed completing the inclusion process, resulting in co-crystals with structures belonging to triclinic *P*1 and monoclinic *C*2 space groups.[67]

Co-grinding of dry anhydrous mixtures of L-serine and oxalic acid has been shown to only result in trace amounts of anhydrous 1:1 serinium oxalate.[68] The addition of water definitely affects the outcome of the mechanical processing, yielding polymorphs of $[L\text{-serH}]_2[ox]\cdot 2H_2O$. Co-grinding of anhydrous oxalic acid and L-serine monohydrate or of anhydrous L-serine and oxalic acid dihydrate results in the formation of polymorph II, *i.e.*, the polymorph that exhibits kinetic stabilization. In contrast, ball milling of L-serine monohydrate and oxalic acid dihydrate mixtures leads to polymorph I, which represents the thermodynamically stable product. Comparison of the experimental results obtained under different conditions suggests that mechanical processing plays the roles of bringing reactants into intimate contact and facilitating the dehydration of crystal hydrates. The availability of water during the co-deformation of molecular crystals enables the co-crystallization to proceed in liquid phase, probably at the points of contact between pressed solid particles.[68]

Co-crystal formation by wet grinding has been shown to benefit drug solubility. In this respect, favorable evidence emerged in the case of tolfenamic acid **8** (HTA, Figure 9.2), *i.e.*, 2-[(2-methyl-3-chlorophenyl)amino] benzoic acid, a well-tolerated, non-steroidal, anti-inflammatory drug belonging to the fenamate family that exhibits a number of desirable properties.[69] HTA **8** has extremely poor solubility in water (~13.6 nM under acidic conditions) and moderate solubility in ethanol (~0.17 M). The mechanical treatment of HTA **8** with sodium acetate (NaAc), sodium carbonate (Na_2CO_3), sodium hydroxide, and imidazole (IM) in the presence of catalytic amounts of water or ethanol resulted in various final products. On the one hand, two salts, namely $NaTA\cdot 0.5H_2O$ and NaTA in their high-temperature form were obtained. On the other hand, co-crystals of salts $NaTA\cdot HTA\cdot 0.5NaAc\cdot 2H_2O$ were prepared. Then, two salt co-crystals, *i.e.*, $NaTA\cdot HTA\cdot H_2O/NaHCO_3$ and IMH-TA·HTA, were also obtained. The different co-crystal forms showed enhanced dissolution rate, particularly $NaTA\cdot HTA\cdot H_2O/NaHCO_3$, NaTA in the high-temperature form, and $NaTA\cdot 0.5H_2O$.[69]

Ball milling has also been used to prepare co-crystals between brivaracetam **9** (BRV) or seletracetam **10** (SEL), two antiepileptic drugs (Figure 9.2), and Ca or Mg chlorides.[70] The co-crystals $BRV_2\cdot MgCl_2\cdot 4H_2O$ and $SEL_2\cdot MgCl_2\cdot 4H_2O$ present structures characterized by octahedral coordination of Mg^{2+} within polymeric chains, giving rise to lipophilic multilayers. Co-crystallization of BRV **9** and SEL **10** with $MgCl_2$ and $CaCl_2$ results in higher thermal stability of the drugs. Since the carbonyl and amide functional groups involved in the formation of ionic BRV and SEL co-crystals are common to a number of pharmaceuticals, the mechanical processing by ball milling show considerable promise for the preparation of drugs with tailored physical and chemical properties.[70]

Suitable considerations regarding the geometry and characteristic lengths associated with preferential hydrogen-bonding interactions of functional groups can lead to the design and subsequent synthesis of co-crystals with desired crystal structures. This is the case of furosemide, *i.e.*, 4-chloro-*N*-furfuryl-5-sulfamoylanthranilic acid (**11**, Figure 9.2), an active pharmaceutical ingredient utilized as loop diuretic. It has been shown that mechanical treatment of furosemide **11** in the presence of 4,4′-bipyridine and a component chosen among dimethyl formamide, dimethyl sulfoxide, methanol, ethanol, 2-propanol, 1-butanol, ethylene glycol, 1,4-butanediol, and hydroquinone, eventually affords the formation of ternary co-crystals with lattice geometry governed by the hydrogen-bonding interactions between the carboxylic sulfonamide and pyridyl functional groups.[71]

Ball milling of caffeine **12** (CF) and citric acid (CTA) mixtures in the presence of catalytic amounts of water has been shown to induce the formation of two co-crystals, namely the known 1:1 CF-CTA co-crystal, hereafter termed co-crystal-1, and a co-crystal with a new structure, hereafter referred to as co-crystal-2.[72] The kinetics of co-crystal formation was studied by X-ray diffraction and numerical simulations providing estimates of the milling intensity. Milling conditions have been shown to affect the transformation rate, and the transformation kinetics has been tentatively described in terms of a three-dimensional diffusion mechanism.[72]

The importance of hydrogen bonding in determining the lattice structure in co-crystals has also been shown using 5-fluorouracil (**13**, Figure 9.2), a drug with biological properties different from those of uracil and application in the treatment of various cancers.[73] In particular, 5-fluorouracil **13** was used as a tecton in combination with urea, thiourea, 2,2′-bipyridine, and 4,4′-bipyridine. Simple grinding with mortar and pestle resulted in the same products yielded by solution methods. Experimental evidence showed that all crystal structures exhibit the formation of hydrogen bonds starting from the N–H moiety from the 5-fluorouracil molecule. Hydrogen bonding establishes connections between two 5-fluorouracil molecules and the second component utilized. The overall geometry of the hydrogen bonding varies as a function of the molecular structure of the second component, which affects the symmetry and stability of the resulting crystalline lattice. Co-crystals 1 and 2 exhibit different crystal structures despite the only difference between urea and thiourea being an S atom instead of an oxygen. The position of the N atoms in 2,2′-bipyridine does not allow their interaction with 5-fluorouracil, whereas hydrogen bonds are formed between the N–H moiety of the 5-fluorouracil molecule and the N atoms of 4,4′-bipyridine. In all cases, the fluorine atoms do not participate in any interactions.[73]

The study of ionic co-crystals between common carbohydrates and alkali or alkaline earth halides dates back to the beginning of chemistry. Time by time, emphasis has been laid on different aspects, according to the specific contemporary interest. They can be conveniently prepared by mechanical treatment of a mixture of an ionic salt and neutral carbohydrate, offering interesting opportunities for fundamental studies addressing crystal engineering and solid state chemistry.[74]

Ionic co-crystals of L- and DL-histidine (His) with $CaCl_2$, $CaBr_2$, and CaI_2 have also been prepared by ball milling.[75] Grinding determines the formation of a structure consisting of L-histidine molecules that bridge Ca^{2+} ions giving rise to enantiopure ribbons, namely homochiral crystals with formula $(\text{L-His})_2{\cdot}CaX_2{\cdot}nH_2O$, where $n = 3$ for X = Cl and Br, and $n = 4$ if X = I. Compound $(\text{L-His})_2{\cdot}CaI_2{\cdot}3H_2O$ can be also formed by partial dehydration of $(\text{L-His})_2{\cdot}CaI_2{\cdot}4H_2O$. Concerning the racemic $(\text{DL-His})_2{\cdot}CaX_2{\cdot}4H_2O$ compounds with X = Cl, Br, and I, molecules of D and L chirality are present in the coordination sphere of Ca^{2+} ions. Similar to the case of enantiopure crystals, this gives rise to ribbon structures with homochiral rims. Intrinsic dissolution rates are higher for histidine-$CaCl_2$ co-crystals than for enantiopure and racemic histidine solids.[75]

The chemical composition has a pivotal role in the formation of co-crystals consequent to the intermolecular interactions that can govern the reciprocal arrangement of components. In light of its importance, systematic investigations have been carried out on solid binary mixtures containing levetiracetam **14** (LEV, Figure 9.2), *i.e.*, (*S*)-2-(2-oxopyrrolidin-1-yl) butanamide, and non-steroidal anti-inflammatory drugs derived from propionic acid such as ibuprofen **15** (IBU), naproxen **16** (NPX), ketoprofen **17** (KTP), and flurbiprofen **18** (FBP) (Figure 9.2). Specifically, the composition was varied over the entire range and the resulting mixtures subjected to ball milling in the absence of solvent. Upon mechanical processing, the mixture LEV + (*S*)-IBU underwent the formation of an equimolar co-crystal. Regarding the mixtures LEV + (*R*,*S*)-IBU, LEV + (*S*)-NPX, LEV + (*R*,*S*)-NPX, LEV + (*R*,*S*)-KTP, and LEV + (*R*,*S*)-FBP, the mechanical treatment led to eutectic/pseudoeutectic mixtures with melting points significantly lower than those of the pure drugs. Enhanced dissolution rates were also observed for all the mixtures investigated. In the case of LEV + (*R*,*S*)-IBU, LEV + (*S*)-NPX, and LEV + (*R*,*S*)-NPX mixtures, the eutectic compositions exhibited the potential of enhancing the pharmacokinetic parameters.[76] Very recently, an (*R*,*S*)-naproxen/(D,L)proline co-crystal system (also using enantiopure substrates) was investigated during *in situ* monitoring of LAG reactions by X-ray diffraction, leading to 17 co-crystals (of which only four were known) and multiple polymorphs with variable stoichiometry (1 : 1, 1 : 2, or 2 : 3), hydrates, and solvates. It was demonstrated that the formation of reaction intermediates depended on the solvent used and that proline was able to form charge-assisted head-to-tail NH··· O hydrogen bonds.[77] A similar investigation was also performed for the preparation of flurbiprofen/proline co-crystals.[78] The ball-milling technology could also be extended to the preparation of inclusion complexes of

IBU **15** and cyclodextrins (β–CD), thus improving the dissolution rate of IBU **15** and consequently its drug absorption. CDs combined with IBU **15** work as useful solubilizers, allowing both solid/liquid oral and parenteral dosage forms.[79] Solid binary systems of drugs and CDs are capable of modifying the physicochemical properties of the drugs, such as the solubility, particle size, crystal habit, and thermal behavior, thereby forming highly water soluble amorphous forms. In a recent paper, an efficient method for IBU/β-CD inclusion complex formation based on ball milling was developed.[80] The efficiency of IBU/β-CD complex formation was quantitatively estimated by means of DSC and TG. The authors achieved highly water-soluble amorphous forms by milling an alkalizing agent ammonia in the presence of appropriate amounts of solvent ($H_2O/C_2H_5OH = 1:4–1:1$).

An alternative methodology to ball milling has also been developed with the aim of processing granular bodies, keeping the damage to particles minimal while assuring effective mixing. Referred to as resonant acoustic mixing (RAM) and extensively applied to the mixing of powders and viscous polymers, it has been recently used to induce the co-crystallization of nicotinamide **19** and carbamazepine **20** mixtures (Figure 9.2) in the presence of catalytic amounts of water.[81] Quite interestingly, co-crystals formed readily at a rate independent of the frequency of oscillation of the mechanical device. *In situ* studies taking advantage of synchrotron radiation highlighted the importance of the number of reactive contacts between powder particles during mixing and comminution processes.[81]

9.4 Hybrid Organic–Inorganic Materials for Pharmaceutical and Biomedical Applications

Mechanochemical milling processes have been reported in recent times as a simple and green synthetic route to different types of advanced materials and nanoparticles without the need for bulk solvents.[11,15,18,82–85] The different methods of milling to achieve hybrid and bio-conjugate materials consist of direct grinding of the precursors in a mortar or in a ball mill (neat grinding), or of LAG performed to activate the reactivity.[82,86] Coordination polymers (CPs) or metal organic frameworks (MOFs) and hybrids materials have gained increased importance in the development of catalytic and energy applications.[83,87–93] Metal–organic frameworks (MOFs) form a class of innovative hybrid materials consisting of metal ion centers, or small clusters, with characteristic coordination geometry chemically bonded to organic bridging ligands. The solid structure can be regarded as a network of organic molecules interconnecting inorganic nodes. The possibility of controlling the coordination geometry and tuning the characteristic length of cavities and channels paved the way to applications in gas storage, sensing, and catalysis. The considerable interest attracted has made MOFs one of the most intensely studied subjects in materials science, with special emphasis on their synthesis. The ever increasing structural complexity of MOFs has also led to a significant surge in studies addressing the functionalization and engineering of already available MOF structures.[94] In this context, considerable attention has been paid to the preparation of MOF structures by mechanical activation of solid reactants. A pioneering study in this regard can be identified in the preparation of Gd(1,4-benzenedi-carboxylate)$_3$(H_2O)$_4$ nanorods by stirring a microemulsion of $GdCl_3$ and bis(methylammonium)benzene-1,4-dicarboxylate in a liquid system comprising cationic cetyltrimethylammonium bromide (CTAB), isooctane, 1-hexanol, and water.[95] Pushed by the interesting results obtained, mechanochemistry is increasingly spreading to the preparation of bio-MOFs with possible pharmaceutical applications.[11] Similar observations hold for the class of divalent metal phosphonates, with an increase in the variety of structures showing different degrees of physical and chemical stability.[96] Bio-inspired metal–organic frameworks (Bio-MOFs) have been recently investigated for applications in biotechnology and nanomedicine with new opportunities in drug encapsulation and controlled release, bioimaging, and biosensing for therapeutics and diagnostics.[97] Bio-MOFs are obtained with metallic cations that are 'essential for life' or metallic cations exhibiting a specific property such as bioactivity, magnetism, or luminescence, for which the amount of metal should be carefully controlled to avoid toxicity side effects. Moreover, the organic linkers of BioMOFs need to be bioactive moieties, endogenous molecules, or non-toxic or safe compounds.[98–102] BioMOFs are divided into two main classes: materials with an active or endogenous linker as building block of the framework and materials with active molecules encapsulated within the pores of the MOF.[98]

Although MOF mechanochemistry synthesis has been used as a powerful technique to obtain solids, the synthesis of BioMOFs using mechanochemistry methodologies has only been recently investigated and a few examples have been realized by manual grinding,[100,102] ball milling, and LAG.[103] Another approach to build BioMOFs consists on the association of a bioactive ligand and a metallic cation. Different bioactive linkers have been used, such as ibuprofen **15**, naproxen **16** (Figure 9.2), and flufenamic acid (**21**, Figure 9.3), which are non-steroidal anti-inflammatory drugs; γ-aminobutyric acid **22**, a neuroleptique drug; or 4-aminosalicylic acid with antibiotic properties, for the preparation of coordination complexes of drugs. Two coordination complexes of the neuroleptic drug gabapentin (**23**, Figure 9.3) were also obtained by grinding gabapentin with $ZnCl_2$ and $CuCl_2$.[102] Novel pharmaceuticals with structures comprising metal centers coordinating flufenamic acid **21** and a bio-inspired framework with muconic acid have been synthesized by mechanical processing. Interestingly, the thermal and on-shelf stabilities were compatible with the relevant temperature range involved in pharmaceutical applications.[100] *trans*-Muconic diacid **24**, an endogenous metabolite, has also been used to generate new coordination polymers with magnesium divalent cations (Figure 9.3). Other coordination networks have been achieved by ball milling and grinding with different divalent cations such as Mn, Ni, Cu, and Zn.[11,100,104] Bonding networks formed by antibiotic 4-aminosalicylic acid and nootropic drug piracetam (**4**, Figure 9.2) coordinating Ag and Ni

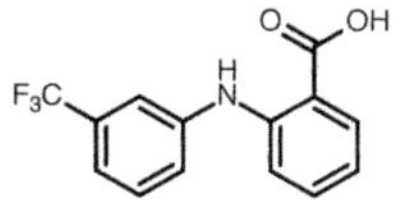

21

Flufenamic acid

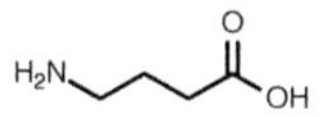

22

γ-Aminobutyric acid

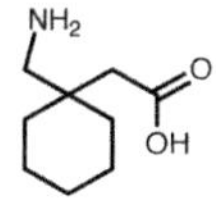

23

Gabapentin

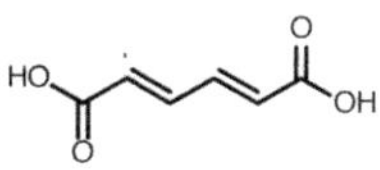

24

trans-Muconic acid

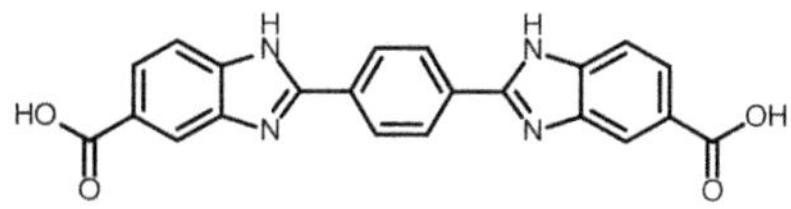

25

1,4-Bis-(5-carboxy-1*H*-benzimidazole-2-yl)benzene

Figure 9.3 APIs and bioactive linkers used for the preparation of BioMOFs by mechanochemistry.

cations, respectively, have also been reported.[104] The considerable versatility of mechanical activation methods clearly emerged in connection with the efficient conversion of Bi_2O_3 into Bi salicylate complexes.[105]

In order to endow BioMOFs with luminescence or magnetic properties for theranostic applications, different lanthanide ions (Ce^{3+}, Nd^{3+}, Er^{3+}, *etc.*...) have been used as metallic cations.[106] Coordination polymers were obtained by manual grinding of lanthanide chlorides and γ-aminobutyric acid **22** as a bioactive linker[106] (Figure 9.3). Along the same line, several coordination networks of gabapentin (**23**, Figure 9.3) with Y(III), Mn(II), and various lanthanides were prepared by mechanochemistry.[106] The coordination networks of pharmaceutical compounds containing lanthanides exhibited thermal stability only at temperatures below ~80 °C, finally decomposing in the temperature range between 200 and 250 °C. Various architectures based on mono-, di-, tri-, and hexa-metallic coordination centers were identified. Gabapentin **23** showed different chelation modes, typically consisting of bidentate coordination-chelation, bridge coordination, and bidentate-bridge coordination, which endowed the coordination complexes with different properties.[106] The mechanical treatment of a Gd salt and the fluorescent linker 1,4-bis(5-carboxy-1*H*-benzimidazole-2-yl)benzene (**25**, Figure 9.3) resulted in the formation of a Gd(III)-based porous structure with one-dimensional open channels filled with hydrogen-bonded water assemblies.[107] The synergy effects for antibacterial applications was investigated for a BioMOF composed of Ag^+ metallic cations with antibacterial properties and a bioactive ligand with antibiotic activity.[104] The network was also obtained by manual grinding.

Moreover, in the same line, silicon-based bio-hybrid nanoparticles with a large bridging organic group were prepared by an original mechanochemical assisted sol–gel reaction.[108] Bis-silylated precursors **29–31-Sil**, containing complex functionalities such as hydantoins or a symmetrical urea, were prepared in a vibration ball mill (VBM). Then, bio-hybrid bridged

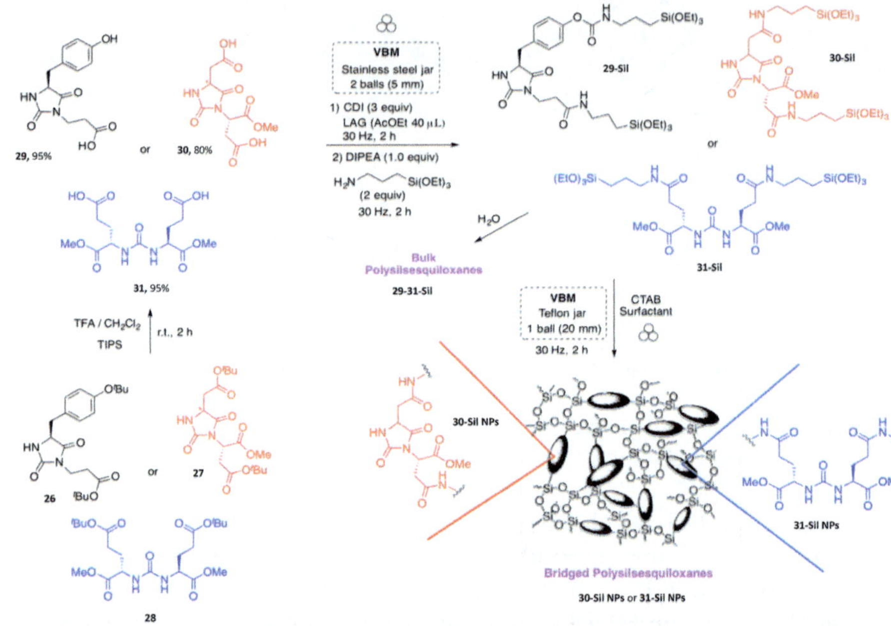

Scheme 9.1 Preparation of bis-silylated precursors **29–31-Sil** and the corresponding bulk materials or the bridged nanosized polysilsesquioxanes **30-Sil NPs** and **31-Sil NPs** by mechanochemistry. Legend: CDI, 1,1′-carbonyldiimidazole; TFA, trifluoroacetic acid; TIPS, triisopropylsilane; CTAB, cetyltrimethylammonium bromide. Reprinted with permission from *ACS Sustainable Chem. Eng.*, 2018, **6**, 511–518. Copyright 2018 American Chemical Society.

silsquioxane nanospheres **30-Sil NPs** and **31-Sil NPs** of 150 ± 15 nm uniform size were attained by mechanochemical 'Stöber-like' sol–gel procedures[108] (Scheme 9.1).

Only a few studies have reported shape and size control under mild mechanochemical milling. As such, the synthesis of amine gold nanoparticles in the size range of 1–4 nm was reported based on a mechanochemical bottom-up approach.[109]

BioMOFs with endogenous ligands were also investigated and the mechanochemistry reaction at room temperature of copper acetate and isonicotinic acid (**32**, Figure 9.4) afforded a 3D network in a short time.[110] LAG grinding of ZnO and fumaric acid **33** resulted in crystals with controlled hydration, depending on the grinding liquid.[111,112] In the presence of ethanol, anhydrous zinc fumarate was obtained, while dihydrate crystals were formed using a mixture of water and ethanol.[111]

Other hydrates were also obtained using different amounts (3 or 4 equiv.) of water during the grinding process. LAG of ZnO with fumaric acid **33** and 4,4′-bipyridyl using DMF or ethanol as the solvent was also carried out to enable the introduction of a secondary ligand in the

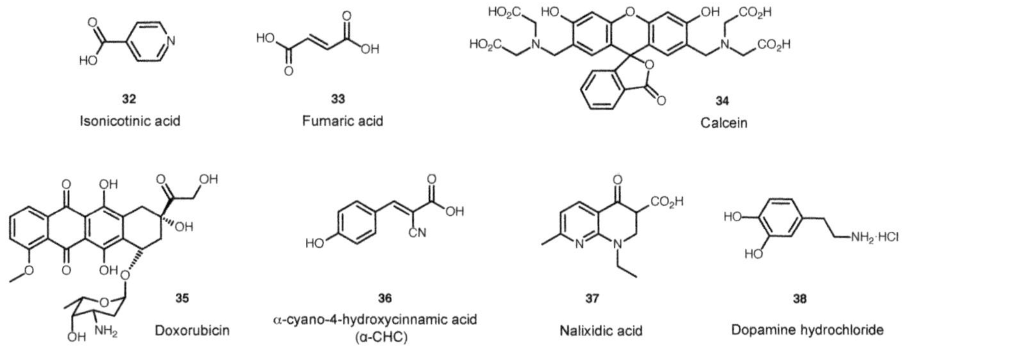

Figure 9.4 APIs and bioactive linkers used for the mechanochemical preparation of inclusion complexes for drug delivery purposes.

BioMOF lattice.[112] Cu_2(1,4-benzenedicarboxylic acid)$_2$(diazabicyclooctane) and Cu_2(1,4-benzenedicarboxylic acid-NH_2)$_2$(diazabicyclooctane) MOFs were prepared by ball milling at room temperature. The materials were loaded with ibuprofen (**15**, Figure 9.2) and their drug delivery properties investigated. Overall, the nanometer-sized MOFs proved able to entrap satisfactorily the model drug and showed good drug release behavior over a one day period with drug loading capacities of about 50% for both structures investigated, which in an unusually high value.[113]

Mechanochemistry has also been recognized as a fast, convenient, and effective method to prepare conventional MOFs such as HKUST-1 (also investigated by *in situ* high-energy synchrotron powder X-ray diffraction[114] and prepared at large scale by extrusion[25]) or MIL-101.[5,93,115,116] Ball milling of copper acetate and benzentricarboxylic acid for 15 min led to the formation of the HKUST solid $Cu_3(BTC)_2$.[117] Mechanochemistry methods were successfully carried out for the formation of MIL-101 obtained with the 1,4-benzendicarboxylic acid linker.[118] It has been reported that MIL-101 exhibits high capacity to load ibuprofen **15** within its pores.[118] The encapsulation of small bioactive molecules was mainly achieved by impregnation, soaking, and diffusions methods under mild conditions.[115] The formation of inorganic oxide-drug nanocomposites using a ball-milling method was also pointed out as an alternative route for drug loading, overcoming issues related to the large amount of solvent handled for drug encapsulation. For instance, the formation of alumina nanocomposites with non-steroidal anti-inflammatory activity was achieved under ball milling in air using a stainless steel planetary mill and by cooling the grinding bowls with circulating water.[119] Highly biocompatible Ag-polysaccharide nanohybrid materials were also prepared by ball milling and tested against human cell lines A549 and SHSY5Y, displaying low toxicity and slow release of silver.[120] Milling (10 Hz for 1 h) was also effective for the preparation of functionalized nanographene oxide (n-GO) targeting selectively breast cancer cells *in vitro via* a Cu(I)-catalyzed azide alkyne cycloaddition (CuAAC) reaction between azide- and maleimide-terminated bifunctional PEG polymer and propargyl-alkyne n-GO. The structural properties of n-GO were preserved upon milling. The hybrid material was further conjugated to a mucin 1 receptor immunoglobulin G antibody (anti-MUC1-IgG) *via* thiol-ene coupling in solution.[121] Water dispersible and biocompatible GO hybridized with oleic acid-coated superparamagnetic iron oxide nanoparticles (SPIOs) for theranostic applications were also prepared by dry ball milling (10–20 Hz for 1 or 2 h).[122] The hydrophobic or π–π interactions between the oleic acid-coated SPIOs and aromatic rings of GO were enhanced upon milling, stabilizing the GO-SPIO hybrids. The *in situ* encapsulation of small bioactive molecules was also reported. Calcein (**34**, Figure 9.4), a hydrophilic model molecule, was encapsulated in the structure formed by $[Zr_6O_4(OH)_4]$ clusters with 1,4-benzodicarboxylic acid struts. Grinding finally induced the amorphization of the MOF, imparting the disordered structure with the unique property of releasing calcein **34** for over 30 days, a period much longer than the two days typically observed in the case of crystalline MOFs.[123] The possibility to achieve the mechanochemical synthesis of zeolitic imidazolate frameworks (ZIFs) was investigated by LAG in the presence of DMF.[124] While the quantitative formation of ZIF-4 was realized using imidazole,[124] the synthesis of amorphous ZIF-8 using 2-methylimidazole as the linker was obtained through the milling of ZnO at 30 Hz for 40 min with a volume of 32 μL of aqueous acetic acid at 1.25 M.[125] The synthesis of nano-sized ZIF-8 with the *in situ* incorporation of caffeine **12**[126] or doxorubicin **35**[127] (Figures 9.2 and 9.4, respectively) was recently reported. ZIF-8 has been the subject of numerous studies including *in situ* monitoring of the mechanochemical synthesis by real-time powder X-ray diffraction.[125] Unexpected amorphization was revealed, evolving toward a recrystallized non-porous material *via* a metastable intermediate. Extrusion methods enabled ZIF-8 continuous large-scale preparation[25] with a throughput of 4 kg h^{-1} and space time yields (STYs) of up to 144×10^3 kg per m^3 per day. ZIF-8 shows considerable promise for drug delivery, being characterized by low density, high surface area and, in principle, tunable coordination frameworks. Any application is, however, hindered by its high cytotoxicity and hydrophobic surface. An obvious way of improving the ZIF-8 properties is to convert the surface hydrophobic character into hydrophilic. In this regard, conventional chemical methods did not result in satisfactory functionalization of the external surface, most often inducing functionalization also on the internal surface, with consequent reduction of the pore size or even obstruction of the pores. In contrast, mechanical activation demonstrated the capability of generating defects exclusively on the external surface, which allowed the formation of unsaturated Zn and N sites able to bind H_2O molecules once the defective ZIF-8 structure was exposed to an aqueous solution.[9] The ZIF-8 structures with more hydrophilic surfaces exhibited superior performances in terms of hydroxyurea delivery and cell cytotoxicity. In particular, a few minutes of ball milling were already able to modify the hydrophobic–hydrophilic balance of ZIF-8, allowing significantly higher cell viability with no drawbacks in the hydroxyurea loading and release capacity.[128] Moreover, ZIF-8 can also be prepared by ILAG. The use of ammonium nitrate, sulfate, and methane sulfonate resulted in the formation of different types of ZIF according to the solvent added during the assisted grinding.[124] The formation of MOFs by mechanical processing of solids has also been the subject of extremely accurate *in situ*, real-time powder X-ray diffraction monitoring.[129] This allowed discovering and capturing a metastable intermediate with unprecedented topology. The study was focused on the synthesis of the so-called zeolitic imidazolate framework (ZIF-8) Zn(2-methylimidazole)$_2$, a MOF consisting of Zn ions coordinated by four imidazolate rings with the same architecture exhibited by Si and Al atoms covalently joined by O atoms in zeolites. *In situ* investigations revealed the unexpected formation of an amorphous phase, followed by recrystallization into a non-porous material upon further grinding. A metastable intermediate with topology never observed before was detected, thus showing that mechanical treatment can give rise to short-lived phases with structures not accessible by conventional chemical methods.[129] The properties toward drug delivery have also been studied for a family of Zr-based MOFs bearing bromo, nitro, and amino functional groups.[10] MOFs were loaded with calcein **34**, a fluorescent molecule, and the anticancer drug α-cyano-4-hydroxycinnamic acid **36** (α-CHC, Figure 9.4). Then, the material was subjected to ball milling to induce its amorphization with the aim of achieving better control of the drug delivery process. As shown by confocal microscopy, the obtained MOFs were able to cross the cell membrane and penetrate cells. Enhanced therapeutic effects of α-CHC **36** on HeLa cells were observed for the case of 20 wt% α-CHC **36** loading into the MOF with the longest linkers. While calcein **34** was released over three days from the crystalline MOFs, the amorphous structures bearing bromo and nitro functionalization were able to release only a fraction of the total amount of calcein **34** loaded. The MOF functionalized with amino groups exhibited, in contrast, the capability of releasing calcein **34** gradually over a period of 15 days. In the case of MOFs loaded with α-CHC **36**, the crystalline and amorphous forms did not reveal differences in their release performance.[130] The preparation of MOFs starting from drugs and inorganic constituents has been shown to have important consequences on solubility and therefore on the bioavailability of drugs. Mechanochemistry has found its relevant place also in this case. For instance, an equimolar mixture of nalidixic acid (**37**, Figure 9.4) and $Zn(NO_3)_2 \cdot 6H_2O$ was manually ground in a mortar with a pestle in the presence of water and ammonia. Nalidixic acid **37** is an antibiotic of the quinolone family with its bioavailability negatively affected by its low solubility. MOFs containing Zn and nalidixic acid were able to double the solubility of the drug.[101]

Magnetic porous MOFs were also fabricated by a multi-step mechanochemical method. Magnetic spinel Fe oxide nanoparticles were incorporated in a Fe(III) carboxylate framework.[15] First, Fe oxide nanoparticles were prepared by mechanical processing of Fe chlorides and NaOH in the presence of NaCl as the dispersing agent. Then, the magnetic nanoparticles were chemically functionalized by neat grinding with benzene-1,3,5-tricarboxylic acid. Finally, they were subjected to ball milling in the presence of hydrated $FeCl_3$, ethanol, and water. The result was a magnetic composite MOF containing encapsulated Fe oxide nanoparticles. The obtained MOF exhibits magnetic properties and remarkably high porosity. It was also tested as a carrier for drug delivery using doxorubicin as the model drug by investigating the loading and release performances.[131]

Surface functionalization should be considered as a crucial post-synthesis strategy not only to improve cell targeting[132] but also to stabilize the nanoparticles in biological media[133] and prevent their aggregation. Polyethylene glycol derivatives are typically the functional groups grafted on the nanoparticle surface to improve their stability and biocompatibility in physiologic media.[134] Surface functionalizations of nanoparticles have been achieved by adsorption[135–137] (using non-covalent interactions) or by covalent conjugation to inorganic surface functional groups.[135] These rational functionalizations are most often carried out as a post-synthesis modification in aqueous solution or solvent according to the solubility of the functional groups. Surprisingly, mechanochemical routes are rarely mentioned as a means to graft functional groups on nanoparticle surfaces. Nevertheless, the grafting of glucose-derived polymers on the surface of iron oxide nanoparticles has been reported.[138] In addition, hybrid protein-iron oxide nanoparticles were obtained *via* two-step mechanochemical synthesis.[139] In a first step, the pre-prepared iron oxide nanoparticles dispersed in 600 μL of water were milled with dopamine hydrochloride (**38**, Figure 9.4) at 200 rpm for 10 min in order to obtain a bio-modified material. In a second step, grinding of the dopamine-Fe_3O_4 suspension with protein BSA dissolved in 600 μL of phosphate buffer solution at 200 rpm afforded bio-conjugates of iron oxide nanoparticles coated with a BSA-dopamine complex. This study demonstrated that the conformation of the BSA protein was preserved despite the applied mechanochemical treatment. In the same way, hemoglobin-dopamine-iron oxide bio-conjugates were achieved by a mechanochemical milling process.[140] The immobilization of enzymes such as laccase, a copper-containing oxidase, on iron oxide particles was also achieved by a mechanochemical approach, and it was found that its activity remained high enough for effective oxygen electro-reduction.[141]

9.5 Conclusions

The recent use of mechanochemistry has opened further development opportunities for biotechnology or biomedical applications, allowing the induction of amorphization of crystals or recrystallization of amorphous phases to phases with various degrees of stability. The milling process[142] provides a suitable alternative for greener and innovative synthesis of nano-biohybrid materials based on bioactive or endogenous molecules, for post-synthesis or *in situ* drug encapsulation, and for surface functionalization or to obtain bioconjugates.

Molecular compounds display a wide range of behaviors and better understanding of mechanically induced transformations at the fundamental level and their systematic investigation will provide new insight into how to efficiently control the various transformations and promote the progress and use of mechanochemical activation in diverse areas of science including not only chemistry, pharmaceutical, materials science, and crystal engineering. Within this context, promising research directions have emerged very recently to obtain mechanistic information[143] through time-resolved *in situ* investigations[144–146] and pair distribution function (PDF) analyses[147] with the ability to trap metastable polymorphs,[148] monitor one-pot syntheses,[149] and disclose the kinetics[150] and milling dynamics of mechanically activated transformations.

Acknowledgements

A.P. is grateful to Fondazione Banco di Sardegna (Italy, project: RICALTRO_CTC_2017_FBS_MONDUZZI) and MIUR for "Finanziamento delle Attività Base di Ricerca (FABR 2017)".

References

1. L. Takacs, *Chem. Soc. Rev.*, 2013, **42**, 7649.
2. V. V. Boldyrev and K. Tkacova, *J. Mater. Synth. Process.*, 2000, **8**, 121.
3. B. Rodriguez, A. Bruckmann, T. Rantanen and C. Bolm, *Adv. Synth. Catal.*, 2007, **349**, 2213.
4. D. Tan and T. Friščić, *Eur. J. Org. Chem.*, 2018, **2018**, 18.
5. S. L. James, C. J. Adams, C. Bolm, D. Braga, P. Collier, T. Friščić, F. Grepioni, K. D. M. Harris, G. Hyett, W. Jones, A. Krebs, J. Mack, L. Maini, A. G. Orpen, I. P. Parkin, W. C. Shearouse, J. W. Steed and D. C. Waddell, *Chem. Soc. Rev.*, 2012, **41**, 413.
6. M. D. Bala and N. J. Coville, *J. Organomet. Chem.*, 2007, **692**, 709.
7. T. Friščić, *Chem. Soc. Rev.*, 2013, **41**, 3493.
8. C. Bolm and J. G. Hernandez, *ChemSusChem*, 2018, **11**, 1410.
9. D. Hasa and W. Jones, *Adv. Drug Delivery Rev.*, 2017, **117**, 147.
10. D. Tan, L. Loots and T. Friscic, *Chem. Commun.*, 2016, **52**, 7760.
11. V. André, S. Quaresma, J. L. F. da Silva and M. T. Duarte, *Beilstein J. Org. Chem.*, 2017, **13**, 2416.
12. V. André and S. Quaresma, Bio-Inspired Metal-Organic Frameworks in the Pharmaceutical World: A Brief Review, in *Metal-Organic Frameworks*, ed. F. Zafar and E. Sharmin, IntechOpen, 2016, ch. 7, DOI: 10.5772/64027.
13. D. Braga, L. Maini and F. Grepioni, *Chem. Soc. Rev.*, 2013, **42**, 7638.
14. A. Lazuen-Garay, A. Pichon and S. L. James, *Chem. Soc. Rev.*, 2007, **36**, 846.
15. P. Baláž, M. Achimovičová, M. Baláž, P. Billik, Z. Cherkezova-Zheleva, J. M. Criado, F. Delogu, E. Dutková, E. Gaffet, F. J. Gotor, R. Kumar, I. Mitov, T. Rojac, M. Senna, A. Streletskii and K. Wieczorek-Ciurowa, *Chem. Soc. Rev.*, 2013, **42**, 7571.

16. S.-E. Zhu, F. Li and G.-W. Wang, *Chem. Soc. Rev.*, 2013, **42**, 7535.
17. E. V. Boldireva, *Chem. Soc. Rev.*, 2013, **42**, 7719.
18. T. Friščić, D. G. Reid, I. Halasz, R. S. Stein, R. E. Dinnebier and M. J. Duer, *Angew. Chem., Int. Ed.*, 2010, **49**, 712.
19. D. Hasa, G. S. Rauber, D. Voinovich and W. Jones, *Angew. Chem., Int. Ed.*, 2015, **54**, 7371.
20. D. Braga, S. d'Agostino, E. Dichiarante, L. Maini and F. Grepioni, *Chem. - Asian J.*, 2011, **6**, 2214.
21. A. V. Trask, W. D. S. Motherwell and W. Jones, *Cryst. Growth Des.*, 2005, **5**, 1013.
22. A. V. Trask, W. D. S. Motherwell and W. Jones, *Cryst. Growth Des.*, 2005, **5**, 1013.
23. C. Medina, D. Daurio, K. Nagapudi and F. Alvarez-Nunez, *J. Pharm. Sci.*, 2010, **99**, 1693.
24. H. Morrison, P. Fung, T. Tran, E. Horstman, E. Carra and S. Touba, *Org. Process Res. Dev.*, 2018, **22**, 1432.
25. D. Crawford, J. Casaban, R. Haydon, N. Giri, T. McNally and S. L. James, *Chem. Sci.*, 2015, **6**, 1645.
26. C. A. Lipinski, *J. Pharmacol. Toxicol. Methods*, 2000, **44**, 235.
27. T. Takagi, C. Ramachandran, M. Bermejo, S. Yamashita, L. X. Yu and G. L. Amidon, *Mol. Pharm.*, 2006, **3**, 361.
28. D. D. Gadade and S. S. Pekamwar, *Adv. Pharm. Bull.*, 2016, **6**, 479.
29. N. Qiao, M. Li, W. Schlindwein, N. Malek, A. Davies and G. Trappitt, *Int. J. Pharm.*, 2011, **419**, 1.
30. N. Blagden, M. de Matas, P. T. Gavan and P. York, *Adv. Drug Delivery Rev.*, 2007, **59**, 617.
31. N. Schultheiss and A. Newman, *Cryst. Growth Des.*, 2009, **9**, 2950.
32. J. E. Patterson, M. B. James, A. H. Forster, R. W. Lancaster, J. M. Butler and T. Rades, *J. Pharm. Sci.*, 2005, **94**, 1998.
33. H. D. Williams, N. L. Trevaskis, S. A. Charman, R. M. Shanker, W. N. Charman, C. W. Pouton and C. J. H. Porter, *Pharmacol. Rev.*, 2013, **65**, 315.
34. B. C. Hancock and M. Parks, *Pharm. Res.*, 2000, **17**, 397.
35. L. Yu, *Adv. Drug Delivery Rev.*, 2001, **48**, 27.
36. R. Laitinen, K. Löbmann, C. J. Strachan, H. Grohganz and T. Rades, *Int. J. Pharm.*, 2012, **453**, 65.
37. K. Löbmann, R. Laitinen, H. Grohganz, K. C. Gordon, C. Strachan and T. Rades, *Mol. Pharmaceutics*, 2011, **8**, 1919.
38. K. Löbmann, C. Strachan, H. Grohganz, T. Rades, O. Korhonen and R. Laitinen, *Eur. J. Pharm. Biopharm.*, 2012, **81**, 159.
39. M. Allesø, N. Chieng, S. Rehder, J. Rantanen, T. Rades and J. Aaltonen, *J. Controlled Release*, 2009, **136**, 45.
40. M. Otsuka, T. Matsumoto and N. Kaneniwa, *Chem. Pharm. Bull.*, 1986, **34**, 1784.
41. K. J. Crowley and G. Zografi, *J. Pharm. Sci.*, 2002, **91**, 492.

42. M. Descamps, J. F. Willart, E. Dudognon and V. Caron, *J. Pharm. Sci.*, 2007, **96**, 1398.
43. S. Desprez and M. Descamps, *J. Non-Cryst. Solids*, 2006, **352**, 4480.
44. J. Pirttimaki, E. Laine, J. Ketolainen and P. Paronen, *Int. J. Pharm.*, 1993, **95**, 93.
45. G. Walker, P. Römann, B. Poller, K. Löbmann, H. Grohganz, J. S. Rooney, G. S. Huff, G. P. S. Smith, T. Rades, K. C. Gordon, C. J. Strachan and S. J. Fraser-Miller, *Mol. Pharmaceutics*, 2017, **14**, 4675.
46. K. Tarp Jensen, F. Hofmann Larsen, C. Cornett, K. Löbmann, H. Grohganz and T. Rades, *Mol. Pharmaceutics*, 2015, **12**, 2484.
47. T. Watanabe, I. Ohno, N. Wakiyama, A. Kusai and M. Senna, *Int. J. Pharm.*, 2002, **241**, 103.
48. K. Gniado, P. MacFhionnghaile, P. McArdle and A. Erxleben, *Int. J. Pharm.*, 2018, **535**, 132.
49. M. Malamataria, S. Somavarapua, M. Bloxhamb and G. Bucktona, *Int. J. Pharm.*, 2015, **495**, 516.
50. C. Brough and R. O. Williams, *Int. J. Pharm.*, 2013, **453**, 157.
51. S. L. James, C. J. Adams, C. Bolm, D. Braga, P. Collier, T. Friščić, F. Grepioni, K. D. M. Harris, G. Hyett, W. Jones, A. Krebs, J. Mack, L. Maini, A. G. Orpen, I. P. Parkin, W. C. Shearouse, J. W. Steed and D. C. Waddell, *Chem. Soc. Rev.*, 2012, **41**, 413.
52. S. A. Myz, T. P. Shakhtshneider, K. Fucke, A. P. Fedotov, E. V. Boldyreva, V. V. Boldyrev and N. I. Kuleshova, *Mendeleev Commun.*, 2009, **19**, 272.
53. S. Karki, T. Friščić, W. Jones and W. D. S. Motherwell, *Mol. Pharmaceutics*, 2007, **4**, 347.
54. S. R. Bysouth, J. A. Bis and D. Igo, *Int. J. Pharm.*, 2011, **411**, 169.
55. K. Yamamoto, S. Tsutsumi and Y. Ikeda, *Int. J. Pharm.*, 2012, **437**, 162.
56. A. Delori, T. Friščić and W. Jones, *CrystEngComm*, 2012, **14**, 2350.
57. M. D. Eddleston, M. Arhangelskis, T. Friščić and W. Jones, *Chem. Commun.*, 2012, **48**, 11340.
58. C. Viedma, J. M. McBride, B. Kahr and P. Cintas, *Angew. Chem., Int. Ed.*, 2013, **52**, 10545.
59. G. R. Desiraju, *J. Am. Chem. Soc.*, 2013, **135**, 9952.
60. M. Moulton and M. J. Zaworotko, *Chem. Rev.*, 2001, **101**, 1629.
61. N. Shan, F. Toda and W. Jones, *Chem. Commun.*, 2002, 2372.
62. N. Shan and W. Jones, *Green Chem.*, 2003, **5**, 728.
63. D. Cincic, T. Friscic and W. Jones, *J. Am. Chem. Soc.*, 2008, **130**, 7524.
64. D. Braga, F. Grepioni, L. Maini, D. Capucci, S. Nanna, J. Wouters, L. Aerts and L. Quere, *Chem. Commun.*, 2012, **48**, 8219.
65. T. J. Chiya and A. Lemmerer, *CrystEngComm*, 2012, **14**, 5124.
66. M. R. Chierotti, K. Gaglioti, R. Gobetto, D. Braga, F. Grepioni and L. Maini, *CrystEngComm*, 2013, **15**, 7598.
67. A. I. Ramos, T. M. Braga, P. Silva, J. A. Fernandes, P. Ribeiro-Claro, M. Silva Lopes, F. A. A. Paz and S. S. Braga, *CrystEngComm*, 2013, **15**, 2822.
68. E. A. Losev and E. V. Boldyreva, *CrystEngComm*, 2014, **16**, 3857.

69. K. Gaglioti, M. R. Chierotti, F. Grifasi, R. Gobetto, U. J. Griesser, D. Hasa and D. Voinovich, *CrystEngComm*, 2014, **16**, 8252.
70. F. Grepioni, J. Wouters, D. Braga, S. Nanna, B. Fours, G. Coquerel, G. Longfils, S. Rome, L. Aerts and L. Quere, *CrystEngComm*, 2014, **16**, 5887.
71. V. K. Srirambhatla, A. Kraft, S. Watt and A. V. Powell, *CrystEngComm*, 2014, **16**, 9979.
72. K. Shomono, K. Kadota, Y. Tozuka, A. Shimosaka, Y. Shirakawa and J. Hidaka, *Eur. J. Pharm. Sci.*, 2015, **76**, 217.
73. N. I. Nadzri, N. H. Sabri, V. S. Lee and N. S. A. Halim, *J. Chem. Crystallogr.*, 2016, **46**, 144.
74. H. Oertling, *CrystEngComm*, 2016, **18**, 1676.
75. O. Shemchuk, L. Degli Esposti, F. Grepioni and D. Braga, *CrystEngComm*, 2017, **19**, 6267.
76. S. M. T. Machado, R. A. E. Castro, T. M. R. Maria and J. Canotilho, *Int. J. Pharm.*, 2017, **533**, 1.
77. N. Tumanova, N. Tumanov, F. Fischer, F. Morelle, V. Ban, K. Robeyns, Y. Filinchuk, J. Wouters, F. Emmerling and T. Leyssens, *CrystEngComm*, 2018, **20**, 7308.
78. N. Tumanova, N. Tumanov, K. Robeyns, F. Fischer, L. Fusaro, F. Morelle, V. Ban, G. Hautier, Y. Filinchuk, J. Wouters, T. Leyssens and F. Emmerling, *Cryst. Growth Des.*, 2018, **18**, 954.
79. B. Cheirsilp and J. Rakmai, *Biol. Eng. Med.*, 2016, **2**, 1.
80. S. Pereva, T. Sarafska, S. Bogdanova and T. Spassov, *J. Drug Delivery Sci. Technol.*, 2016, **35**, 34.
81. A. A. L. Michalchuk, K. S. Hope, S. R. Kennedy, M. V. Blanco, E. V. Boldyreva and C. R. Pulham, *Chem. Commun.*, 2018, **54**, 4033.
82. T. Friščić, *J. Mater. Chem.*, 2010, **20**, 7599.
83. M. J. Muñoz-Batista, D. Rodriguez-Padron, A. R. Puente-Santiago and R. Luque, *ACS Sustainable Chem. Eng.*, 2018, **6**, 9530.
84. G.-J. Lee, E.-K. Park, S.-A. Yang, J.-J. Park, S.-D. Bu and M.-K. Lee, *Sci. Rep.*, 2017, **7**, 46241.
85. J. M. González-Domínguez, V. León, M. I. Lucío, M. Prato and E. Vázquez, *Nat. Protoc.*, 2018, **13**, 495.
86. T. Friščić, *Chem. Soc. Rev.*, 2012, **41**, 3493.
87. C. Xu, S. De, A. M. Balu, M. Ojeda and R. Luque, *Chem. Commun.*, 2015, **51**, 6698.
88. Z. Dong, R. Zhang, D. Ji, N. A. Chernova, K. Karki, S. Sallis, L. Piper and M. S. Whittingham, *Adv. Sci.*, 2016, **3**, 1500229.
89. C. Liu, F. Li, L.-P. Ma and H.-M. Cheng, *Adv. Mater.*, 2010, **22**, E28–E62.
90. L. Ouyang, Z. Cao, H. Wang, R. Hu and M. Zhu, *J. Alloys Compd.*, 2017, **691**, 422.
91. W. Ouyang, E. Kuna, A. Yepez, M. A. Balu, A. A. Romero, C. J. Colmenares and R. Luque, *Nanomaterials*, 2016, **6**, 93.
92. K. Ralphs, C. Hardacre and S. L. James, *Chem. Soc. Rev.*, 2013, **42**, 7701.
93. M. Klimakow, P. Klobes, A. F. Thünemann, K. Rademann and F. Emmerling, *Chem. Mater.*, 2010, **22**, 5216.

94. T. D. Bennett and A. K. Cheetham, *Acc. Chem. Res.*, 2014, **47**, 1555.
95. W. J. Rieter, K. M. L. Taylor, W. L. H. An and W. Lin, *J. Am. Chem. Soc.*, 2006, **128**, 9024.
96. M. Wilke, S. Bach, T. E. Gorelik, U. Kolb, W. Tremel and F. Emmerling, *Z. Kristallogr. - Cryst. Mater.*, 2017, **232**, 209.
97. A. C. McKinlay, R. E. Morris, P. Horcajada, G. Férey, R. Gref, P. Couvreur and C. Serre, *Angew. Chem., Int. Ed.*, 2010, **49**, 6260.
98. W. Chen and C. Wu, *Dalton Trans.*, 2018, **47**, 2114.
99. V. André and S. Quaresma, in *Metal-Organic Frameworks*, ed. F. Zafar, InTech, Croatia, 2016, pp. 135–156, DOI: 10.5772/64027.
100. S. Quaresma, V. André, A. Fernandes and M. T. Duarte, *Inorg. Chim. Acta*, 2017, **455**, 309.
101. V. André, F. Galego and M. Martins, *Cryst. Growth Des.*, 2018, **18**, 2067.
102. D. Braga, F. Grepioni, L. Maini, R. Brescello and L. Cotarca, *CrystEngComm*, 2008, **10**, 469.
103. E. H. H. Chow, F. C. Strobridge and T. Friščić, *Chem. Commun.*, 2010, **46**, 6368.
104. D. Braga, F. Grepioni, V. André and M. T. Duarte, *CrystEngComm*, 2009, **11**, 2618.
105. V. André, A. Hardeman, I. Halasz, R. S. Stein, G. J. Jackson, D. G. Reid, M. J. Duer, C. Curfs, M. T. Duarte and T. Friščić, *Angew. Chem., Int. Ed.*, 2011, **50**, 7858.
106. S. Quaresma, V. André, A. M. M. Antunes, L. Cunha-Silva and M. T. Duarte, *Cryst. Growth Des.*, 2013, **13**, 5007.
107. T. Kundu, S. Mitra, P. Patra, A. Goswami, D. Diaz Diaz and R. Banerjee, *Chem. - Eur. J.*, 2014, **20**, 10514.
108. M. Lupacchini, A. Mascitti, L. Tonucci, N. d'Alessandro, E. Colacino and C. Charnay, *ACS Sustainable Chem. Eng.*, 2018, **6**, 511.
109. M. J. Rak, N. K. Saadé, T. Friščić and A. Moores, *Green Chem.*, 2014, **16**, 86.
110. A. Pichon, A. Lazuen-Garay and S. L. James, *CrystEngComm*, 2006, **8**, 211.
111. F. C. Strobridge, N. Judaš and T. Friščić, *CrystEngComm*, 2010, **12**, 2409.
112. T. Friščić and L. Fábián, *CrystEngComm*, 2009, **11**, 743.
113. N. Nadizadeh, M. R. Naimi-Jamal and L. Panahi, *J. Solid State Chem.*, 2018, **259**, 35.
114. T. Stolar, L. Batzdorf, S. Lukin, D. Zilic, C. Motillo, T. Friscic, F. Emmerling, I. Halasz and K. Uzarevic, *Inorg. Chem.*, 2017, **56**, 6599.
115. I. B. Vasconcelos, T. G. da Silva, G. C. G. Militão, T. A. Soares, N. M. Rodrigues, M. O. Rodrigues, N. B. da Costa Jr, R. O. Freire and S. A. Junior, *RSC Adv.*, 2012, **2**, 9437.
116. A. K. Das, R. S. Vemuri, I. Kutnyakov, B. P. McGrail and R. K. Motkuri, *Sci. Rep.*, 2016, **6**, 28050.
117. S. S. Y. Chui, S. M. F. Lo, J. P. H. Charmant, A. G. Orpen and I. D. Williams, *Science*, 1999, **283**, 1148.
118. R. Babarao and J. Jiang, *J. Phys. Chem. C*, 2009, **113**, 18287.

119. T. P. Shakhtshneider, S. A. Myz, M. A. Dyakonova, V. V. Boldyrev, E. V. Boldyreva, A. I. Nizovskii, A. V. Kalinkin and R. Kumar, *Acta Phys. Pol., A*, 2011, **120**, 272.
120. R. A. D. Arancon, A. M. Balu, A. A. Romero, M. Ojeda, M. Gomez, J. Blanco, J. L. Domingo and R. Luque, *Environ. Res.*, 2017, **154**, 204.
121. N. Rubio, K.-C. Mei, R. Klippstein, P. M. Costa, N. Hodgins, J. T.-W. Tzu-Wen Wang, F. Festy, V. Abbate, R. C. Hider, K. L. A. Chan and K. T. Al-Jamal, *ACS Appl. Mater. Interfaces*, 2015, 7, 18920.
122. K.-C. Mei, Y. Guo, J. Bai, P. M. Costa, H. Kafa, A. Protti, R. C. Hider and K. T. Al-Jamal, *ACS Appl. Mater. Interfaces*, 2015, 7, 14176.
123. C. Orellana-Tavra, E. F. Baxter, T. Tian, T. D. Bennett, N. K. H. Slater, A. K. Cheetham and D. Fairen-Jimenez, *Chem. Commun.*, 2015, **51**, 13878.
124. P. J. Beldon, L. Fábián, R. S. Stein, A. Thirumurugan, A. K. Cheetham and T. Friščić, *Angew. Chem., Int. Ed.*, 2010, **49**, 9640.
125. A. D. Katsenis, A. Puškarić, V. Štrukil, C. Mottillo, P. A. Julien, K. Užarević, M.-H. Pham, T.-O. Do, S. A. J. Kimber, P. Lazić, O. Magdysyuk, R. E. Dinnebier, I. Halasz and T. Friščić, *Nat. Commun.*, 2015, **6**, 6662.
126. N. Liédana, A. Galve, C. Rubio, C. Téllez and J. Coronas, *ACS Appl. Mater. Interfaces*, 2012, **4**, 5016–5021.
127. H. Zheng, Y. Zhang, L. Liu, W. Wan, P. Guo, A. M. Nystrom and X. Zou, *J. Am. Chem. Soc.*, 2016, **138**, 962.
128. E. Shearier, P. Cheng, Z. Zhu, J. Bao, Y. H. Hu and F. Zhao, *RSC Adv.*, 2016, **6**, 4128.
129. A. D. Katsenis, A. Puškarić, V. Štrukil, C. Mottillo, P. A. Julien, K. Užarević, M. H. Pham, T. O. Do, S. A. J. Kimber, P. Lazić, O. Magdysyuk, R. E. Dinnebier, I. Halasz and T. Friščić, *Nat. Commun.*, 2015, **6**, 6662.
130. C. Orellana-Tavra, R. J. Marshall, E. F. Baxter, I. Abanades Lazaro, A. Tao, A. K. Cheetham, R. S. Forgan and D. Fairen-Jimenez, *J. Mater. Chem. B*, 2016, **4**, 7697.
131. M. Bellusci, P. Guglielmi, A. Masi, F. Padella, G. Singh, N. Yaacoub, D. Peddis and D. Secci, *Inorg. Chem.*, 2018, **57**, 1806.
132. D. Schmid, C. G. Park, C. A. Hartl, N. Subedi, A. N. Cartwright, R. B. Puerto, Y. Zheng, J. Maiarana, G. J. Freeman, K. W. Wucherpfennig, D. J. Irvine and M. S. Goldberg, *Nat. Commun.*, 2017, **8**, 1747.
133. S.-A. Yang, S. Choi, S. M. Jeon and J. Yu, *Sci. Rep.*, 2018, **8**, 185.
134. J. V. Jokerst, T. Lobovkina, R. N. Zare and S. S. Gambhir, *Nanomedicine*, 2011, **6**, 715.
135. B. Pelaz, P. del Pino, P. Maffre, R. Hartmann, M. Gallego, S. Rivera-Fernández, J. M. de la Fuente, G. U. Nienhaus and W. J. Parak, *ACS Nano*, 2015, **9**, 6996.
136. A. J. Raghavendra, K. Fritz, S. Fu, J. M. Brown, R. Podila and J. H. Shannahan, *Sci. Rep.*, 2017, 7, 8382.
137. Z. Bujnakova, M. Balaz, E. Dutkova, P. Balaz, M. Kello, G. Mojzisova, J. Mojzis, M. Vilkova, J. Imrich and M. Psotka, *J. Colloid Interface Sci.*, 2017, **486**, 97.

138. W. Zhang, X. Li, R. Zou, H. Wu, H. Shi, S. Yu and Y. Liu, *Sci. Rep.*, 2015, **5**, 11129.
139. D. Rodriguez-Padrón, A. R. Puente-Santiago, A. M. Balu, A. A. Romero and R. Luque, *Chem. Commun.*, 2017, **53**, 7635.
140. D. Rodríguez-Padrón, A. D. Jodlowski, G. de Miguel, A. R. Puente-Santiago, A. M. Balu and R. Luque, *Green Chem.*, 2018, **20**, 225.
141. D. Rodríguez-Padrón, A. R. Puente-Santiago, A. Caballero, A. M. Balu, A. A. Romero and R. Luque, *Nanoscale*, 2018, **10**, 3961.
142. T. P. Shakhtshneider and V. V. Boldyrev, in *Mechanochemical Synthesis and Mechanical Activation of Drugs*, ed. E. Boldyreva and V. Boldyrev, Reactivity of molecular solids, John Wiley & Sons, Chichester, UK, 1999, pp. 271–312.
143. I. A. Tumanov, A. F. Achkasov, E. V. Boldyreva and V. V. Boldyrev, *CrystEngComm*, 2011, **13**, 2213.
144. T. Friščić, I. Halasz, P. J. Beldon, A. M. Belenguer, F. Adams, S. A. J. Kimber, V. Honkimäki and R. E. Dinnebier, *Nat. Chem.*, 2013, **5**, 66.
145. I. Halasz, S. A. J. Kimber, P. J. Beldon, A. M. Belenguer, F. Adams, V. Honkimäki, R. C. Nightingale, R. E. Dinnebier and T. Friščić, *Nat. Protoc.*, 2013, **8**, 1718.
146. L. Batzdorf, F. Fischer, M. Wilke, K.-J. Wenzel and F. Emmerling, *Angew. Chem., Int. Ed.*, 2015, **54**, 1799.
147. N. T. Yen, K. Rademann and F. Emmerling, *CrystEngComm*, 2015, **17**, 9029.
148. H. Kulla, S. Greiser, S. Benemann, K. Rademann and F. Emmerling, *Cryst. Growth Des.*, 2017, **17**, 1190.
149. H. Kulla, S. Haferkamp, I. Akhmetova, M. Rollig, C. Maierhofer, K. Rademann and F. Emmerling, *Angew. Chem., Int. Ed.*, 2018, **57**, 5930.
150. E. Colacino, M. Carta, G. Pia, A. Porcheddu, P. C. Ricci and F. Delogu, *ACS Omega*, 2018, **3**, 9196.

CHAPTER 10

Microwave Dielectric Heating for Solvent-free Organic Transformations

ELENA CINI AND MAURIZIO TADDEI*

Dipartimento di Biotecnologie, Chimica e Farmacia, Università degli Studi di Siena, 53100, Siena, Italy
*E-mail: maurizio.taddei@unisi.it

10.1 Introduction

It is well known that the main features of microwave (MW) dielectric heating in organic synthesis are the high purity of reaction mixtures, increased product yields, low reaction times, and reduction of unwanted side products. The acceleration effects of MW heating were thought to arise from both thermal and specific MW effects.[1,2] Thermal effects are the consequence of internal frictions between polar molecules during changes in their orientation towards the MW the electric field; MW effects are related to specific electrostatic interactions of MWs with the reaction transition states.[3] Although Kappe has demonstrated that most MW-assisted chemical reactions proceed under simple thermal effects, some recent results have highlighted the possibility that electromagnetic effects (*i.e.*, non-thermal effects) might speed up certain processes.[4,5]

MW dielectric heating is effective with media having a sufficient dielectric loss tangent; however, the use of a solvent is not mandatory for heat transport. Indeed, the use of solvent-free conditions is especially propitious to

Green Chemistry Series No. 61
Green Synthetic Processes and Procedures
Edited by Roberto Ballini

Published by the Royal Society of Chemistry, www.rsc.org

MW activation as they provide an opportunity to avoid the risk of high pressure development with the possibility of carrying out reactions on a preparative scale simplifying manipulation.[6]

Solvent-free MW-assisted synthesis (sometimes used as the acronym SFMAS) represents an alternative strategy to reduce the E-factor and the environment impact of some chemical processes as many organic solvents are ecologically harmful.[7]

The three main general protocols to carry out solvent-free microwave-assisted reactions are:

1. Heating reactants adsorbed on solid supports such as alumina, silica, montmorillonite clay and zeolite;[8,9]
2. Heating reactant and reagents in the presence of phase-transfer catalysis (PTC) conditions;[10]
3. In neat reactions.[11–13]

In the first case, reactants are impregnated on the solid support and the contribution of the solid active surface is not limited to be a good reaction medium. In most cases, the support catalyses the reactions too.[14] Alumina, among other solid supports, is an interesting reagent as it can be modified to acts as an acidic or basic catalyst (additive). The main drawback of this strategy is that organic solvents are often required to adsorb the reagents on the support and to remove the product, decreasing the overall sustainability of the process. In the second technique, which is generally applied to anionic reactions such as alkylations, the reagents are mixed with a tetraalkylammonium salt as the phase-transfer catalyst. The third approach is the simple use of mixtures of the neat reagents, often ground and mixed before heating.

In this chapter, we report a roundup of the most interesting results achieved in solvent-free organic synthesis using focused MW irradiation, underlining the recent catalytic processes that limit or avoid the use of solvents in all the reaction steps including preparation, work-up and purification. The aim of this chapter is also to underline the improvements of SFMAS in classic organic transformations rather than for more complex structures synthesis and to collect different results and points of view from the previous reviews by Loupy[13,15–17] and Varma[18–24] and the more recent chapters dedicated to SFMAS.[25–33]

10.2 Alcohol Oxidation

One of the significant transformations of organic synthesis is the oxidation of alcohols to aldehydes or ketones, since the resultant functional groups are used in a variety of biologically interesting compounds or represent the starting point for more complicated organic transformations such as multicomponent reactions. A large number of oxidants have been developed to achieve this and SFMAS has been efficiently applied to a wide range of oxidation reactions under different conditions. Varma described the solvent-free oxidation of benzoins[34] using oxone on wet alumina under MW irradiation and, ten years later, an optimised procedure using NBS on Al_2O_3 was described by Khurana[35] on alcohols different than benzoins. Other neat microwave-assisted oxidation protocols involve the use of MnO_2 or CrO_3 in the presence of solid alumina, clay or silica.[19,21,36]

Despite the inherent efficiency of these methods, the development of catalytic processes employing more eco-friendly oxidants under controlled conditions has been recently explored. In 2016, Cu(II) and Co(II) tetrazole-saccharinate complexes were described as effective catalysts for the oxidation of secondary alcohols under solvent-free MW-assisted conditions in the presence of *tert*-butyl hydroperoxide (TBHP), an environmentally friendly oxidant.[37] One year later, a contribution of the same authors reported an organocatalysed oxidation of benzyl alcohols by a tetrazole-amino-saccharin in the presence of TBHP under SFMAS conditions (Scheme 10.1).[38]

This reaction proceeds through a radical mechanism, with the tetrazole-amino-saccharin assisting both the cleavage of the O–O bond in TBHP and the H-abstraction from the alcohol.

Another promising technology for environmentally friendly oxidations is the catalytic oxidation with an appropriate MW absorbent catalyst. Martins and co-workers prepared the series of first-row-transition-metal ferrite magnetic nanoparticles (NPs) MFe_2O_4 (M = Mn^{2+}, Fe^{2+}, Co^{2+}, Ni^{2+}, Cu^{2+} or Zn^{2+}) by a co-precipitation method.[39] Those NPs were used as catalysts (5%) for the MW-assisted oxidation of various alcohols in solvent-free medium using TBHP, with $MnFe_2O_4$, $CoFe_2O_4$ and $CuFe_2O_4$ affording the best results (Scheme 10.2).

The reaction mechanism involves again radical formation, accounted by reaction of TBHP with the metal ions on the surface of the NPs. It is noteworthy that these metal catalysts are stable and can be recovered after the reaction with unchanged composition.

The solvent-free microwave-assisted oxidative borrowing hydrogen process[40,41] will be discussed in Section 10.3.3.

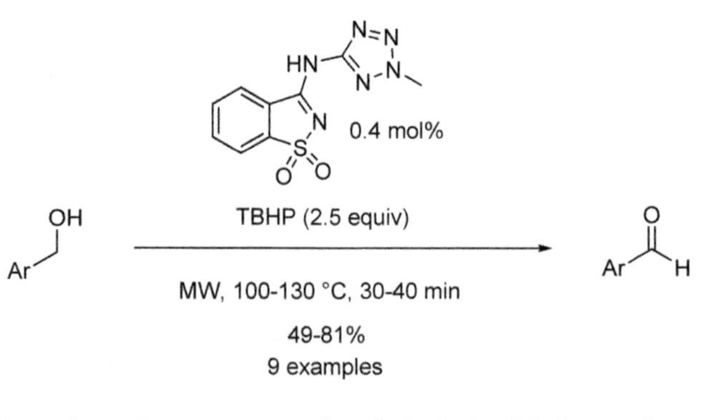

Scheme 10.1 Solvent-free organocatalysed alcohol oxidation under MW dielectric heating.

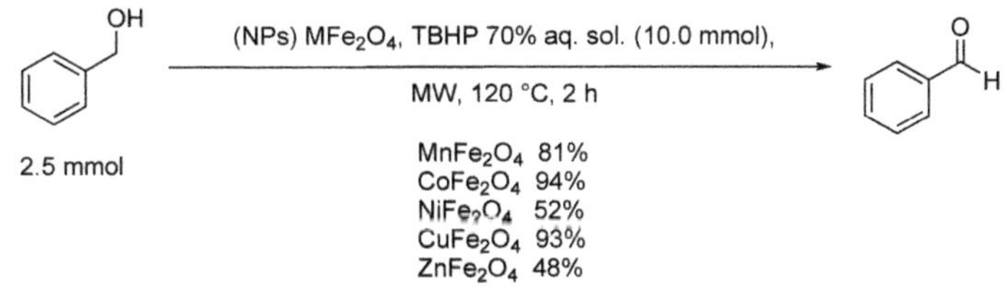

Scheme 10.2 Solvent-free microwave assisted alcohol oxidation using NPs MFe_2O_4 under MW dielectric heating.

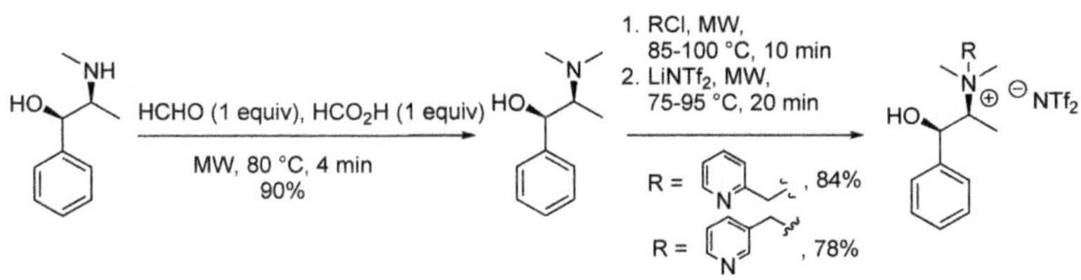

Scheme 10.3 SFMAS of chiral ionic liquids.

10.3 Alkylation Reactions

The increased efficiency in these transformations employing alkyl halides under MW irradiation was often attributed to specific effects stabilising polar transition states, as carbon–halogen covalent bond cleavage can be facilitated by an increase in the polarity of the system during the progress of the reaction between the ground and transition states.[42] The existence of non-thermal MW effects was recently revaluated by Kappe and co-workers for the quaternisation of phosphine with benzyl bromide; acceleration effects of MW heating were found to be caused by high temperature gradients and overheating of the reaction mixture rather than by stabilisation of the transition states.[43]

10.3.1 Alkylation with Alkyl Halides

Several heterocyclic and other substrates, such as pyrazols,[44] other azoles,[45] and also *N*-substituted amides and lactams,[46] are rapidly *N*-alkylated with alkyl halides under solvent-free phase-transfer catalytic conditions using MWs.

SFMAS was also effective in the quaternisation of 1-alkyl-imidazoles for the preparation of classic imidazolium-based ionic liquids (ILs)[47–50] and for the preparation of chiral ionic liquids (CILs) based on (1*R*,2*S*)-(−)-ephedrine (Scheme 10.3).[42,51]

More recently, a series of antibacterial molecules containing a tetrahydroisoquinoline nucleus and a glycinamide linker were prepared by alkylating the corresponding chloroacetyl precursor under neat MW-assisted

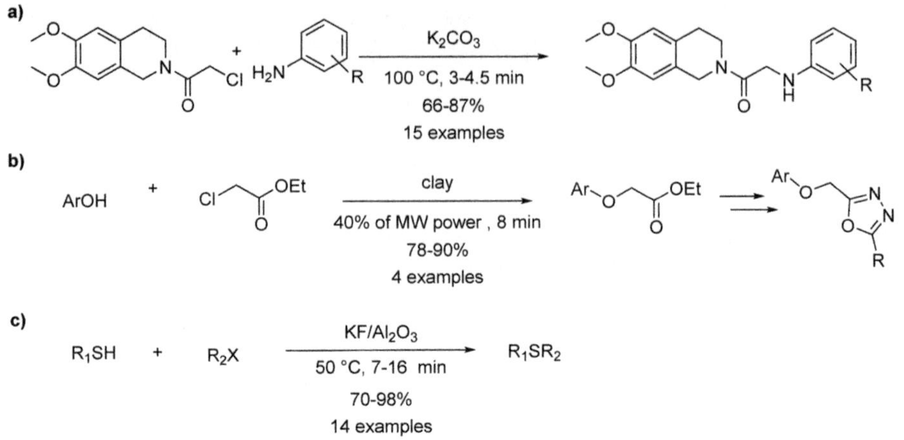

Scheme 10.4 Solvent-free alkylation of tetrahydroisoquinolines (a), phenols (b) and thiols (c) under MW dielectric heating.

conditions.[52] Equimolar amounts of starting materials were mixed without a phase-transfer catalyst and reacted for 3–4.5 min at 100 °C. At the end of the reaction, the product was precipitated with water from the crude reaction mixture and no further purification was required while, under convectional heating, the reaction afforded the products in less than 40% yield (Scheme 10.4a).

SFMAS was also effective in the base-free alkylation of phenols with ethylchloroacetate using clay as the solid support[53] as the key step for the synthesis of substituted 1,3,4-oxadiazoles (Scheme 10.4b) and in the alkylation of either aromatic and aliphatic thiols with several alkyl halides in the presence of KF/Al_2O_3[54] (Scheme 10.4c). In this last contribution, comparative experiments under conventional heating and under ultrasound irradiation as an alternative energy source resulted in lower yields and longer reaction times, demonstrating the crucial effect of the MWs on the reaction.

10.3.2 Alkylation by Ring-opening Reactions

The regioselective ring-opening reactions of strained cycles such as aziridines and epoxides by *N*- and *O*-nucleophiles or by carbanionic species is a valuable alternative to achieve highly functionalised *N*-, *O*- and *C*-alkylated compounds preventing the use of halides.

Alkylation of long chain alcohols with 1,1,1-trifluoro-2,3-epoxypropane[55] was successful under SFAMS conditions and an analogous ring opening of Cbz-aziridines[56] with benzylamine has also been described (Scheme 10.5). In that last article, the authors underlined the importance of using a solid support in the regioselectivity of the reaction, being bentonite the most suitable when compared to alumina or silica.

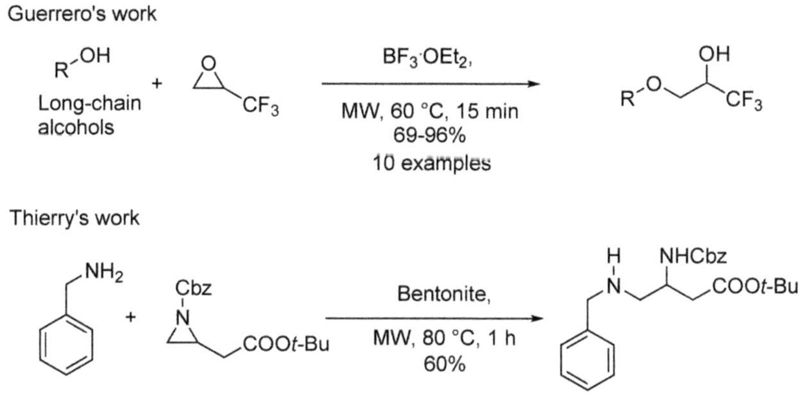

Scheme 10.5 Solvent-free ring opening reactions under MW dielectric heating.

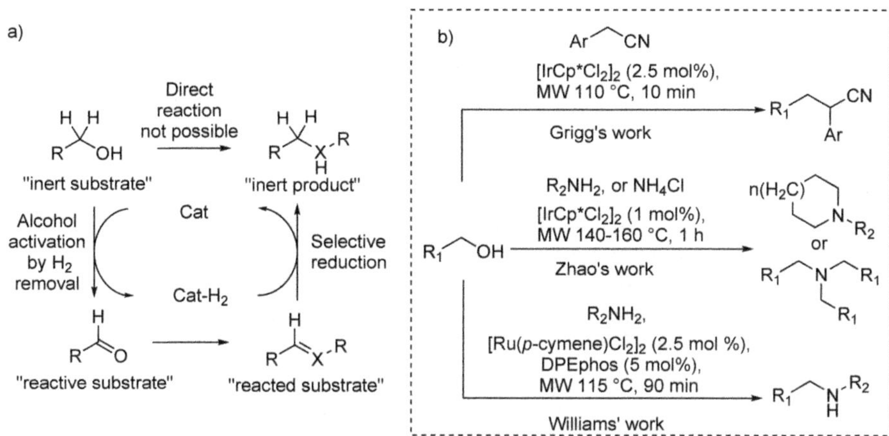

Scheme 10.6 (a) General scheme of the BH mechanism. (b) Application of SFMAS to BH reactions.

10.3.3 Alkylation with Alcohols

Direct catalytic alkylation with alcohols using a borrowing hydrogen (BH) protocol, an environmentally friendly and atom economical solution, has emerged as the most promising alternative to alkyl halides.[40,41] This procedure overcomes the use of stoichiometric reducing agents or toxic sacrificial nucleophiles and usually generates water as the only waste product.[57]

Alcohols are temporarily converted into the corresponding carbonyl compounds by the metal-catalysed removal of hydrogen. The so-formed aldehyde or ketone can undergo a wide range of transformations to give *in situ* reactive substrates such as imines, alkenes and α-functionalised carbonyl compounds; then, the metal returns the hydrogen to the carbonyl intermediate affording the saturated final product (Scheme 10.6a).

MW heating enables this strategy without the need for a solvent, while reducing the reaction times. The first *C*- and *N*-alkylation reactions under SFMAS BH conditions were described by Grigg, Zhao and Williams; whose seminal work is summarised in Scheme 10.6b.[58–60]

The reaction mechanism involves three key sequential steps: (i) dehydrogenation of the primary alcohol to the corresponding aldehyde and a metal hydride, (ii) Knoevenagel condensation (Grigg's work) or nucleophilic attack of the amine (Williams' and Zhao's work) on the *in situ* formed aldehyde, and (iii) hydrogenation of the unsaturated intermediate (arylacrylonitrile or imine) by the metal hydride to afford the monoalkylated arylacetonitrile or amine.

William's protocols was found also suitable for the base-free alkylation of amides and sulphonamides in the absence of a base; similar results in the alkylation of amides were achieved by Trudell and co-workers with $(Cp^*IrCl_2)_2$ as the catalyst.[61]

Fe^{III}(Pc)]Cl (Pc: phthalocyaninato), an inexpensive catalyst typically used as an industrial additive for ink and rubber manufacturing, was also reported to display good catalytic activity in the direct SFMAS of *N*-alkylated amines with benzylic alcohols in the presence of *t*BuOK.[62] The authors proposed a BH-type catalytic cycle but they also underlined that the mechanism was not clear.

Direct alkylation with alcohols can also proceed through a different mechanism depending on the substrate and/or the catalyst. When benzylic/allylic or secondary alcohols were used as alkylating agents, Lewis acid or transition metal complex-catalysed dehydrative formation of a carbocation or a coordinated allylic-metal complex may proceed as the key step in the alkylation mechanism rather than a borrowing hydrogen event.[63]

In 2009, the MW-assisted alkylation of 1,3-dicarbonyls in the presence of $FeCl_3/SiO_2$ using alcohols directly without previous transformation into the corresponding halides, carboxylates, carbonates, phosphonates or related compounds was reported (Scheme 10.7).[64]

This method was restricted to primary/secondary benzyl alcohols and allylic alcohols such as cinnamyl alcohol, while aliphatic alcohols did not react under these conditions, confirming a Friedel–Craft-like mechanism. However, this is a big improvement in the search for better catalysts for

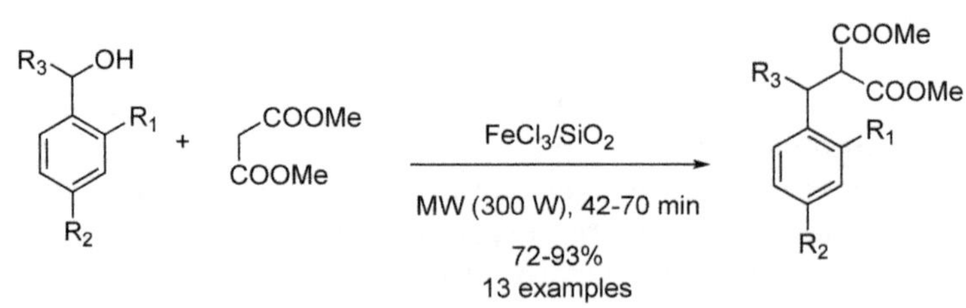

Scheme 10.7 Solvent-free Lewis acid-catalysed alkylation of dimethylmalonate with benzylic alcohols under MW dielectric heating.

carbon–carbon bond formation in terms of operational simplicity, economic viability and applicability at large scale. It is noteworthy that the reaction was carried out simply by grinding a $FeCl_3/SiO_2$ (10 mol%) mixture with the substrates, dimethyl malonate and benzyl alcohol, before MW dielectric heating. After completion, a simple extraction with environmentally risky diethyl ether provided the products; no need for chromatography was mentioned. Finally, one of the latest contribution in this field reported the MW-assisted alkylation of *m*-cresol with isopropyl alcohol (IPA) catalysed by strong acid resins such as CT-151 and Nafion NR-50 without any solvent, a challenging procedure to obtain mono-alkylphenols.[65]

10.4 Application of SFMAS to Carbonyl Chemistry

10.4.1 Imine Synthesis

Aldehydes and ketones typically react with amines to form imines. Imine synthesis represents a nice example of atom economy producing water as the only side product of the reaction. Unfortunately, this functional group transformation often requires long reaction times and refluxing acid conditions. SFMAS has successfully shifted this transformation towards more sustainable conditions. In one of the earliest protocols described, a mixture 3 : 1 of salicylaldehyde and several aryl amines were heated in a MW domestic oven (power 616–800 W) for 4 min to give the corresponding imine without the need of any acid catalyst (it is likely that the phenol group acts as the acid catalyst).[66]

Ferrocene-derived Schiff bases were analogously prepared without any water-trapping additive using a 1 : 1 ratio of reactants in a focused MW system Prolabo Synthewave1 402 (reaction conditions 120 °C, 3 min). The protocol allowed the purification of the products by simple recrystallisation from the crude reaction mixture.[67,68]

Porcheddu and co-workers developed in 2015 the SFMAS of arylimines based on a hydrogen-transfer (HT)[41,69] methodology using a Ru-based catalyst in the presence of crotononitrile as a sacrificial hydrogen acceptor.[70] After a screening study of several catalysts, $RuH_2CO(PPh_3)_3$ was selected; moreover, the use of Xantphos as the ligand led to a big improvement of the yield (Scheme 10.8). Interestingly, once total conversion was achieved, the addition of a second alcohol in the presence of further crotononitrile, acting as a hydrogen acceptor, and catalytic amounts of TFA under MW dielectric heating (130 °C, 3 h) afforded the corresponding quinolines in good yield (Scheme 10.8)

The aliphatic alcohol undergoes a ruthenium-assisted oxidative dehydrogenation reaction, as benzyl alcohol did in the previous step. The nucleophilic attack of the corresponding enol to the protonated imine, followed by hetero-annulation and aromatisation, are the final key steps of the proposed mechanism. A large variety of aromatic alcohols are amenable to this protocol, but the reaction is sensitive to electronic and steric effects.

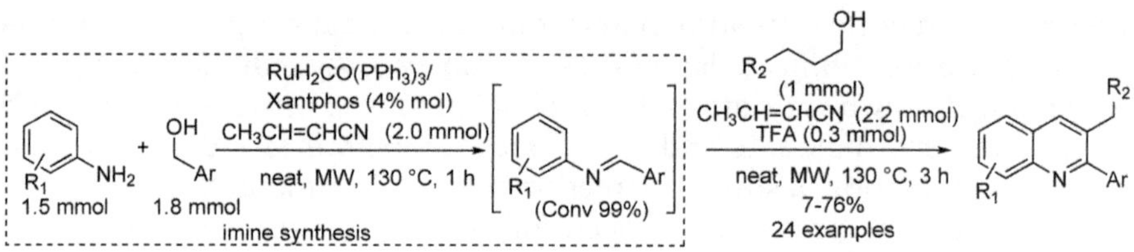

Scheme 10.8 Application of HT-based imine synthesis to the telescopic SFMAS of quinolines.

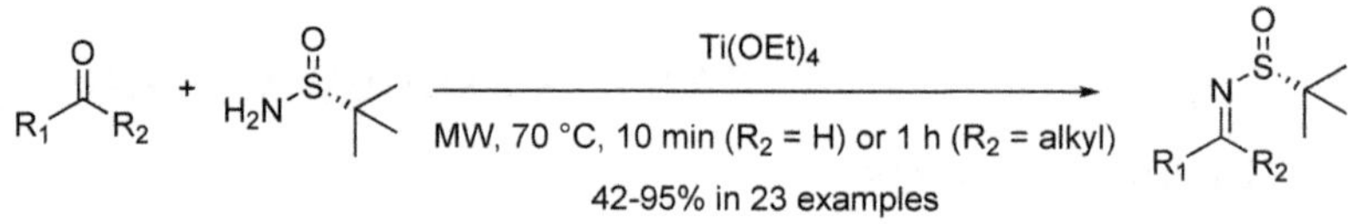

Scheme 10.9 SFMAS of enantiomerically pure *N*-(*tert*-butylsulphinyl)imines.

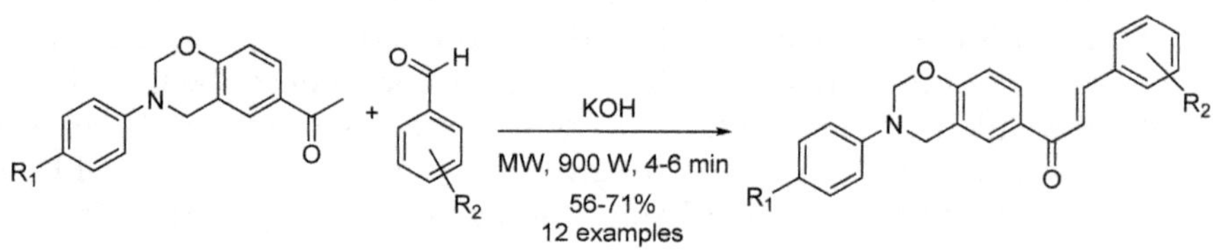

Scheme 10.10 SFMAS of 3,4-dihydro-2*H*-1,3-benzoxazines.

Chiral sulphinylimines were prepared in enantiomerically pure form under SFMAS conditions using $Ti(OEt)_4$ as the activator and water scavenger.[71] Both *N*-(*tert*-butylsulphinyl)aldimine and (*tert*-butylsulphinyl)ketamine were synthesised at 70 °C and isolated in pure form without the need for chromatography purification (Scheme 10.9). In both cases, control experiments under convectional heating afforded worse results.

This protocol tolerates other substituents different from the *tert*-butyl group on the sulphur atom. *N*-(*p*-Tolylsulphinyl)imines, very good substrates for asymmetric synthesis, can also be prepared following this procedure.[72]

10.4.2 Claisen–Schmidt and Knoevenagel Reactions

The Claisen–Schmidt condensation is one of the most useful reactions in organic chemistry. A representative class of chalcones was prepared by such Claisen–Schmidt reaction, some of which were reported to exhibit anti-TB[73] or antitumoural[74] activity. An effective neat procedure to prepare such derivatives under MW dielectric heating was described by Akhter and co-workers in 2011 (Scheme 10.10).[75]

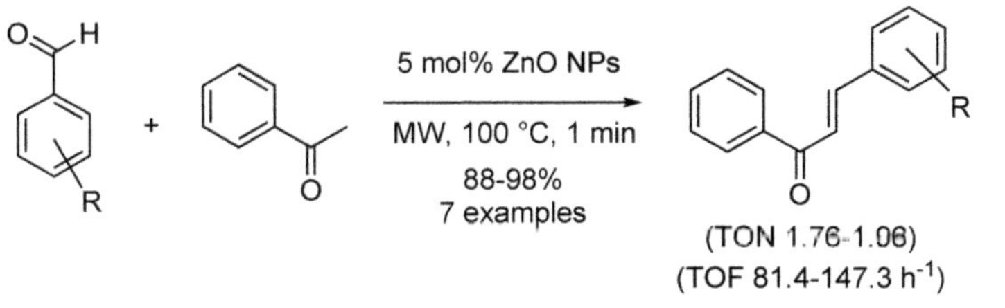

Scheme 10.11 Solvent-free ZnO NPs-catalysed Claisen–Schmidt condensation under MW dielectric heating.

In this protocol, acetophenones, KOH and aromatic aldehydes are first ground and then subjected to MW dielectric heating with four pulses of 1–1.5 min at 900 W in a domestic oven, as suggested by the absence of a recorded temperature. After completion of the reaction, crushed ice was added to the reaction mixture and the products precipitated by acidification with dilute HCl.

In 2015, Tamuly and co-workers published a microwave-assisted Claisen–Schmidt condensation neat protocol based on the use of bio-derived ZnO NPs. This eco-friendly nanocatalyst was prepared from the peel of *Musa balbisiana*, a common medicinal plant, and it was described to convert acetophenone and several benzaldehydes (bearing both electron withdrawing and electron donating groups, EWGs and EDGs, respectively) into the corresponding chalcones in only 1 min at 100 °C without any additive, solvent or base (Scheme 10.11).[76]

After filtration from the crude reaction mixture, the ZnO NPs could be recovered and, after washing with hot water/ethanol to remove any adsorbed products, reused up to five times with the same catalytic performance.

Other SFMAS of chalcones *via* Claisen–Schmidt reactions and other examples of crossed aldol condensations were achieved with NaOH–Al_2O_3,[77] H_2SO_4–SiO_2,[78] NH_4Cl–Al_2O_3[79] and neutral Al_2O_3.[80] In all these examples, convectional heating was reported to give worse results in terms of yield and reaction time; the same beneficial accelerating effect of SFMAS conditions was also observed for the Knoevenagel and Doebner condensations.[81–88]

Recently, a green alternative to several condensation protocols has been developed using a low cost and environmentally friendly catalyst such as boric acid. Even though other boric derivatives such as phenyl boronic acid display stronger Lewis acid behaviour, boric acid has successfully coupled several aldehydes to activated methylene compounds under neat MW-assisted conditions including substrates with acid or base-sensitive functional groups (Scheme 10.12).[89]

This method is suitable also for thiohydantoin, hydantoin and rhodanine derivatives, which are interesting scaffolds for drug discovery. The general procedure is noteworthy for its simplicity; the mixture of the three reagents is irradiated under optimised conditions and, at the end of the reaction, the

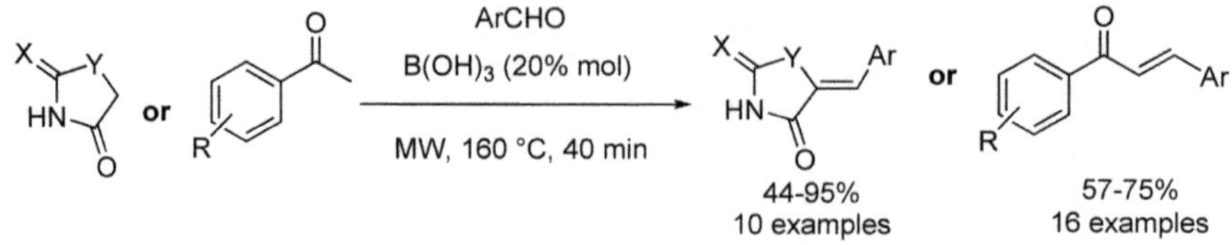

Scheme 10.12 Solvent-free boric acid-catalysed condensation of activated methylene compounds and aldehydes under MW dielectric heating.

product is precipitated by addition of EtOH or H_2O/EtOH without the need for chromatography to obtain pure products. The proposed mechanism involves the formation of a boron enolate that attacks the aldehyde, affording a six-membered ring adduct that undergoes a final dehydration through intramolecular coordination of the boron.

10.4.3 Other Nucleophilic Additions on Aldehydes and Ketones

SFMAS has been efficiently applied also to Wittig reactions, another standard methodology for the synthesis of alkenes, and many of the recent results on this topic have been reported in a recently published review.[29] Several other nucleophilic additions to carbonyl compounds have been realised by SFMAS, such as the synthesis of thio analogues of ketones using Lawesson's reagent,[90,91] the Lewis acid-free cyanosilylation of aldehydes and ketones[11] and the synthesis of bis(aryl)methanes.[92–94]

10.4.4 Conjugate Addition: Michael and Related Reactions

Conjugate addition to α,β-unsaturated carbonyl compounds and analogous substrates represents a widely applied organic transformation; however, depending on the nature of the nucleophile, it often requires harsh reaction conditions, long reaction times, and the use of toxic reagents and solvents. Big improvements in the success and sustainability have been achieved by MW dielectric heating under solvent-free conditions.

10.4.4.1 C-Nucleophiles

An environmentally benign approach for the Michael addition of diethyl malonate to α,β-unsaturated carbonyl compounds was described by Shen and co-workers in 2011.[95] The solvent-free MW-assisted basic reaction on ferrocenyl-substituted chalcones proceeds rapidly, whereas up to several days were required for this transformation without MWs. After screening several bases, such as K_2CO_3, NaOH or KF, on alumina, the use of KOH resulted in 57–70% yield of the desired products in only 1–5 min (MW power 90–180 W).

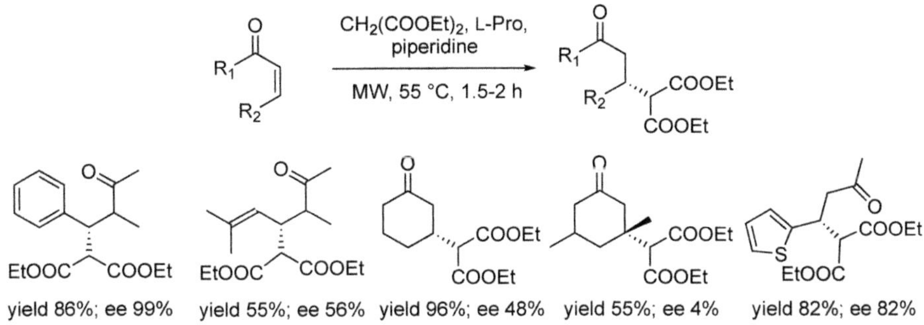

Scheme 10.13 Solvent-free organocatalytic Michael reaction under MW dielectric heating.

A remarkable eco-friendly enantioselective Michael reaction protocol based on organocatalysis was developed by Procopio and co-workers.[7] Starting from different 1-alkyl-3-monosubstituted enones and in the presence of 15 mol% of L-proline, diethyl malonate afforded the corresponding optically enriched Michael adducts after MW dielectric heating at 55 °C for 1.5 h in the presence of piperidine as the base. Although the ee's were strongly influenced by the enone substituents, this method represents a big improvement over protocols employing convectional heating (Scheme 10.13).

The authors highlighted that the use of solvents such as THF, DMSO, CH_2Cl_2 or CH_3CN afforded worse results than those of comparative experiments performed under oil bath heating.

SFMAS was also effective with other *C*-nucleophiles such as nitromethane[96,97] and TMSCN[98] It is noteworthy that the addition of TMSCN to chalcones, 3-nonen-2-one, and benzalacetone under SFMAS conditions resulted in the regioselective formation of β-cyanoketones without the need for a Lewis acid catalyst, often required to prevent 1,2-addition.[98]

10.4.4.2 N-*Nucleophiles*

Several anilines[99] and azaheterocycles[100] including nucleosides[101] have been successfully used as nucleophiles on acrylate derivatives in the presence of CH_3COOH,[99] DABCO and TBAB[100] or KF/Al_2O_3 [101] under SFMAS conditions. SFMAS was effective also in the synthesis of *N*-(2-(pyridin-2-yl)ethyl)sulphonamides by a similar 1,4-aza-conjugate addition between sulphonamides and 2-vinylpyridines (Scheme 10.14).[102]

This alternative to classic Michael reactions requires neat mixing of the two starting materials (1 : 1.5 ratio of vinylpyridine/sulphonamide) without any adjuvant, reagent or solvent. After MW dielectric heating for 45 min at 170 °C, a streamlined work-up procedure afforded the pure products without the need for chromatography. This protocol may be applied to 2- and 4-vinylpyridines, while 3-vinylpiridines failed completely to react, thus proving the conjugate addition-based mechanism of this transformation.

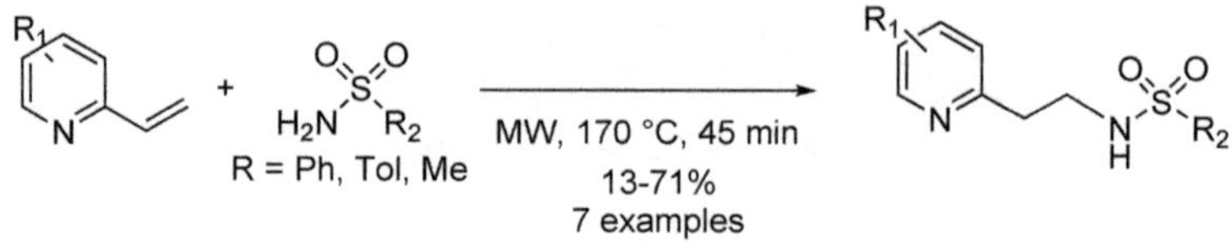

Scheme 10.14 Solvent-free microwave assisted 1,4-conjugate addition of sulphonamides to 2-vinylpyridines under MW dielectric heating.

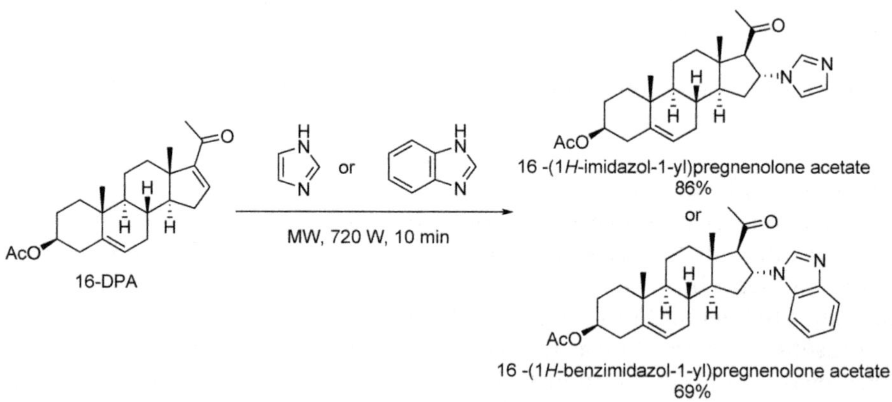

Scheme 10.15 Solvent-free 1,4-conjugate addition to steroid frameworks under MW dielectric heating.

Moreover, SFMAS was used for the preparation of 16α-(1*H*-imidazol-1-yl) pregnenolone acetate and 16α-(1*H*-benzimidazol-1-yl)pregnenolone acetate by catalyst-free Michael addition to the α,β-unsaturated precursor (16-DPA).[103] The authors highlighted that the same reaction under convectional heating (8 h at 140 °C) afforded the products with less than 40% yield, while the use of aprotic solvents such as CH_3CN and DMF resulted in no addition (Scheme 10.15). This reaction is completely selective giving the *trans*-isomers as the only products.

10.4.4.3 Phospha–Michael Reaction

Stockland, Jr and co-workers optimised the addition of diarylphosphine oxides on several Michael acceptors such as chalcones, cinnamates, crotonates, and other internal alkenes by SFMAS, overcoming the use of a radical initiator, strong base, or metal catalyst (Scheme 10.16).[104]

MW dielectric heating of the ground reagent mixture simply gave high yields of the products. The reaction times and temperatures depend on the alkene substituents; highly hindered olefins need harsher reaction conditions (up to 60 min at 200 °C), while alkenes containing multiple activating groups (such as malononitrile derivatives) lead to good results in shorter times at lower temperatures when compared to oil bath heating methods.

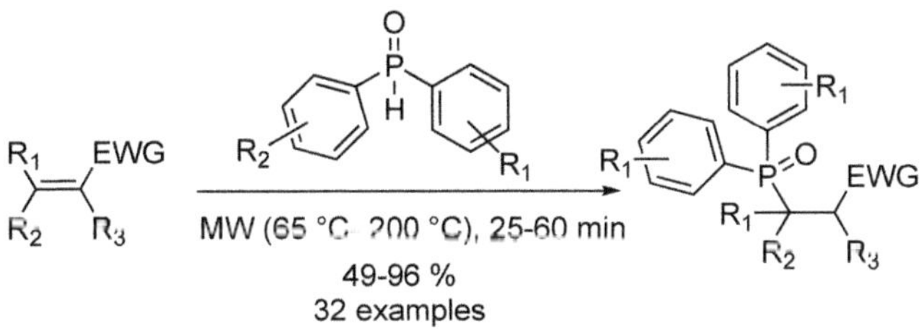

Scheme 10.16 Solvent-free phospha-Michael reaction under MW dielectric heating.

10.4.4.4 O-Nucleophiles

Oxa-Michael reactions have several drawbacks due to the nucleophile low reactivity and the high rate of retro-Michael reactivity. SFAMS has been described as one of the possible approaches to overcome these limits. For example, a MW-assisted intramolecular oxa-Michael addition in the presence of K_2CO_3 was described as an efficient and environmentally friendly protocol for the synthesis of flavanones starting either from 2′-hydroxychalcones or from 2′-hydroxyacetophenones in good yield in 3–5 min.[105]

10.4.5 Mannich and Mannich-type Reactions

Several researchers have targeted the development of more sustainable Mannich reactions. Amongst the last results in this field, the $CeCl_3$-catalysed SFMAS of β-aminoketones from cyclohexanone-substituted aldehydes and primary amines[106] gave better results respect to the previously described reactions of primary or secondary amides, aldehydes and amines in the presence of basic alumina as an amidic nitrogen activator.[107]

Moreover, several SFMAS protocols for the Betti reaction have been developed. Chavan and co-workers described in 2010 the SFMAS of 1-amidoalkyl-2-naphthols from β-naphthols, aromatic aldehydes and urea or primary amides in the presence of $InCl_3$,[108] while in 2014 the same neat transformation was achieved using nano-Fe_3O_4-modified carbon nanotubes as the catalyst.[109]

The Kabchnick–Field (also known as phospha-Mannich reaction) is the condensation of primary or secondary amines, carbonyl compounds and dialkyl phosphite compounds to generate α-aminophosphonate with the involvement of an imine intermediate. SFMAS is the best of choice for this transformation, as proven by a review published in 2012 on this topic.[110]

The following efforts in the development of new environmentally benign protocols for the Kabchnick–Field reaction have led to new contributions (Scheme 10.17).[111,112]

In Keglevich's work, the SFMAS of α-aminophosphinates starting for alkyl phenyl-*H*-phosphinates, paraformaldehyde and primary or secondary amines was explored. The outcome of the reaction depends on the substrate

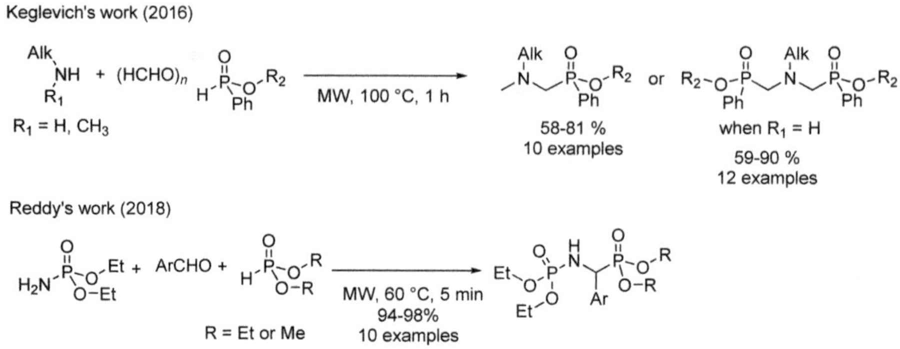

Scheme 10.17 Solvent-free Kabachnik–Fields reaction under MW dielectric heating.

and stoichiometric ratios of the reagents. With secondary amines and 1 equiv. of paraformaldehyde, the formation of alkyl monoaminomethyl-phosphinates was achieved while, starting from primary amines in the presence of excess paraformaldehyde, the major product was the corresponding bis-aminomethyl-phosphinate.[101]

In Reddy's work, the Kabachnik–Fields reaction was employed to generate a series of antimicrobial compounds based on the dialkoxyphopsphorylmethyl phosphoramidates skeleton. Unfortunately, in both the cases, chromatography was necessary to afford the pure final products.[102]

Finally, a neat Petatesis reaction, also referred as a boronic-acid Mannich reaction, was successfully realised by irradiating a 1:1:1 mixture of the appropriate boronic acid, salicylaldehyde and the respective amine in a sealed tube at 120 °C for 2 h.[113]

10.5 Application of SFMAS to the Multicomponent Synthesis of Heterocycles

Multicomponent reactions (MCRs) represent an attractive approach to sustainable chemicals. The requirements of saving solvents (a single purification and isolation step is necessary) and energy (as unnecessary steps such as protection–deprotection reactions can be avoided) can be achieved with MCRs.[108,114]

These reactions are often promoted by MW heating. The employment of neat conditions, often in the presence of solid supports, provides further outstanding benefits in the sustainability of MCRs, as testified by the large number of articles on SFMA-MCRs.[14]

One of the most popular MCRs is the Biginelli reaction (BR), in which an aldehyde, a ß-ketoester or a diketone and urea (or thiourea) react under acid conditions to give dihydropyrimidones, interesting compounds with a wide range of biological activities. Several BR protocols by SFMAS have been developed differing mainly in the catalyst. Alumina sulphuric acid

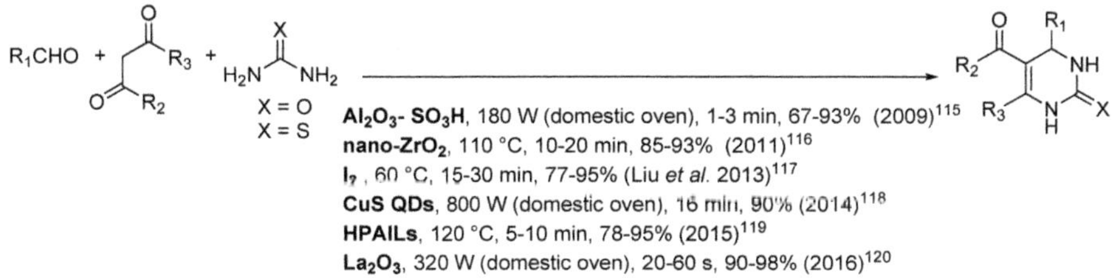

Scheme 10.18 Recent protocols for solvent-free Biginelli reactions under MW dielectric heating.

(Al_2O_3-SO_3H, an environmentally friendly heterogeneous recyclable catalyst),[115] nanocrystalline zirconium(IV) oxide (nc-ZrO_2, synthesised by a low temperature eco-friendly combustion method using alanine as a new organic fuel),[116] iodine,[117] recyclable CuS quantum dots (QDs),[118] HIPALs (hyteropolyanion-based ionic liquids),[119] and even lanthanum oxide[120] are just some of the catalyst used in these recently improved Biginelli protocols (Scheme 10.18).

Several other heterocycles have been synthesised by SFMA-MCRs demonstrating the beneficial effects of MWs in this kind of transformations, most of them affording the desired heterocycles in short times and high purity without chromatography. A remarkable example is the four-component synthesis of heavily substituted pyrroles catalysed by *p*-toluenesulphonic acid-doped polystyrene (PS-PTSA) (Scheme 10.19).[121]

The role of PS-PTSA is to accelerate the formation of the β-enamine carbonyl compound between **a** and **c**, the nitrostyrene from **b** and **d**, and the final cyclisation with elimination of nitrous acid. The reaction requires simple irradiation of the reagent/catalyst mixture and simple filtration with ethyl acetate at the end of the reaction, with consequent recovery of the catalyst that can be reused up to seven times. Unfortunately, column chromatography is often necessary to purify the final products.

The synthesis of a library of 1,2,3-triazoloimines was described starting from Si/Ge-substituted propynals, trimethylsilylazide and primary amines under catalyst-free SFMAS conditions.[122]

Fused-imidazo heterocycles have been prepared *via* the Goebke–Bienayme–Blackurn reaction (Scheme 10.20).[123] This methodology employs $Yb(OTf)_3$ as a reusable water-soluble catalyst and, after a short reaction time (5 min), the products were obtained by simple trituration with pentane followed by washing with water.

A possible mechanism for this reaction was proposed, based on the coordination of Yb-Lewis acid with the two nitrogens of the arylamine and the carbonyl of aldehydes, resulting in the formation of an arylimine that undergoes a cycloaddition reaction with the isonitrile. A similar synthetic strategy, published by these authors, affords furane-fused chromene-ones or quinolin-ones.[124]

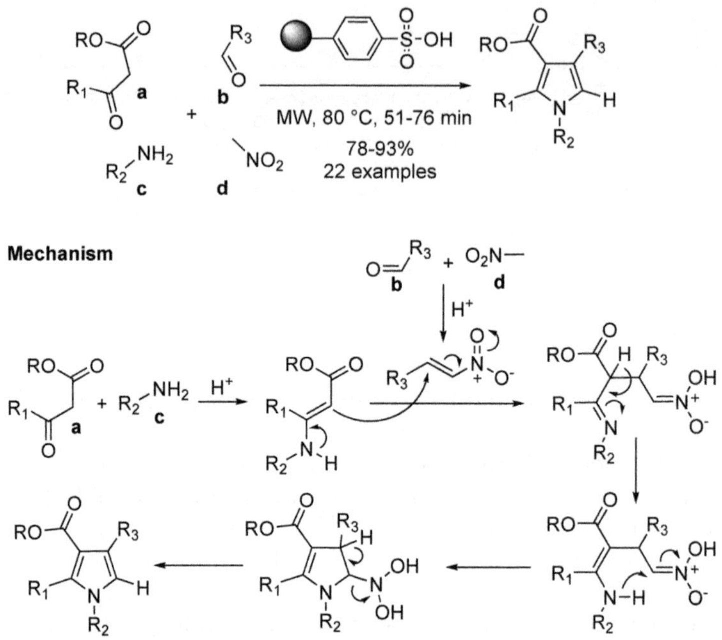

Scheme 10.19 SFMAS of heavily substituted pyrroles.

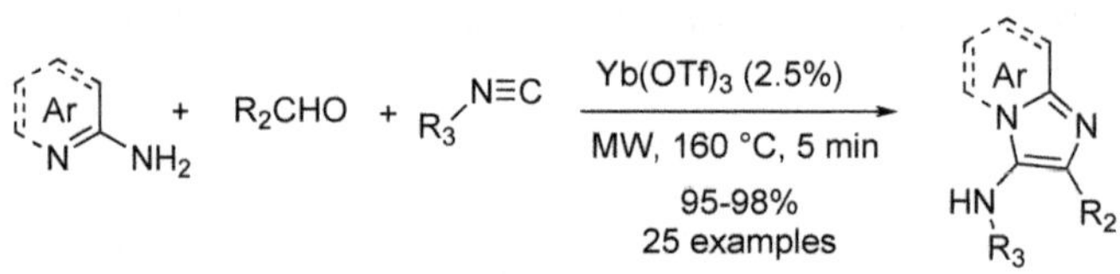

Scheme 10.20 SFMAS of fused-imidazo heterocycles.

The use of rare-earth metal triflates was effective in other SFMA-MCRs, where $Sc(OTf)_3$ catalysed also an environmentally benign protocol affording benzimidazolyl-imidazo[1,2-*a*]-pyridines.[125]

Among six-membered ring heterocycles synthesis, a recently published study described cation-exchanged montmorillonites as efficient and reusable catalyst for the solvent-free Hantzsch synthesis of dihydropyridines under MW irradiation.[126]

Boruah described the $Pd(OAc)_2$-catalysed annulation of β-bromovinyl aldehyde **a** or **b** (Scheme 10.21) with a substituted acetylene and benzylamine in the presence of PPh_3, Na_2CO_3 and neutral alumina. In this protocol, the reagents after vigorous mixing were irradiated in a closed vessel at 140 °C for 10 min (Scheme 10.21).[127]

The proposed mechanism involves the formation of an organopalladium imine intermediate between the vinylic halide, palladium and benzylamine; this complex reacts with acetylene, with palladium adding to the more

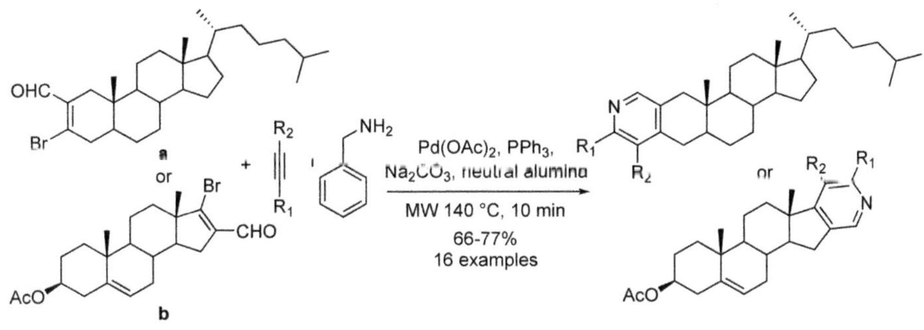

Scheme 10.21 SFMAS of steroidal A- and D-ring fused pyridines.

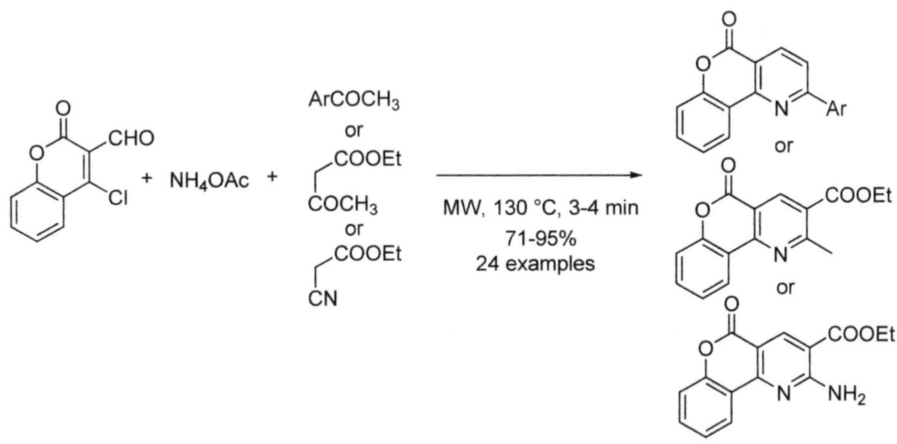

Scheme 10.22 SFMAS of pyrido [3,2-*c*]coumarins.

hindered end of the alkyne forming a seven-membered palladacyclic ammonium salt. The following elimination leads to Pd(0) and the intermediate benzylpiridinium, which is readily transformed into pyridine and benzyl halide. This reaction is also effective with non-steroidal β-bromovinyl aldehydes.

Bhuyan[128] and co-workers developed the SFMAS of pyrido [3,2-*c*]coumarins; differently substituted products were obtained depending on the substrate (Scheme 10.22) as a consequence of slightly different mechanism pathways.[128]

Convectional heating comparative experiments either in the presence or absence of a solvent afforded the desired products in less than 60% yield.

Despite the unusual combination of heating and stereoselectivity, an environmentally benign procedure for the stereoselective catalyst-free synthesis of 1,4-pyranonaphthoquinones by sequential four-component reaction was developed in 2012 (Scheme 10.23).[129]

The reaction is performed with a telescopic protocol: first a mixture of arylamine and the acetylene are grinded for 30 s at room temperature to form the

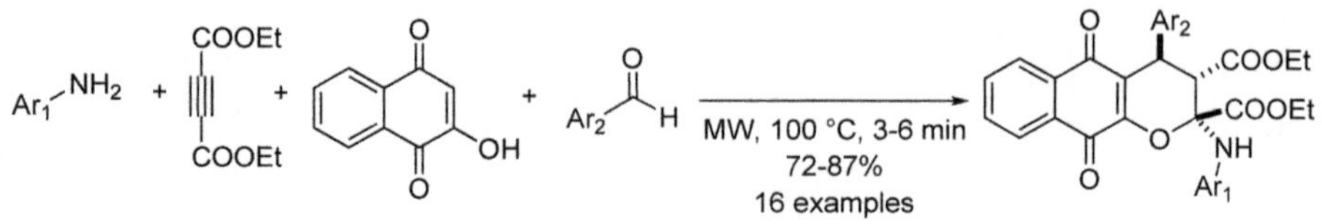

Scheme 10.23 SFMAS of 1,4-pyranonaphthoquinones.

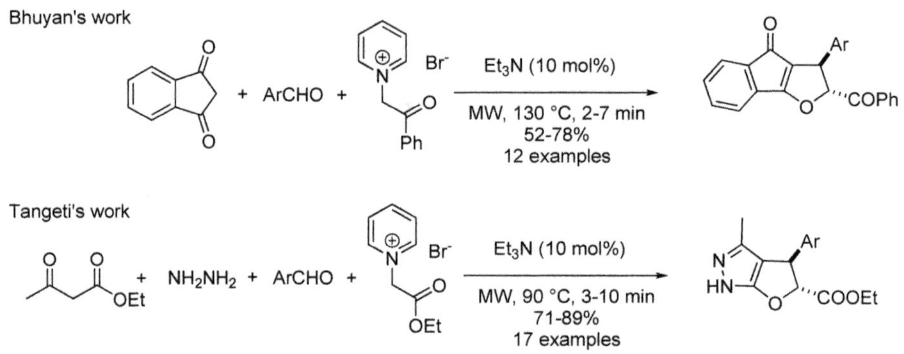

Scheme 10.24 SFMAS of fused heterocycles.

corresponding Michael adduct, then the 2-hydroxynaphthalene-1,4-dione and aldehyde are added, and the resulting mixture is irradiated for 3 min at 100 °C to afford the final products. The relative configuration was assigned by NMR and the authors highlighted that no chromatography purification was required as the products were isolated by crystallisation and no other isomers were observed. Water is the only by-product generated in this procedure.

Two similar protocols based on SFMA-MCRs have been described for the synthesis of *trans*-subsitued-2,3-dihydrofurans fused with indene[130] or pyrazol rings[131] (Scheme 10.24). The proposed mechanism involves a cascade Knoevenagel condensation/Michael addition/cyclisation sequence catalysed by 10 mol% of Et_3N. Unfortunately, in both protocols, chromatography was a necessary purification step.

10.6 Other Cyclocondensations

The SFMAS of pyrido[2,3-*d*]pyridine proceeds under $Pd(OAc)_2$/Xanphos catalysis *via* cascade imination, Buchwald–Hartwig cross-coupling and cycloaddition of β-bromovinyl aldehydes with 6-amino-1,3-dialkyluracils.[132] The catalyst- and solvent-free synthesis of imidazo[1,2-*a*]pyridines from 2-aminopyridine and α-bromoacetophenone was also successfully effected by MW dielectric heating with an improvement of yields from 80% to 90% and a decrease of reaction time from 60 to 12 min.[133] The enantioconservative SFMAS of chiral oxazolines from nitriles and enantiopure amino

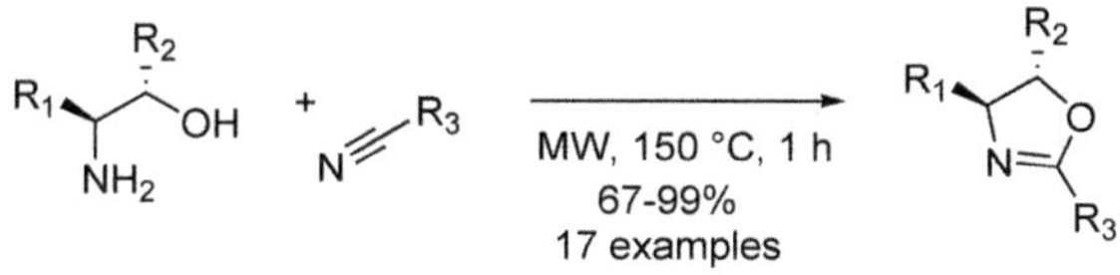

Scheme 10.25 SFMAS of enantiopure oxazolines.

alcohols was reported in 2014.[134] Under SFMAS conditions, the harsh protocol required for the so-called Witte and Seeliger reaction was improved to 1 h of MW dielectric heating at 150 °C with final quantitative oxazoline formation without the need for chromatography and loss of stereochemistry information (Scheme 10.25).

10.7 Rearrangements and Transpositions

Since the first example of MW-assisted Beckman rearrangement in 1999,[135] several other examples of neat rearrangements and transpositions under MW dielectric heating have been reported.[136,137] One of the latest contribution is the *N*-fluorobenzenesulphonimide/$FeCl_3{\cdot}6H_2O$ (NFSI/$FeCl_3{\cdot}6H_2O$)-catalysed pinacol rearrangement of hydrobenzoins (Scheme 10.26).[138]

The outcome of the pinacol rearrangement on hydrobenzoin substrates can be an aldehyde or a ketone as a result of phenyl or hydride migration, respectively. Here, the regioselectivity is addressed by tuning the electronic nature of the phenyl groups. The migration of the phenyl group, with selective formation of the aldehyde, was achieved by electron-rich substituents, while the ketone produced by hydride migration was obtained by reducing the electron density on one of the rings according to the results predicted by DFT calculations (Scheme 10.26a).

The use of the readily available and environmentally friendly NFSI/$FeCl_3$ catalyst under MW irradiation in the absence of solvents circumvents the employment of harsh conditions usually required for this transformation, being also well tolerated by several functional groups. With this protocol, starting from the appropriate hyrobenzoins, the aldehydes generated by the regioselective pinacol rearrangement afforded in five overall steps (±)-Latifine and (±)-Cherylline, two 4-aryltetrahydroisoquinoline alkaloids isolated from *Amaryllidacecae* plants. Only one chromatography purification was required at the final step of the synthesis (Scheme 10.26b).

10.8 Cross-coupling Reactions

10.8.1 Suzuki Coupling

Palladium-catalysed Suzuki coupling is considered the most efficient method to produce bi- and poly-aryl compounds. Unfortunately, this reaction employs high reaction temperatures and long reaction times[139] and

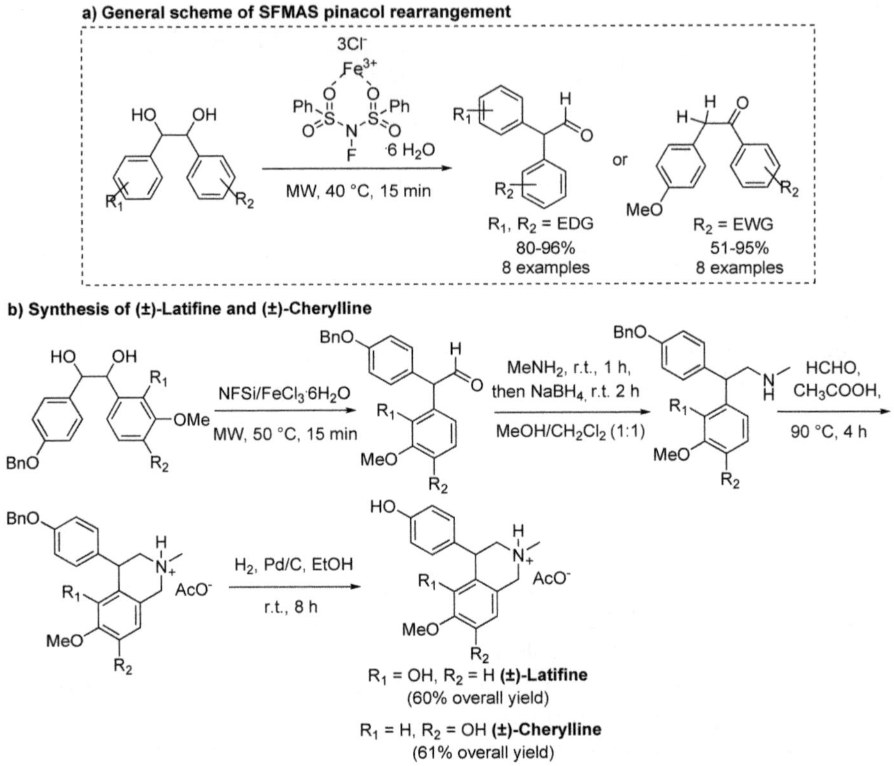

Scheme 10.26 Solvent-free microwave-assisted pinacol rearrangement (a) and its application to the synthesis of (±)-Latifine and (±)-Cherylline (b).

moreover, it generally proceeds in the presence of homogeneous Pd catalysts, such as $Pd(PPh_3)_4$, with consequently difficult purification of the products and catalyst recovery.[140] To improve these drawbacks, several protocols using MWs for activation have emerged.[141]

The first strategy involved the use of classic homogeneous palladium catalysts in the presence of a solid support such as $KF–Al_2O_3$[142,143] or basic alumina[144] under solvent-free conditions. Successively, research has evolved towards the development of new catalysts to successfully couple the Suzuki reaction and SFMAS conditions. In 2009, the use of PEPPSI-iPr (Pyridine Enhanced Pre-catalyst Preparation, Stabilisation and Initiation) was described to fully convert several boronic acids and aryl halides into the corresponding biaryl derivatives in the presence of K_2CO_3 (Scheme 10.27).[139] The authors underlined that grounding the catalyst, base and boronic acid in a mortar before the addition of halides is important for the success of the reaction, while a simply filtration and work-up afforded pure products without any further purification.

Recently, solid supported heterogeneous Pd-catalysts have gained much attention because they represent a green turning point in cross-coupling

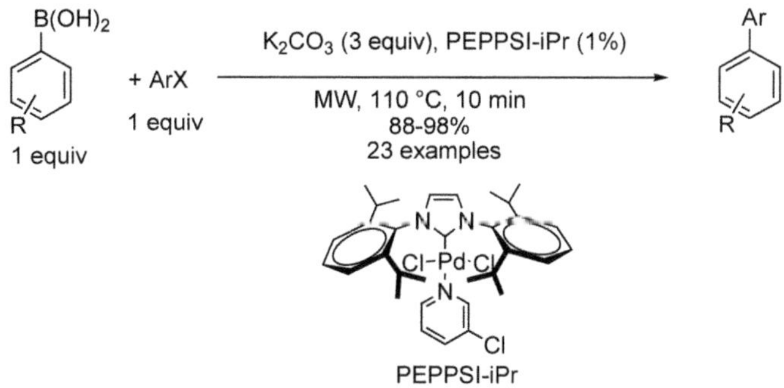

Scheme 10.27 Solvent-free Suzuki coupling with PEPPSI-iPr under MW dielectric heating.

reactions owing to their easy recovery and reuse. The choice of the best solid support is crucial for the success and sustainability of the reaction: it needs to be both eco-friendly and low cost, as well as having high thermal and chemical stability.[145] Chitin, chitosan, cellulose, silica, zeolite and carbon are widely used biomaterials[146] to develop new transition metal-supported catalysts in the green chemistry field.

In this regard, Pd/MCM-41 and Pd/SBA-15 were efficiently used in a microwave-assisted Suzuki coupling protocol in the presence of K_2CO_3 as the base (reaction conditions: 120 °C, 2 min) and in the absence of solvent.[140] Later, chitosan and chitosan-*Ulva* sp. (green alga)-supported Pd(II) catalysts were used by Baran and co-workers to obtain biaryl compounds under basic SFMAS conditions in only 4 min at 50 °C.[146,147] The same authors developed a practical, economical and eco-friendly starch-supported palladium catalyst that in only 5 min gave remarkable yields (reaction conditions: 1 equiv. of aryl halide, 1 equiv. of phenyl boronic acid, 3 equiv. of K_2CO_3, 4×10^{-3} mol% of catalyst, MW, 50 °C).[145] They also proved that the catalyst could be reused for at least 10 times.

10.8.2 Sonogashira and Heck Coupling

The first contribution of Sonogashira coupling under SFMAS conditions was reported in 2000. Several terminal alkynes were coupled with aryl or alkenyl iodide on palladium-doped alumina in the presence of triphenylphosphine and cuprous iodide to provide high yields of product in 2.5 min.[148] Later, also Ni(0)-doped KF/Al_2O_3 was described to work in this kind of transformation.[149] More recently, Pd-MCM-41 and Pd–Y, as well as for the Suzuki coupling, were successfully applied to MW-assisted Sonogashira coupling in the presence of DBU as the base (Scheme 10.28) under neat conditions.[150]

Although this protocol works with different phenyl iodides, its main drawback is the product purification by silica gel column chromatography and the substrate-dependent results. For example, the diaryl alkyne yield decreased

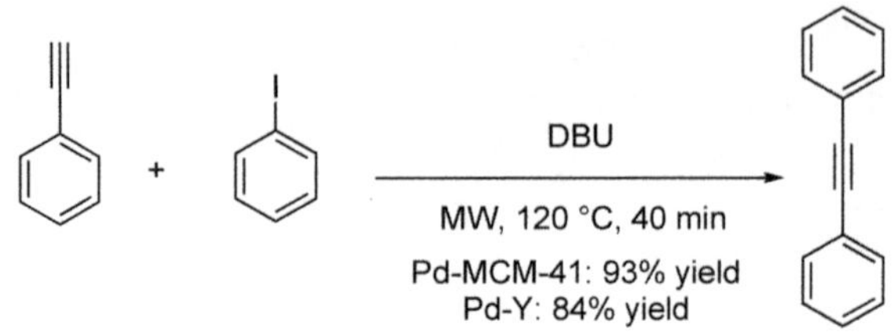

Scheme 10.28 Solid-supported Pd catalysts in solvent-free Sonogashira coupling under MW dielectric heating.

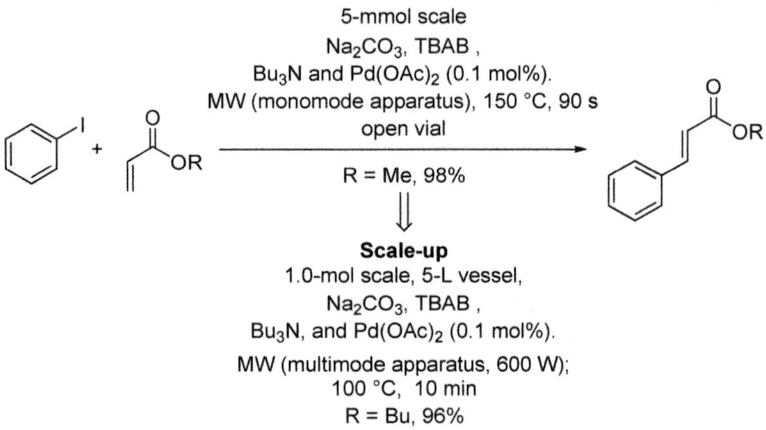

Scheme 10.29 Solvent-free Heck coupling under MW dielectric heating.

from the 80% obtained starting from *p*-chloro-iodobenzene to the 30% from *p*-bromo-iodobenzene, and to less than 10% with *p*-iodo-phenol.

SFMAS also promotes Heck couplings (Scheme 10.29). In a typical experiment at mmol scale, all the reagents were mixed in a round-bottomed flask with a reflux condenser and the mixture was irradiated at 150 °C for 90 s. Afterwards, H_2O and exaction with EtOAc afforded the final alkene which was further purified by crystallisation.[151]

The process was scaled up using a multimode apparatus. A further improvement was realised with Pd NPs supported on laponite clay (PdNPsLaponite), which efficiently catalysed the Heck coupling under SFMAS conditions in the presence of NEt_3 in 5–10 min.[152]

10.9 Conclusion

The employment of SFMAS conditions has proven to be one of the most advantageous protocols to achieve high-performance transformations with a lower negative impact on the environment. This outcome is due not only to the elimination of the solvent but also to limitation of side-product formation, which prevents the production of waste materials and makes the product purification easier, often averting the use of column chromatography. Moreover, the use of MWs is frequently crucial for the formation of the expected product and the reduction of the reaction time compared to that of the corresponding conventionally heated protocols is impressive, also compared to other non-conventional energy sources such as mechanochemistry (ball milling) and ultrasound-based methods. Finally, being nowadays the scale-up from milligrams to kilograms feasible, MW technology is becoming widely applied in pharmaceutical companies; the coupling with solvent-free protocols could therefore emerge as one of the most promising tools for the development of more sustainable Active Pharmaceutical Ingredients synthetic methods.

Acknowledgements

The authors thank the Italian Ministry of Education, University, and Research (MIUR) for the financial support through the program "Dipartimenti di Eccellenza – 2018–2022".

References

1. S. Ravichandran and E. Karthikeyan, *Int. J. ChemTech Res.*, 2011, **3**, 466.
2. M. A. Surati, S. A. Jauhari and K. R. Desai, *Arch. Appl. Sci. Res.*, 2012, **4**, 645.
3. L. Perreux, A. Loupy and M. Delmotte, *Tetrahedron*, 2003, **59**, 2185.
4. D. Obermayer, B. Gutmann and C. O. Kappe, *Angew. Chem., Int. Ed.*, 2009, **48**, 8321.
5. J. Zhou, W. Xu, Z. You, Z. Wang, Y. Luo, L. Gao, C. Yin, R. Peng and L. Lan, *Sci. Rep.*, 2016, **6**, 25149.
6. S. A. Khanum, S. Shashikanth, B. S. Sudha, S. A. Deepak and H. S. Shetty, *Pest Manage. Sci.*, 2004, **60**, 1119.
7. A. Procopio, A. De Nino, M. Nardi, M. Oliverio, R. Paonessa and R. Pasceri, *Synlett*, 2010, 1849.
8. R. S. Varma, *Tetrahedron*, 2002, **58**, 1235.
9. U. R. Pillai, E. Sahle-Demessie and R. S. Varma, *J. Mater. Chem.*, 2002, **12**, 3199.
10. B. Rissafi, A. El Louzi, A. Loupy, A. Petit, M. Soufiaoui and S. F. Tétouani, *Eur. J. Org. Chem.*, 2002, 2518.
11. H. Iida, H. Hamana and K. Matsumoto, *Synth. Commun.*, 2007, **37**, 1801.
12. A. Loupy, *Actual. Chim.*, 2004, 18.
13. A. Loupy, A. Petit, J. Hamelin, F. Texier-Boullet, P. Jacquault and D. Mathe, *Synthesis*, 1998, 1213.
14. M. R. Naimi-Jamal, M. Mirzaei, M. Bolourtchian and A. Sharifi, *Synth. Commun.*, 2006, **36**, 2711.
15. K. Bougrin, A. Loupy and M. Soufiaoui, *J. Photochem. Photobiol., C*, 2005, **6**, 139.

16. A. Loupy, *C. R. Chim.*, 2004, **7**, 103.
17. A. Loupy, L. Perreux, M. Liagre, K. Burle and M. Moneuse, *Pure Appl. Chem.*, 2001, **73**, 161.
18. R. S. Varma, *Pure Appl. Chem.*, 2001, **73**, 193.
19. C. R. Strauss and R. S. Varma, *Top. Curr. Chem.*, 2006, **266**, 199.
20. V. Polshettiwar and R. S. Varma, *Acc. Chem. Res.*, 2008, **41**, 629.
21. V. Polshettiwar and R. S. Varma, in *Eco-Friendly Synthesis of Fine Chemicals*, ed. R. Ballini, RSC Green Chemistry Series, Royal Society of Chemistry, Cambridge, UK, 2009, ch. 8, vol. 3, pp. 275–292.
22. R. S. Varma, *Green Chem. Lett. Rev.*, 2007, **1**, 37.
23. R. S. Varma, *ACS Symp. Ser.*, 2000, **767**, 292.
24. R. S. Varma, *Green Chem.*, 1999, **1**, 43.
25. M. B. Gawande, V. D. B. Bonifacio, R. Luque, P. S. Branco and R. S. Varma, *ChemSusChem*, 2014, **7**, 24.
26. G. Perin, D. Alves, R. G. Jacob, A. M. Barcellos, L. K. Soares and E. J. Lenardão, *ChemistrySelect*, 2016, **1**, 205.
27. S. M. Prajapati, K. D. Patel, R. H. Vekariya, S. N. Panchal and H. D. Patel, *RSC Adv.*, 2014, **4**, 24463.
28. J. G. Hernandez, C. G. Avila-Ortiz and E. Juaristi, Useful Chemical Activation Alternatives in Solvent-free Organic Reactions, in *Comprensive Organic Synthesis*, ed. G. A. Molandery and P. Knochel, Elsevier, Oxford, 2nd edn, 2004, ch. 11, vol. 9, pp. 287–314.
29. A. Sarkar, S. Santra, S. K. Kundu, A. Hajra, G. V. Zyryanov, O. N. Chupakhin, V. N. Charushin and A. Majee, *Green Chem.*, 2016, **18**, 4475.
30. A. De La Hoz, A. Diaz-Ortiz and P. Prieto, *RSC Green Chem. Ser.*, 2016, **47**, 1.
31. M. A. P. Martins, C. P. Frizzo, D. N. Moreira, L. Buriol and P. Machado, *Chem. Rev.*, 2009, **109**, 4140.
32. C. Leonelli and T. J. Mason, *Chem. Eng. Process.*, 2010, **49**, 885.
33. K. Kranjc and M. Kocevar, *Curr. Org. Chem.*, 2010, **14**, 1050.
34. R. S. Varma, R. Dahiya and D. Kumar, *Mol. Online*, 1998, **2**, 82.
35. J. M. Khurana and R. Arora, *ARKIVOC*, 2008, 211.
36. R. S. Varma and V. V. Namboodiri, *Pure Appl. Chem.*, 2001, **73**, 1309.
37. L. M. T. Frija, E. C. B. A. Alegria, M. Sutradhar, M. L. S. Cristiano, A. Ismael, M. N. Kopylovich and A. J. L. Pombeiro, *J. Mol. Catal. A: Chem.*, 2016, **425**, 283.
38. L. M. T. Frija, M. L. Kuznetsov, B. G. M. Rocha, L. Cabral, M. L. S. Cristiano, M. N. Kopylovich and A. J. L. Pombeiro, *Mol. Catal.*, 2017, **442**, 57.
39. N. Martins, L. Martins, C. Amorim, V. Amaral and A. Pombeiro, *Catalysts*, 2017, **7**, 222.
40. M. H. S. A. Hamid, P. A. Slatford and J. M. J. Williams, *Adv. Synth. Catal.*, 2007, **349**, 1555.
41. S. Bähn, S. Imm, L. Neubert, M. Zhang, H. Neumann and M. Beller, *ChemCatChem*, 2011, **3**, 1853.
42. G. V. Thanh, B. Pegot and A. Loupy, *Eur. J. Org. Chem.*, 2004, 1112.
43. A. Marinkovic, D. Mijin, J. Mirkovic, V. Maslak and C. O. Kappe, *J. Serb. Chem. Soc.*, 2014, **79**, 759.

44. I. Almena, E. Díaz-Barra, A. de la Hoz, J. Ruiz and A. Sánchez-Migallón, *J. Heterocycl. Chem.*, 1998, **35**, 1263.
45. E. R. Pérez, A. Loupy, M. Liagre, A. M. De Guzzi Plepis and P. J. Cordeiro, *Tetrahedron*, 2003, **59**, 865.
46. D. Bogdal, *Molecules*, 1999, **4**, 333.
47. R. S. Varma and V. V. Namboodiri, *Pure Appl. Chem.*, 2001, **73**, 1309.
48. G. Gong, W. Li, B. Xiao and Y. Li, *Huaxue Yu Nianhe*, 2013, **35**, 41.
49. P. Ranjan, B. S. Kitawat and M. Singh, *RSC Adv.*, 2014, **4**, 53634.
50. R. S. Varma, *ACS Symp. Ser.*, 2003, **856**, 82.
51. T. K. T. Truong and G. Vo-Thanh, *Tetrahedron*, 2010, **66**, 5277.
52. S. Chander, P. Ashok, R. Singh, P. N. Jha and S. Murugesan, *Curr. Microwave Chem.*, 2015, **2**, 44.
53. S. A. Khanum, S. Shashikanth and B. S. Sudha, *Heteroat. Chem.*, 2004, **15**, 37.
54. K. N. Nguyen, F. Duus and T. X. T. Luu, *J. Sulfur Chem.*, 2016, **37**, 349.
55. J. Rayó, L. Muñoz, G. Rosell, M. Bosch and Á. Guerrero, *Synthesis*, 2010, 3117.
56. J. Thierry and V. Servajean, *Tetrahedron Lett.*, 2004, **45**, 821.
57. E. Cini, E. Petricci, G. I. Truglio, M. Vecchio and M. Taddei, *RSC Adv.*, 2016, **6**, 31386.
58. C. Löfberg, R. Grigg, M. A. Whittaker, A. Keep and A. Derrick, *J. Org. Chem.*, 2006, **71**, 8023.
59. A. J. A. Watson, A. C. Maxwell and J. M. J. Williams, *J. Org. Chem.*, 2011, **76**, 2328.
60. W. Zhang, X. Dong and W. Zhao, *Org. Lett.*, 2011, **13**, 5386.
61. T. D. Apsunde and M. L. Trudell, *Synthesis*, 2014, **46**, 230.
62. M. Minakawa, M. Okubo and M. Kawatsura, *Bull. Chem. Soc. Jpn.*, 2015, **88**, 1680.
63. C. Liu, S. Liao, Q. Li, S. Feng, Q. Sun, X. Yu and Q. Xu, *J. Org. Chem.*, 2011, **76**, 5759.
64. M. R. Shushizadeh and M. Kiany, *Chin. Chem. Lett.*, 2009, **20**, 1068.
65. F. Teodorescu, A. Enache and M. Sandulescu, *ARKIVOC*, 2017, 58.
66. H. Yang, W.-H. Sun, Z. Li and L. Wang, *Synth. Commun.*, 2002, **32**, 2395.
67. G. Öztürk, A. Petit and C. Kouklovsky, *Synth. Commun.*, 2008, **38**, 2707.
68. G. Öztürk, M. Çolak and N. Demirel, *Chirality*, 2011, **23**, 374.
69. A. J. A. Watson, A. C. Maxwell and J. M. J. Williams, *Org. Biomol. Chem.*, 2012, **10**, 240.
70. M. G. Mura, S. Rajamäki, L. De Luca, E. Cini and A. Porcheddu, *Adv. Synth. Catal.*, 2015, **357**, 576.
71. J. F. Collados, E. Toledano, D. Guijarro and M. Yus, *J. Org. Chem.*, 2012, **77**, 5744.
72. F. A. Davis, Y. Zhang and D. Li, *Tetrahedron Lett.*, 2007, **48**, 7838.
73. L. D. Chiaradia, P. G. A. Martins, M. N. S. Cordeiro, R. V. C. Guido, G. Ecco, A. D. Andricopulo, R. A. Yunes, J. Vernal, R. J. Nunes and H. Terenzi, *J. Med. Chem.*, 2012, **55**, 390.
74. M. Go, X. Wu and X. Liu, *Curr. Med. Chem.*, 2005, **12**, 483.

75. M. Akhter, S. Habibullah, S. M. Hasan, M. M. Alam, N. Akhter and M. Shaquiquzzaman, *Med. Chem. Res.*, 2011, **20**, 1147.
76. C. Tamuly, I. Saikia, M. Hazarika, M. Bordoloi, N. Hussain, M. R. Das and K. Deka, *RSC Adv.*, 2015, **5**, 8604.
77. J. P. Singh, M. Dulawat, N. Jaitawat, S. S. Chundawat, A. Devpura and S. S. Dulawat, *Indian J. Chem., Sect. B: Org. Chem. Incl. Med. Chem.*, 2012, **51B**, 1623.
78. K. Ranganathan, R. Arulkumaran, D. Kamalakkannan, R. Sundararajan, S. P. Sakthinathan, S. Vijayakumar, R. Suresh, G. Vanangamudi, K. Thirumurthy, P. Mayavel and G. Thirunarayanan, *Int. J. Pharma Med. Biol. Sci.*, 2012, **1**, 62.
79. R. Pal, *IOSR-JAC*, 2013, **3**, 74.
80. R. S. Dhivare and S. S. Rajput, *Int. J. ChemTech Res.*, 2016, **9**, 325.
81. A. K. Mitra, A. De and N. Karchaudhuri, *Synth. Commun.*, 1999, **29**, 2731.
82. A. Loupy, S.-J. Song, S.-M. Sohn, Y.-M. Lee and T.-W. Kwon, *J. Chem. Soc., Perkin Trans. 1*, 2001, **10**, 1220.
83. S. S. Pandit, B. N. Jejurkar and R. L. Harde, *J. Indian Chem. Soc.*, 2007, **84**, 1035.
84. R. S. Bhatti, M. Singh and J. S. Sandhu, *Rasayan J. Chem.*, 2008, **1**, 738.
85. D. Kumar and J. S. Sandhu, *Synth. Commun.*, 2010, **40**, 1915.
86. R. Dhivare and S. S. Rajput, *Heterocycl. Lett.*, 2016, **6**, 443.
87. D. Kumar and J. S. Sandhu, *Synth. Commun.*, 2010, **40**, 510.
88. E. Balducci, E. Attolino and M. Taddei, *Eur. J. Org. Chem.*, 2011, 311.
89. E. Brun, A. Safer, F. Carreaux, K. Bourahla, J.-M. L'Helgoua'ch, J.-P. Bazureau and J. M. Villalgordo, *Molecules*, 2015, **20**, 11617.
90. R. S. Varma and D. Kumar, *Org. Lett.*, 1999, **1**, 697.
91. R. Rico-Gomez, F. Najera, J. M. Lopez-Romero and P. Canada-Rudner, *Heterocycles*, 2000, **53**, 2275.
92. M. Xia, S. Wang and W. Yuan, *Synth. Commun.*, 2004, **34**, 3175.
93. S. A. Sadaphal, A. H. Kategaonkar, V. B. Labade and M. S. Shingare, *Chin. Chem. Lett.*, 2010, **21**, 39.
94. J. Azizian, M. R. Mohammadizadeh, F. Teimouri, A. A. Mohammadi and A. R. Karimi, *Synth. Commun.*, 2006, **36**, 3631.
95. Z.-L. Shen, D.-G. Gu, J.-M. Yang and S.-J. Ji, *Synth. Commun.*, 2011, **41**, 851.
96. O. Correc, K. Guillou, J. Hamelin, L. Paquin, F. Texier-Boullet and L. Toupet, *Tetrahedron Lett.*, 2004, **45**, 391.
97. S. Li, Z. Xie, X. Bian and C. Wang, *J. Chem. Res.*, 2007, 660.
98. H. Iida, T. Moromizato, H. Hamana and K. Matsumoto, *Tetrahedron Lett.*, 2007, **48**, 2037.
99. K. M. Amore, N. E. Leadbeater, T. A. Miller and J. R. Schmink, *Tetrahedron Lett.*, 2006, **47**, 8583.
100. L. Dubois, F. C. Acher and I. McCort-Tranchepain, *Synlett*, 2012, **23**, 791.
101. A. Zare, A. Hasaninejad, M. H. Beyzavi, A. R. M. Zare, A. Khalafi-Nezhad, F. Asadi, L. Baramaki, S. Jomhori-Angali and R. Ghaleh-Golabi, *Synth. Commun.*, 2009, **39**, 139.

102. P. Ghattas, J. W. Carlin, S. L. Murkli and D. L. Jacobs, *Synth. Commun.*, 2014, **44**, 520.
103. S. Gogoi, R. C. Boruah, P. Saikia, A. Addlagatta and V. Saddanapu, *PCT Int. Appl. Pat.*, WO2015170336A1, 2015.
104. H. K. Lenker, M. E. Richard, K. P. Reese, A. F. Carter, J. D. Zawisky, E. F. Winter, T. W. Bergeron, K. S. Guydon and R. A. Stockland, *J. Org. Chem.*, 2012, **77**, 1378.
105. R. Mondal, A. Das Gupta and A. K. Mallik, *Tetrahedron Lett.*, 2011, **52**, 5020.
106. U. Sankappa Rai, A. M. Isloor, P. Shetty, N. Isloor, S. Malladi and H.-K. Fun, *Eur. J. Med. Chem.*, 2010, **45**, 6090.
107. A. Sharifi, F. Mohsenzadeh and M. R. Naimi-Jamal, *J. Chem. Res., Synop.*, 2000, 394.
108. N. L. Chavan, P. N. Naik, S. K. Nayak and R. S. Kusurkar, *Synth. Commun.*, 2010, **40**, 2941.
109. J. Safari and Z. Zarnegar, *J. Ind. Eng. Chem.*, 2014, **20**, 2292.
110. G. Keglevich and E. Balint, *Molecules*, 2012, **17**, 12821.
111. E. Balint, A. Tajti, A. Dzielak, G. Hagele and G. Keglevich, *Beilstein J. Org. Chem.*, 2016, **12**, 1493.
112. K. M. K. Reddy, S. M. Sadik, K. Peddanna, N. B. Reddy, G. Sravya, G. V. Zyryanov and C. S. Reddy, *Phosphorus, Sulfur Silicon Relat. Elem.*, 2018, **193**, 329.
113. P. Nun, J. Martinez and F. Lamaty, *Synthesis*, 2010, 2063.
114. B. Jiang, X. Wang, F. Shi, S. J. Tu, T. Ai, A. Ballew and G. Li, *J. Org. Chem.*, 2009, **74**, 9486.
115. H. R. Shaterian, A. Hosseinian and M. Ghashang, *Phosphorus, Sulfur Silicon Relat. Elem.*, 2009, **184**, 197.
116. M. R. M. Bhojegowd, A. Siddaramanna, A. B. Siddappa, C. G. Thimmanna and M. A. Pasha, *Chin. J. Chem.*, 2011, **29**, 1863.
117. Q. Liu, N. Pan, J. Xu, W. Zhang and F. Kong, *Synth. Commun.*, 2013, **43**, 139.
118. G. R. Chaudhary, P. Bansal and S. K. Mehta, *Chem. Eng. J.*, 2014, **243**, 217.
119. R. Fu, Y. Yang, W. Lai, Y. Ma, Z. Chen, J. Zhou, W. Chai, Q. Wang and R. Yuan, *Synth. Commun.*, 2015, **45**, 477.
120. A. Kuraitheerthakumaran, S. Pazhamalai and M. Gopalakrishnan, *Arabian J. Chem.*, 2016, **9**, S461.
121. V. V. Shinde, S. D. Lee, Y. S. Jeong and Y. T. Jeong, *Tetrahedron Lett.*, 2015, **56**, 859.
122. A. S. Medvedeva, M. M. Demina, T. V. Kon'kova, T. L. H. Nguyen, A. V. Afonin and I. A. Ushakov, *Tetrahedron*, 2017, **73**, 3979.
123. A. J. Ansari, S. Sharma, R. S. Pathare, K. Gopal, D. M. Sawant and R. T. Pardasani, *ChemistrySelect*, 2016, **1**, 1016.
124. M. Kumar, T. Kaur, V. K. Gupta and A. Sharma, *RSC Adv.*, 2015, **5**, 17087.
125. B. Maiti, K. Chanda, M. Selvaraju, C. C. Tseng and C. M. Sun, *ACS Comb. Sci.*, 2013, **15**, 291.

126. D. D. Pham, N. T. Le and G. Vo-Thanh, *ChemistrySelect*, 2017, **2**, 12041.
127. K. Shekarrao, D. Nath, P. P. Kaishap, S. Gogoi and R. C. Boruah, *Steroids*, 2013, **78**, 1126.
128. S. Kolita and P. J. Bhuyan, *ChemistrySelect*, 2018, **3**, 1411.
129. B. Devi Bala, S. Michael Rajesh and S. Perumal, *Green Chem.*, 2012, **14**, 2484.
130. S. Majumder, M. Sharma and P. J. Bhuyan, *Tetrahedron Lett.*, 2013, **54**, 6868.
131. V. S. Tangeti, G. V. Siva Prasad, J. Panda and K. R. Varma, *Synth. Commun.*, 2016, **46**, 878.
132. P. Saikia, G. Sharma, S. Gogoi and R. C. Boruah, *RSC Adv.*, 2015, **5**, 23210.
133. D. Kong, X. Wang, Z. Shi, M. Wu, Q. Lin and X. Wang, *J. Chem. Res.*, 2016, **40**, 529.
134. R. Hassani, A. Requet, S. Marque, A. Gaucher, D. Prim, Y. Kacem and B. Ben Hassine, *Tetrahedron: Asymmetry*, 2014, **25**, 1275.
135. A. Loupy and S. Regnier, *Tetrahedron Lett.*, 1999, **40**, 6221.
136. J.-A. Ahn, D.-H. Chang, Y. J. Park, Y. R. Yon, A. Loupy and C.-H. Jun, *Adv. Synth. Catal.*, 2006, **348**, 55.
137. P. Ravi and S. P. Tewari, *Propellants, Explos., Pyrotech.*, 2013, **38**, 147.
138. H. Shi, C. Du, X. Zhang, F. Xie, X. Wang, S. Cui, X. Peng, M. Cheng, B. Lin and Y. Liu, *J. Org. Chem.*, 2018, **83**, 1312.
139. P. Nun, J. Martinez and F. Lamaty, *Synlett*, 2009, 1761.
140. W. Chang, G. Chae, S. R. Jang, J. Shin and B. J. Ahn, *J. Ind. Eng. Chem.*, 2012, **18**, 581.
141. L. Bai and J. Wang, *Curr. Org. Chem.*, 2005, **9**, 535.
142. M. Melucci, G. Barbarella and G. Sotgiu, *J. Org. Chem.*, 2002, **67**, 8877.
143. B. Basu, P. Das, M. H. Bhuiyan and S. Jha, *Tetrahedron Lett.*, 2003, **44**, 3817.
144. P. Saha, S. Naskar, P. Paira, A. Hazra, K. B. Sahu, R. Paira, S. Banerjee and N. B. Mondal, *Green Chem.*, 2009, **11**, 931.
145. T. Baran, *J. Colloid Interface Sci.*, 2017, **496**, 446.
146. T. Baran, T. Inanan and A. Menteş, *Carbohydr. Polym.*, 2016, **145**, 20.
147. T. Baran, I. Sargin, M. Kaya and A. Menteş, *J. Mol. Catal. A: Chem.*, 2016, **420**, 216.
148. G. W. Kabalka, L. Wang, V. Namboodiri and R. M. Pagni, *Tetrahedron Lett.*, 2000, **41**, 5151.
149. J. Yan, Z. Wang and L. Wang, *J. Chem. Res.*, 2004, 71.
150. W. Chang, J. Shin, G. H. Chae, S. R. Jang and B. J. Ahn, *J. Ind. Eng. Chem.*, 2013, **19**, 739.
151. N. Leadbeater, V. Williams, T. Barnard and M. Collins Jr, *Synlett*, 2006, 2953.
152. A. V. Martínez, F. Invernizzi, A. Leal-Duaso, J. A. Mayoral and J. I. García, *RSC Adv.*, 2015, **5**, 10102.

CHAPTER 11

Advances in Catalysis for More Sustainable Synthesis of Phenolics

T. TABANELLI AND F. CAVANI*

Alma Mater Studiorum Università di Bologna, Dipartimento di Chimica Industriale "Toso Montanari", Viale Risorgimento 4, Bologna, 41136, Italy
*E-mail: fabrizio.cavani@unibo.it

11.1 Phenol Methylation with Methanol: The First Industrial Example of Green Functionalisation of Phenol

The unique chemical properties of phenol and phenolic compounds derive from the presence of the hydroxyl group. The conjugation with the aromatic ring gives phenol higher acidity compared to other aliphatic or aromatic alcohols; indeed, the unshared electron pair located on the –OH group can be delocalised over the aromatic ring, leading to negative charge excess in the *ortho* and *para* positions. This behaviour facilitates the alkylation and other functionalisations with suitable electrophiles, but also increases the capability of phenol hydroxyl group to act as a nucleophile in order to obtain different derivatives.

For all these reasons, phenol has been widely investigated as the starting material for the production of a variety of commercially significant compounds (Scheme 11.1).

Green Chemistry Series No. 61
Green Synthetic Processes and Procedures
Edited by Roberto Ballini

Published by the Royal Society of Chemistry, www.rsc.org

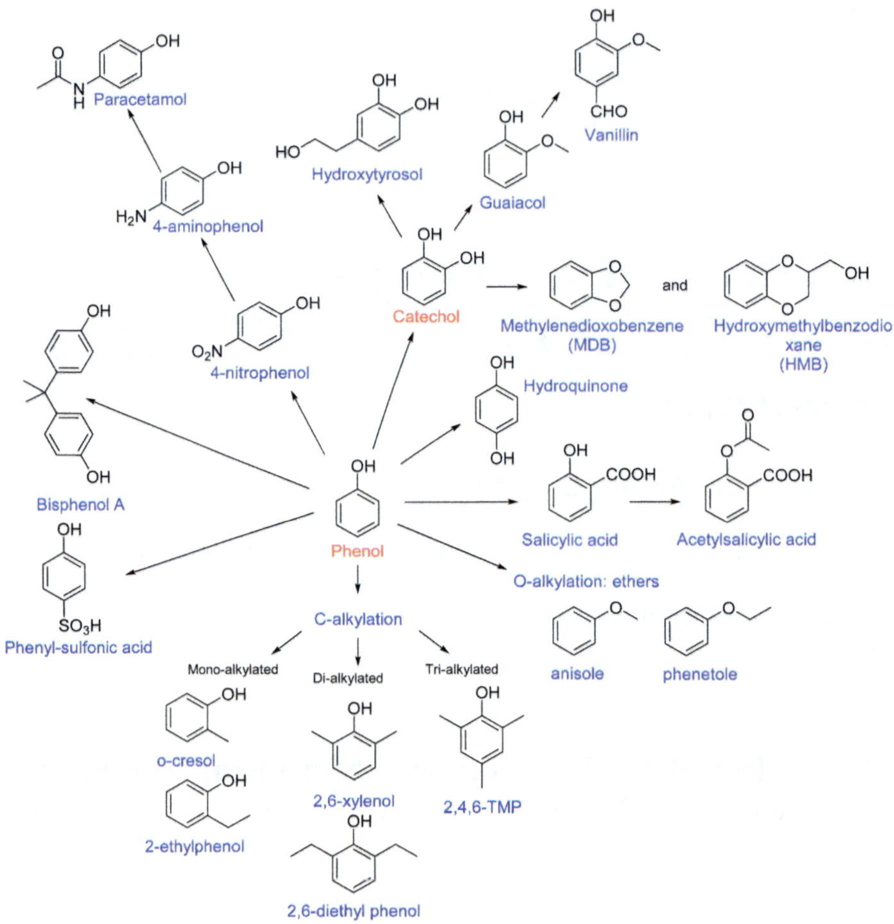

Scheme 11.1 The phenol 'tree', showing some of the most important derivatives obtained from phenol.

Among the alkylation processes, phenol methylation is the most investigated due to the importance of the related methylated products. In fact, anisole (methylphenyl ether) is an important precursor for the synthesis of fragrances, as well as an octane booster for gasoline. *C*-alkylated compounds find a wide range of applications: (a) *o*-cresol is an important precursor for various dyes, pharmaceuticals (*e.g.* carvacrol) and antioxidants;[1] however, it is mainly used in polymer manufacture, especially for the production of epoxy-*o*-cresol novolak resins. (b) *p*-Cresol is used as intermediate for the production of antioxidants and preservatives used as additives in foods and plastics. (c) 2,6-Xylenol (2,6-dimethylphenol), similarly to *o*-cresol, is used by the polymer industry to produce poly(phenylene oxide) resins (PPE), a high-temperature, high-performance thermoplastic characterised by a high impact resistance, as well as thermal stability and fire

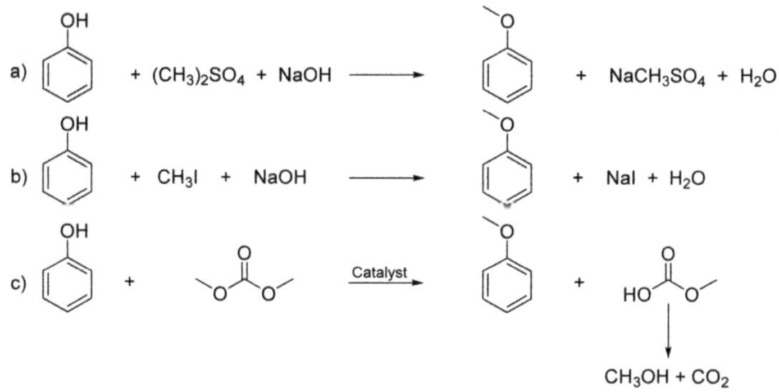

Scheme 11.2 Phenol *O*-methylation reactions with different alkylating agents.

resistance. (d) 2,4,6- and 2,3,6-Trimethyl phenol (2,4,6-TMP and 2,3,6-TMP) are important intermediates for the synthesis of pharmaceuticals and food additives (vitamin supplements), particularly they are used for the production of vitamin E.

Conventional methylation agents are dimethylsulphate and methyl iodide and chloride, which are still used especially for the *O*-methylation of phenolics (*e.g.* catechol) in the liquid phase. Methanol is the most used reagent for gas-phase alkylation, whereas dimethyl carbonate may be considered an innovative methylating agent (Scheme 11.2). Alkyl halides are known carcinogens, while dimethylsulphate is characterised by a very high toxicity and may be fatal if inhaled.

11.1.1 Phenol *C*-methylation with Methanol: Classes of Catalysts

The catalysts used for the gas-phase methylation of phenol with methanol may be divided into several classes, based on their catalytic behaviour. Indeed, in this case, the distinction between acid and basic catalysts should be based on their ability to act as proton acceptors or donors towards phenol, the molecule containing the most acidic proton in the two reactants. Due to the strong proton-donor properties of phenol, which forms a phenolate anion upon interaction with a large number of oxides, including those conventionally considered Brønsted acids, most of the heterogeneous catalysts listed below should be regarded as basic.

Catalysts based on alkali or alkaline earth oxides. Their most important component is MgO (possibly doped with other elements). MgO-based catalysts are active in the temperature range of 420–460 °C. They are extremely *ortho*-selective, and both phenol conversion and 2,6-xylenol-to-*o*-cresol ratio depend on the catalyst properties (possible presence of a second component, surface area...). Moreover, only traces of anisole are formed. MgO-based catalysts are used in the General Electric process for 2,6-xylenol production.[2]

Catalysts based on transition metal oxides with redox properties, such as Cr, Fe, V, and Mn oxides (and mixed oxides with these components). Their behaviour is very similar to that observed with conventional basic systems, that is, characterised by a high regio- and chemo-selectivity (only *ortho-C*-alkylation, no *O*-alkylation), whereas they are active at considerably lower temperatures, *e.g.* in the 350–400 °C range. In this case, the extent of methanol decomposition is much greater than with MgO-based catalysts. Therefore, various strategies are adopted to try to limit methanol decomposition, *e.g.* feeding water or changes in catalyst composition, for instance by making mixed oxides with Mg. The process based on V/Fe mixed oxides is used commercially by Asahi.[2]

Amphoteric catalysts. In this case, the catalyst may behave as a base towards phenol and as an acid towards methanol. Thus, the catalytic behaviour shown, in terms of regio- and chemo-selectivity, might be intermediate between the two 'main' classes of catalysts. Alumina is the main component of this class of catalysts; to a certain extent, their behaviour is similar to that of catalysts belonging to the main classes, with the preferred (but not exclusive) formation of the *ortho*-methylated compound. On the other hand, there is little or no decomposition of methanol into light compounds, while little chemo-selectivity is shown, because of the formation of anisole (the product of *O*-methylation) and of various polyalkylated compounds in addition to 2,6-xylenol. Indeed, with oxides like titania and alumina, the generation of Me–OH groups by phenol dissociation may play an important role in the acid-type activation of methanol. In fact, with these oxides, there are acid sites that are sufficiently strong to directly activate methanol.

Acid catalysts. Weak acid sites catalyse almost exclusively the formation of anisole whereas, with medium-strength Brønsted-type catalysts, the typical reactivity behaviour shows negligible decomposition of methanol into light compounds and the formation of hydrocarbons that are responsible for catalyst deactivation. These catalysts are active over a wide range of temperatures; anisole is the favoured product at low temperatures (*i.e.* 250–350 °C) but, at higher temperatures, mono- and polyalkylated phenolic compounds are formed. The regio-selectivity may be limited, with the formation of all possible isomers but, depending on the catalyst features, either *o*-cresol or even *p*-cresol is the preferred compound. With strongly acidic catalysts, side reactions occurring on methanol lead to rapid catalyst deactivation due to the formation of aliphatic and aromatic compounds with the coproduction of water.

Table 11.1 shows some examples of the gas-phase methylation of phenol with methanol using different catalysts.

The main peculiarity of the industrial methylation of phenol catalysed by either MgO-based systems or V/Fe mixed oxides is its very high regio- and chemo-selectivity. Both systems are selective for the *C*-alkylation of phenol *ortho* positions, leading to the formation of *o*-cresol and 2,6-xylenol, while both *p*-cresol and anisole (or alkylated anisole) are formed as minor by-products.

Table 11.1 Examples of catalysts active for phenol *O*- and *C*-alkylation with methanol.

Phenol conv.	Main product, yield	Catalyst	Reference
84%	Anisole 77%	Sm/Cs phosphate	3
85%	*o*-Cresol 30%, 2,6-xylenol 37%	γ-Al_2O_3	4
70%	Anisole 23%	$AlPO_4$–Al_2O_3–F^-/SO_4^{2-}	5
nd	2,6-Xylenol nd	Fe_xO_y/Si/Cr/K	6
72%	*o*-Cresol 27%	Cu/Al/O	7
99%	2,6-Xylenol 52%	Mg/Al/O	8
65%	*o*-Cresol 29%, 2,6-xylenol 29%	V_2O_5	9
10%	Anisole 10%	Cs-X	10
95%	*o*-Cresol 24%, 2,6-xylenol 65%	V_2O_5/TiO_2	11
99%	Mg/Al/O: anisole 35%, *o*-cresol 40%, 2,6-xylenol 25% Mg/Fe/O, Mg/Cr/O: *o*-cresol 88%	Mg/Al/O, Mg/Fe/O, Mg/Cr/O	12
53%	Anisole 45%	ThO_2	13
95%	*o*-Cresol 39%, 2,6-xylenol 56%	Cu, Mg, Ca, Ba,Zn, Mn/Fe/O	14
15%	Anisole 10%	H-Y, H-ZSM5	15, 16
12%	Anisole 8%	SAPO-5, AlPO-5, SAPO-11	17
47%	Cresols 24%	H-X, H-Y, H-ZSM5	18, 19
50%	Anisole 49%	$La_2(HPO_4)_3$, $BaSO_4$, $SrSO_4$	20
55%	*o*-Cresol 45%	MgO	21
95%	2,6-Xylenol 72%	Mn_3O_4	22
96%	2,6-Xylenol 78%	Mn/O modif.	23
78%	*o*-Cresol 38%, 2,6-xylenol 30%	$Ca_3(PO_4)_2$	24
56%	Anisole 13%, cresols 23%	H-USY, H-ZSM5	25
61%	*p*-Cresol 22%	H-Y/K, Na	26
36%	*o*-Cresol 33%	CeO_2	27
65%	Anisole 62%	Mn,K/SO_4/γ-Al_2O_3	28
90%	2,6-Xylenol 63%	V/Cr/O	29
60%	Anisole 38%	$AlPO_4$–TiO_2	30
30%	*o*-Cresol	Ce/Mg/O	31

Historically, the first hypothesis to explain such high regio-selectivity was formulated by Kotanigawa[14,32] on a Fe_2O_3 catalyst. It was suggested that the orthogonal orientation of adsorbed phenol on Lewis acidic sites prevents the aromatic ring from direct interaction with the strong acidic sites on the surface, while the proton bound to the basic sites facilitates the generation of a carbonium ion, which in turn reacts with an adjacent phenolate species in the more available *ortho* positions. This hypothesis is still used in the literature sometimes to interpret catalytic results.[33]

Pioneers in the study of reactant adsorption on metal oxides in the heterogeneously catalysed alkylation of hydroxyarenes with alcohols were Klemm and Taylor,[34–37] who first hypothesised the importance of the adsorption geometry on the regio-selectivity for phenol and 1-naphthol methylation. Indeed, in addition to the usual electronic effects that address the position in ring methylation, entropy effects were hypothesised to be significant, in particular the effects of orientation of the substrate with respect to the catalyst surface. In the case of phenol methylation, the dissociative vertical adsorption as phenate species, as well as the confinement of the methylating agent in a layer over the catalytic surface were proposed to foster substitution at the –OH *ortho* positions. On the other hand, a flatwise configuration over the surface may promote the substitution in other positions in the ring (*e.g. para* position). The absence of *O*-methylation products was attributed to the reversibility of ether formation, and to the shielding effect promoted by the catalyst surface on the O atom. The chemisorption of phenol on alumina was also investigated by the same authors by means of IR spectroscopy.[38] In particular, working in the temperature range of 300–450 °C, the formation of aluminium phenoxide surface species was observed with perturbation of the vibrational bands. This evidenced that the phenol molecule is oriented with the ring almost coplanar to the surface, but tilted upward in such a way that the *para* position is far from the surface and methylation is more difficult.

The geometric model was later proposed by Tanabe *et al.*[21,39,40] They demonstrated that the aromatic ring of the phenolate anion interacts with the acid sites of the catalyst, attracting the benzene ring closer to the surface and finally facilitating the methylation of both *ortho* and *para* positions of the ring. On the other hand, the same effect is not observed over a catalytic surface having basic characteristics, such as magnesium oxide. In this case, the phenate ions are adsorbed perpendicular to the surface with only the *ortho* positions available for reaction with the adsorbed, activated methanol. Indeed, this is the first example in the literature of phenol methylation with methanol performed on an alkaline-earth oxide catalyst. The 'Tanabe' model has been cited ever since the 1980s by the majority of papers dealing with phenol or diphenol methylation over basic catalysts.[41] For example, Gopinath *et al.*[42] suggested a mechanism in which methanol is activated over the surface of the catalyst by the proton released by the adsorbed phenol, promoting electrophilic aromatic substitution in the available *ortho* positions.

Later, a different mechanism was proposed by Grabowska *et al.*[43] studying the reaction between 1-naphthol and methanol catalysed by Fe oxide, in order to rationalise the high *ortho-C*-methylation selectivity observed. In particular, the authors suggested that the coordinative properties of Fe^{3+} might promote the simultaneous adsorption of both the activated, deprotonated naphthol and methanol over the same site, finally promoting only the methylation of the closer *ortho* positions. However, this model is also essentially an adsorptive/geometric one.

In summary, in the case of both heterogeneous basic catalysis and catalysts made of transition-metal oxides (Fe, V...), geometric adsorptive effects are considered the reason for the high regio-selectivity of this reaction. This is one of the very few, well-demonstrated examples in heterogeneous catalysis where the distribution of products is dramatically influenced by how reactants adsorb and interact with the catalyst surface.

However, over acid catalysts, both the chemo and regio-selectivity decrease. Indeed, in a confined environment, direct *C*-alkylation at both the *ortho* and *para* positions is observed, as well as the production of anisole (and methylated anisoles) through *O*-alkylation.[25] The *O*/*C*-alkylation ratio depends on the catalyst acid strength.

11.1.2 Phenol *C*-methylation with Methanol: The Reaction Mechanism Revisited

In the Kirk–Othmer Encyclopedia, it is stated that:[44] "*The methylation of phenol is more selective to o-cresol with base catalysts than with acid catalysts. On basic catalysts, phenol adsorbs dissociatively on a pair of basic and acidic sites to form a protonic site and an adsorbed phenolate species, respectively. This proton site activates methanol to produce a carbonium ion, which reacts with the benzene ring of an adjacently adsorbed phenolate species at the ortho position*". The acid-catalysed mechanism on basic catalysts has been accepted in the literature for a long time; even in papers where the mechanism is debated, the conclusion is still the same.[33,45] It should be pointed out, however, that the protonation of methanol by the H^+ abstracted by Mg–O from Ph–O–H has never been fully demonstrated. Moreover, the following should be taken into account:

1. With the industrial catalysts based on either MgO or V/Fe/O, one major problem is the extent of methanol decomposition into CO, H_2, CO_2, and CH_4. This can hardly be explained by taking an acid mechanism for methanol activation into account.
2. The total absence of anisole experimentally observed with basic catalysts cannot be explained by the mechanisms reported in the literature. In fact, the carbonium ion generated by the activation of methanol on the protonic site should react with the phenolic oxygen to generate anisole.
3. A conventional acid mechanism does not explain why the same catalyst that catalyses phenol alkylation with methanol is poorly active in phenol alkylation with *t*-butanol.[46]

The interaction of methanol with metal cations with Lewis-type acid properties yields an undissociated CH_3OH_{ads} Lewis-bound species,[47,48] while dissociated methoxy species, accompanied by gaseous evolution of water, are preferentially formed on basic (ionic) oxides (Bi_2O_3, Fe_2O_3, NiO, ZnO). On the other hand, covalent and Lewis-acidic oxides (WO_3, SiO_2, V_2O_5, Nb_2O_5, MoO_3) and amphoteric materials (CeO_2, TiO_2) produce both undissociated and dissociated methoxy species.

MgO is known to catalyse methanol dehydrogenation to formaldehyde.[49] The development of formate species occurs by means of a Cannizzaro-type reaction, with intermolecular disproportionation: a nucleophilic attack by

O^{2-} surface species to carbonyl is accompanied by a hydride transfer to a second molecule of adsorbed formaldehyde, which is thus converted to a methoxy species.[50,51] The disproportionation mechanism may also occur on metal oxides such as ZnO. The role of the dehydrogenating properties of MgO-based systems is also demonstrated by the fact that 2,6-xylenol can even be obtained by reaction between cyclohexanol/one and methanol.[52] All these considerations lead to the hypothesis that formaldehyde may be the true reactant for phenol methylation, which is generated *in situ* by methanol dehydrogenation. However, only a few authors have suggested a possible role of aldehydes in the alkylation of substrates using alcohols as alkylating agents; moreover, no experimental evidence has been provided.[31,36,42,53–55]

In a series of recent papers, Cavani *et al.* demonstrated that the reaction mechanism of phenol methylation on MgO and V/Fe/O catalysts involves formaldehyde as the true (hydroxy)methylating agent, while the formation of *o*-cresol occurs by means of the mechanism shown in Scheme 11.3.[56–63] The adsorbed phenolate reacts with formaldehyde, with the formation of a complex in which the catalyst forms a bridge between the phenolic hydroxyl group and the *ortho*-hydroxymethyl group formed by the reaction of phenol and formaldehyde. This complex decomposes into *o*-methylenecyclohexadienone, which immediately reacts with additional methanol to produce *o*-cresol, water, and formaldehyde; therefore, the formation of the methyl group is explained by an H-transfer mechanism. This mechanism clearly explains why catalysts consisting of V/Fe/O show such outstanding performance: they are extremely reactive in methanol dehydrogenation to formaldehyde,

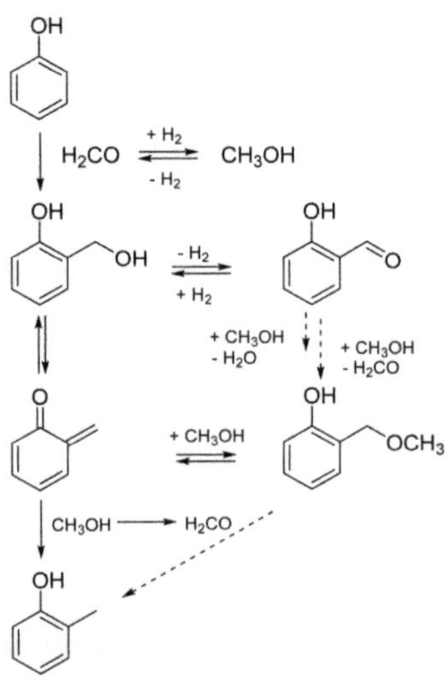

Scheme 11.3 Reaction mechanism for phenol methylation with methanol and with basic/dehydrogenating catalysts.

the rate-determining step of the process. On the other hand, formaldehyde is so reactive that it immediately reacts with the adsorbed phenolate to yield the intermediate salicyl alcohol.

It is worth noting that the intermediate formaldehyde readily reacts with phenolate and it is also likely to contribute to the reduction *via* H-transfer of reaction intermediates, to produce *o*-cresol with the co-production of H_2O and CO_2. Therefore, this extremely efficient cascade reaction, with *in situ* generation of a very reactive intermediate, makes it possible to overcome the problem of formaldehyde toxicity.

11.1.3 Phenolic Alkylation with Organic Carbonates

Organic carbonates (OCs) have been extensively investigated as 'greener' alkylating agents for a wide range of substrates, including aromatics, replacing more traditional, harmful, and undesirable compounds, namely alkyl iodides and dimethyl sulphate, that are known to be carcinogens and extremely toxic and hazardous compounds, respectively (Table 11.2).[64–66]

Nevertheless, only a few organic carbonates are commercially available. Also, for this reason, only the methylation reaction with the simplest linear OC, dimethyl carbonate (DMC), has been investigated in detail. Indeed, it was demonstrated that DMC is able to react with a plethora of nucleophilic substrates under both batch and continuous flow conditions. Of particular interest are the methylation of amines, thiols, sulfones, derivatives of aryl and aroxy-acetic acids and of course of phenolics compounds.[67]

As reported in Table 11.2, DMC represents a benign alternative for methylations, overcoming the problems related to the toxicity and corrosiveness of other alkylating agents. Moreover, the use of alkyl halides and sulphates leads also to the need for stoichiometric amounts of a base, with the related drawbacks associated with the waste salt separation and disposal. Even more

Table 11.2 Comparison of the toxicological and ecotoxicological properties of DMC and traditional alkylating agents.

Property	Dimethylsulphate (DMS)	Methyl iodide (CH_3I)	Dimethylcarbonate (DMC)
Oral acute toxicity (rats)	LD50 440 mg kg^{-1}	LD50 76 mg kg^{-1}	LD50 13.8 g kg^{-1}
Acute toxicity inhalation	LC50 1.5 mg L^{-1} (4 h)	LC50 1.3 mg L^{-1} (4 h)	LC50 140 mg L^{-1} (4 h)
Biodegradability (OECD-301C)	Rapid hydrolysis	16% (28 days)	>90% (28 days)
Mutagenic properties	Mutagenic	No data available	None
Carcinogenic properties	Carcinogen	Suspected carcinogen	None

important, the utilisation of OCs as alkylating agents leads to the formation of unstable alkoxy-carbonate anions as co-products of the reaction. This intermediate rapidly decomposes into CO_2 and alcohol (methanol in the case of DMC), leading to an irreversible reaction pathway that finally contributes to increase the yield into the desired products. Furthermore, this reaction shows a very high and specific chemo-selectivity that may be considered complementary compared to the behaviour of methanol, as explained in the preceding chapters. Indeed, considering phenol methylation, the reaction with DMC is highly selective towards *O*-alkylation, promoting the formation of anisole (phenylmethyl ether).[68,69] Another interesting example of the peculiar behaviour of OCs is represented by the reaction of DMC and substrates characterised by the co-presence of both hydroxyl and carboxylic acid functional groups on the phenyl ring. In this case, the reaction is selective towards the methylation of the carboxylic acid, leading to the formation of the corresponding ester.[70,71]

Lastly, DMC is also highly efficient for the selective methylation of other heteroatoms. In particular, it is reported to be highly selective in the mono-*N*-methylation of anilines,[72,73] in the *S*-methylation of thiophenols and in the methylation of more activated carbon atoms (*e.g.* in arylacetonitriles).[74]

The list of catalysts for methylation with DMC mainly includes basic or amphoteric materials, namely: hydroxides, alkoxides and carbonates (*e.g.* Cs_2CO_3), hydrotalcites, basic zeolites, alkali metal halides, tertiary amines, phosphines and phosphoniumhalides, *etc.*[75] However, also acidic systems, such as some acidic commercial zeolites and supported $AlPO_4$, have been reported to be active catalysts for this kind of alkylation. Of particular interest is the selective *O*-methylation of catechol to yield guaiacol and veratrole in the gas-phase.[76]

However, despite the promising results obtained in methylation reactions with DMC, only a few studies are focused on the alkylation using longer-chain carbonates. For instance, Li *et al.* reported the ethylation of benzene with diethyl carbonate (DEC) on different acid zeolites (*e.g.* HZSM-5), leading to the production of a mixture of mono and poly-alkylated compounds.[77] In the case of phenol alkylation, DEC was investigated as ethylating agent at high temperature (360 °C) on mesoporous nano-silicas (MCM-41). Over these materials, the chemo-selectivity was not total towards *O*-alkylation, leading to the production of a mixture of compounds in which phenylethyl ether and 2-ethylphenol were the main components (at a molar ratio of 2 : 1). Furthermore, impregnating the catalytic system with TiO_2 increased both the phenol conversion and *C*-alkylation selectivity.[78] Similarly, the incorporation of acidic Al sites in the silicate structure of MCM-41 also led to an increase of poly-alkylated products, occurring along with the reduction of the catalyst Si/Al ratio.[79]

Diallylcarbonate is reported to be an innovative and benign allylation agent for a wide range of substrates, including phenolics derived from lignins.[80–82]

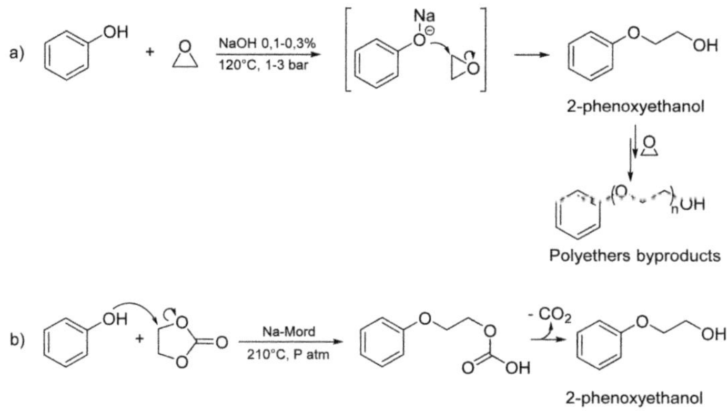

Scheme 11.4 Synthesis of 2-phenoxyethanol from phenol *O*-hydroxyalkylation with (a) ethylene oxide or (b) ethylene carbonate.

It is worth noting that only the reactivity of symmetric carbonates has been considered so far. In the open literature, only a few papers have reported the utilisation of asymmetric carbonates as alkylating agents. For instance, Perosa *et al.* reported on the reactivity of various methyl alkyl carbonates with aromatic substrates in the liquid phase. In this way, the selective methylation of different aromatic compounds has been achieved, thus demonstrating the higher reactivity of the shorter alkyl chain of the asymmetric carbonate.[83]

Finally, also cyclic alkylene carbonates have been considered for the hydroxyalkylation of aromatic nucleophiles. For example, the selective synthesis of 2-phenoxyethanol was achieved using ethylene carbonate (EC) as an alternative reagent for phenol, working at a relatively high temperature (*ca.* 200 °C) in the liquid phase over basic heterogeneous catalytic systems (*e.g.* Na-mordenites). In this way, the formation of the undesired oligomeric by-product was limited, along with an increase of the selectivity of the overall process compared to the industrial one that uses ethylene oxide.[84] The two different reaction pathways are shown below (Scheme 11.4).

2-Phenoxyethanol is an important chemical intermediate for the production of pharmaceuticals, plasticisers and fragrances with an annual global capacity of around 170 000 tons.

Lastly, both EC and propylene carbonate (PC) have also been investigated in the alkylation of aromatic amines.[85]

11.2 Production of Vanillin: Various Industrial Routes and Sources

Vanillin (4-hydroxy-3-methoxybenzaldehyde) is the most important organoleptic component giving the characteristic aroma. Vanillin and its homolog, vanillal (4-hydroxy-3-ethoxybenzaldehyde, or ethylvanillin), have a wide range of applications as flavouring agents in the food industry (approximately 60%) and as additives in perfumery (approximately 30–35%).[86–89]

Mainly, there are two commercially available types of vanillin: (a) the chemical pure vanillin, obtained from the derivatisation of guaiacol or lignosulphonates from the paper and pulp industry; and (b) natural vanilla, extracted from the cured pods of the flowers of *Vanilla planifolia*, an orchid from Mexico, or other botanical sources. Indeed, the natural 'vanilla extract' is characterised by the presence of several hundred different compounds in addition to vanillin.

The synthetic vanillin market represents over 97% of the overall demand; its market size totalled close to 27 Million USD in 2015 and is expected to grow at a rate of more than 7% over the coming years. Natural vanillin supplies less than 2% of the total demand for vanillin and is produced from glucovanillin when the beans of the orchid are subjected to a complex curing process: the Mexican process (sun method) or the Madagascar process (Bourbon method). After this curing process, the dry beans obtained contain vanillin as the most abundant component at 2–3 wt%. The price of chemically synthesised 'nature-identical' vanillin is lower than the price of cured vanilla pods. Of increasing importance are biotechnological transformations from natural substrates such as the biofermentation of natural ferulic acid obtained from rice bran.[88,90–93]

Synthetic pathways include chemical routes derived from guaiacol; the latter can be obtained from benzene either *via* the classical route with an *o*-chloronitrobenzene intermediate, or by hydroxylation of phenol into cathecol with H_2O_2 and methylation of catechol to guaiacol.

In the 1870s, Haarmann and Reimer were the firsts to produce vanillin, using coniferin as the raw material. Instead, the Reimer–Tiemann protocol toward aromatic aldehydes reacts guaiacol with $CHCl_3$ in KOH. Synthetic vanillin derived from eugenol (4-allyl-2-methoxyphenol) found in clove oil – an industrial method still in use – was the main source for vanillin until the 1920s. After the 1930s, the major synthetic vanillin supply was from lignin waste (black liquor) produced by the sulphite pulping process. Vanillin from wood is claimed to have the lowest CO_2 footprint, producing, for instance, 90% less CO_2 than vanillin from guaiacol.[94–97] Since the 1980s, vanillin production from petrochemical guaiacol has been gaining increasing importance.

Lignin is the raw material for the production of vanillin based on lignosulphonate. In particular, the lignin present in sulphite liquor from the pulp and cellulose industry has been considered. The chemical composition of lignin varies greatly depending on the plant species. Industrially, softwood spent sulphite liquors are oxidised in alkaline media by air in the presence of a catalyst. Vanillin is then extracted with a solvent, and a vacuum distillation yields technical-grade vanillin, which is re-crystallised to obtain food-grade vanillin.

In the last years, the increasing demand for natural flavours has led to a growing interest in the production of vanillin from natural sources using microorganisms or enzymes; in fact, the term 'natural' can be applied in both the United States and EU when a product is derived from a natural raw material *via* a biological process. Options include the biotransformation of stilbenes (commonly found in spruce bark), of ferulic acid (3-(4-hydroxy-3-methoxyphenyl)-propenoic acid), a compound found in the cell walls of grass and corn hulls (today mainly isolated from rice bran), and of eugenol (the main component of clove essential oil).[89–91]

Currently, chemical routes account for the majority of vanillin production. The availability of the key reactant, guaiacol, along with low manufacturing costs compared to natural and biotechnological processes, are the main reasons for the growth in the synthetic vanillin market size. The older *o*-chloronitrobenzene (ONCB) process, including nitroanisole and anisidine as intermediates, generates pollutants and toxic by-products; therefore, it has been discontinued in many countries. Not only does the ONCB process use benzene as the starting reactant, but it is also used as the solvent and has a highly negative eco-balance, with large amounts of tars and COD in liquid effluents, and VOCs emitted. Moreover, traces of chlorine impurities may derive from the starting material, despite the purification steps. The greener process starts with the reaction of guaiacol with glyoxylic acid, as summarised in Scheme 11.5; this process, developed by Rhodia, yields a high-quality low-pollution product.

An alternative pathway has been investigated, but not exploited commercially; in this route, characterised by a very high atomic efficiency (all C atoms of the reactants are incorporated into the final product) and co-production of water only, guaiacol is reacted with an aqueous formalin solution in the presence of a solid acid catalyst.[98–100] The hydroxymethylation reaction yields *ortho* and *para* vanillic alcohols, which can then be oxidised to vanillin with O_2 in the presence of a supported noble metal catalyst promoted by Bi^{3+}. This procedure, however, has some drawbacks, such as the use of very diluted aqueous solutions in the hydroxymethylation step, the low selectivity to *p*-vanillol on formaldehyde basis, and the use of a toxic reactant.

Variants of the process include (a) the use of TS-1, a titanium silicalite, as the heterogeneous catalyst for the hydroxylation of phenol to catechol and hydroquinone, developed by EniChem Synthesis, now implemented at

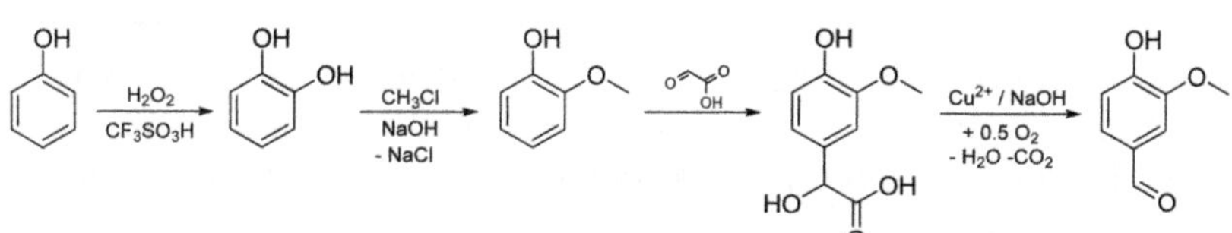

Scheme 11.5 Rhodia process for the synthesis of vanillin.

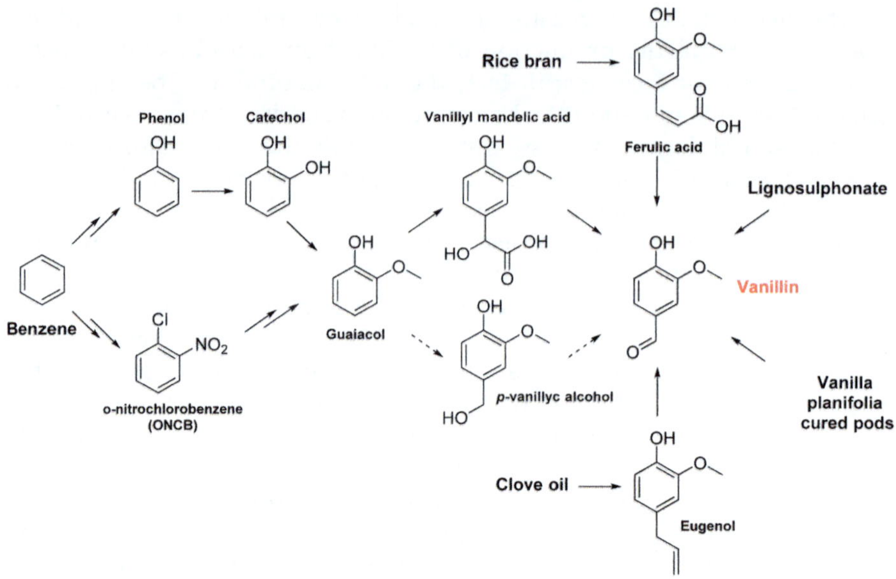

Scheme 11.6 Overall synthetic pathways for the production of vanillin.

the Ravenna (I) factory by Camlin FS Europe;[101] and (b) methanol instead of methyl chloride (or dimethylsulphate) for the mono-methylation of catechol to guaiacol.[102] The reaction with methanol is carried out in the gas phase, as developed by Ube.

Scheme 11.6 summarises all the different pathways used in industry today for the synthesis of vanillin.

11.3 Hydroxytyrosol: The Continuous Pursuit for Sustainable Production

Hydroxytyrosol (3,4-dihydroxyphenylethanol, HyT) is a phenolic derivative considered the most powerful antioxidant compound after gallic acid. Indeed, the regular consumption of HyT has several beneficial effects – antioxidant, anti-inflammatory, anticancer, and protective of skin and eyes – and so it has been widely researched as a food additive.[103,104]

HyT can be found in olives and olive leaves; therefore, it is generally recovered from olive oil mill wastewater, in addition to tyrosol and oleuropein, from which HyT can be produced by hydrolysis.[105,106] However, the recovery of these value-added compounds from oil mill wastewaters by extraction or membrane filtrations has several drawbacks, namely the use of huge amounts of organic solvents, low extraction yield, and expensive multi-step purification steps.[107,108]

For these reasons, in recent years, several efforts have been made to find suitable synthetic pathways towards the efficient production of HyT, and different possibilities starting from various substrates have been suggested (Scheme 11.7).

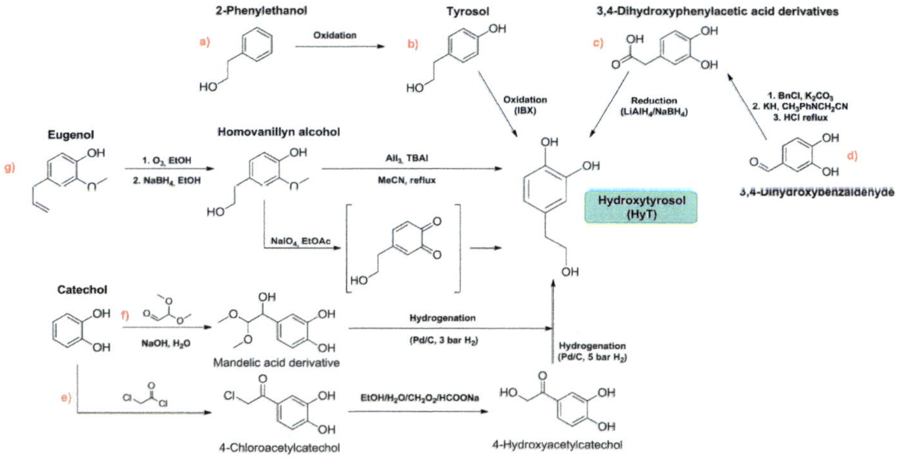

Scheme 11.7 Main synthetic strategies for hydroxytyrosol production. See text for description of the labels.

One possibility is to promote the oxidation of tyrosol or 2-phenylethanol (Scheme 11.7a and b), introducing new hydroxyl groups on aromatic rings. However, this reaction has been reported to need either iodine-based oxidants (*e.g.* 2-iodoxybenzoic acid, IBX), followed by a reduction workup with sodium dithionite ($Na_2S_2O_4$),[109] or a complex multistep approach to promote a Baeyer–Villiger oxidation mechanism, using halogenated solvents, 3-chloroperoxybenzoic acid as the oxidant, and diphenyl selenide as the catalyst.[110] Another interesting possibility is that of using supported or genetically modified enzymes, namely tyrosinase and toluene monooxygenases, to promote the selective introduction of new –OH groups in an aromatic ring.[111–113] Nevertheless, these procedures suffer from low productivity and the need for diluted streams, low reaction temperatures, and buffered pH conditions.

A completely different approach is the chemical reduction of 3,4-dihydroxyphenylacetic acid (or its esters and derivatives) by stoichiometric reductants such as hydrides ($LiAlH_4$, $NaBH_4$) in the presence of tetrabutylammonium boronate (Scheme 11.7c).[106] Although this procedure rapidly leads to high yields of the desired product, the high cost of the starting materials precludes its development as a satisfactory industrial-scale source of HyT.

More recently, the one-carbon homologation of a protected form of 3,4-dihydroxybenzaldehyde was proposed (Scheme 11.7d). However, this pathway also suffers from several drawbacks: (i) the starting material is expensive; (ii) it is a multistep reaction pathway that requires the protection of the aromatic –OH groups and leads to the formation of a protected 3,4-dihydroxyphenylacetic acid derivative, then following the same reaction steps as seen before; (iii) it requires the use of halogenated reactants and solvents.

A few years ago, Quideau *et al.* proposed an interesting route starting from clove-derived eugenol *via* oxidative cleavage of the double bond, and subsequent reduction of the acid (Scheme 11.7g).[114] This is an interesting approach, because of the commercial availability of eugenol and its relatively low market price of around US$ 5 per kg. Moreover, eugenol is produced in large quantities and is widely used in dentistry, as an essence in cosmetics and as a flavouring in food products.[115] Nevertheless, this procedure requires the use of stoichiometric oxidants and reducing agents, with the generation of a non-negligible amount of waste.

Lastly, HyT can be obtained starting from catechol, which is the cheapest starting material compared to the other alternatives discussed above. Interestingly, a DSM patent in 2009 proposed the synthesis of HyT *via* a three-step reaction pathway (Scheme 11.7e): first, 4-chloroacetylcatechol is produced by the alkylation of catechol with 4-chloroacetylchloride in the presence of phosphorus oxychloride; this compound is then isolated and dissolved in a solution (H_2O/ethanol/formic acid) of sodium formate to produce 4-hydroxyacetylcatechol; lastly, the latter is hydrogenated to HyT in batch, using 10% w/w Pd/C and 5 bar of hydrogen.[116]

However, the latter approach needs halogenated reagents, entails difficult workout procedures, and leads to the formation of inorganic waste. Very recently, a step forward towards the sustainable synthesis of HyT starting from catechol was reported by Ziosi *et al.* (Scheme 11.7f).[117] In their two-step procedure, catechol is hydroxyalkylated in a basic aqueous medium with an equimolar amount of 2,2-dimethoxyacetaldehyde to produce the corresponding mandelic *para*-hydroxyalkylated derivative with 70% selectivity. This intermediate is then hydrogenated using a 10% w/w Pd/C catalyst (150 °C, 3 bar H_2), obtaining the total conversion of the substrate and 68% selectivity to the target product. The latter approach represents an innovative example of a HyT synthetic pathway that does not require the use of halogenated reagents or solvents and prevents the production of stoichiometric waste.

11.4 1,2-Methylendioxybenzene and Related Compounds

1,2-Methylendioxybenzene (MDB, 1,3-benzodioxole) is an important catechol derivative and heterocycle compound. MDB plays a key role in the chemical industry. Indeed, many compounds containing a methylenedioxy core are bioactive and find a wide range of applications in a number of sectors, namely: (i) in agrochemical and household products as pesticides (for example, piperonyl butoxide or PBO, an insecticide synergist),[118] (ii) in pharmaceuticals (Paroxetine, Tadalafil, also known as Cialis®, *etc.*),[119–121] and (iii) in fragrances (piperonal, also called heliotropin, helional, *etc.*).[122] Moreover, both safrole and iso-safrole, important intermediates in the synthesis of many valuable compounds (including PBO and piperonal) can be produced starting from MDB.[123,124] MDB and its major derivatives are shown in Scheme 11.8.

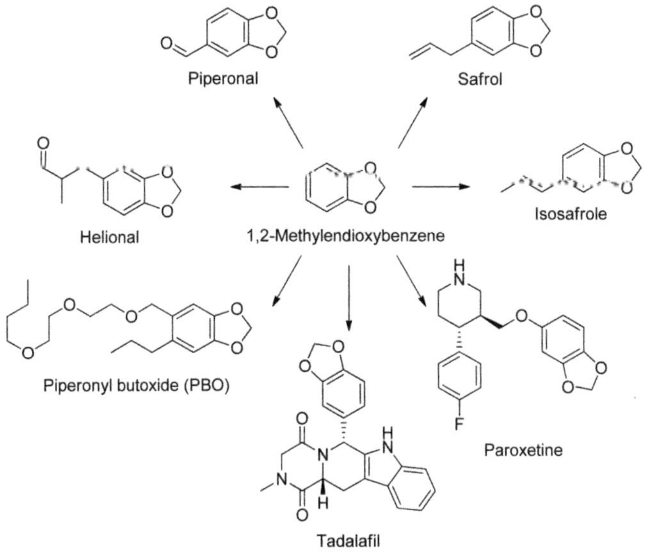

Scheme 11.8 MDB main derivatives.

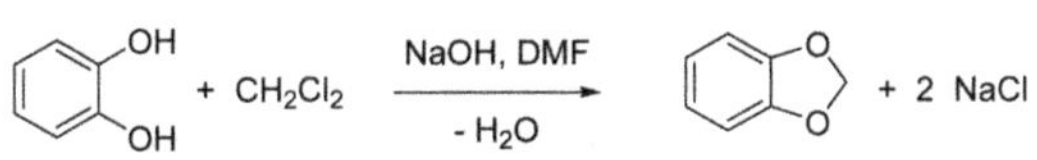

Scheme 11.9 Industrial production of MDB.

Industrially, MDB is usually synthesised in batch reactors, working in liquid phase under alkaline conditions by reacting the activated catecholate anions with a dihalomethane (mainly dichloro or dibromomethane).[125] This reaction needs the presence of aprotic polar solvents to improve the yield to the desired product; for this reason, dimethylsulphoxide, *N*-methylpyrrolidone or dimethylformamide is often used (Scheme 11.9).

This synthetic route has some significant drawbacks, namely the use of toxic solvents or halogenated reagents that lead to the production of equimolar amounts of inorganic salt waste that require disposal.

An innovative approach for the sustainable production of MDB may be found in the continuous-flow, gas-phase alkylation of catechol. Indeed, this approach, based on a fixed bed of a heterogeneous catalyst may lead to the improved production of MDB with the concomitant benefit of simpler purification of the product and less waste. Giugni *et al.* recently proposed an eco-friendly synthesis of MDB in the gas phase by reacting catechol with formaldehyde acetals (*e.g.* diethoxymethane) in the presence of appropriate heterogeneous catalysts.[126] In particular, Ti-silicalite (TS-1) and Sn-doped MCM-41 proved to be active in the reaction, affording an MDB yield of up to 20% with selectivity values higher than 80%.

11.5 Conclusions

Phenolic compounds are used in several sectors of the chemical industry. Over the past few years, many alternative synthetic pathways have been suggested for the replacement of older processes with more sustainable ones. Examples include the replacement of halogenated reactants with alcohols and aldehydes for ring functionalisation, and the substitution of homogeneous catalysts with heterogeneous systems, which are more easily removed from the reaction medium, regenerated, and reused, finally simplifying the product purification steps.

In this context, the more emblematic reaction is that of phenol ring-methylation with methanol: a process that has been used in industry for a number of years, but whose mechanism has only been revealed recently.

References

1. H. Fiege, Cresols Xylenols, in *Ullmann's Encyclopedia of Industrial Chemistry*, 2005, pp. 1.
2. H. Fiege, H.-W. Voges, T. Hamamoto, T. Iwata, H. Miki, Y. Fujita, H.-J. Buysch, D. Garbe and W. Paulus, Phenol Derivatives, in *Ullmann's Encyclopedia of Industrial Chemistry*, 2005, pp. 1.
3. G. S. Devi, D. Giridhar and B. M. Reddy, *J. Mol. Catal. A: Chem.*, 2002, **181**, 173.
4. E. Santacesaria, D. Grasso, D. Gelosa and S. Carrà, *Appl. Catal.*, 1990, **64**, 83.
5. F. M. Bautista, J. M. Campelo, A. Garcia, D. Luna, J. M. Marinas, A. Romero, J. A. Navio and M. Macias, *Appl. Catal., A*, 1993, **99**, 161.
6. H. Grabowska, W. Kaczmarczyk and J. Wrzyszcz, *Appl. Catal.*, 1989, **47**, 351.
7. S. Velu and C. S. Swamy, *Appl. Catal., A*, 1996, **145**, 141.
8. S. Velu and C. S. Swamy, *Appl. Catal., A*, 1994, **119**, 241.
9. V. V. Rao, V. Durgakumari and S. Narayanan, *Appl. Catal.*, 1989, **49**, 165.
10. S. C. Lee, S. W. Lee, K. S. Kim, T. J. Lee, D. H. Kim and J. C. Kim, *Catal. Today*, 1998, **44**, 253.
11. V. V. Rao, K. V. R. Chary, V. Durgakumari and S. Narayanan, *Appl. Catal.*, 1990, **61**, 89.
12. S. Velu and C. S. Swamy, *Appl. Catal., A*, 1997, **162**, 81.
13. S. Karuppannasamy, K. Narayanan and C. N. Pillai, *J. Catal.*, 1980, **66**, 281.
14. T. Kotanigawa, M. Yamamoto, K. Shimakowa and Y. Yoshida, *Bull. Chem. Soc. Jpn.*, 1971, **44**, 1961.
15. L. Garcia, G. Giannetto, M. R. Goldwasser, M. Guisnet and P. Magnoux, *Catal. Lett.*, 1996, **37**, 121.

16. M. Marczewski, J.-P. Bodibo, G. Perot and M. Guisnet, *J. Mol. Catal.*, 1989, **50**, 211.
17. V. Durgakumari, S. Narayanan and L. Guczi, *Catal. Lett.*, 1990, **5**, 377.
18. P. Beltrame, P. L. Beltrame, P. Carniti, A. Castell and L. Forni, *Appl. Catal.*, 1987, **29**, 327.
19. S. Balsamà, P. Beltrame, P. L. Beltrame, P. Carniti, L. Forni and G. Zuretti, *Appl. Catal.*, 1984, **13**, 161.
20. R. Pierantozzi and A. F. Nordquist, *Appl. Catal.*, 1986, **21**, 263.
21. K. Tanabe and T. Nishizaki, in *Proceed. 6th ICC*, ed. G. C. Bond *et al.*, The Chemical Society, London, 1977, p. 863.
22. C. Bezouhanova and M. A. Al-Zihari, *Appl. Catal., A*, 1992, **83**, 45.
23. K.-T. Li, I. Wang and K.-R. Chang, *Ind. Eng. Chem. Res.*, 1993, **32**, 1007.
24. F. Nozaki and I. Kimura, *Bull. Chem. Soc. Jpn.*, 1977, **50**, 614.
25. R. F. Parton, J. M. Jacobs, D. R. Huybrechts and P. A. Jacobs, *Stud. Surf. Sci. Catal.*, 1989, **46**, 163.
26. S. Namba, T. Yashima, Y. Itaba and N. Hara, *Stud. Surf. Sci. Catal.*, 1980, **5**, 105.
27. S. Sato, K. Koizumi and F. Nozaki, *Appl. Catal., A*, 1995, **133**, L7.
28. M. C. Samolada, E. Grigoriadou, Z. Kiparissides and I. A. Vasalos, *J. Catal.*, 1995, **152**, 52.
29. S. Narayanan, V. V. Rao and V. Durgakumari, *J. Mol. Catal.*, 1989, **52**, L29.
30. J. M. Campelo, A. Garcia, D. Luna, J. M. Marinas and M. S. Moreno, *Stud. Surf. Sci. Catal.*, 1988, **41**, 249.
31. S. Sato, K. Koizumi and F. Nozaki, *J. Catal.*, 1998, **178**, 264.
32. T. Kotanigawa, *Bull. Chem. Soc. Jpn.*, 1974, **47**, 950.
33. T. Mathew, S. Shylesh, B. M. Devassy, M. Vijayaraj, C. V. V. Satyanarayana, B. S. Rao and C. S. Gopinath, *Appl. Catal., A*, 2004, **273**, 35.
34. J. Shabtai, L. H. Klemm and D. R. Taylor, *J. Org. Chem.*, 1968, **33**, 1494.
35. J. Shabtai, L. H. Klemm and D. R. Taylor, *J. Org. Chem.*, 1968, **33**, 1489.
36. L. H. Klemm, J. Shabtai and D. R. Taylor, *J. Org. Chem.*, 1968, **33**, 1480–1488.
37. J. Shabtai, L. H. Klemm and D. R. Taylor, *J. Org. Chem.*, 1970, **35**, 1075.
38. D. R. Taylor and K. H. Ludlum, *J. Phys. Chem.*, 1972, **76**, 2882.
39. H. Hattori, K. Shimazu, N. Yoshii and K. Tanabe, *Bull. Chem. Soc. Jpn.*, 1976, **49**, 969.
40. K. Tanabe, *Stud. Surf. Sci. Catal.*, 1985, **20**, 1.
41. J. Wrzyszcz, H. Grabowska, W. Miśta, L. Syper and M. Zawadzki, *Appl. Catal., A*, 1998, **166**, 249.
42. T. Mathew, M. Vijayaraj, S. Pai, B. B. Tope, S. G. Hegde, B. S. Rao and C. S. Gopinath, *J. Catal.*, 2004, **227**, 175.
43. H. Grabowska, W. Miśta, L. Syper, J. Wrzyszcz and M. Zawadzki, *J. Catal.*, 1996, **160**, 134.
44. B. V. Vora, J. A. Kocal, P. T. Barger, R. J. Schmidt and J. A. Johnson, Alkylation, *Kirk-Othmer Encycl. Chem. Technol.*, 2003, **2**, 169.

45. T. Mathew, B. B. Tope, N. R. Shiju, S. G. Hegde, B. S. Rao and C. S. Gopinath, *Phys. Chem. Chem. Phys.*, 2002, **4**, 4260.
46. T. Mathew, B. S. Rao and C. S. Gopinath, *J. Catal.*, 2004, **222**, 107.
47. G. Busca, *Catal. Today*, 1996, **27**, 457.
48. J. C. Lavalley, *Catal. Today*, 1996, **27**, 377.
49. P. Mars, J. J. Scholten and P. Zwietering, *Adv. Catal.*, 1963, **14**, 35.
50. X. D. Peng and M. A. Barteau, *Langmuir*, 1989, **5**, 1051.
51. N. D. Lazo, D. K. Murray, M. L. Kieke and J. H. Haw, *J. Am. Chem. Soc.*, 1992, **114**, 8552.
52. F. Wang and T. Tsai, *Catal. Today*, 1998, **44**, 259.
53. A. H. Padmasri, A. Venugopal, V. Durga Kumari, K. S. Rama Rao and P. Kanta Rao, *J. Mol. Catal. A: Chem.*, 2002, **188**, 255.
54. A. R. Shyam, R. Dwivedi, V. S. Reddy, K. V. R. Chary and R. Prasad, *Green Chem.*, 2002, **4**, 558.
55. A. R. Gandhe and J. B. Fernandes, *J. Mol. Catal. A: Chem.*, 2005, **226**, 171.
56. M. Bregolato, V. Bolis, C. Busco, P. Ugliengo, S. Bordiga, F. Cavani, N. Ballarini, L. Maselli, S. Passeri, I. Rossetti and L. Forni, *J. Catal.*, 2007, **245**, 285.
57. M. Bolognini, F. Cavani, D. Scagliarini, C. Flego, C. Perego and M. Saba, *Catal. Today*, 2002, **75**, 103.
58. N. Ballarini, F. Cavani, L. Maselli, A. Montaletti, S. Passeri, D. Scagliarini, C. Flego and C. Perego, *J. Catal.*, 2007, **251**, 423.
59. F. Cavani, L. Maselli, S. Passeri and J. A. Lercher, *J. Catal.*, 2010, **269**, 340.
60. V. Crocellà, G. Cerrato, G. Magnacca, C. Morterra, F. Cavani, S. Cocchi, S. Passeri, D. Scagliarini, C. Flego and C. Perego, *J. Catal.*, 2010, **270**, 125.
61. N. Ballarini, F. Cavani, L. Maselli, S. Passeri and S. Rovinetti, *J. Catal.*, 2008, **256**, 215.
62. N. Ballarini, F. Cavani, S. Passeri, L. Pesaresi, A. F. Lee and K. Wilson, *Appl. Catal., A*, 2009, **366**, 184.
63. T. Tabanelli, S. Cocchi, B. Gumina, L. Izzo, M. Mella, S. Passeri, F. Cavani, C. Lucarelli, J. Schütz, W. Bonrath and T. Netscher, *Appl. Catal., A*, 2018, **552**, 86.
64. G. Lamoureux and C. Agüero, *ARKIVOC*, 2009, **2009**, 251.
65. J. Mccann, E. Choi, E. Yamasaki, B. N. Ames, *Proc. Natl. Acad. Sci. U. S. A.*, 1975, **72**, 5135.
66. L. A. Poirier, G. D. Stoner and M. B. Shimkin, *Cancer Res.*, 1975, **35**, 1411.
67. P. Tundo and M. Selva, *Acc. Chem. Res.*, 2002, **35**, 706.
68. S. Ouk, S. Thiébaud, E. Borredon and P. Le Gars, *Green Chem.*, 2002, **4**, 431.
69. L. S. Zhen, Z. J. Xuan, M. M. Wei, X. H. Bao and N. Sun, *Green Chem.*, 2005, **7**, 97.

70. W. C. Shieh, S. Dell and O. Repič, *J. Org. Chem.*, 2002, **67**, 2188.
71. M. Selva and P. Tundo, *J. Org. Chem.*, 2006, **71**, 1464.
72. K. Takeda, A. Nakajima, M. Takeda and E. Yoshii, *Org. Synth.*, 1999, **76**, 169.
73. M. Selva, P. Tundo and A. Perosa, *J. Org. Chem.*, 2003, **68**, 7374.
74. P. Tundo, F. Trotta, G. Moraglio and F. Ligorati, *Ind. Eng. Chem. Res.*, 1988, **27**, 1565.
75. J. H. Clements, *Ind. Eng. Chem. Res.*, 2003, **42**, 663.
76. R. Luque, J. Manuel Campelo, T. D. Conesa, D. Luna, J. M. Marinas and A. A. Romero, *New J. Chem.*, 2006, **30**, 1228.
77. Y. Li, B. Xue and Y. Yang, *Fuel Process. Technol.*, 2009, **90**, 1220.
78. B. Xue, J. Xu, P. Liu, L. Lv, C. Xu, Y. Li and K. Zhang, *J. Mol. Catal. A: Chem.*, 2012, **357**, 50.
79. S. Udayakumar, A. Pandurangan and P. K. Sinha, *Appl. Catal., A*, 2004, **272**, 267.
80. O. Kreye, L. C. Over, T. Nitsche, R. Z. Lange and M. A. R. Meier, *Tetrahedron*, 2015, **71**, 293.
81. L. C. Over and M. A. R. Meier, *Green Chem.*, 2015, **18**, 197.
82. F. Manjolinho, M. F. Grünberg, N. Rodríguez and L. J. Gooßen, *Eur. J. Org. Chem.*, 2012, 4680.
83. A. Perosa, M. Selva, P. Tundo and F. Zordan, *Synlett*, 2000, **2**, 272.
84. P. Ziosi, T. Tabanelli, G. Fornasari, S. Cocchi, F. Cavani and P. Righi, *Catal. Sci. Technol.*, 2014, **4**, 4386.
85. M. Selva, M. Fabris, V. Lucchini, A. Perosa and M. Noè, *Org. Biomol. Chem.*, 2010, **8**, 5187.
86. B. M. Upton and A. M. Kasko, *Chem. Rev.*, 2016, **116**, 2275.
87. I. Delidovich, P. J. C. Hausoul, L. Deng, R. Pfützenreuter, M. Rose and R. Palkovits, *Chem. Rev.*, 2016, **116**, 1540.
88. S. Khoyratty, H. Kodja and R. Verpoorte, *Ind. Crops Prod.*, 2018, **125**, 433.
89. S. Ramachandra Rao and G. Ravishankar, *J. Sci. Food Agric.*, 2000, **80**, 289.
90. R. H. H. Van den Heuvel, M. W. Fraaije, C. Laane and W. J. H. Van Berkel, *J. Agric. Food Chem.*, 2001, **49**, 2954.
91. H. Priefert, J. Rabenhorst and A. Steinbüchel, *Appl. Microbiol. Biotechnol.*, 2001, **56**, 296.
92. J. Schrader, M. M. Etschmann, D. Sell, J. M. Hilmer and J. Rabenhorst, *Biotechnol. Lett.*, 2004, **26**, 463.
93. T. Li and J. P. N. Rosazza, *Appl. Environ. Microbiol.*, 2000, **66**, 684.
94. V. E. Tarabanko, D. V. Petukhov and G. E. Selyutin, *Kinet. Catal.*, 2004, **45**, 569.
95. E. A. B. da Silva, M. Zabkova, J. D. Araújo, C. A. Cateto, M. F. Barreiro, M. N. Belgacem and A. E. Rodrigues, *Chem. Eng. Res. Des.*, 2009, **87**, 1276.
96. H.-R. Bjørsvik and F. Minisci, *Org. Process Res. Dev.*, 1999, **3**, 330.
97. V. E. Tarabanko and N. Tarabanko, *Int. J. Mol. Sci.*, 2017, **18**, 2421.

98. F. Cavani, L. Dal Pozzo, L. Maselli and R. Mezzogori, *Stud. Surf. Sci. Catal.*, 2002, **142A**, 565.
99. F. Cavani, M. Corrado and R. Mezzogori, *J. Mol. Catal. A: Chem.*, 2002, **182**, 447.
100. M. Armandi, B. Bonelli, E. Garrone, M. Ardizzi, F. Cavani, L. Dal Pozzo, L. Maselli, R. Mezzogori and G. Calestani, *Appl. Catal., B*, 2007, **70**, 585.
101. M. G. Clerici and P. Ingallina, *Catal. Today*, 1998, **41**, 351.
102. M. Ardizzi, N. Ballarini, F. Cavani, E. Chiappini, L. Dal Pozzo, L. Maselli and T. Monti, *Appl. Catal., B*, 2007, **70**, 597.
103. L. Martínez, G. Ros and G. Nieto, *Medicines*, 2018, **5**, 13.
104. A. Segura-Carretero and J. Curiel, *Int. J. Mol. Sci.*, 2018, **19**, 2899.
105. C. Romero, M. Brenes, K. Yousfi, P. García, A. García and A. Garrido, *J. Agric. Food Chem.*, 2004, **52**, 479.
106. K. L. Tuck, H. W. Tan and P. J. Hayball, *J. Agric. Food Chem.*, 2000, **48**, 4087.
107. N. Sabatini, *Recent Pat. Food, Nutr. Agric.*, 2010, **2**, 154.
108. A. Roig, M. L. Cayuela and M. A. Sánchez-Monedero, *Waste Manage.*, 2006, **26**, 960.
109. R. Bernini, E. Mincione, F. Crisante, M. Barontini and G. Fabrizi, *Tetrahedron Lett.*, 2009, **50**, 1307.
110. G. Piersanti, M. Retini, J. L. Espartero, A. Madrona and G. Zappia, *Tetrahedron Lett.*, 2011, **52**, 4938.
111. M. Brouk and A. Fishman, *Food Chem.*, 2009, **116**, 114–121.
112. G. Botta, B. M. Bizzarri, A. Garozzo, R. Timpanaro, B. Bisignano, D. Amatore, A. T. Palamara, L. Nencioni and R. Saladino, *Bioorg. Med. Chem.*, 2015, **23**, 5345.
113. M. Guazzaroni, C. Crestini and R. Saladino, *Bioorg. Med. Chem.*, 2012, **20**, 157.
114. D. Deffieux, P. Gossart and S. Quideau, *Tetrahedron Lett.*, 2014, **55**, 2455.
115. P. Milind and K. Deepa, *Int. J. Res. Ayurveda Pharm.*, 2011, **2**, 47.
116. M. Breuninger and M. Joray, *Eur. Pat.*, EP2114846 B1, DSM Nutritional Products AG, 2009.
117. P. Ziosi, C. Paolucci, F. Santarelli, T. Tabanelli, S. Passeri, F. Cavani and P. Righi, *ChemSusChem*, 2018, **11**, 2202.
118. J. E. Casida, *J. Agric. Food Chem.*, 1970, **18**, 753.
119. N. Micale, M. Zappalà and S. Grasso, *Farmaco*, 2002, **57**, 853.
120. M. Pollack, R. Zaninelli, A. Goddard, J. McCafferty, K. Bellew, D. Burnham and M. Iyengar, *J. Clin. Psychiatry*, 2001, **62**, 350.
121. K. M. Smith and F. Romanelli, *J. Am. Pharm. Assoc.*, 2005, **45**, 63.
122. J. Panten and H. Surburg, Flavors and Fragrances, 1. General Aspects, in *Ullman's Encyclopedia of Industrial Chemistry*, 2015.
123. S. Wang, J. Liu, C. Qian and X. Chen, *Res. Chem. Intermed.*, 2012, **38**, 147.

124. H. M. Alvarez, D. P. Barbosa, A. T. Fricks, D. A. G. Aranda, R. H. Valdés and O. A. C. Antunes, *Org. Process Res. Dev.*, 2006, **10**, 941.
125. W. Bonthrone and J. W. Cornforth, *J. Chem. Soc. C*, 1969, **0**, 1202–1204.
126. A. Giugni, D. Impalà, O. Piccolo, A. Vaccari and A. Corma, *Appl. Catal., B*, 2010, **98**, 72.

CHAPTER 12

Transition Metal Catalysis in Micellar Media: Much More Than a Simple Green Chemistry Promise

ALESSANDRO SCARSO AND GIORGIO STRUKUL*

Università Ca' Foscari Venezia, Dipartimento di Scienze Molecolari e Nanosistemi, via Torino 155, 30100 Venezia, Italy
*E-mail: strukul@unive.it

12.1 Introduction

Spurred by the nowadays well-established environmental sensitivity, the new perspectives in organic synthesis and catalysis are strongly bound to the twelve principles of Green Chemistry. The latter is not just a young subdiscipline of Chemistry but also a new reference standard for research in the general field of synthetic organic chemistry. In the synthesis of fine chemicals or pharmaceuticals, solution chemistry is the standard approach, often performed with the aid of soluble catalysts using traditional batch reactors. Among the constituents of a liquid phase reaction, the solvent is by far the major component, representing about 80% of the total mass; regrettably, however, its recovery and recycle suffer from only moderate efficiency. This in turn imposes the problem of disposing large

Green Chemistry Series No. 61
Green Synthetic Processes and Procedures
Edited by Roberto Ballini

Published by the Royal Society of Chemistry, www.rsc.org

amounts of exhausted solvents, particularly in the pharmaceutical industry where the final product is very often the result of numerous steps. According to an evaluation carried out on the Swiss chemical industry (based mainly on fine and pharma chemicals), at the end of a process, solvent recovery is barely above 30%, the remaining being simply burnt to recover heat.[1] This situation entails several economic as well as environmental drawbacks related to both new solvent supply and increased gas emissions into the atmosphere. In this context, the environmental acceptability of the solvent used becomes a key issue. This is the main reason why, in industrial practice, methanol, acetone, and toluene are the most common reaction media.

In a chemical reaction, the solvent fulfills several functions at the same time: (i) it ensures the contact between reagents of different polarities, (ii) it controls heat transfer, and (iii) it favors reactant interactions leading to the formation of the final product. Among the different possible solvents, water could be the most desirable option as it conjugates the lowest impact on the environment with other advantages. In fact, it is economic, non-toxic, non-flammable, it has tuneable acidity, high heat capacity and large heat of evaporation (which allows easy control of exothermic reactions), high polarity, the coexistence of both hydrogen bond donors and acceptors, it produces neither greenhouse emissions nor it requires synthesis, and finally its E-factor is assumed to be equal to zero.[2] In one word, it can be considered the green solvent *par excellence*. However, against these beautiful features, water suffers from the fundamental disadvantage of not being able to dissolve hydrophobic molecules, such as most organic molecules and organometallic catalysts used in homogeneous phase industrial processes for the synthesis of many commodities, fine chemicals, pharmaceuticals, *etc.* This behavior is due mainly to the so-called classical 'hydrophobic effect',[3] an example of entropic effect where the solvation by water of a hydrophobic organic reactant or the hydrophobic portion of a molecule spontaneously leads to the aggregation of the apolar surfaces, thus requiring a reduced number of water molecules for solvation and liberating several other water molecules with a large overall increase of entropy.

Nevertheless, organic chemistry in water has been successfully applied in several cases because the hydrophobic effect may result also in 'compaction' of the reagents in the transition state, increasing the reaction rates as observed for the first time by Breslow some forty years ago in a Diels–Alder reaction.[4] The most common approach in the use of water as the solvent has been the introduction of appropriate hydrophilic tags on the ligands of intrinsically hydrophobic catalysts and the use of water–organic two phase systems. In this way, the soluble catalyst remains confined in water and can be easily recycled at the end of the reaction by simple decantation, thus solving one of the major weak spots of homogeneous catalysis. This approach has been the basis for the well-known Ruhrchemie-Rhone Poulenc industrial process for the production of butyraldehyde *via* hydroformylation.[5]

Still, how does Mother Nature manage the problem? All enzymatic reactions occur in water and enzymes are amphiphilic macromolecules based on proteins containing both hydrophilic functional groups and large hydrophobic portions. Their compatibility with water is ensured by the hydrophobic effect, which forces the protein part of the enzyme to fold exposing the hydrophilic groups on the external surface in contact with water, thus endowing the enzyme with solubility. At the same time, the hydrophobic portions are confined inside and assume the enzyme characteristic structure, held together by weak interactions (H-bonds, dipole–dipole interactions, *etc.*), that are responsible for the extraordinary catalytic activity and selectivity of these macromolecules capable of picking up the right substrate in a complex mixture and transforming it into the right product. The inside part of an enzyme, where the active site is typically located, is therefore a hydrophobic nanophase separated from water and a suitable reaction medium for hydrophobic reagents.

Soaps, which have been used by man for about 2000 year to remove fats, operate by the same principle. Records on their preparation are reported by Pliny the Elder in his *Naturalis Historia*,[6] where he describes a strange practice by Gauls, who used to melt together tallow and wood ashes, obtaining upon cooling a solid material called *sapo*. Surfactants, the active ingredients of soaps, are amphiphilic molecules with a polar hydrophilic head and a hydrophobic tail that, above a certain concentration (the critical micellar concentration or cmc), self-assemble in solution to give nanometric aggregates (micelles) where the hydrophilic heads are concentrated on the surface in contact with water while the hydrophobic tails point inside. Similarly to enzymes, they constitute an organic nanophase dispersed in water capable of dissolving hydrophobic molecules (such as fats). Surfactants are a wide class of generally cheap chemicals largely available commercially as anionic (**A**), cationic (**C**), zwitterionic (**Z**), and neutral (**N**) molecules (see Figure 12.1 for some examples).

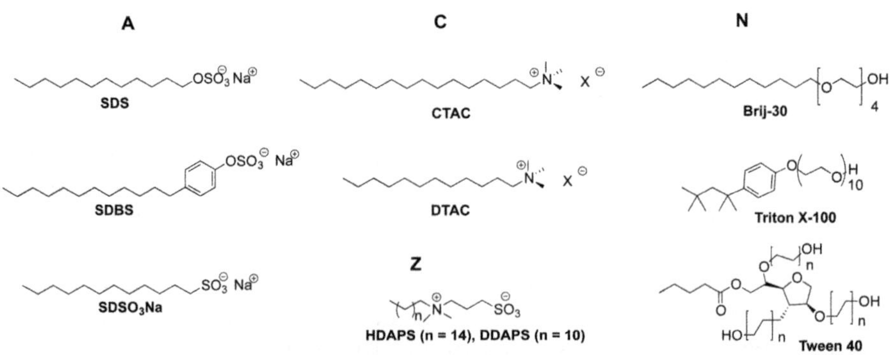

Figure 12.1 Some examples of commercially available surfactants: **A** (anionic), **C** (cationic), **Z** (zwitterionic), and **N** (neutral).

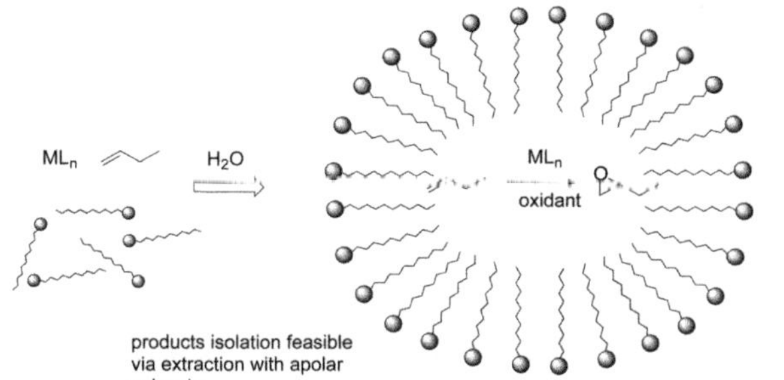

Figure 12.2 Scheme representing a micelle as a nanoreactor through self-assembly and dissolution of reactants, products, and catalyst in an epoxidation process. The advantages of this approach are also indicated.

Micelles can be seen as nanoreactors[7] capable, in principle, of dissolving hydrophobic substrates and catalysts already known to work in organic solvents without the need for modifying the ligands to make the system compatible with water, thus being potentially applicable to a large array of processes. Dissolution inside micelles generally leads to higher local concentrations with respect to ordinary solvents, implying a possible increase in reactivity; moreover, the chance to extract reaction products with water-immiscible organic solvents at the end of the reaction opens the way to possible catalyst recycling with evident economic advantages and better perspectives for practical applications (Figure 12.2).

Clearly, the need for product extraction from aqueous phases raises some possible criticism about the green character of catalysis in water, because organic solvents are involved anyway and the resulting water becomes a water stream contaminated by organics subjected to restrictions before disposal that must be purified by common techniques, such as stripping under vacuum or adsorption on activated carbon.[8] However, as long as water provides extra performance in terms of activity and selectivity, the concerns about the use of limited amounts of traditional solvents at the end of the reaction are at least mitigated and, in the case of successful recycling, reduced to a minimum.

The general subject of micellar catalysis has been reviewed several times over the years since the 1970s.[7a,9] Furthermore, some recent papers have appeared on some new, specific trends in micellar catalysis such as the application to multicomponent reactions,[10] the use of metal nanoparticles associated with micellar media,[11,12] the combination with visible light photo-redox catalysis,[13] or the development of new surfactants endowed with new catalytic head-groups.[14] The present contribution will focus specifically on the use of classical transition metal catalysts, picking from the authors' personal experience, and the advantages of their use in micellar media in terms of activity, selectivity, and ultimately green chemistry advances with specific reference to the synthesis of fine chemicals and pharmaceuticals.

12.2 Early Studies on Enantioselective Oxidation

When we first entered this topic 15 years ago, the most common approach to transition metal catalysis in water was inspired by the Ruhrchemie-Rhone Poulenc process for the synthesis of butyraldehyde by hydroformylation,[5] *i.e.*, *via* the introduction of hydrophilic tags on the ligands surrounding the transition metal center in order to make it soluble in water. Thus, having to decide what prototype reaction to investigate to test the limits of micellar conditions, we opted for enantioselective oxidation for two main reasons: first, because it is a key step in the synthesis of a variety of active pharmaceutical ingredients (APIs) such as omeprazole (anti-ulcer), taxol (anti-cancer), indinavir (anti-HIV), among others; second, because enantioselective oxidation with hydrogen peroxide as the oxidant and platinum(II) complexes as catalysts was a system that we had thoroughly studied for the previous 25 years. In fact, complexes of general formula $[(P\text{-}P)Pt(R)]_n(A)_n$, where P-P is a diphosphine; R = $-CF_3$, $-C_6F_5$, $-OH$; A is a weakly coordinating anion such as BF_4^-, ClO_4^-, or triflate; and n = 1 or 2, are stable catalysts for a variety of oxidation reactions such as epoxidation,[15] sulfoxidation, hydroxylation of aromatics,[16] and the Baeyer–Villiger oxidation of ketones,[17] showing peculiar selectivity properties and good activity. Two examples are reported in Figure 12.3.

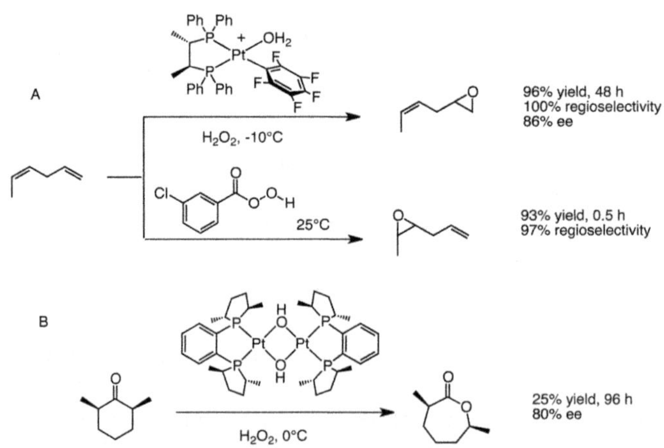

Figure 12.3 Two typical examples of enantioselective oxidation with hydrogen peroxide as the oxidant using chiral platinum(II) complexes as the catalysts: (A) epoxidation of a diene and regioselectivity comparison with *m*-chloroperbenzoic acid, and (B) desymmetrization of a *meso* cyclohexanone through Baeyer–Villiger oxidation.

These systems allowed the first successful use of aqueous hydrogen peroxide as the oxidant in a variety of oxidations without any waste of oxidant in side-reactions using soluble transition metal complexes as catalysts. These properties, along with the high product yields, high product selectivity, and mild conditions required, make these systems a good example of Green Chemistry *ante litteram*, as their discovery dates back to the early 1980s, *i.e.*, several years before the twelve principles of Green Chemistry were established. This excellent performance is strongly linked to the choice of Pt(II) centers as catalysts for a number of reasons, some of which were clear only retrospectively: (i) ligands on Pt(II) are less labile than on, *e.g.*, Pd(II) or Rh(I), making the catalyst more stable under strong oxidizing conditions and allowing the safe use of phosphines as ligands with the consequent advantage of better tuning of the steric and electronic properties at the metal center; (ii) diphosphine Pt(II) complexes are inert to oxidative addition, reducing the number of oxidant activation steps and improving selectivity; (iii) Pt(II) does not undergo one-electron redox pathways imparting the oxidant with better stability; and (iv) Pt(II) is a slow catalyst in β-hydride addition/eliminations, preventing the occurrence of Wacker chemistry as is the case of, *e.g.*, Pd(II). In addition to that and with respect to the specific use of these systems in micellar catalysis, we can also add: (v) the observed high activity and selectivity at all levels (chemo-, regio-, enantio-) are important for testing the potential of micellar systems and evaluate all the *pro*'s and *con*'s with respect to a traditional solvent; (vi) water is already present in the system both as the solvent for hydrogen peroxide and as the reaction product; (vii) the catalyst is stable in water, a condition that does not occur with, *e.g.*, early transition metals being strongly Lewis acidic that readily decompose in the presence of water.

12.3 Oxidation in Micellar Media

The first process investigated in soapy water was the enantioselective sulfoxidation of a variety of arylmethylsulfides with hydrogen peroxide using a $[(P\text{-}P)Pt(\mu\text{-}OH)_2]^{2+}$ catalyst, where P-P is a chiral diphosphine ligand.[18] The synthetic protocol is very simple: reagents and catalyst are placed in water, where (with the exception of H_2O_2) they are not soluble and, upon addition of a suitable surfactant at a concentration above the cmc, they readily dissolve. The reaction can be carried out in air at room temperature with excellent yield in short times. As shown in Figures 12.4 and 12.5, the choice of surfactant as well as its concentration are critical to maximize the reaction rate and enantioselectivity.

As can be seen under optimized conditions and in comparison with the reaction carried out in an organic solvent, high conversions are preserved while the enantioselectivity of the system is improved.

Similar results were observed in the enantioselective epoxidation of terminal alkenes[15e,19] and the Baeyer–Villiger oxidation of cyclic ketones[20] also

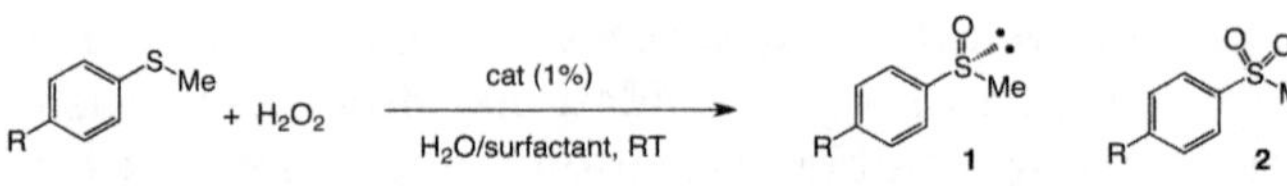

cat:

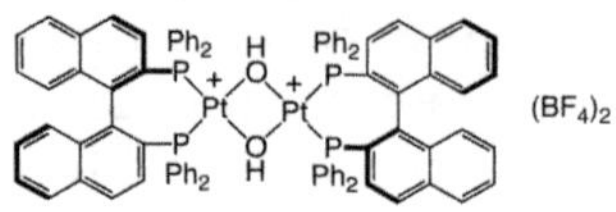

R	conv (%)	1/2 ratio	ee (%)	solvent
H	99	80	16	CH_2Cl_2
p-NO_2	41	25	26	CH_2Cl_2
H	16	31	5	H_2O/CTABr
H	20	56	0	H_2O/Triton-X100
H	98	> 200	40	H_2O/SDS

Figure 12.4 Oxidation of thioanisoles with hydrogen peroxide using a chiral platinum(II) complex as the catalyst: performance in different reaction media showing how H_2O/SDS improves the chemo- and enantioselectivity.

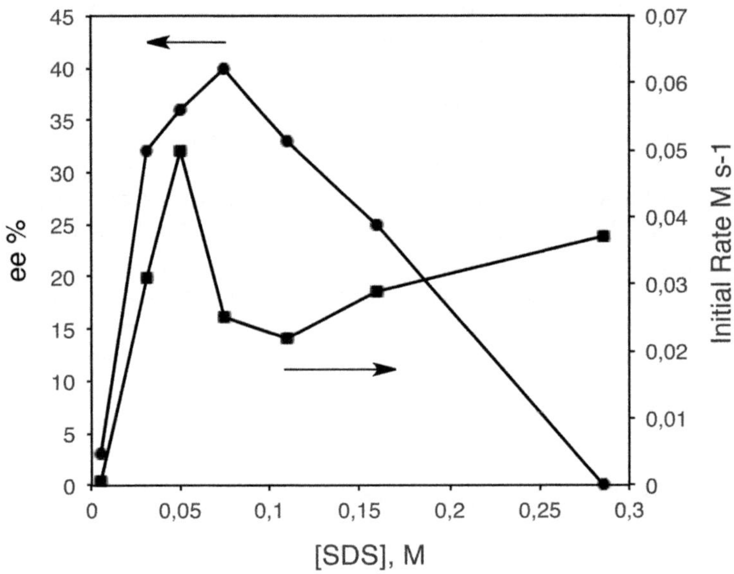

Figure 12.5 Effect of surfactant concentration on the initial rate of reaction (squares) and sulfoxide enantioselectivity (circles).

using hydrogen peroxide as the oxidant. In the former case (Figure 12.6), the use of Triton-X100 neutral surfactant allowed to significantly increase the enantioselectivity and also to separate and recycle the chiral Pt(II) catalyst for at least three times without loss of product yield and ee. A spectacular increase in both the yield and enantioselectivity was observed in the Baeyer–Villiger oxidation of substituted cyclobutanones using a Co(salen) complex previously reported by Katsuki and found virtually inactive in the same reaction.[21] As can be seen in Figure 12.7, the product yield increased from 15% to

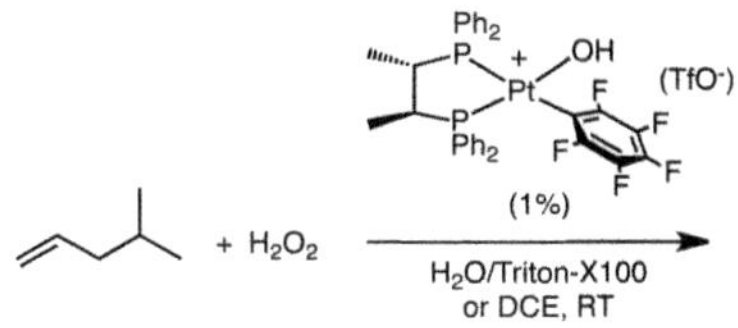

solvent	yield %	ee %
DCE	56	58
H2O/Triton-X100	51	84

Figure 12.6 Effect of a micellar medium *vs.* DCE on the enantioselectivity of the epoxidation of a terminal alkene using a chiral platinum(II) catalyst.

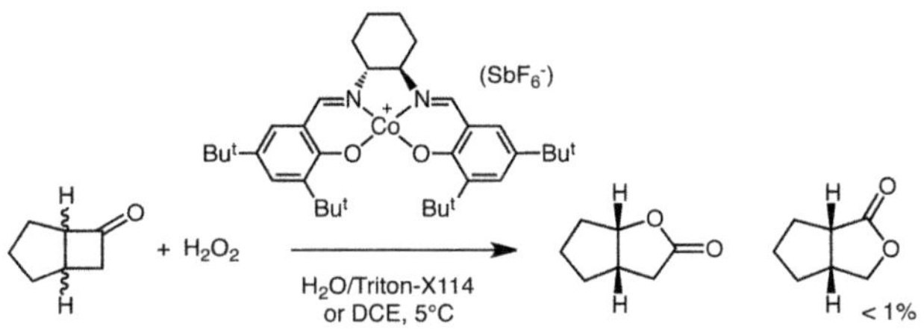

solvent	yield %	ee %
DCE	15	0
H2O/Triton-X114	80	88

Figure 12.7 Spectacular amplification of the activity and enantioselectivity by a micellar medium in the Baeyer–Villiger oxidation of a substituted cyclobutanone using a chiral cobalt(III) complex as the catalyst.

80%, while the enantioselectivity increased from zero to 88%, an effect that is only due to the micellar medium.

12.4 Other Reactions

After this seminal work, many other reactions have been tested over the years, *e.g.*, Friedel–Crafts,[22] Stille,[23] Diels–Alder,[24] hydration,[25] hydroformylation,[26] nucleophilic aromatic substitutions,[27] indol functionalization,[28] alkyne hydroboration,[29] hydroaminomethylation,[30] isomerization,[31] hydrogenation,[32] C–H functionalization,[33] C–O bond formation,[34] and metathesis.[35] A comprehensive review of these processes is beyond the scope of this contribution; however, a few selected examples illustrating the synthetic potential and practical utility of micellar catalysis will be briefly summarized.

Figure 12.8A shows the asymmetric transfer hydrogenation of aromatic ketones to obtain the corresponding chiral alcohols, catalyzed by unmodified hydrophobic ruthenium, rhodium, and iridium chiral amido complexes.[36] This reaction was successfully performed with significant enhancement of activity, chemoselectivity, and enantioselectivity (up to 99% ee) in aqueous media containing micelles. The hydrophobic catalyst, embedded in micelles constructed from the surfactant cetyltrimethylammonium bromide (CTAB), could be separated from the organic phase along with the products and was recycled for at least six times.

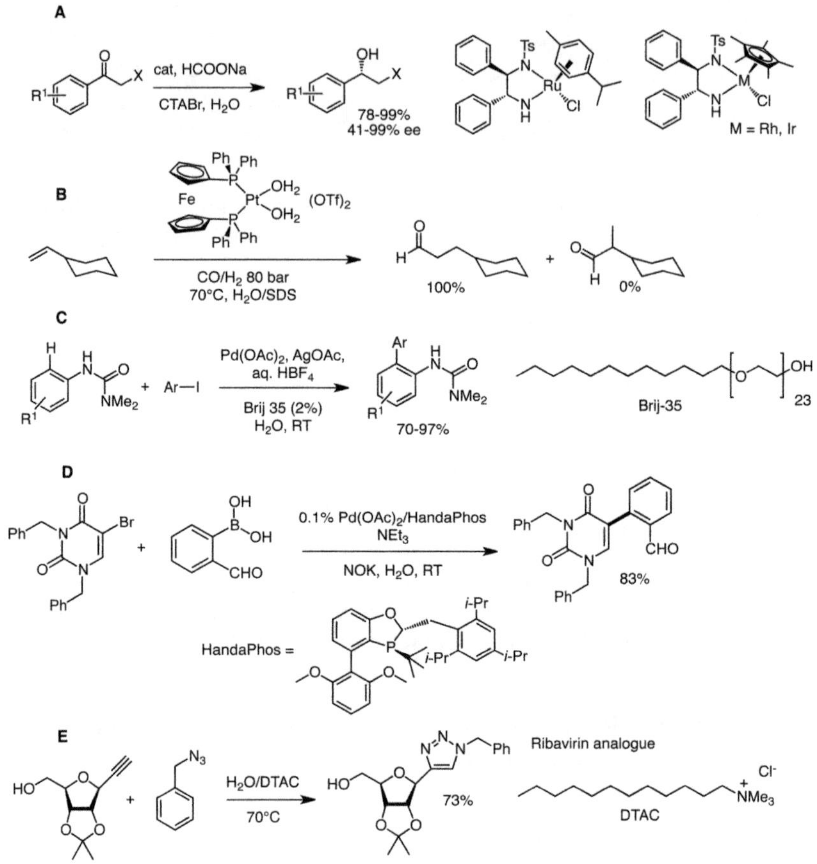

Figure 12.8 Some examples of catalytic reactions performed in micellar media: (A) asymmetric transfer hydrogenation of aromatic ketones using amido ruthenium, rhodium, and iridium catalysts; (B) hydroformylation of alkenes performed by cationic platinum triflate complexes; (C) Pd-catalyzed arylation of anilides with aryl iodides under mild conditions; (D) Suzuki–Miyaura cross-coupling carried out with ppm levels of the catalyst; and (E) synthesis of a ribavirin analogue *via* 1,3-dipolar cycloaddition between an alkynylated riboside and benzyl azide.

The hydroformylation of a variety of terminal and internal alkenes was efficiently performed by cationic platinum triflate complexes of the type $[P_2Pt(H_2O)_2](OTf)_2$ under mild conditions in water/SDS as the micellar medium (Figure 12.8B).[26] The use of surfactants is essential to ensure dissolution in water of both the catalyst and substrate, the former being positioned on the anionic surface of the micelles. A large variety of diphosphine ligands could be used; the best results were observed with the ferrocenyldiphosphine shown in Figure 12.8B that afforded linear aldehydes as the exclusive products. The catalyst could be separated by extraction of the organic products with hexane and recycled for at least four times with only a modest loss of activity and no effect on the selectivity.

In 2010, the group of Lipshutz reported for the first time a Pd-catalyzed C–H activation/cross-coupling at room temperature consisting in the arylation of anilides with aryl iodides to give biaryl derivatives in good yields.[37] The combination of $Pd(OAc)_2$ and HBF_4 allowed for reactions of aryl ureas with aryl iodides under very mild conditions using micellar catalysis (Figure 12.8C). This reaction was made possible using an *in situ* generated cationic palladium catalyst in the absence of phosphine ligands. Essential to drastically improve the yields of some 24 different substrates with respect to non-micellar systems was the use of the commercially available neutral surfactant Brij-35.

More recently, the same group reported a large series of Suzuki–Miyaura cross-couplings involving highly functionalized reaction partners for the synthesis of pharmaceuticals using monophosphine HandaPhos as the ligand for the Pd catalyst and operating in water containing NOK.[38] This enabled Pd-catalyzed reactions to be run at ppm levels of this pre-catalyst instead of the usual 5 mol% employed in the traditional synthesis carried out in organic solvents, thus reducing the level of residual Pd in the products to about 1 ppm (Figure 12.8D). The catalyst could be recycled up to four times without loss of activity and product yield.

Finally, Uziel and coworkers reported in 2009 the synthesis of carbonated analogues of ribavirin (an antiviral drug active against both DNA and RNA viruses) starting from carbohydrate derivatives obtained by an alkynylation reaction mediated by indium(0).[39] The C-ribosides were then engaged in a Huisgen 1,3-dipolar cycloaddition reaction in micellar medium (Figure 12.8E). It is noteworthy that here, the surfactant does not only provide the reaction medium but also catalyzes the click reaction. Under these conditions, the formation of 1,2,3-triazoles with control of the regioselectivity was observed.

12.5 Designer Surfactants and Metallosurfactants

Given the importance of C–C bond forming reactions in drug synthesis, over the years, the group of Lipshutz has reported a series of especially designed unconventional neutral surfactants capable of improving the catalyst performance of a series of Pd-catalyzed transformations such as Stille, Suzuki–Miyaura, metathesis, allylic amination. *etc.* Some of these surfactants are shown in Figure 12.9.

They are all based on a PEG unit as the hydrophilic head, a natural product as the hydrophobic pocket (β-sitosterol in the case of PSS and NOK and tocopherol in the case of PTS and TPGS-750-M) and a variable length dicarboxylic acid as a spacer in between.

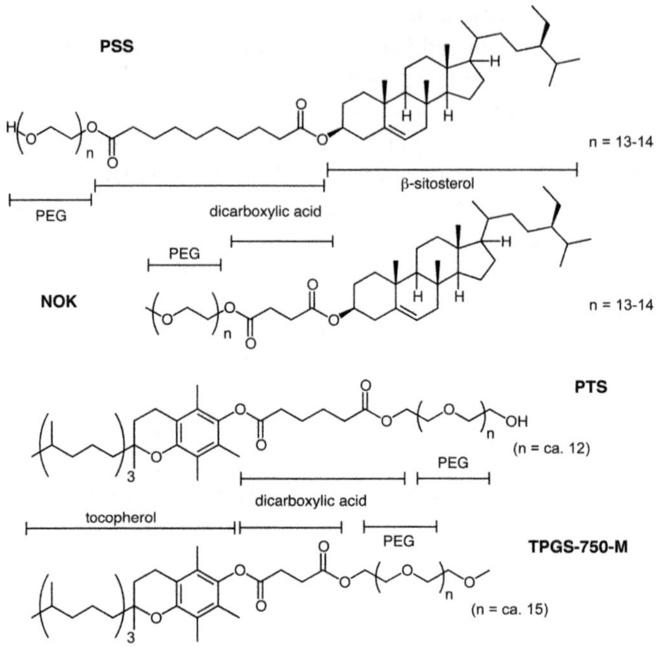

Figure 12.9 Designer surfactants reported by the group of Lipshutz.

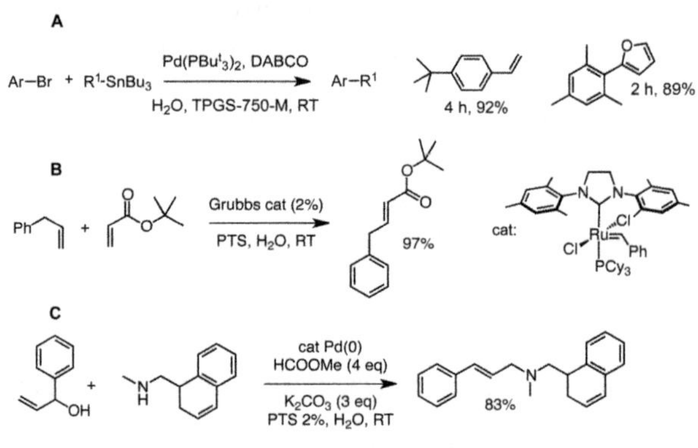

Figure 12.10 Some examples of the use of designer surfactants in synthetically useful reactions: (A) Stille nucleophilic substitution, (B) metathesis between allylbenzene and acrylate esters, and (C) allylic amination.

Some examples on the use of these designer surfactants are shown in Figure 12.10 and they address some important features with respect to their possible practical applications. In fact, Stille nucleophilic substitution of arylbromides with organotin derivatives allowed the preparation of a variety of alkylaromatics working in water containing TPGS-750-M in short times and high yield under considerably milder conditions with respect to those of organic solvent systems, where similar yields could be observed only at 80 °C, thus saving energy (Figure 12.10A).[40] Metathesis of allylbenzene and acrylate esters could be easily performed with Grubbs catalyst in PTS/water, proving a flexible system compatible with a variety of functional groups and allowing easier product isolation with respect to organic solvent systems (Figure 12.10B).[41] Finally, the antifungal Naftifine could be prepared *via* allylic amination under very mild conditions using traditional Pd catalysts operating in PTS/water media, allowing the facile amination of primary, secondary, and tertiary allylic alcohols (Figure 12.10C).[42]

The issue of catalyst permanent confinement in the micellar medium, sometimes a problem preventing catalyst separation and recycling, was addressed designing metallo-surfactants, where the catalyst is part of the surfactant structure, thus being not extracted from the micellar medium during product separation with organic solvents. Two examples are shown in Figure 12.11. A ring-closing metathesis reaction could be realized using the ruthenium catalyst shown in Figure 12.11A, where the metal is linked to the hydrophobic core (a ubiquinol unit) connected to a PEG chain with a dicarboxylic acid spacer using the same strategy adopted for the designer surfactants shown in Figure 12.9.[43] This strategy allows to recycle the catalyst up to ten times without loss of activity and selectivity. The opposite approach is shown in Figure 12.11B, where the rhodium catalyst is placed close to the hydrophilic head of a sulfonamide surfactant containing a variable length alkyl chain.[44] This was used for the asymmetric hydrogenation of a series of ketones to give the corresponding chiral alcohols with high conversion and enantioselectivity, observing that the latter could be increased upon increasing the surfactant alkyl chain length.

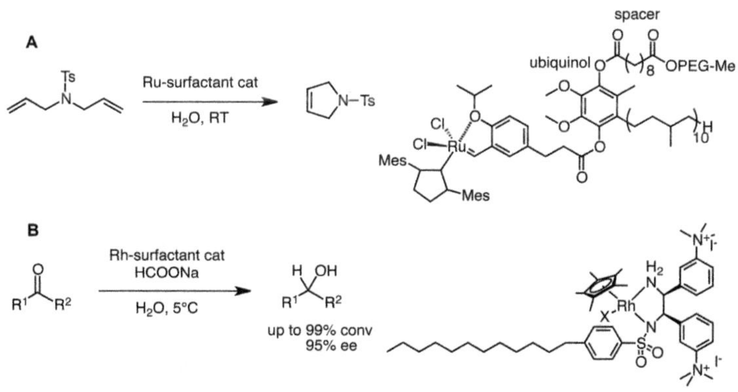

Figure 12.11 Two examples of the use of metallo-surfactants: (A) ring-closing metathesis using a ruthenium-surfactant catalyst, and (B) enantioselective hydrogenation of ketones using a rhodium-surfactant catalyst.

12.6 Importance of Catalyst–Surfactant Interactions

As evident from the examples reported in the previous sections, the use of micellar media not only entails the advantages arising from water being an environmentally benign solvent and the possibility to separate and recycle the catalyst, but also from the increase in yields and selectivity at all levels: chemo-, regio-, and enantio-. All that being related solely to the presence of micelles where the reaction takes place. In this respect, an important issue is the nature of the catalyst–surfactant interactions and how these can drive the course of the organic transformation under study. While it is easy to demonstrate by simple calculations that the reactant and catalyst concentrations inside the micelle are at least one order of magnitude higher than in a regular solution, which would account for the increased reaction rates, insight into catalyst–surfactant interactions has been almost completely neglected in the current literature in this field.

In a series of oxidation reactions, we reported a systematic study of catalyst positioning in a micellar medium using two-dimensional NMR techniques (NOESY, DOSY) to identify the interactions of the catalyst with different segments of the surfactant molecule. The surfactant was in all cases SDS.

In the sulfoxidation performed with the class of bis-cationic (diphosphine)Pt dimers shown in Figure 12.4,[18] the aromatic rings of the phosphine were clearly found to interact with the first methylene group adjacent to the SO_4^- hydrophilic head, suggesting that the bis-cationic catalyst, somehow expectedly, was likely sitting on the anionic surface of the surfactant, as indicated in Figure 12.12A. In the epoxidation of alkenes carried out with the (diphosphine)Pt(C_6F_5) catalyst shown in Figure 12.6,[19] the aromatic rings of

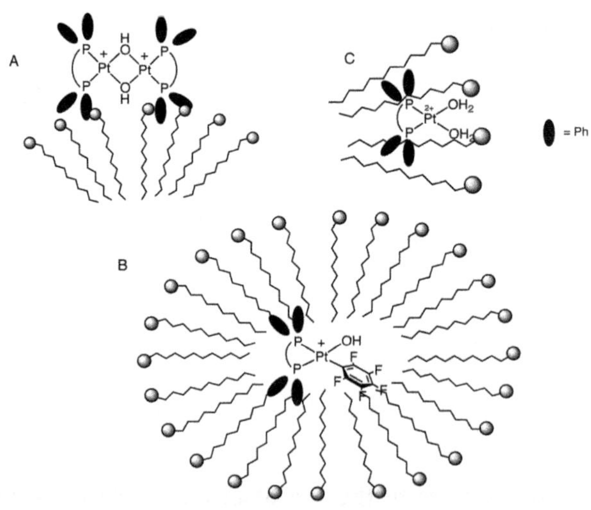

Figure 12.12 Positioning of bis-cationic (diphosphine)Pt complexes in SDS micelles in different oxidation reactions with hydrogen peroxide: (A) sulfoxidation, (B) epoxidation, and (C) Baeyer–Villiger oxidation.

the phosphine were found to interact with the terminal methyl group of the alkyl chain of SDS, suggesting that the catalyst was deeply buried inside the micelle, as represented in Figure 12.12B. Finally, in the Baeyer–Villiger oxidation of ketones with $[(diphosphine)Pt(OH_2)_2]^{2+}$,[45] the phenyl groups of the ligand were found to interact with the methylene groups of the central part of the alkyl chain of SDS, suggesting a third different catalyst positioning mechanism, as indicated in Figure 12.12C.

These observations clearly indicate that, even if the complexes are very similar (they all contain platinum, tetraaryl diphosphines, and they are all cationic) and catalyze oxidation reactions where hydrogen peroxide is the oxidant, in their interactions with SDS micelles (indeed a very simple structure), they occupy different positions and are surrounded by environments of different polarity. In fact, at variance with any homogeneous reaction medium where catalyst solvation is a random phenomenon, a micelle is an ordered, anisotropic aggregate where a polarity gradient exists from the surface to the core. This spontaneous positioning in differently polar environments will influence the electronic demand of the reaction transition state, often resulting in an increase in activity and, more importantly, in selectivity.

Enantioselective reactions allow the evaluation of the extent of this effect on the transition state. In fact, the ee is the result of the difference in energy between the two diastereomeric transition states leading to the formation of *R* and *S* products ($\Delta\Delta G^{\neq}$). The difference in ee observed when comparing the same reaction carried out in either an ordinary solvent or a micellar medium represents the ability of either medium to exert a better discriminating influence on the stabilization of the two diastereomeric transition states. Hence, in Figure 12.6, the 58% ee observed in DCE means $\Delta\Delta G^{\neq} = -3.23$ kJ mol^{-1}, while the 84% ee observed in Triton-X100 means $\Delta\Delta G^{\neq} = -5.95$ kJ mol^{-1}, the difference (2.72 kJ mol^{-1}) being due to the peculiar 'pocket' found in the micelle by the substrate when interacting with the catalyst. Similarly, in the Baeyer–Villiger oxidation reported in Figure 12.7, Triton-X114 is able to increase the $\Delta\Delta G^{\neq}$ by -6.82 kJ mol^{-1}, while no difference was observed in DCE, a huge discrimination effect solely due to the micelle.

These properties of micelles emphasize the already mentioned parallelism with enzymes (see Section 12.1), where the kinetic acceleration on the reaction under consideration is the result of the interactions (steric and/or electronic) exerted on the transition state by the polypeptide functional groups surrounding the 'pocket' where the active site is located.

A typical property of enzymes is substrate selectivity. This consists in the ability to pick up from the external reaction medium only the substrate that has to be transformed out of a large number of molecules (the well-known 'lock and key' concept). This peculiar property is generally not shared by ordinary catalysts, especially homogeneous, except for zeolites where the regular structure of channels and pores is able to select substrates on the basis of their size and shape (shape selectivity). Micelles can do the same, as has been demonstrated in a few cases even if on a different discriminating basis. For example, in the Cr(salen)-catalyzed Diels–Alder reaction between cyclopentadiene and a series of α,β-unsaturated aldehydes of the type R–CH=CH–CHO (where R was a series of linear aliphatic chains from $-CH_3$ to $-C_7H_{15}$),[24b] it was found that the product yield was completely insensitive to the R-chain length when the reaction was performed in chloroform. Conversely, when the catalyst was embedded in SDS in water, the yield in the case of the longest aldehyde ($R = C_7H_{15}$) was 3.5 times higher than that for the shortest aldehyde ($R = CH_3$). The aldehydes were not tested individually but fed all together into the reactor in a real competitive test, thus proving certain substrate selectivity properties of the micellar system. Similarly, when the same aldehydes were hydrogenated with a Pd metal catalyst generated *in situ* by reduction of palladium acetate with hydrogen, again in a fully competitive test, when the reaction was tested in THF, the initial rate was found to be 3.6 times higher in the case of the shortest aldehyde with respect to the longest one. Conversely, when the Pd particles were embedded in SDSN (sodium dodecylsulfonate), not only were the longer aldehydes preferred to shorter ones, but the initial rate for $R = C_7H_{15}$ was found to be 330 times higher than that for $R = CH_3$.[46] These results indicate that even simple structures like those of SDS or SDSN micelles are able to impart the catalyst with substrate selectivity, the driving force likely being the substrate hydrophobicity, *i.e.*, its affinity with the micelle inside, thus marking a not too surprising similarity with enzymes. This important property of micelles has huge practical potential, but has not been adequately addressed in the literature and deserves further insight.

12.7 Improvement of the E-factor and Practical Applications

In general terms, micellar media can be defined as green because they make possible the use of water as the reaction medium. They offer extra advantages such as the ability to improve the activity and selectivity of a variety of catalytic organic transformations and also entail the concept of substrate selectivity, as shown in the previous section. The crucial point however, implied in the title of this contribution, is: How green are these processes? Can we provide some metrics? And finally, are they amenable for practical application?

Speaking about metrics, it has to be pointed out that recent advances in the field are suggesting the replacement of classical concepts such as the Atom Economy or E-factor with other, more informative ones.[47] In the pharmaceutical industry, particularly useful to evaluate the industrial production of APIs is the Process Mass Intensity (PMI) proposed by the ACS Green Chemistry Institute.[48] The PMI is the ratio between the quantity of raw materials input (in kilograms) and the quantity of bulk API output (also in kilograms). An important difference with respect to the E-factor is that the water used in the processes is assumed as a raw material in the PMI. However, in the open literature, no examples of PMI comparisons between classical and micellar routes have been reported to date. Hence, for the purpose of this contribution, the E-factors will be considered in the following.

The group of Lipshutz analyzed a series of C–C bond-forming reactions (Heck, Suzuki–Miyaura, Sonogashira) involved in the synthesis of some APIs and evaluated the E-factor of the traditional, organic solvent-based synthetic protocols according to published procedures as compared to the parallel use of micellar media.[49] As the examples in Figure 12.13 illustrate, the E-factors for each of the above reactions can be drastically reduced, in some cases by greater than an order of magnitude. Since these were run in water on a much smaller, laboratory scale, the corresponding drop in E-factor would likely be even more pronounced at larger scales, where small losses will have lesser consequences. Moreover, in most cases, cross-coupling reactions take place at room temperature while, in traditional organic media, heating might be necessary to initiate and/or drive reactions to completion. Heating, especially in high boiling solvents such as DMF, usually causes the unavoidable formation of varying amounts of side products, with potential complications in the subsequent product purification and decrease in yields of the desired products. Conversely, reactions run in water at room temperature are typically very clean with better product yields, as seen in each of the cases reported in Figure 12.13. Furthermore, running reactions in water greatly simplifies the set up, execution, and especially the workup procedures. When the reaction is complete, no

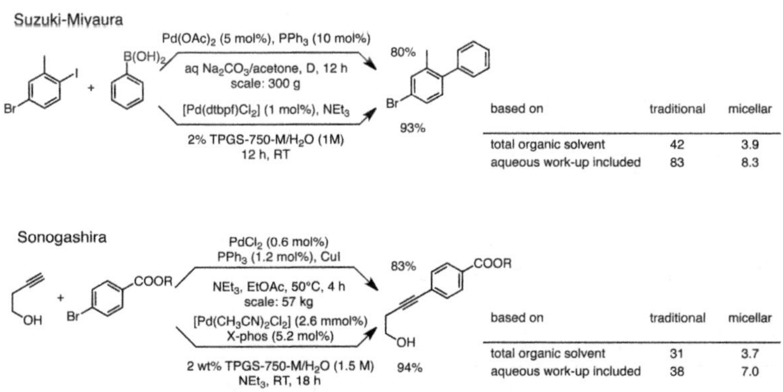

Figure 12.13 Some examples of C–C bond forming reactions carried out either in traditional organic or micellar media: comparison of the E-factors calculated in both cases for Heck, Suzuki–Miyaura, and Sonogashira reactions.

additional water needs to be added. Product extraction is simply performed by adding a limited amount of an organic solvent to the reaction vessel with subsequent in-flask extraction and product removal.

This procedure makes catalyst recycling quite straightforward. The surfactant and catalyst remain in the aqueous phase, in the reaction vessel, and are ready for reuse. As calculated by the authors, even a single recycle of the medium can result in a further reduction of the E-factor of over 50% (down to 2.8 for Heck and to 1.9 for Suzuki–Miyaura).

These results point further out the huge practical potential of micellar catalysis and justify its green character on a quantitative basis, even if the metrics are applied to a single reaction and not to an entire process. However, having potential is not the real thing, and so the answer to the third question raised at the beginning of this section came from industry. In 2016, an industrial research group from Novartis Pharma published the industrial synthesis of an API by comparing the organic traditional route to a micellar one by evaluating the yields, costs, and cycle time.[50] The synthetic procedure (Figure 12.14), carried out at kilogram scale, consisted of five different steps: (1) a nucleophilic aromatic substitution, (2) a

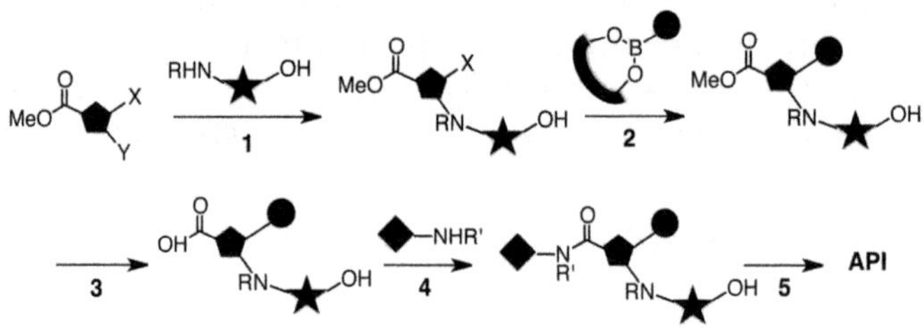

Step	Conditions	Yield (%)	Cycle time (hr)	Cost difference
1 S_NAr	reactant 1.2 eq, i-PrOH/toluene, 80°C, 5h	87	104	+4%
	reactant 1.0 eq, TPGS-750-M/water, RT, 12-16 h	75	61	
2 Suzuki-Miyaura	reactant 1.4 eq, Pd, t-AmOH, 85°C, 6h	=	61	-38%
	reactant 1.0 eq, Pd, TPGS-750-M/water, RT, 4-6 h	=	24	
3 hydrolysis	NaOH, MeOH/water, RT, 8h, MeTHF	70	137	
	NaOH, RT, 4-6 h	87	53	
4 amide bond formation	reactant 1.2 eq, acetonitrile, 50°C, 10h	76	105	-7%
	reactant 1.0 eq, TPGS-750-M/water, RT, 10-12 h	80	76	
5 final deprotection	HCl, MeOH/water, 0°C to RT, 16 h	92	62	0%
	HCl, MeOH/water, 0°C to RT, 16 h	92	62	
Overall	organic traditional	42.5	469	-17%
	micellar	48	276	

Figure 12.14 Multistep synthesis of an active pharmaceutical ingredient: comparison between the organic traditional (highlighted in grey) and micellar routes.

Suzuki–Miyaura C–C bond formation, (3) an ester hydrolysis, (4) an amide bond formation, and (5) a final deprotection to eventually release the API. The specific nature of the targeted drug substance and of the intermediates through the sequence were masked with stars, diamonds, circles, *etc.* for commercial sensitivity reasons.

The table in Figure 12.14 summarizes both the reaction conditions as well as product yields, the cycle times (the time necessary to carry out the operation), and the cost savings for each individual step comparing the organic traditional and micellar routes. As can be seen, the latter was performed in TPGS-750-M (where appropriate) at substantially milder conditions, using fewer reactants, in shorter times, overall reducing costs by 17% and the total cycle time to three fifths, which means that the multipurpose facilities employed could be used for other operations for that equivalent time.

Another important issue concerns the quality of the final compound and is of paramount importance in the pharmaceutical industry. In all steps described, very high quality of the crude intermediate products was claimed. The intermediates could simply be isolated and purified by crystallization, the purity of the targeted API turned out to be 99.5% or greater, even better than the routine purity obtained in the original process conducted in organic solvents.

12.8 Conclusions

As we have tried to demonstrate in this outline, the micellar approach, whose major feature is the possibility to operate in water as the reaction medium, is already a major tool in the hands of synthetic organic chemists and is on the way to become an important green technology with ubiquitous fields of application.

Surfactants are self-assembling auxiliaries, widely available on the market, that need no special technique to be used but are simply added to water. They allow the use of transition metal complexes developed for work in organic solvents, already optimized by ordinary synthetic elaboration without any special modification to make them compatible with water. Their use does not require any significant plant design or operational changes in the industrial practice while simplifying the separation processes.

The activity and selectivity are strongly bound to the different catalyst (and substrate) positioning within the micelles because of the existence of a polarity gradient, affording a local 'pocket' that can influence the outcome of the reaction with different levels of selectivity (chemo-, stereo-, enantio-). In enantioselective transformations, the steric constrain and anisotropy of the medium can amplify the free energy difference of the diastereomeric transition states, bearing interesting similarities with enzymes including their substrate selectivity.

The range of reactions in which micellar media can be successfully used is already very wide. Micelles can improve yields, selectivity at all levels, reaction conditions, product separation, and catalyst recycling, ultimately decreasing the costs of operation and energy input. The surfactant choice is a key issue that, in specific cases, can be optimized with especially designed surfactants and metallo-surfactants. Practical examples provide some metrics demonstrating that micellar catalysis can indeed reduce the E-factor and in industrial practice also improve the yields, reduce the energy consumption, shorten the cycle times, and ultimately reduce production costs. In short, catalysis in micellar media is much ahead of a mere Green Chemistry promise and can be already considered a profitable industrial opportunity.

References

1. (a) C. Seyler, C. Capello, S. Hellweg, C. Bruder, D. Bayne, A. Huwiler and K. Hungerbühler, *Ind. Eng. Chem. Res.*, 2006, **45**, 7700; (b) C. Capello, U. Fischer and K. Hungerbühler, *Green Chem.*, 2007, **9**, 927.
2. R. A. Sheldon, *Green Chem.*, 2005, **7**, 267.
3. F. Biedermann, W. M. Nau and H.-J. Schneider, *Angew. Chem., Int. Ed.*, 2014, **53**, 11158.
4. D. C. Rideout and R. Breslow, *J. Am. Chem. Soc.*, 1980, **102**, 7816.
5. C. W. Kohlpainter, R. W. Fischer and B. Cornils, *Appl. Catal., A*, 2001, **221**, 219.
6. Pliny the Elder, *Naturalis Historia*, vol. 28, ch. 47.
7. (a) G. La Sorella, G. Strukul and A. Scarso, *Green Chem.*, 2015, **17**, 644; (b) M. T. De Martino, L. K. E. A. Abdelmohsen, F. P. J. T. Rutjes and J. C. M. van Hest, *Beilstein J. Org. Chem.*, 2018, **14**, 716.
8. D. G. Blackmond, A. Armstrong, V. Coombe and A. Wells, *Angew. Chem., Int. Ed.*, 2007, **46**, 3798.
9. (a) E. J. Fendler and J. H. Fendler, *Adv. Phys. Org. Chem.*, 1970, **8**, 271; (b) J. H. Fendler and E. J. Fendler, *Catalysis in Micellar and Macromolecular Systems*, Academic Press, New York, 1975, p. 1; (c) T. Dwars, E. Paetzold and G. Oehme, *Angew. Chem., Int. Ed.*, 2005, **44**, 7174; (d) S. Taşcioğlu, *Tetrahedron*, 1996, **52**, 11113; (e) J. B. F. N. Engberts, *Pure Appl. Chem.*, 1992, **64**, 1653; (f) K. Holmberg, *Eur. J. Org. Chem.*, 2007, 731; (g) N. J. Buurma, *Annu. Rep. Prog. Chem., Sect. B: Org. Chem.*, 2012, **108**, 316–333; (h) D. Sinou, *Top. Curr. Chem.*, 1999, **206**, 42; (i) F. Mancin, P. Scrimin, P. Tecilla and U. Tonellato, *Coord. Chem. Rev.*, 2009, **253**, 2150.
10. D. Paprocki, A. Madej, D. Koszelewski, A. Brodzka and R. Ostaszewski, *Front. Chem.*, 2018, **6**, 502.
11. B. H. Lipshutz, S. Ghorai and M. Cortes-Clerget, *Chem. - Eur. J.*, 2018, **24**, 6672.
12. (a) B. H. Lipshutz, *Curr. Opin. Green Sustain. Chem.*, 2018, **11**, 1; (b) G. La Sorella, L. Sperni, P. Canton, L. Coletti, F. Fabris, G. Strukul and A. Scarso, *J. Org. Chem.*, 2018, **14**, 7438.

13. M.-J. Bu, G.-P. Lu, J. Jiang and C. Cai, *Catal. Sci. Technol.*, 2018, **8**, 3728.
14. S. Polarz, M. Kunkel, A. Donner and M. Schlötter, *Chem. - Eur. J.*, 2018, 18842–18856.
15. See *e.g.*: (a) G. Strukul and R. A. Michelin, *J. Chem. Soc., Chem. Commun.*, 1984, 1538; (b) G. Strukul and R. A. Michelin, *J. Am. Chem. Soc.*, 1985, **107**, 7563; (c) A. Zanardo, F. Pinna, R. A. Michelin and G. Strukul, *Inorg. Chem.*, 1988, **27**, 1966; (d) M. Colladon, A. Scarso, P. Sgarbossa, R. A. Michelin and G. Strukul, *J. Am. Chem. Soc.*, 2007, **129**, 7680; (e) M. Colladon, A. Scarso, P. Sgarbossa, R. A. Michelin and G. Strukul, *J. Am. Chem. Soc.*, 2006, **128**, 14006 and references therein.
16. A. Marsella, S. Agapakis, F. Pinna and G. Strukul, *Organometallics*, 1992, **11**, 3578.
17. See *e.g.*: (a) M. Del Todesco Frisone, F. Pinna and G. Strukul, *Organometallics*, 1993, **12**, 148; (b) A. Gusso, C. Baccin, F. Pinna and G. Strukul, *Organometallics*, 1994, **13**, 3442; (c) C. Paneghetti, R. Gavagnin, F. Pinna and G. Strukul, *Organometallics*, 1999, **18**, 5056; (d) R. A. Michelin, P. Sgarbossa, A. Scarso and G. Strukul, *Coord. Chem. Rev.*, 2010, **254**, 646 and references therein.
18. A. Scarso and G. Strukul, *Adv. Synth. Catal.*, 2005, **347**, 1227.
19. M. Colladon, A. Scarso and G. Strukul, *Adv. Synth. Catal.*, 2007, **349**, 797.
20. G. Bianchini, A. Cavarzan, A. Scarso and G. Strukul, *Green Chem.*, 2009, **11**, 1517.
21. T. Katsuki and T. Uchida, *Tetrahedron Lett.*, 2001, **42**, 6911.
22. (a) K. Rajendar Reddy, K. C. Rajanna and K. Uppalaiah, *Tetrahedron Lett.*, 2013, **54**, 3431; (b) F. Rosati, J. Oelerich and G. Roelfes, *Chem. Commun.*, 2010, **46**, 7804.
23. G. P. Lu, C. Cai and B. H. Lipshutz, *Green Chem.*, 2013, **15**, 105.
24. (a) K. Bica, P. Gärtner, P. J. Gritsch, A. K. Ressmann, C. Schröder and R. Zirbs, *Chem. Commun.*, 2012, **48**, 5013; (b) F. Trentin, A. Scarso and G. Strukul, *Tetrahedron Lett.*, 2011, **52**, 6978.
25. (a) A. Rühling, H.-J. Galla and F. Glorius, *Chem. - Eur. J.*, 2015, **21**, 12291; (b) A. Cavarzan, A. Scarso and G. Strukul, *Green Chem.*, 2010, **12**, 790; (c) F. Trentin, A. M. Chapman, A. Scarso, P. Sgarbossa, R. A. Michelin, G. Strukul and D. F. Wass, *Adv. Synth. Catal.*, 2012, **354**, 1095.
26. M. Gottardo, A. Scarso, S. Paganelli and G. Strukul, *Adv. Synth. Catal.*, 2010, **352**, 2251.
27. N. A. Isley, R. T. H. Linstadt, S. M. Kelly, F. Gallou and B. H. Lipshutz, *Org. Lett.*, 2015, **17**, 4734.
28. (a) T. Kitanosono, M. Miyo and S. Kobayashi, *Tetrahedron*, 2015, **71**, 7739; (b) T. Kitanosono, M. Miyo and S. Kobayashi, *ACS Sustainable Chem. Eng.*, 2016, **4**, 6101.
29. J. S. da Costa, R. K. Braun, P. A. Horn, D. S. Lüdtke and A. V. Moro, *RSC Adv.*, 2016, **6**, 59935.
30. A. Behr and A. Wintzer, *Chem. Eng. Technol.*, 2015, **38**, 2299.
31. L. Sperni, A. Scarso and G. Strukul, *Inorg. Chim. Acta*, 2017, **455**, 535.

32. P. Stathis, D. Stavroulaki, N. Kaika, K. Krommyda and G. Papadogianakis, *Appl. Catal., B*, 2017, **209**, 579.
33. M. Alvarez, R. Gava, M. R. Rodríguez, S. G. Rull and P. J. Pérez, *ACS Catal.*, 2017, **7**, 3707.
34. (a) D. Ahanthem and W. S. Laitonjam, *Asian J. Org. Chem.*, 2017, **6**, 1492; (b) X.-H. Li, C. Mi, X.-H. Liao and X.-G. Meng, *Catal. Lett.*, 2017, **147**, 2508.
35. (a) L. Laville, C. Charnay, F. Lamaty, J. Martinez and E. Colacino, *Chem. - Eur. J.*, 2012, **18**, 760; (b) D. Astruc, A. K. Diallo, S. Gatard, L. Liang, C. Ornelas, V. Martinez, D. Méry and J. Ruiz, *Beilstein J. Org. Chem.*, 2011, **7**, 94.
36. F. Wang, H. Liu, L. Cun, J. Zhu, J. Deng and Y. Jiang, *J. Org. Chem.*, 2005, **70**, 9424.
37. T. Nishikata, A. R. Abela and B. H. Lipshutz, *Angew. Chem., Int. Ed.*, 2010, **49**, 781.
38. S. Handa, M. P. Andersson, F. Gallou, J. Reilly and B. H. Lipshutz, *Angew. Chem., Int. Ed.*, 2016, **55**, 1.
39. R. A. Youcef, M. Dos Santos, S. Roussel, J.-P. Baltaze, N. Lubin-Germain and J. Uziel, *J. Org. Chem.*, 2009, **74**, 4318.
40. G.-P. Lu, C. Cai and B. H. Lipshutz, *Green Chem.*, 2013, **15**, 105.
41. B. H. Lipshutz, G. T. Aguinaldo, S. Ghorai and K. Voigtritter, *Org. Lett.*, 2008, **10**, 1325.
42. T. Nishikata and B. H. Lipshutz, *Org. Lett.*, 2009, **11**, 2377.
43. (a) B. H. Lipshutz and S. Ghorai, *Org. Lett.*, 2009, **11**, 705; (b) B. H. Lipshutz and S. Ghorai, *Tetrahedron*, 2010, **66**, 1057.
44. J. Li, Y. Tang, Q. Wang, X. Li, L. Cun, X. Zhang, J. Zhu, L. Li and J. Deng, *J. Am. Chem. Soc.*, 2012, **134**, 18522.
45. A. Cavarzan, G. Bianchini, P. Sgarbossa, L. Lefort, S. Gladiali, A. Scarso and G. Strukul, *Chem. - Eur. J.*, 2009, **15**, 7930.
46. G. La Sorella, P. Canton, G. Strukul and A. Scarso, *ChemCatChem*, 2014, **6**, 1575.
47. D. J. C. Constable, A. D. Curzons and V. L. Cunningham, *Green Chem.*, 2002, **4**, 521.
48. ACS Green Chemistry Institute, www.acs.org/content/acs/en/greenchemistry/research-innovation/tools-for-green-chemistry.html.
49. B. H. Lipshutz, N. A. Isley, J. C. Fennewald and E. D. Slack, *Angew. Chem., Int. Ed.*, 2013, **52**, 10952.
50. F. Gallou, N. A. Isley, A. Ganic, U. Onken and M. Parmentier, *Green Chem.*, 2016, **18**, 14.

CHAPTER 13

Supported ILs and Materials Based on ILs for the Development of Green Synthetic Processes and Procedures

SILVIA MONTOLIO, BELÉN ALTAVA, EDUARDO GARCÍA-VERDUGO* AND SANTIAGO V. LUIS*

Dpt, of Inorganic and Organic Chemistry, University Jaume I, Av. Sos Baynat, s/n, E-12080 Castellón, Spain
*E-mail: cepeda@uji.es, luiss@uji.es

13.1 Introduction

Ionic Liquids (ILs) have received much attention as an alternative to traditional organic solvents.[1] Their unique properties and high structural modularity facilitate their use for multiple applications, including sustainable chemical processes.[2,3] Nevertheless, their use faces challenges such a their high cost or (eco)toxicological concerns.[4] This, along with some specific properties such as high viscosities and boiling points hamper often their use, in particular when requiring easy separation and recovery in pure form.[5]

The use of materials based on ILs can overcome many of the above-mentioned drawbacks. Immobilization of ILs on solid supports provides a

Green Chemistry Series No. 61
Green Synthetic Processes and Procedures
Edited by Roberto Ballini

Published by the Royal Society of Chemistry, www.rsc.org

substantial reduction of the amount of IL required for a given application. This decreases accordingly the associated costs, facilitates handling and manipulation, and greatly mitigates the potential environmental risks. Both inorganic and organic polymers have been used as supports.[6–8] ILs have been supported either by coating the surface of an insoluble material with a thin layer of the corresponding IL or by covalent attachment of structural fragments related to ILs.[9] Materials prepared by absorption of ILs on a porous support are known as Supported Ionic Liquid Phases (SILPs) and have proved applicable in different fields, but their essential limitation is the potential leaching of the IL phase.[10,11] This can be avoided by covalent attachment of structural fragments related to ILs to functional groups in the support to provide Supported Ionic Liquid-Like Phases (SILLPs). This can be achieved through postfuncionalization of a preformed polymer using '*grafting from*' or '*grafting to*' methodologies,[12] or by polymerization of a monomer containing IL-like moieties (Figure 13.1). Polymerization can be carried out directly or in combination with a support.[13,14]

This chapter will focus on materials based on organic polymers, although the concepts discussed can also be of application to materials based on inorganic supports. A broad range of applications in catalysis,[15] responsive materials,[16] sensors,[17] CO_2 capture,[18] separation,[11] self-healing materials,[19] ion-transport materials, energy, and bio-related topics have been reported.[20] Our goal, however, is to highlight how polymeric frameworks containing IL-like units endow these materials with new properties and advantages of relevance for the development of green synthetic processes and procedures, in particular when catalysis is involved.

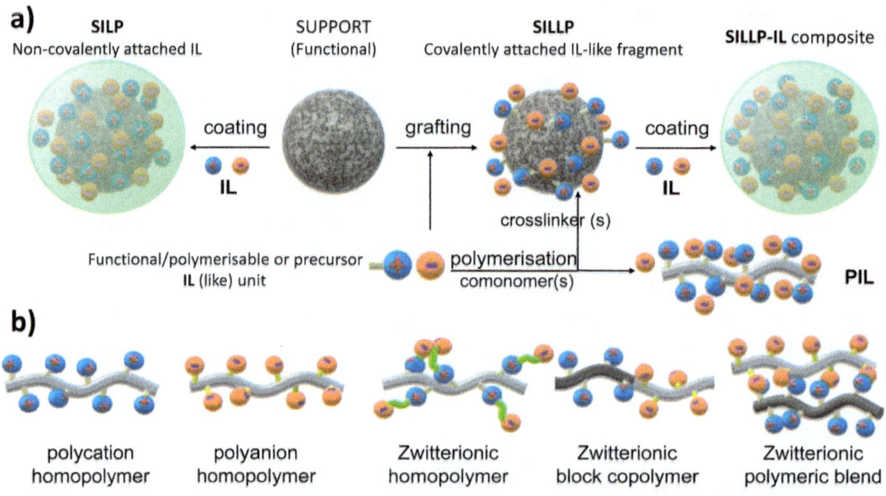

Figure 13.1 (a) General approaches for the preparation of materials based on ILs. (b) Some structural arrangements available for linear PILs.

13.2 Synthesis and Structures of PILs and Related IL-like Polymeric Phases

Polymeric ionic liquids (PILs) are macromolecules formed by ionic liquid units connected through a polymeric backbone.[7,8,21,22] They combine the unique characteristics of ILs with those of polymers,[23] achieving enhanced mechanical stability, processability, flexibility, and durability, and improved dimensional control over their structure.[24,25] All possible combinations of anions, polyanions, cations, polycations, and zwitterionic species have been analyzed (Figure 13.1).[26] Like ILs, polymers with IL-like moieties can form specific 3D supramolecular structures driven by extended networks of interactions including anion–cation, H-bonds, π–π, or C–H···π.[27] Thus, new materials with particular properties can be predesigned, at the molecular level, to form specific structures at the micro- or nanoscale.[28,29]

The simplest approach for the preparation of PILs and SILLPs is the use of preformed polymers with appropriate properties and functionality, introducing IL-like units by grafting (Figure 13.1). Amines, phosphines, imidazoles, and other heterocycles can react with alkylating groups in the support to generate the IL-like fragment.[30] Other groups in the support, such as $-NH_2$, CH_2Cl, $-CH_2SH$, $-CO_2H$, *etc.* have been used for the introduction of IL-like units, and the creation of imidazolium rings from polymer-bound amines has been reported.[31] The progress of the reaction can be easily monitored through standard methodologies.[30,32] Poly(acrylamide-homocysteine thiolactone) (PAHT) polymers provide easy access to multifunctional PILs through thiolactone ring-opening, followed by thiol-ene reactions.[33] In this approach, the architecture and morphological features of the starting polymer are well known, which control some of the properties of the resulting materials. For inorganic supports, free hydroxyl groups on the surface are used for the attachment, employing sol–gel approaches and alkoxyorganosilyl derivatives containing the desired IL-like functionality or its precursor.[9] To reduce potential leaching in SILPs, the coating of a material with an IL layer can be enhanced by introducing in its surface additional IL-like fragments.[34]

An alternative strategy is the polymerization of specific monomers containing IL-like fragments in the presence, when appropriate, of other monomers such as vinyl, (meth)acrylic, or styrenic derivatives. IL-like moieties can be incorporated in the polymer backbone or in its side chains. Many applications correspond to soluble PILs containing imidazolium, phosphonium, or ammonium groups that can be classified as polyelectrolytes.[21,35] When variable amounts of a crosslinking agent are added (*i.e.*, divinylbenzene, diacrylate), the corresponding insoluble polymers (SILLPs) are obtained.[30]

A large number of publications describe polymerizations to obtain PILs.[36] Many involve conventional free radical polymerization (FRP),[37] but an increasing number of PILs are prepared using controlled/living radical polymerization (CRP), including nitroxide mediated polymerization (NMP), atom transfer radical polymerization (ATRP), and reversible addition-fragmentation chain transfer (RAFT) polymerization.[38–40] This, allows preparing PILs with controlled molecular weights, low dispersity (PDI < 1.5), multifunctionality, and with tight control of the macromolecular architecture.[41,42] Such polymeric architectures include homopolymers, gradient or block copolymers, star and hyperbranched polymers,[43] or polymer brushes.[40] Other approaches such as photopolymerization,[44] electropolymerization,[45] microwave-assisted dispersion polymerization,[46] or polycondensation have also been described.[47]

13.3 Physico-chemical Characteristics of PILs and Related IL-like Polymeric Phases

In SILPs, the IL-phase coated on the surface of the material is expected to retain its main properties. In PILs/SILLPs and related materials, appropriate transfer of the essential properties of the ILs to the solid phase needs to be guaranteed. The physico-chemical characteristics of a polymer are associated with their composition and morphology and the IL-like fragments covalently attached will modify and tune many of them.[36]

13.3.1 Polarity

The polarity of ILs can be tuned through changes in the structure of the cation or the anion.[48] Solvatochromic probes permit a rapid polarity analysis in ILs,[49] and can also be applied for IL-based materials. Different SILLPs were obtained by reaction of alkylimidazoles with commercial crosslinked polystyrene-divinylbenzene (PS-DVB) Merrifield resins and further counteranion exchange (Figure 13.2) and Reichardt's probe was used to evaluate their polarity. According to the $E_T(N)$ values, there is a big change in polarity between the starting Merrifield resin (0.292) and SILLPs displaying values between 0.6 and 0.7,[30] in good agreement with those reported for butylimidazolium salts.[50] A wider variation in polarity as a function of the anion was observed when pyrene was employed as the fluorescent probe using the I_3/I_1 ratio.[51] As seen in Figure 13.2, the more coordinating anion (Cl^-) is associated with a higher value of polarity (higher value of I_3/I_1), whereas the presence of the less coordinating NTf_2^- anion affords a lower value of I_3/I_1,[52] with values again in reasonable agreement with those obtained for related ILs.[53]

The polarity of ILs has also been determined from their ATR-FT-IR spectra by studying the vibrational modes of IL-associated water.[54] Again, the values obtained in SILLPs agree well with those reported for similar ILs,[55] confirming that water is coordinated to SILLPs in a similar way. The effect of the anion on the polarity in the microenvironment of the IL-like species in SILLPs

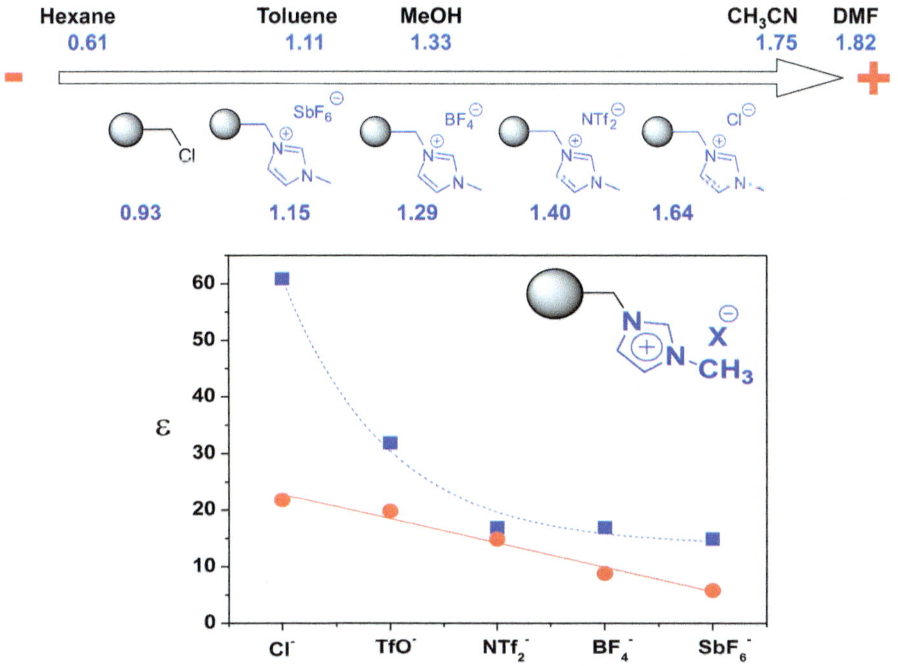

Figure 13.2 Top: Polarity of SILLPs using Reichardt's dye as the solvatochromic probe and comparison of E_T(N) values with those for common solvents. Bottom: Comparison of dielectric constants calculated for SILLPs using FT-IR spectroscopy (blue squares) and pyrene as the solvathocromic probe (red dots).[30,52]

follows the trend: $Cl^- > TfO^- > NTf_2^- \approx BF_4^- \approx SbF_6$, with the butylimidazolium fragment providing a slightly lower polarity than the methylimidazolium moiety. However, a quantitative comparison of data obtained from different methodologies is not easy (Figure 13.2). In SILLPs, low polarity regions associated with the polymeric backbone can coexist with hydrophilic ones related to IL-like fragments. Therefore, the higher affinity of pyrene for the aromatic structure of the polymeric backbone can lead to experimental values of I_1/I_3 (and thus of ε) lower than expected. Thus, although quantitative data must be handled with caution, the former results suggest the efficient transference of the polarity of ILs to SILLPs.

Polarity in SILPs was studied using the solvation parameter model for GC stationary phases using ILs coated on solid supports.[56] The anion significantly influenced the hydrogen bond basicity and dipolarity. The hydrogen-bond basicity increased in the order: $PF_6^- < NTf_2^- < BF_4^- < TfO^- < Cl^-$. Additional functional groups in the IL-phase, including aromatic fragments, were found to also influence different properties, leading to tunable stationary phases for GC applications.[57–59]

In summary, the polarity of these materials can be fine-tuned through changes in the structure of the IL-like fragment and this is important,

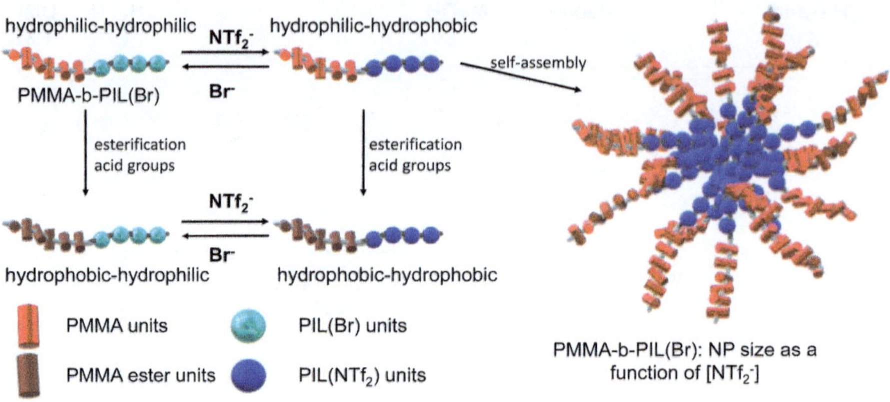

Figure 13.3 Chemical manipulation of the polarity in IL-based block copolymers by defining their self-assembly state through external stimuli.[62]

for instance, for applications involving CO_2 capture and separation, an environmentally relevant and challenging task.[60] Moreover, PILs are polyelectrolytes able to self-assemble by non-covalent interactions to produce highly ordered and complex hierarchical structures organized at the nanolevel, such as micelles or vesicles, and changes in polarity can be critical in this regard.[27,61] Thus, the dual block PILs shown in Figure 13.3 present sensitivity to anions in water.[62] The IL-based block can change reversibly between hydrophobic and hydrophilic as a function of the counterion and the chemical modification of the hydrophilic poly(methacrylic acid) (PMMA) block affords a hydrophobic poly(methyl methacrylate acid) fragment. While soluble unimers were observed with bromide counteranions, micellar aggregates of varying size were detected for variable amounts of NTf_2^- added.

13.3.2 Swelling

Swelling in a good solvent defines the accessibility to functional sites in polymers and is essential for many applications, determining the efficiency in many catalytic applications.[63] Initial reports showed that cross-linked PIL beads prepared from vinylimidazolium cations were able to swell in organic solvents such as acetone, but did not quantify the degree of swelling.[64] PILs produced by photopolymerization of 1-octyl acrylate-3-methylimidazolium chloride in the presence of 23 wt% water formed hydrogels in water, swelling to almost 200 times their dry volume while, in EtOH, DMSO, MeOH, and DMF, the swelling was approximately 50 times.[65] Less polar solvents such as 1-octanol, hexanes, EtOAc, and $CHCl_3$ produced minimal or no swelling.

PS-DVB resins are hydrophobic, but the introduction of IL-like subunits provides important changes in polarity favoring the swelling in polar solvents, including water. Gel-type PS-DVB-SILLPs with imidazolium

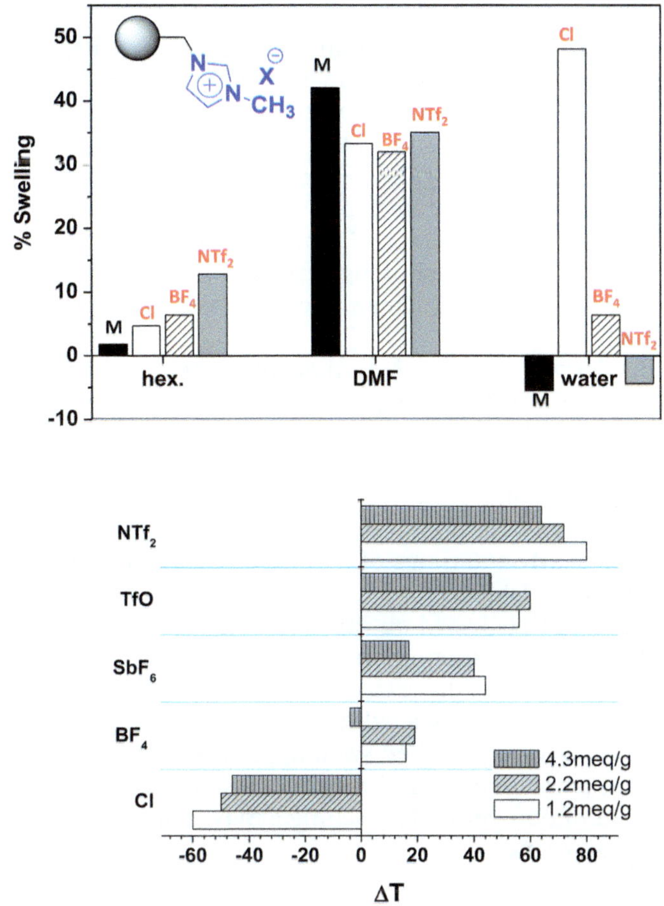

Figure 13.4 (Top) Influence of the anion on the swelling of gel-type SILLPs measured as the percentage of volume increase. M: original Merrifield resin (4.3 mmol Cl g^{-1}). (Bottom) Effect of loading and nature of the anion on the thermal stability of SILLPs measured by TGA. ΔT is the difference in thermal stability between the starting Merrifield resin and the corresponding SILLP.[30]

moieties showed significant changes in swelling depending on the nature of these fragments.[30] Figure 13.4 shows data for SILLPs prepared from a gel-type Merrifield resin (4.3 mmol Cl g^{-1}). Although in water the starting Merrifield shrank, the SILLP (Cl^- anion) resulting from the reaction with 1-methylimidazole increased in size up to a 43%. However, the SILLP with NTf_2^- shrank in water. In hexanes, the starting resin shrank but the SILLP with NTf_2^- showed appreciable swelling in this medium, opening the possibility of carrying out processes in this apolar medium. As expected, substitution of the *N*-methyl group by the more lipophilic *N*-decyl produced an appreciable decrease of the swelling in water (from 47% to 15% with Cl^- anion).

Polyelectrolyte gels containing a ternary mixture of vinyl imidazolium monomers, crosslinkable vinyl gemini imidazolium salts, and tetradecyl acrylate swelled in solvents with low dielectric constants, such as CCl_4, $CHCl_3$, CH_2Cl_2, THF, and toluene (~10–50 times, w/w), but not in more polar solvents like CH_3CN, DMSO, and DMF.[66] A more dramatic effect was found for crosslinked polyacrylate gels containing lipophilic and bulky tetraalkylammonium tetraphenylborate groups, swelling over 120 times (w/w) in CH_2Cl_2 and THF, 99 times in $CHCl_3$, 70 times in chlorobenzene, and 11 times in hexane. Reducing the crosslinking degree allowed achieving swelling in CH_2Cl_2 of up to 500 times.[67] Such an effect of crosslinking is rather general in these systems.[68]

Simple PILs derived from methyl imidazolium tethered to a PS-backbone also displayed significant swelling in molecular solvents despite not being covalently crosslinked. Some solvents that were fully miscible with the IL-monomer were unable to swell or dissolve this material, revealing strong non-covalent crosslinking. However, solvents with dipole moments >2 D able to weaken the cation–anion and/or π–cation interactions provided significant swelling to nearly 400 times of its initial mass in propylene carbonate and over 200 times in DMSO.[69]

Nanoporous materials based on PILs are obtained by radical polymerization of crosslinker-type IL monomers in the presence of a similar IL acting as a porogenic solvent that can be extracted after polymerization and recycled.[70] The monomer/porogen ratios used define the properties of the materials. The polymers showed very limited swelling in water due to the hydrophobic anion used, but presented a non-linear increase of the swelling in acetone with the amount of porogenic IL, reaching a maximum of 500 vol% for a 1 : 5 monomer/porogen ratio.

13.3.3 Thermal Behavior

Thermal stability governs the suitability of supported IL-phases for many applications. The nature of both the cation and anion in ILs is important in this regard. Imidazolium-based cations exhibit higher thermal stabilities than tetraalkylammonium counterparts.[71] For imidazolium-based ILs, the thermal stabilities follow the reversal of anion nucleophilicity: Cl^-, Br^-, I^- < BF_4^- < $CF_3SO_3^-$ < NTf_2^- < PF_6^-.[72] PILs, SILPs, and SILLPs present similar trends.[73] TGA data showed that the introduction of imidazolium subunits in PS-DVB matrices could significantly increase the thermal stability of the resulting SILLPs (Figure 13.4).[30] The anion is the main parameter determining the thermal stability (higher for less nucleophilic anions), while changes in the *N*-aliphatic chain of the imidazolium have a much less pronounced influence. These effects slightly increase with the loading. The polymerization method is also important: for a family of imidazolium PILs, step-growth polymers revealed to be more thermally stable than related chain-growth.[74]

The thermoresponsiveness reflects the balance between polymer–polymer and polymer–solvent interactions.[16] Upper critical solution temperature (UCST)- and lower critical solution temperature (LCST)-type phase transitions can provide 'smart' materials. Tunable LCST behavior was reported for a copolymer of poly(*N*-isopropylacrylamide) (PNIPAM) containing IL-moieties. Below the LCST, the polymer chains exist as individual coils due to hydrogen bonding of the amide groups with water; however, above this temperature, conformational equilibria are activated and hydrophobic interactions from the backbone become dominant, leading to the aggregation of hydrophobic moieties from different polymeric chains.[75] In a PNIPAM and poly(1,1,3,3-tetramethylguanidine acrylate) copolymer, the LCST behavior was controlled by the hydrophilic IL content and could be regulated by treatment with CO_2/N_2.[42] In a poly(tripentyl(vinylbenzyl)phosphonium) PIL, the LCST phase transition decreased with the concentration of NaCl, which also reduced its swelling.[76] These stimulus-responsive polymers offer a wide range of applications in drug delivery, sensing, separation, *etc.*[16a] Thus, a copolymer of NIPAM and vinylbutylimidazolium with alanine counteranion was used for the selective extraction of tocopherols with good recyclability.[77] In catalysis, they enabled catalyst separation just by changing the temperature of the solution.[78]

13.4 Application of PILs and Related IL-like Polymeric Phases in Catalytic Processes

PILs, SILPs, and SILLPs have demonstrated their potential as catalysts, supports for catalysts, or precursors for catalytic species.[15] A series of selected examples, involving organocatalysis, metal complexes, or metal nanoparticles (MNPs) are summarized below highlighting the strong relationships between the structure and catalytic activity. Despite the important applications for biocatalysts in ILs,[79] the number of examples encompassing biocatalysis in supported ILs is scarce, some of them being considered in Section 13.5 (*vide infra*). The improved catalytic performance and facilitation of the work up, in particular the separation, recovery, and reuse of the catalytic system are the main targets for polymer-supported systems.[80] In this regard, separation processes are much simpler with crosslinked insoluble systems or with magnetic materials. For soluble systems, membrane filtration or solvent-induced precipitation strategies have been used for catalyst recovery.[81] The polymeric catalysts (*i.e.*, PILs) can be soluble in the hot reaction solution, which becomes homogeneous, but insoluble at lower temperature, while products remain dissolved. For all the systems considered, it is essential to take into account the potential of using the polymeric backbone to enhance the activity or stability or to modify the selectivity.[82,83] In general, the catalytic efficiency of a polymer-supported catalyst is very sensitive to the method employed for its immobilization and to the microporous or macroporous structure of the support.[84,85]

13.4.1 Catalytic Systems Not Involving Metallic Species

Structural units present in ILs and accordingly in their polymeric analogues can provide activation mechanisms for organocatalysis. The cation can interact through coulombic forces and/or as a hydrogen bond donor, while the anion can participate through coulombic forces, as a base or as a nucleophile. Besides, imidazol(in)ium fragments can be transformed into *N*-heterocyclic carbenes (NHCs) in the presence of a strong base.

Thus, lineal PILs obtained by FRP of 1-vinyl-3-alkylimidazolium bromide and anion exchange with $LiNTf_2$ were deprotonated with a strong base to afford poly(NHC)s that, upon reversible carboxylation with CO_2, led to poly(NHC–CO_2) adducts (Figure 13.5a).[86] Both species catalyzed, under homogeneous conditions, transesterification (R-I) and benzoin condensation (R-II) reactions, being recovered by induced precipitation. Unlike poly(NHC)s, precatalytic poly(NHC–CO_2)s were air stable, less moisture sensitive, and easier to handle, allowing their reuse for up to eight runs with a minor decrease in activity. The appropriate choice of the alkyl substituent (R) revealed to be crucial. Related poly([NHC(H)][HCO_3]) s were studied as (pre)catalysts in similar processes with improved air stability over poly(NHC)s.[87] Similar poly([NHC(H)][HCO_3])s derived from chloromethylated PS quantitatively reacted with various electrophiles, including CS_2 or isothiocyanate, to afford poly(NHC-CS_2) or poly(NHC-isothiocyanate) and also catalyzed the benzoin condensation (Figure 13.5b).[88] The same reaction was catalyzed by poly(NHC)s generated *in situ* from PILs prepared by cross-coupling copolymerization of *N*-alkyl-3,4(4-bromophenyl)imidazoles with a fluorenyl boronic acid diester, followed by *N*-methylation (Figure 13.5c).[89] Higher efficiency, reactivity, and recyclability were observed relative to their monomeric analogues, which were ascribed to cooperative effects between the repeating catalytic units along the polymer chain.

Poly(vinyl ethylimidazolium bromide) obtained by RAFT polymerization and the related PIL with NTf_2^- anion catalyzed the Diels–Alder reactions of isoprene with different dienophiles in quantitative yield in 30 min with selectivities ranging from 75% to 100%.[90] The polymer prepared by grafting tetra-ethylene glycol-bridged 1-vinylimidazolium mesylate to the surface of crosslinked polyethylene glycol methacrylate acted as the catalyst for the multicomponent solventless synthesis of complex heterocycles with only a slight decrease in activity after six reuses (Figure 13.5d, from 96% to 90% yield).[91] The catalytic generation of H_2 or the reduction of nitroaromatics using $NaBH_4$ has also been reported.[92] One of the scarce examples described up to now on the catalytic use of hyperbranched PILs is the water-soluble ionically tagged polyethylene imine that promotes the dephosphorylation of diethyl dinitrophenyl phosphate ester in water,[93] associated with a cooperative effect of substrate activation by the imidazolium units in the proximity of nucleophilic amine groups. Increased reactivity was obtained in the presence of water-insoluble 2-methyl-4-nitroimidazole.

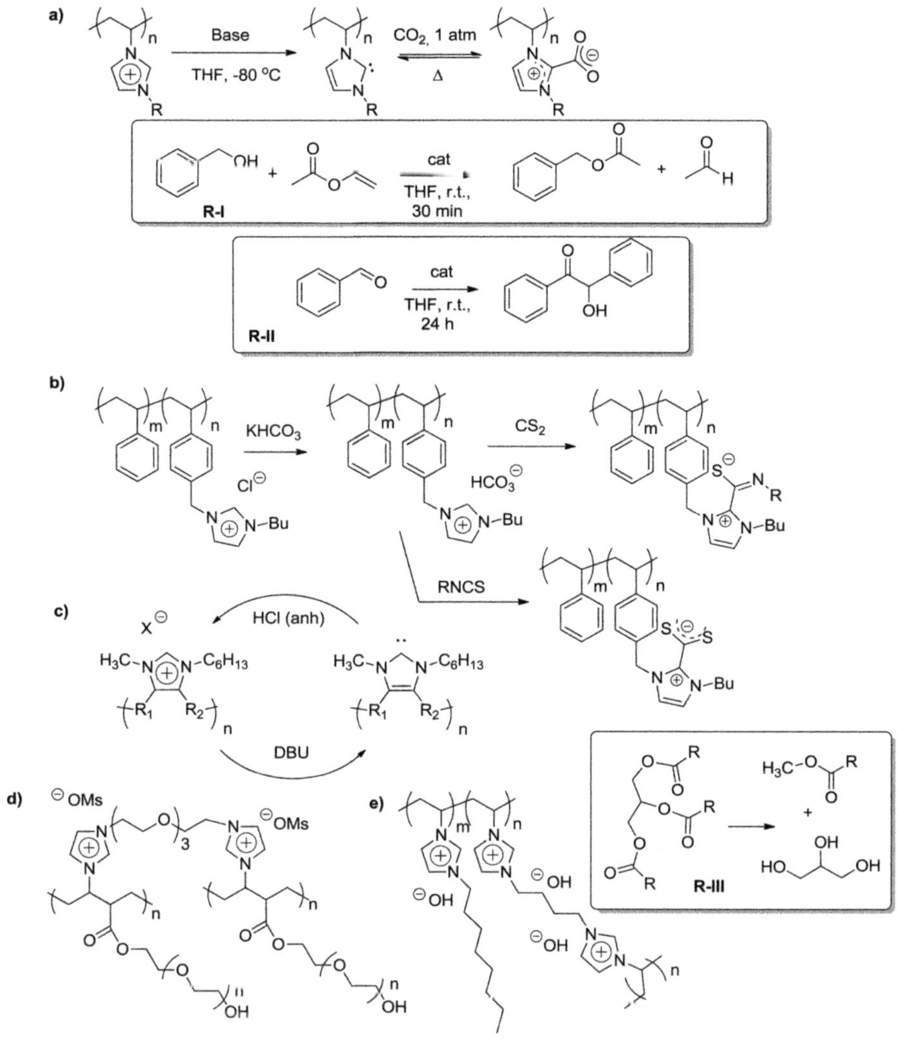

Figure 13.5 Some examples of organocatalytic IL-based materials (a–e) for general organic transformations.

The basicity and acidity of PILs and SILLPs can be enhanced following different strategies for the development of acidic/basic catalysts. Thus, quaternization of poly(vinylimidazole)-*co*-DVB or poly(vinylpyridine)-*co*-DVB nanoporous polymers with propane sultone and treatment with HSO_3CF_3 afforded a strong solid acid able to catalyze the hydration and hydroamination of alkynes with results better than those using classical acids.[94] Most applications have been reported for biodiesel synthesis (R-III). Anionic microgel PILs synthesized from tetrabutylphosphonium 4-styrenesulfonate with *N*,*N*′-methylenebisacrylamide as the crosslinker were treated with a controlled amount of HSO_3CF_3 providing catalytic sulfonic acid fragments that catalyzed the esterification of palmitic acid and transesterification of tripalmitin with methanol, as well as the formation of biodiesel from cooking oil.[95] The calculated kinetic constants were slightly higher than those obtained with toluenesulfonic acid. Microgels could be recovered by centrifugation and reused for several cycles. Magnetic materials containing IL-like fragments with strong acidic or basic character have been reported, including their efficient use for at least five runs.[96] Results with the acidic materials highlighted the importance of diffusional factors for efficiency, and the easy modification of IL-based polymeric materials was used to control both the hydrophobicity and acidity in order to optimize the catalyst, achieving final yields of 90%. The improvement of diffusion was also at the origin of the preparation of a basic superhydrophobic and porous IL-based material (Figure 13.5e) that was assayed successfully for biodiesel synthesis from soybean oil.[97]

Processes for CO_2 capture and conversion to high-value products is of utmost importance in Green and Sustainable Chemistry and IL-based materials have attracted increasing attention in this field. Polyimidazolium salts prepared by reaction of imidazole with bis(halomethyl)arenes were efficient catalysts (5 mol%) for the solventless cycloaddition of CO_2 to styrene oxide (SO) and other epoxides (R-IV, Figure 13.6).[98] Yields of up to 87% were achieved for SO after 24 h at 100 °C and 1 atm CO_2 and could be increased to ≥99% at higher CO_2 pressures. The best results were obtained with the material derived from bis(chloromethyl)biphenyl that could be reused ten times without significant loss in catalytic activity (Figure 13.6a). Interestingly, the use of a tripodal tris(halomethyl)arene affording a hyperbranched material but also some covalent crosslinking was the least porous and active material, which reflects the importance of diffusional factors. A triazolium hyperbranched PIL decorated with pyrene groups at the terminal positions has been efficiently supported on rGO through π–π interactions and used for the cycloaddition of CO_2 to SO for six consecutive runs without appreciable changes in activity.[99] The PS-*co*-PVBnBuIm(Cl or HCO_3) PILs shown in Figure 13.5b also displayed organocatalytic activity for the reaction between CO_2 and propylene oxide.[88]

Self-assembly of polycationic with polyanionic PILs affords materials that have been termed poly(ionic liquid) complexes (PILCs).[100] One of such systems was obtained from zwitterionic PILs derived from imidazolium homopolymers containing an easily hydrolyzable ester group (Figure 13.6b).[101] Although the lineal PIL–COOH prepared by removal of the ester group was soluble, organic solvents containing NH_3 triggered *in situ* interpolyelectrolyte complexation affording a material with a mesoporous structure (6–20 nm in size) and stable in organic solvents such as acetone, THF, or acetonitrile, even after several cycles of wet–drying or refluxing. These PILCs were successfully used as catalysts for the cycloaddition of CO_2 to epoxides with high productivities (70.5 g g_{cat}^{-1}, 1 MPa, 160 °C, 6–24 h). The catalyst was reused four times with average conversions of 94% and selectivities of ≥97%.

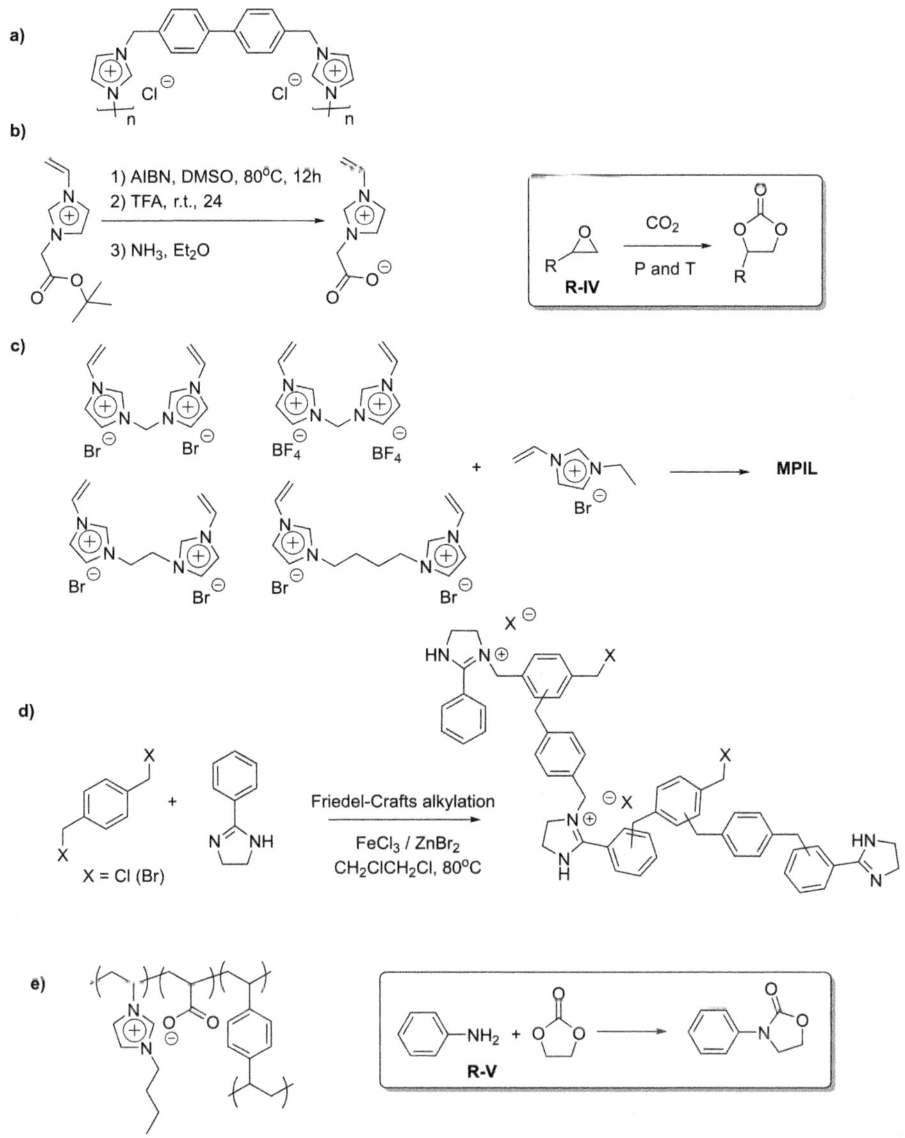

Figure 13.6 Examples of IL-based materials catalyzing the cycloaddition of CO_2 to epoxides and related processes.

The overall performance of these mesoPILCs was attributed to their excellent structural stability and unique nanoscale morphology.

Monolithic hierarchical meso-macroporous crosslinked materials (MPILs) were prepared by copolymerization of divinyl bisimidazolium derivatives with alkylvinylimidazolium bromides (Figure 13.6c).[102] The pore volume was favored by the presence of some water and initially increased with the alkyl length in the IL, although the monolith suffered a great shrinkage during drying in the case of *N*-octyl substituents, with the collapse of the meso-structure. Their high catalytic activity for the cycloaddition of CO_2 to epoxides was attributed to the large surface area and specific hierarchical porous structure enabling good dispersion of the active sites and accelerated mass transfer. Their good CO_2-philicity facilitated the concentration of CO_2 at the catalytic sites inside the pores of the polymeric framework. Additional examples on the synthesis and catalytic activity of hierarchical meso-macroporous MPILs can be found in the literature.[103,104] Porous hyper-crosslinked polymers containing imidazolium fragments have been synthetized by Friedel–Crafts reaction of bis(halomethylbenzenes) and further reaction of *N*-methylimidazole with the residual halomethyl fragments,[105] or from bis(halomethylbenzenes) and 2-phenylimidazoline (Figure 13.6d).[106] These materials showed much higher activities for conversion of CO_2 into cyclic carbonates than the related homogeneous imidazolium derivatives. Their CO_2 capture capacity closely correlated with the micropore volume (BET surface area of up to 926 $m^2\ g^{-1}$) and imidazolium loading, whilst introduction of additional functional groups reduced the polymer porosity and the capacity for CO_2 capture. In all these porous polymers, catalytic improvements were associated to high loadings of IL-like fragments, large surface areas, accessibility of catalytic sites, acceleration of mass transfer, hydrophobic features, and compatibility for substrates.

The same cycloaddition of CO_2 to epoxides was studied using highly crosslinked polymeric nanoparticles (CLPNs) incorporating IL-like fragments. CLPNs were obtained as spherical NPs (10–100 nm) by radical copolymerization of 4-vinylbenzyl-tributylphosphonium chloride and ethylene glycol dimethacrylate (EGDMA) as the crosslinker in alcohols.[107] The solvent and monomer/crosslinker ratio allowed tuning their properties. These materials showed excellent catalytic properties and the reaction with epichlorohydrin took place with 100% yield and 100% selectivity after 3 h, using 0.1 g CLNPs and 3 mL of epoxide (3.0 MPa CO_2, 140 °C). The recovery of the CLPNs required precipitation with an antisolvent or centrifugation, being reused efficiently for six times. In a related field, the copolymer prepared by radical copolymerization of *N*-vinyl imidazolium and sodium acrylate using DVB as the crosslinker was tested as the catalyst for the reaction of ethylene carbonate with aniline (R-V) (Figure 13.6e).[108]

Chirality is a key element in chemistry and a great deal of effort has been devoted to design and synthesize chiral ionic liquids (CILs).[109] The availability of a simple and robust modular synthetic strategy for the synthesis of configurationally and structurally diverse CILs[110] allowed the preparation of a broad family of chiral PILs (CPILs) by chemical modification of preformed poly(chloromethylstyrene) or polymerization of the related vinylic monomers.[111] Some of them provided efficient catalytic activity for the aldol reaction in water, catalyzing the consecutive aldol-dehydration processes and thus behaving as a synthetic mimic of

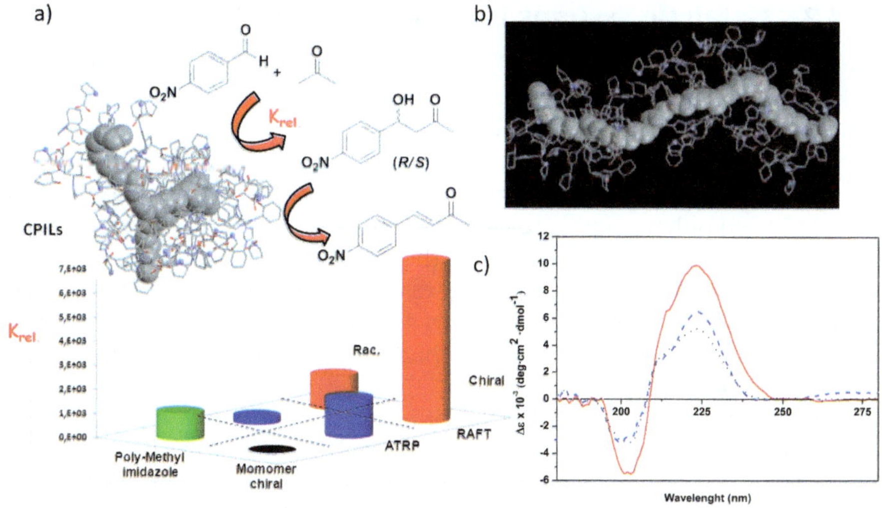

Figure 13.7 (a) Relative kinetic constants for the aldol reaction using CPILs prepared by CRP. (b) Optimized structure obtained from computational calculations for the polymer prepared from the enantiopure monomer. (c) Increase in the circular dichroism signal intensity for the polymer obtained by RAFT.[111]

the aldolase-dehydrogenase enzymatic system (Figure 13.7). At suitable concentration and acetone/water ratio, the preferential enantioselective dehydration of the *R* aldol was achieved.

In water/acetone 1:4, these CPILs were up to *ca.* 6.6×10^3 times more active than their corresponding monomeric counterpart, while in water the only active catalyst was the polymeric one. This can be related to certain chirality transfer to the main polymeric chain, producing a more properly organized polymeric structure. Both the nature of the component CIL and the preparation methodology, in particular the exact CRP method used for polymerization, were found to influence the performance. CILPs prepared by RAFT polymerization of the chiral monomer were more active and enantioselective than the analogous ones prepared by ATRP or grafting. The catalytic nanoparticles prepared by complexation in basic organic media of an imidazolium PIL and poly(L-prolinamide-*co*-MAA) were also sensitive to the preparation methodology. By working with excess PIL to locate this component on the outer shell, high activity and enantioselectivity were realized in water for the aldol reaction and for several multicomponent reactions.[26b]

Quinine-based catalysts were efficiently encapsulated in hydrogels obtained by radical polymerization of vinyl ethylimidazolium bromide and *N*,*N*′-methylenebisacrylamide and tested for the asymmetric nitroaldol reaction.[112] Although the catalytic activity was reduced compared to the homogeneous analogue, a similar enantioselectivity was achieved when using the dried gel. Recycling experiments revealed an appreciable reduction in activity while maintaining the enantioselectivity.

13.4.2 Catalytic Systems Involving Metallic Species

Poly(NHC)s represent excellent ligands for transition metal complexes,[88] and additional functionalities present in the cation, anion, or polymeric backbone can also favor the formation of polymer-supported catalytic metallic complexes. The use of a sulfonated phosphine as the counteranion in nano-PILs incorporating IL moieties in the periphery of hyperbranched polyether cores afforded polymers able to form catalytically active complexes with transition metals such as rhodium and with sufficient solubility in polar organic solvents (Figure 13.8a).[113] They were tested for 1-hexene hydroformylation in methanol (R-VI). Higher selectivities were observed than for the unbound complex, which was attributed to the high local phosphine concentration in the polymer periphery.

PILCs prepared by slow addition of a solution of poly(3-cyanomethyl-1-vinylimidazolium bis(trifluoromethylsulfonyl)imide) and poly(acrylic acid) to a solution of NH_3 in ethanol were obtained as micro-mesoporous particles with high surface areas (330 m^2 g^{-1}) and large pore volumes (1.10 cm^3 g^{-1}).[100] Several processing parameters, including the ratio of components, their concentration, and the solvent(s) used, controlled the resulting porous structure. Cu@PILC species were obtained refluxing these PILCs with $CuCl_2$ in ethanol overnight. Catalytic tests were performed for the aerobic C–H selective oxidation of various hydrocarbons under mild conditions (50 °C), using *N*-hydroxyphthalimide (NHPI) (R-VII) as a combined oxygen-transfer system (Figure 13.8b). Diphenylmethane was oxidized to diphenylketone in 10 h with 91% conversion and 96% selectivity, while the individual components were inefficient. The Cu@PILC was recovered by centrifugation, maintaining the original morphology, and efficiently reused for five cycles. For the hyper-crosslinked polymers containing imidazolinium fragments shown in Figure 13.6d, the catalytic activity for the cycloaddition of CO_2 to epoxides was significantly enhanced in the presence of $ZnBr_2$, allowing to work at low CO_2 pressures (0.15 atm, flue gas simulation, long reaction times).[106]

In the same way ILs stabilize transition metal NPs with anions coordinating to their surface and forming an outer shell that precludes their further reaction and aggregation,[114] PILs containing a large number of IL-like fragments can wrap around the MNP surface allowing their immobilization and stabilization.[6,115] The combination of MNPs with PILs and SILLPs usually exhibits excellent synergistic effects, enhancing the activity and durability of the resulting composites.[116] The influence of the structure and morphology of SILLPs synthesized by grafting the corresponding imidazole onto commercial Merrifield-type resins on the catalytic activity of immobilized AuNPs has been studied in detail (Figure 13.8c). The variables considered included the loading of imidazolium units, resin morphology (gel-type *vs.* macroporous), alkyl substitution at the imidazolium moiety (methyl, butyl, or decyl), counteranion (Cl^- or NTf_2^-), and particle size.[117] Both the microgel/macroporous structure of the polymer and alkyl-substitution on the imidazole could

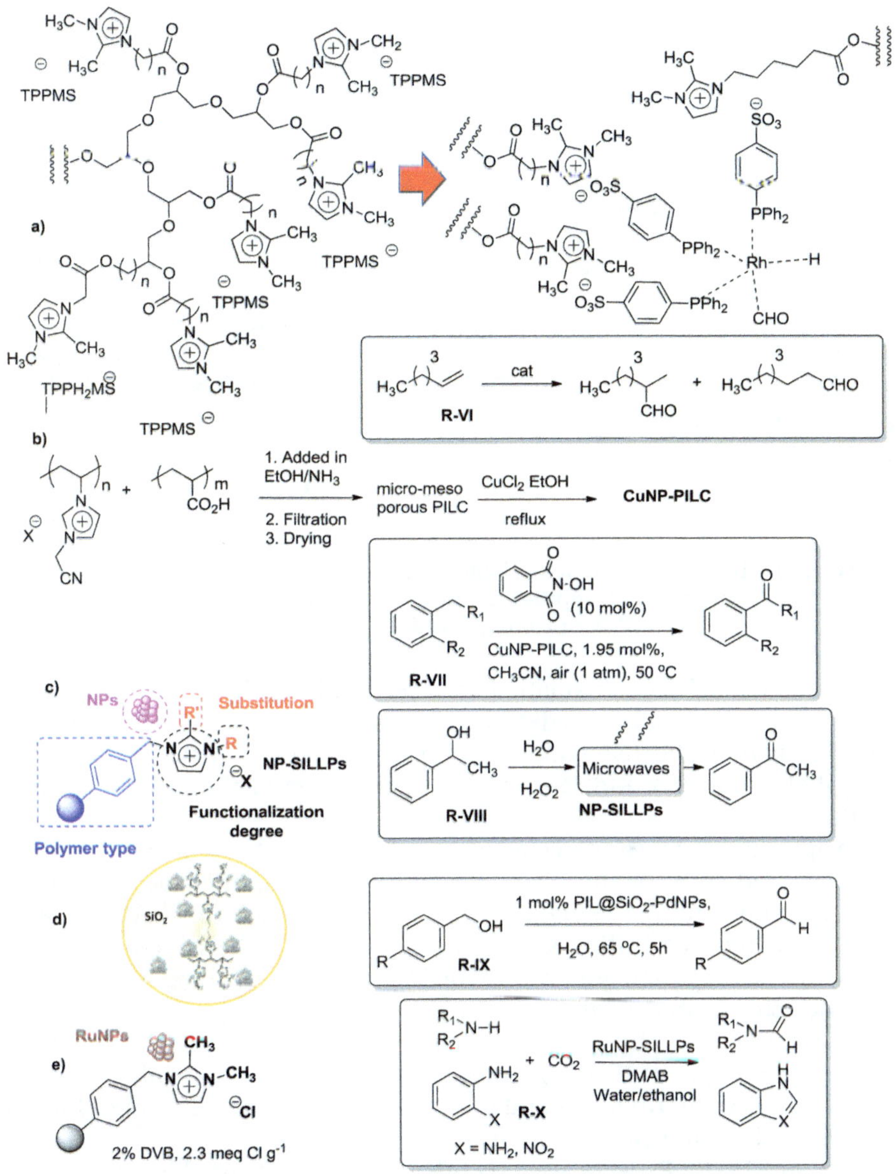

Figure 13.8 Examples of IL-based materials (a–e) used for catalytic systems involving metallic species.

be used to tune the AuNP size distribution. Similar NP sizes were obtained for gel-type SILLPs (1% crosslinking) with imidazolium substituents R = CH_3, R′ = H and R = C_4H_9, R′ = H, but a five-fold increase in size was observed for polymers with R = $C_{10}H_{21}$, R′ = CH_3 (Figure 13.9). Interestingly, the TEM images also showed changes in their morphology and shape, from spherical

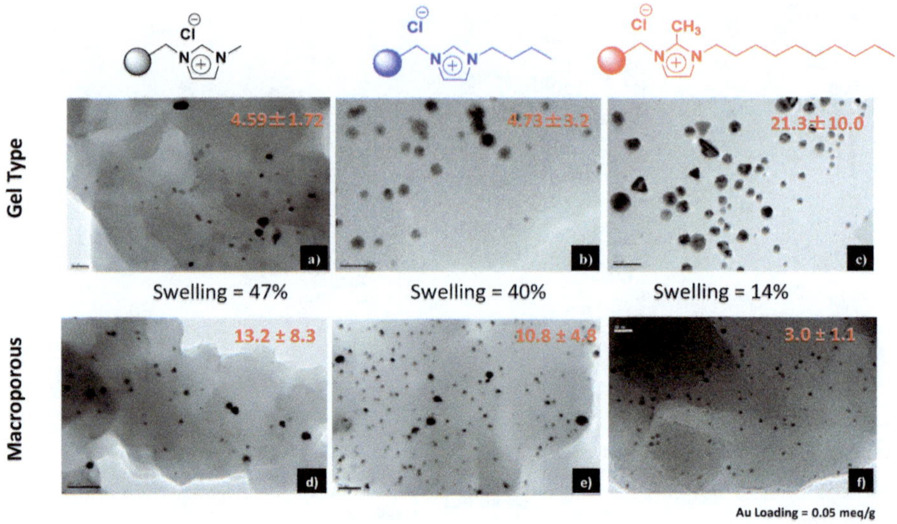

Figure 13.9 Influence of structural and morphological parameters on the formation of SILLP-supported AuNPs: (a)–(c) PS-DVB 1% crosslinked, and (d)–(f) PS-DVB 20% crosslinked.

particles to triangles or prismatic particles. The long aliphatic substituents on the imidazolium subunits decrease the swelling in water, leading to a significantly higher effective concentration of the precursor inside the more hydrophobic polymer and favoring the aggregation of the initially formed AuNPs.

The opposite trend was detected for macroporous SILLPs for which the NP size decreased with the increasing size of the aliphatic residue, with decyl *N*-substitution leading to smaller particles. Here, access to the IL-like sites is not regulated by swelling and no changes exist in the effective concentration of the supported gold precursor. Nevertheless, the long hydrophobic aliphatic chains can contribute to regulate the diffusion processes hampering NP aggregation. Thus, appropriate control of the polymer morphology, loading, and nature of IL-like groups, allows obtaining immobilized AuNPs with narrow size distributions and long-term (months) stability (even for 2–3 nm NPs) under atmospheric conditions.

These AuNP-SILLPs were used as efficient catalysts for the oxidation of phenyl ethanol in water using H_2O_2 as the oxidant and under microwave irradiation (R-VIII) (Figure 13.8c).[118] For gel-type materials, increased activity was observed with larger *N*-alkyl chains, in spite of containing AuNPs of larger sizes. This was associated with fine-tuning of the hydrophobicity–hydrophilicity balance, favoring the absorption of the substrate from a polar solvent and also the potential formation of 'hot-spots' on the NPs through the efficient absorption of microwave irradiation. Moreover, better performance of the macroporous resins in comparison with that of gel-type ones was observed, again associated with facilitated diffusion of the reagents to the active sites. Changes in the counteranion and IL-like unit loading were found to have a smaller effect.

Poly(DVB-*co*-vinylimidazole) polymers and the corresponding SILLPs obtained by alkylation of imidazole groups with an alkyl iodide have been shown to stabilize PdNPs of very small size (<2 nm) and catalyze the selective hydrogenation of *p*-chloronitrobenzene to *p*-chloroaniline.[119] Materials with *N*-butyl substitution displayed almost total selectivity at full conversion. Since the average size of the PdNPs was the same in all the catalysts, the results suggested that the chemoselectivity depended on the polarity of the environment provided by the IL-like fragments, which directly affects the electronic and surface properties of the PdNPs. Imidazolium-based porous organic ionic polymers also provided well-dispersed PdNPs with small sizes and narrow distributions displaying excellent catalytic performances in the hydrogenation of nitroarenes.[120] An improvement of the catalytic activity was observed and ascribed to the higher specific surface area and smaller granular particles formed. In the case of the PdNPs supported on poly(DVB-*co*-bisimidazolium) studied for the same reaction, the catalysts were reused efficiently for 10 runs and the results suggested the participation of NHC ligands in the stabilization of the PdNPs.[121] As will be discussed below in more detail, PIL/SILLP-supported PdNPs have also been used as catalysts for a variety of C–C coupling reactions.[41b,122]

PdNPs have also been supported on hybrid micelle-like amphiphilic nanoreactors composed of a hydrophobic PIL core and a hydrophilic mesoporous silica shell, displaying good dispersion in water (Figure 13.8d).[123] This system enhanced the catalytic activity of PdNPs and showed good recyclability in the selective oxidation of aromatic alcohols to aldehydes (R-IX). Such catalytic enhancement was associated to the accumulation of the hydrophobic substrates in the PIL core in close proximity to PdNPs (Figure 13.8d).

In the area of CO_2 transformation, RuNPs supported on SILLPs prepared from Merrifield resins (Figure 13.8c, RuNPs) were used to catalyze the reaction of CO_2 with amines and diamines in the presence of dimethylamine borane (DMAB) to form *N*-formamines and benzimidazoles in water/ethanol (R-X).[124] The catalyst was recovered and reused five times with only a slight decrease in yield and minor metal leaching.

13.5 Multicatalytic Systems and Continuous Flow Processes

In this context, multicatalytic systems can be achieved either by developing multifunctional IL-based materials or by combining different catalytic PILs or SILLPs. This last approach is favored under continuous flow conditions that, by themselves, represent a simple implementation toward greener chemical processes based on supported ILs.[125] Most reported examples involve IL-like fragments covalently attached to a polymeric matrix.[126]

As mentioned above, SILLPs represent an excellent medium for the generation and stabilization of PdNPs catalytically active for different C–C forming reactions (Heck, Stille, or Sonogashira). IL-like moieties stabilize the NPs and simultaneously reduce leaching through a release-and-catch mechanism.[127] Additionally, the combination in a single reactor of SILLPs with different functionalities that can act simultaneously or consecutively as supported reagents (*i.e.*, bases or acids), precatalysts, or scavengers for undesired reaction side-products, led to polymeric cocktails with improved properties. Thus, the Heck reaction (R-XI) could be carried out under continuous-flow conditions using a packed bed reactor with a polymeric cocktail containing basic SILLPs and PdNP-SILLPs and supercritical carbon dioxide ($scCO_2$) as the solvent (Figure 13.10a). The same reactor could be reused efficiently not only for successive Heck reactions but also for alternative C–C coupling processes, which represents a significant improvement in terms of green chemical processes.[128]

Imidazolium chloride SILLPs catalyzed the solvent-free cyanosilylation of carbonyl compounds using trimethylsilyl cyanide. Productivities of up to 45.1 g g^{-1}_{SILLP} h^{-1} were achieved under flow conditions, with no indication of decrease in activity/selectivity after running continuously for more than two days.[129] The exchange of chloride by (hydroxide or acetate as counteranion) allowed developing a continuous flow solvent-free Henry nitroaldol reaction (R-XII) using a monolithic mini-reactor (Figure 13.10b).[130] In all cases, the continuous flow processes allowed obtaining relevant TON values. PS-DVB-SILLPs have been studied for the cycloaddition of CO_2 to epoxides using a continuous flow fixed-bed reactor that maintained its efficiency for more than 130 h.[131]

The immobilization of biocatalysts in SILLPs and their application in flow processes is also possible. Thus, CALB-SILLP systems were efficient and stable catalysts for the continuous flow synthesis of citronellyl propionate in $scCO_2$ (R-XIII, Figure 13.11a).[132] High IL/enzyme ratios enhanced the stabilization.[133,134] Yields of 93% were obtained at 80 °C for more than 10 h without any enzyme deactivation, being superior to those found using silica-supported CALB. This was associated with an optimal microenvironment for lipase activity and improved mass-transfer phenomena.

The high enantioselectivity of processes based on enzymes allowed the continuous kinetic resolution (KR) of *sec*-alcohols with a mini-reactor containing CALB-SILLP. The maximum achievable yield of 50% for the expected enantiopure ester was obtained for 6 days using $scCO_2$ (50 °C, 10 MPa).[134] Excellent results for the corresponding dynamic KR (DKR) were obtained using a multicatalytic system under flow conditions (R-XIV). For this, a columnar minireactor was packed with a mixture of CALB-SILLP and zeolite

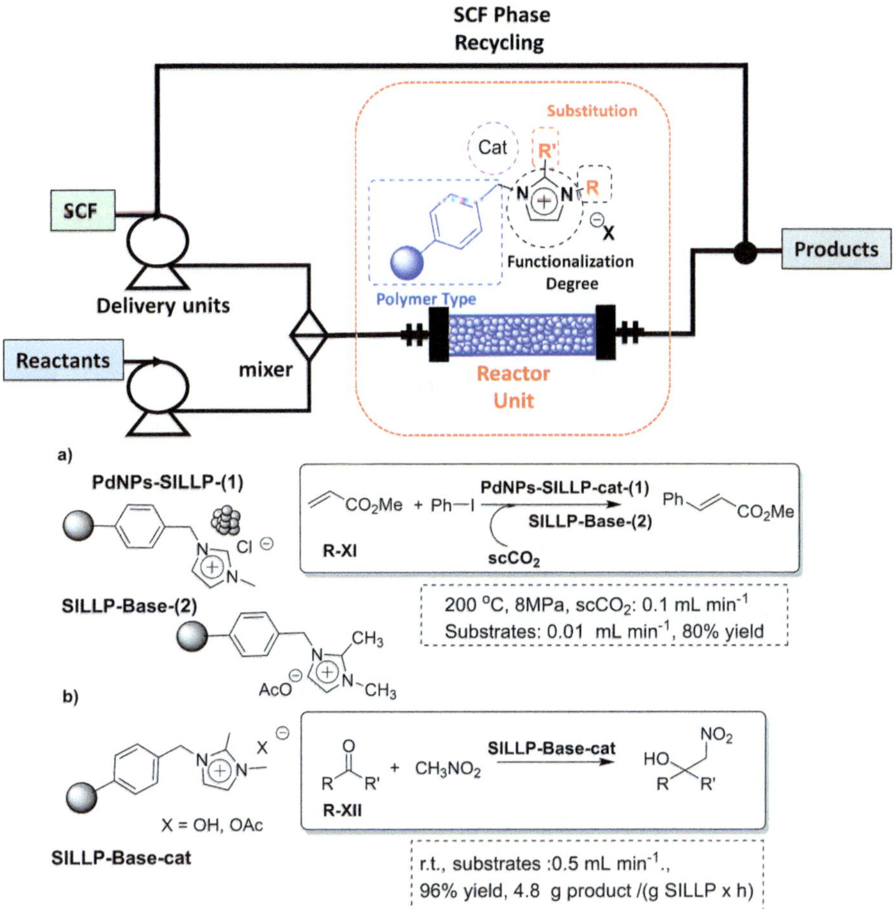

Figure 13.10 Continuous flow processes using (a) a SILLP cocktail in $scCO_2$[128] and (b) a basic SILLP.[130]

CP811E-150 coated with a small amount of an IL (Figure 13.11b). This allowed the effective isolation of both catalysts and the continuous racemization of the undesired enantiomer of the alcohol not transformed by the enzyme. Yields higher than 92% with enantioselectives ≥99.9% were observed under the same conditions used for the KR experiments.

Appropriate mass-transfer processes at the interface (substrates to the active site or products from it) are always key requirements for efficiency, which can be controlled by tuning the IL-like moieties. Thus, CALB immobilized on macroporous PS-DVB-SILLP bearing 1-decyl-2-methyimidazolium cations was the optimal biocatalyst for continuous flow methyl oleate synthesis (biodiesel) from triolein ($scCO_2$, 18 MPa, 45 °C, Figure 13.11b, R-III in Figure 13.5), achieving an average 85% yield of biodiesel after 45 cycles of 4–8 h.[135]

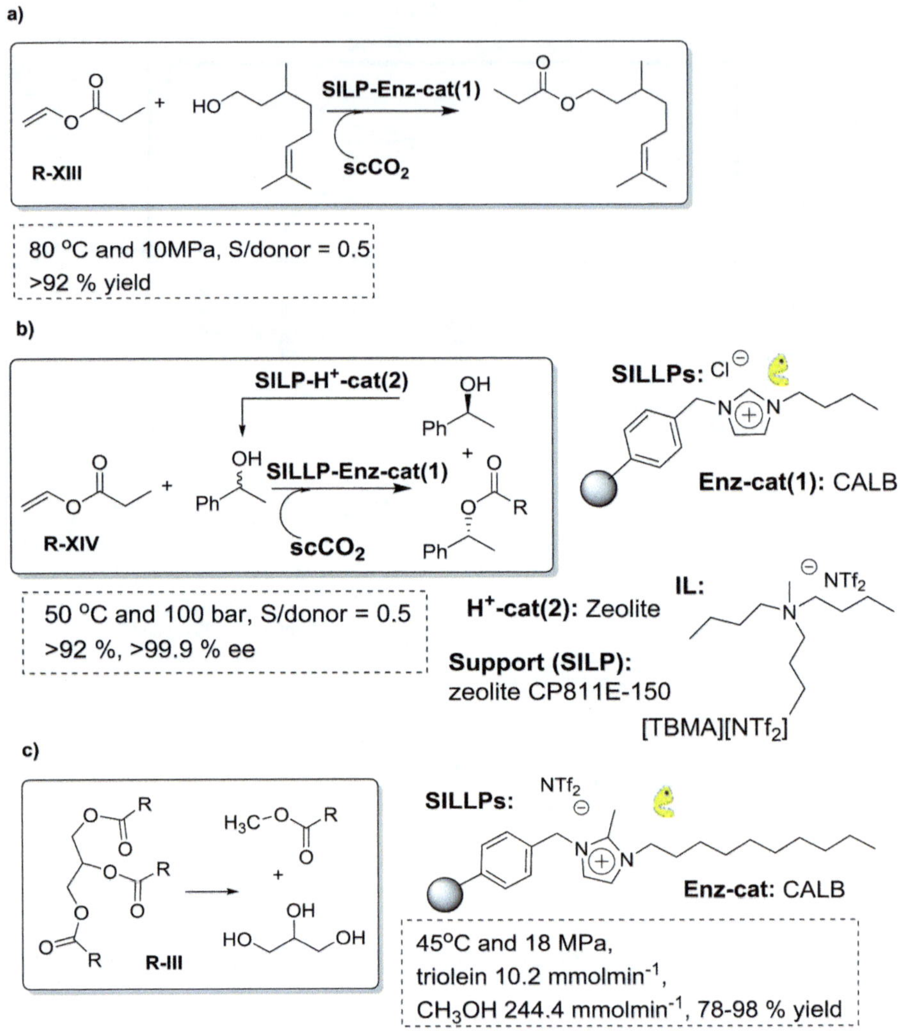

Figure 13.11 Continuous flow enzymatic reactions in SILLPs: (a) transesterification, (b) kinetic resolution, (c) biodiesel synthesis.

Following this approach, the multistep synthesis of aminoalcohols has been reported. The first step was the continuous flow epoxidation of an alkene in a reactor packed with CALB supported on an imidazolium PS-DVB-SILLP with long *N*-alkyl chains (chloride anion) using dimethyl carbonate as the solvent and source for the *in situ* formation of the reactive peroxyacid. After eliminating the excess H_2O_2, two consecutive transformations were coupled under flow conditions in DMC. In a first reactor, the ring opening of the epoxide with an amine took place catalyzed by an imidazolium sulfonate zwitterionic SILLP treated with $Sc(OTf)_2$ and, in a second reactor, the KR of the *rac*-aminoalcohol using a commercial CALB (Novozyme 435) and DMC as the acyl donor. The flow at the exit of the second reactor showed a 49% yield of the acetylated aminoalcohol with 99% ee.[136]

13.6 Outlook and Prospectives

The provided examples highlight the potential of IL-based materials for the development of efficient green chemical processes. Analysis of the recent results in the field suggest that future trends in the area will involve the integration of more complex systems based on IL-based materials. These should include additional multicatalytic and flow processes, with the perspective of achieving automated synthesis on-demand,[137] the exploitation of the possibilities of biotransformations and their combination with chemocatalytic reactions. They must also consider the integration of transformation and separation processes.[138] In this regard, the studies carried out on the use of these materials, both for the separation and transformation of CO_2, open new avenues for their extensive application. Finally, the application of new advanced manufacturing technologies, in particular 3D printing, for instance, for the development of specific microreactors represents an additional opportunity in this field.[139]

References

1. P. T. Anastas, W. Leitner, P. G. Jessop, C.-J. Li, P. Wasserscheid and A. Stark, *Handbook of Green Chemistry - Green Solvents*, Wiley-VCH Verlag GmbH, New York, 2010.
2. R. Hayes, G. G. Warr and R. Atkin, *Chem. Rev.*, 2015, **115**, 6357.
3. P. Wasserscheid and T. Welton, *Ionic Liquids in Synthesis*, Wiley-VCH, Weinheim, 2008.
4. A. Jordan and N. Gathergood, *Chem. Soc. Rev.*, 2015, **44**, 8200.
5. N. Sun, H. Rodríguez, M. Rahman and R. D. Rogers, *Chem. Commun.*, 2011, **47**, 1405.
6. P. Zhang, P. Wu and B. Han, *Adv. Mater.*, 2014, **26**, 6810.
7. W. J. Qian, J. Texter and F. Van, *Chem. Soc. Rev.*, 2017, **46**, 1124.
8. A. Eftelchari and T. Saito, *Eur. Polym. J.*, 2017, **90**, 245.
9. R. Fehrmann, A. Riisager and M. Haumann, *Supported Ionic Liquids*, Wiley-VCH, Weinheim, 2014.
10. B. Xin and J. Hao, *Chem. Soc. Rev.*, 2014, **43**, 7171.
11. E. Rynkowska, K. Fatyeyeva and W. Kujawski, *Rev. Chem. Eng.*, 2018, **34**, 341.
12. S. Zhang, K. Dokko and M. Watanabe, *Chem. Sci.*, 2015, **6**, 3684.
13. M. Gruttadauria, L. F. Liotta, A. M. P. Salvo, F. Giacalone, V. La Parola, C. Aprile and R. Noto, *Adv. Synth. Catal.*, 2011, **353**, 2119.
14. P. Agrigento, S. M. Al-Amsyar, B. Sorée, M. Taherimehr, M. Gruttadauria, C. Aprile and P. Pescarmona, *Catal. Sci. Technol.*, 2014, **4**, 1598.
15. (a) V. Campisciano, F. Giacalone and M. Gruttadauria, *Chem. Rec.*, 2017, **17**, 918; (b) F. Giacalone and M. Gruttadauria, *ChemCatChem*, 2016, **8**, 664.
16. (a) Y. Kohno, S. Saita, Y. Men, J. Yuan and H. Ohno, *Polym. Chem.*, 2015, **6**, 2163; (b) Q. Zhang, L. B. Liu and C. G. Pan, *J. Mater. Chem.*, 2018, **53**, 7.

17. S. V. Muginova, D. A. Myasnikova, S. G. Kazarian and T. N. Shekhovtsova, *Anal. Sci.*, 2017, **33**, 261.
18. (a) S. Zulfiqar, M. I. Sarwar and D. Mecerreyes, *Polym. Chem.*, 2015, **6**, 6435; (b) L. C. Toméa and I. M. Marrucho, *Chem. Soc. Rev.*, 2016, **45**, 2785; (c) M. Sadeghpour, R. Yusoff and M. K. Aroua, *Rev. Chem. Eng.*, 2017, **33**, 183.
19. P. Guo, H. Zhang, X. Liu and J. Sun, *ACS Appl. Mater. Interfaces*, 2018, **10**, 2105.
20. J. Yuan, D. Mecerreyes and M. Antonietti, *Prog. Polym. Sci.*, 2013, **38**, 1009.
21. J. Yuan and M. Antonietti, *Polymer*, 2011, **52**, 1469.
22. D. Mecerreyes, *Prog. Polym. Sci.*, 2011, **36**, 1629.
23. N. Nishimura and H. Ohno, *Polymer*, 2014, **55**, 3289.
24. O. Green, S. Grubjesic, S. Lee and M. Firestone, *Polym. Rev.*, 2009, **49**, 339.
25. F. Chen, J. Guo, D. Xu and F. Yan, *Polym. Chem.*, 2016, **7**, 1330.
26. (a) Z. Y. Guo, X. Hai, Y. T. Wang, Y. Shu, X. W. Chen and J. H. Wang, *Biomacromolecules*, 2018, **19**, 53; (b) X. J. Li, C. N. Lv, X. B. Jia, M. Q. Cheng, K. Wang and Z. G. Hu, *ACS Appl. Mater. Interfaces*, 2017, **9**, 827; (c) K. M. Meek and Y. A. Elabd, *Macromol. Rapid Commun.*, 2016, **37**, 1200.
27. W. Xu, P. A. Ledin, V. V. Shevchenko and V. V. Tsukruk, *ACS Appl. Mater. Interfaces*, 2015, **7**, 12570.
28. B. Kerscher, F. Schüler, A. Appel, K. Schadt and R. Mülhaupt, *Adv. Polym. Sci.*, 2013, **262**, 431.
29. X. P. Gao, F. Lu, B. Dong, A. L. Wu, N. Sun and L. Q. Zheng, *J. Mater. Chem. A*, 2016, **4**, 13316.
30. V. Sans, N. Karbass, M. I. Burguete, V. Compañ, E. García-Verdugo, S. V. Luis and M. Pawlak, *Chem. - Eur. J.*, 2011, **17**, 1894.
31. A. Castro-Grijalba, E. M. Reyes-Gallardo, R. G. Wuilloud, R. Lucena and S. Cardenas, *RSC Adv.*, 2017, **7**, 42797.
32. B. Altava, M. I. Burguete, E. Garcia-Verdugo, S. V. Luis and M. J. Vicent, *Tetrahedron*, 2001, **57**, 8675.
33. S. Montolio, O. Zagorodko, R. Porcar, M. I. Burguete, S. V. Luis, H. Tenhu and E. Garcia-Verdugo, *Polym. Chem.*, 2017, **8**, 4789.
34. (a) A. Garcia-Bernabe, A. Rivera, A. Granados, S. V. Luis and V. Compañ, *Electrochim. Acta*, 2016, **213**, 887; (b) B. Altava, V. Compan, A. Andrio, L. F. del Castillo, S. Molla, M. I. Burguete, E. Garcia-Verdugo and S. V. Luis, *Polymer*, 2015, **72**, 69.
35. M. D. Green and T. E. Long, *Polym. Rev.*, 2009, **49**, 291.
36. A. S. Shaplov, D. O. Ponkratov and Y. S. Vygodskii, *Polym. Sci., Ser. B*, 2016, **58**, 73.
37. D. O. Ponkratov, E. I. Lozinskaya, P. S. Vlasov, P. H. Aubert, C. Plesse, F. Vidal, Y. S. Vygodskii and S. S. Shaplov, *Electrochim. Acta*, 2018, **281**, 777.
38. K. Matyjaszewski, *Macromolecules*, 2012, **45**, 4015.

39. E. Rizzardo and D. Solomon, *Aust. J. Chem.*, 2012, **65**, 945.
40. T. Morinaga, S. Honma, T. lshizuka, T. Kamijo, T. Sato and Y. Tsujii, *Polymers*, 2016, **8**, 146.
41. (a) K. Nakabayashi, A. Umeda, Y. Sato and H. Mari, *Polymer*, 2016, **96**, 81; (b) S. Ghasemi and Z. A. Harandi, *RSC Adv.*, 2018, **8**, 14570.
42. Y. N. Zhou, L. Leí, Z. H. Luo and S. P. Zhu, *Macromolecules*, 2017, **50**, 8378.
43. (a) A. L. Wang, H. Xu, X. Liu, R. Gao, S. Wang, Q. Zhou, J. Chen, X. F. Liu and L. Y. Zhang, *Polym. Chem.*, 2017, **8**, 3177; (b) Q. J. Chen and Z. S. An, *Chin. J. Polym. Sci.*, 2017, **35**, 54.
44. (a) E. Andrzejewska, *Polym. Int.*, 2017, **66**, 366; (b) A. C. Tibbits, Y. S. Yan and C. J. Kloxin, *Macromol. Rapid Commun.*, 2017, **38**, 1700113; (c) S. M. Chen, T. L. Wang, P. Y. Chang, C. H. Yang and Y. C. Lee, *React. Funct. Polym.*, 2016, **108**, 103; (d) H. Gu, F. Van and J. Texter, *Macromol. Rapid Commun.*, 2016, **37**, 1218.
45. A. M. Devasurendra, C. Zhang, J. A. Young, L. M. V. Tillekeratne, J. L. Anderson and J. R. Kirchhoff, *ACS Appl. Mater. Interfaces*, 2017, **9**, 24955.
46. Y. I. Dong, J. B. Yin, J. H. Yuan and X. P. Zhao, *Polymer*, 2016, **97**, 408.
47. (a) S. M. Morozova, A. S. Shaplov, E. I. Lozinskaya, P. S. Vlasov, H. Sardon, D. Mecerreyes and Y. S. Vygodskii, *High Perform. Polym.*, 2017, **29**, 691; (b) M. Suckow, M. Roy, K. Sahre, L. Haussler, N. K. Singha, B. Voit and F. Bohme, *Polymer*, 2017, **111**, 123.
48. C. Chiappe, M. Malvaldi and C. S. Pomelli, *Pure Appl. Chem.*, 2009, **81**, 767.
49. P. G. Jessop, D. A. Jessop, D. Fua and P. Lam, *Green Chem.*, 2012, **14**, 1245.
50. (a) C. Reichardt, *Green Chem.*, 2005, **7**, 339; (b) S. V. Dzyuba and R. A. Bartsch, *Tetrahedron Lett.*, 2002, **43**, 4657.
51. D. C. Dong and M. A. Winnik, *Can. J. Chem.*, 1984, **62**, 2560.
52. M. I. Burguete, F. Galindo, E. García-Verdugo, N. Karbass and S. V. Luis, *Chem. Commun.*, 2007, **29**, 3086.
53. (a) K. A. Fletcher, I. A. Storey, A. E. Hendricks, S. Pandey and S. Pandey, *Green Chem.*, 2001, **3**, 210; (b) D. S. Karpovich and G. J. Blanchard, *J. Phys. Chem.*, 1995, **99**, 3951.
54. T. Köddermann, C. Wertz, A. Heintz and R. Ludwig, *Angew. Chem., Int. Ed.*, 2006, **45**, 3697.
55. L. Cammarata, S. G. Kazarian, P. A. Salter and T. Welton, *Phys. Chem. Chem. Phys.*, 2001, **3**, 5192.
56. J. L. Anderson, J. Ding, T. Welton and D. W. Armstrong, *J. Am. Chem. Soc.*, 2002, **124**, 14247.
57. N. K. Sharma, M. D. Tickell, J. L. Anderson, J. Karr, V. Pino, B. F. Wicker, D. W. Armstrong, J. H. Davis Jr and A. J. Russell, *Chem. Commun.*, 2006, 646.
58. J. L. Anderson and D. W. Armstrong, *Anal. Chem.*, 2003, **75**, 4851.

59. L. G. Mendoza, J. Gonzalez-Alvarez, C. F. Gonzalo, P. Arias-Abrodo, B. Altava, S. V. Luis, M. I. Burguete and M. D. Gutierrez-Alvarez, *Talanta*, 2015, **143**, 212.
60. (a) D. Nikolaeva, I. Azcune, E. Sheridan, M. Sandru, A. Genua, M. Tanczyk, M. Jaschik, K. Warmuzinski, J. C. Jansen and I. F. J. Vankelecom, *J. Mater. Chem. A*, 2017, **5**, 19808; (b) H. Y. Ran, J. X. Wang, A. A. Abdeltawab, X. C. Chen, G. R. Yu and Y. H. Yu, *J. Energy Chem.*, 2017, **26**, 909; (c) H. Tang, J. Tang, S. Ding, M. Radosz and Y. Shen, *J. Polym. Sci., Part A: Polym. Chem.*, 2005, **43**, 1432; (d) J. Tang, W. Sun, H. Tang, M. Radosz and Y. Shen, *Macromolecules*, 2005, **38**, 2037.
61. (a) K. Manojkumar, D. Mecerreyes, D. Taton, Y. Gnanou and K. Vijayakrishna, *Polym. Chem.*, 2017, **8**, 3497; (b) Z. L. Hou, T. Huang, C. Y. Cai, T. Resheed, C. Y. Yu, Y. F. Zhou and D. Y. Van, *Chin. J. Polym. Sci.*, 2017, **35**, 602.
62. K. Vijayakrishna, S. Jewrajka, A. Ruiz, R. Marcilla, J. Pomposo, D. Mecerreyes, D. Taton and Y. Gnanou, *Macromolecules*, 2008, **41**, 6299.
63. Y. Zhang, Y. Zhang, B. Chen, L. Qin and G. Gao, *ChemistrySelect*, 2017, **2**, 9443.
64. M. J. Muldoon and C. M. Gordon, *J. Polym. Sci., Part A: Polym. Chem.*, 2004, **42**, 3865.
65. D. Batra, D. N. T. Hay and M. A. Firestone, *Chem. Mater.*, 2007, **19**, 4423.
66. A. Pourjavadi, M. Doulabi and S. H. Hosseini, *Polymer*, 2012, **53**, 5737.
67. T. Ono, T. Sugimoto, S. Shinkai and K. Sada, *Nat. Mater.*, 2007, **6**, 429.
68. D. Han, X. X. Li, J. Peng, L. Xu, J. Q. Li, H. B. Li and M. L. Zhai, *RSC Adv.*, 2016, **6**, 69052.
69. W. J. Horne, M. A. Andrews, K. L. Terrill, S. S. Hayward, J. Marshall, K. A. Belmore, M. S. Shannon and J. E. Bara, *ACS Appl. Mater. Interfaces*, 2015, **7**, 8979.
70. I. Azcune, I. Garcia, P. M. Carrasco, A. Genua, M. Tanczyk, M. Jaschik, K. Warmuzinski, G. Cabañero and I. Odriozola, *ChemSusChem*, 2014, **7**, 3407.
71. H. L. Ngo, K. LeCompte, L. Hargens and A. B. McEwen, *Thermochim. Acta*, 2000, **357–358**, 97.
72. A. E. Visser, W. M. Reichert, H. D. Willauer, G. A. Broker, R. D. Rogers and J. G. Huddleston, *Green Chem.*, 2001, **3**, 156.
73. (a) M. M. Obadia, S. Fagour, Y. S. Vygodskii, F. Vidal, A. Serghei, A. S. Shaplov and E. Drockenmuller, *J. Polym. Sci., Part A: Polym. Chem.*, 2016, **54**, 2191; (b) Y. S. Vygodskii, O. A. Mel'nik, E. I. Lozinskaya, A. S. Shaplov, I. A. Malyshkina, N. D. Gavrilova, K. A. Lyssenko, M. Y. Antipin, D. G. Golovanov, A. A. Korlyukov, N. Ignatev and U. Welz-Biermann, *Polym. Adv. Technol.*, 2007, **18**, 50; (c) W. Ogihara, S. Washiro, H. Nakajima and H. Ohno, *Electrochim. Acta*, 2006, **51**, 2614.
74. K. Roeleveld, F. David and F. Lynen, *J. Chromatogr. A*, 2016, **1451**, 135.
75. K. Jain, R. Vedarajan, M. Watanabe, M. Ishikiriyama and N. Matsumi, *Polym. Chem.*, 2015, **6**, 6819.

76. A. Okafuji, Y. Kohno and H. Ohno, *Macromol. Rapid Commun.*, 2016, **37**, 1130.
77. Y. Lu, G. Yu, W. Wang, Q. Ren, B. Li and S. Zhu, *Macromolecules*, 2015, **48**, 915.
78. H. Zayas, A. Lu, D. Valade, F. Amir, Z. Jia, K. O'Reilly and M. Monteiro, *ACS Macro Lett.*, 2013, **2**, 327.
79. (a) P. Lozano, J. M. Bernal, S. Nieto, C. Gomez, E. Garcia-Verdugo and S. V. Luis, *Chem. Commun.*, 2015, **51**, 17361; (b) P. Lozano, J. M. Bernal, E. Garcia-Verdugo, G. Sanchez-Gomez, M. Vaultier, M. I. Burguete and S. V. Luis, *Green Chem.*, 2015, **17**, 3706.
80. B. Altava, M. I. Burguete, E. Garcia-Verdugo and S. V. Luis, *Chem. Soc. Rev.*, 2018, **47**, 2722.
81. D. E. Bergbreiter, *ACS Macro Lett.*, 2014, **3**, 260.
82. B. Altava, M. I. Burguete, J. M. Fraile, J. I. García, S. V. Luis, J. A. Mayoral and M. J. Vicent, *Angew. Chem.*, 2000, **112**, 1563.
83. B. Altava, M. I. Burguete, E. García-Verdugo, S. V. Luis and M. J. Vicent, *Green Chem.*, 2006, **8**, 717.
84. M. I. Burguete, J. M. Fraile, E. García-Verdugo, S. V. Luis, V. Martínez-Merino and J. A. Mayoral, *Ind. Eng. Chem. Res.*, 2005, **44**, 8580.
85. B. Altava, M. I. Burguete, E. García-Verdugo, S. V. Luis, M. J. Vicent and J. A. Mayoral, *React. Funct. Polym.*, 2001, **48**, 25.
86. J. Pinaud, J. Vignolle, Y. Gnanou and D. Taton, *Macromolecules*, 2011, **44**, 19008.
87. P. Coupillaud, J. Pinaud, N. Guidolin, J. Vignolle, M. Fèvre, E. Veaude-crenne, D. Mecerreyes and D. Taton, *J. Polym. Sci., Part A: Polym. Chem.*, 2013, **51**, 4530.
88. P. Coupillaud, J. Vignolle, D. Mecerreyes and D. Taton, *Polymer*, 2014, **55**, 3404.
89. A. B. Powel, Y. Suzuki, M. Ueda, C. W. Bielawski and A. H. Cowley, *J. Am. Chem. Soc.*, 2011, **133**, 5218.
90. N. Pothanagandhi, N. A. Sivaramakrishna and K. Vijayakrishna, *Tetrahedron*, 2016, **106**, 132.
91. M. V. Reddy, K. R. Byeon, S. H. Park and D. W. Kim, *Tetrahedron*, 2017, **73**, 5289.
92. (a) O. F. Ozturk, S. Demirci, S. B. Sengel and N. Sahiner, *Polym. Adv. Technol.*, 2018, **564**, 56; (b) A. Chinnappan, A. H. Tamboli, W. J. Chung and H. Kim, *Chem. Eng. J.*, 2016, **285**, 554.
93. J. G. L. Ferreira, L. M. Ramos, A. L. de Oliveira, E. S. Orth and A. D. Neto, *J. Org. Chem.*, 2015, **80**, 5979.
94. D. J. Tao, F. J. Liu, L. Wang and L. L. Jiang, *Appl. Catal., A*, 2018, **564**, 56.
95. S. Chen, Y. Peng, Q. Wu, A. Chang, A. Qu, J. Shen, J. Xie, Z. H. Farooqi and W. Wu, *Polym. Chem.*, 2016, **7**, 5463.
96. (a) H. Zhang, H. Li, H. Pan, A. P. Wang, S. Souzanchi, C. B. Xu and S. Yang, *Appl. Energy*, 2018, **223**, 416; (b) H. Zhang, H. Li, H. Pan, A. P. Wang, C. B. Xu and S. Yang, *Energy Convers. Manage.*, 2017, **153**, 462.

97. B. Jiang, Y. Wang, L. Zhang, Y. Sun, H. Yang, B. Wang and N. Yang, *Energy Fuels*, 2017, **31**, 5203.
98. W. Zhong, F. D. Bobbink, D. Felix, Z. F. Fei and P. J. Dyson, *ChemSusChem*, 2017, **10**, 2728.
99. Y. Wang, X. Guan, F. Y. Chen, S. S. Zhu, Y. S. Ye, H. Y. Peng, X. P. Zhou, X. L. Xie and Y. W. Mai, *Catal. Sci. Technol.*, 2017, **7**, 4173.
100. Q. Zhao, P. Zhang, M. Antonietti and J. Yuan, *J. Am. Chem. Soc.*, 2012, **134**, 11852.
101. S. Soll, P. Zhang, Q. Zhao, Y. Wang and J. Yuan, *Polym. Chem.*, 2013, **4**, 5048.
102. X. Wang, Y. Zhou, Z. Guo, G. Chen, J. Li, Y. Shi, Y. Liu and J. Wang, *Chem. Sci.*, 2015, **6**, 6916.
103. X. Wang, J. Li, G. Chen, Z. Guo, Y. Zhou and J. Wang, *ChemCatChem*, 2015, **7**, 993.
104. C. Gao, G. Chen, X. Wang, J. Li, Y. Zhou and J. Wang, *Chem. Commun.*, 2015, **51**, 4969.
105. J. Wang, W. Sng, G. Yi and Y. Zhang, *Chem. Commun.*, 2015, **51**, 12076.
106. J. Li, D. Jia, Z. Guo, Y. Liu, Y. Lyu, Y. Zhou and J. Wang, *Green Chem.*, 2017, **19**, 2675.
107. (a) Y. Xiong, Y. Wang, H. Wang and R. Wang, *Polym. Chem.*, 2011, **2**, 2306; (b) Y. Xiong, H. Wang, R. Wang, Y. Yan, B. Zheng and Y. Wang, *Chem. Commun.*, 2010, **46**, 3399.
108. Y. Zhang, B. Wang, E. H. M. Elageed, L. Qin, B. Ni, X. Liu and G. Gao, *ACS Macro Lett.*, 2016, **5**, 435.
109. (a) S. Payra, A. Saha and S. Banerjee, *Curr. Organocatal.*, 2017, **4**, 4; (b) A. Sing and H. K. Chopra, *Curr. Org. Synth.*, 2017, **14**, 488; (c) K. Bica and P. Gaertner, *Eur. J. Org. Chem.*, 2008, 3235; (d) A. Winkel, P. V. G. Reddy and R. Wilhelm, *Synthesis*, 2008, 999.
110. (a) E. Busto, V. Gotor-Fernández, N. Ríos-Lombardía, E. García-Verdugo, I. Alfonso, S. García-Granda, A. Menéndez-Velázquez, M. I. Burguete, S. V. Luis and V. Gotor, *Tetrahedron Lett.*, 2007, **48**, 5251; (b) N. Ríos-Lombardía, E. Busto, V. Gotor-Fernández, V. Gotor, R. Porcar, E. García-Verdugo, S. V. Luis, I. Alfonso, S. García-Granda and A. Menéndez-Velázquez, *Chem. - Eur. J.*, 2010, **16**, 836.
111. E. Karjalainen, D. F. Izquierdo, V. Martí-Centelles, S. V. Luis, H. Tenhu and E. García-Verdugo, *Polym. Chem.*, 2014, **5**, 1437.
112. J. Grosseheilmann, J. Bandomir and U. Kragl, *Chem. - Eur. J.*, 2015, **21**, 18957.
113. E. Schwab and S. Mecking, *Organometallics*, 2005, **24**, 3758.
114. (a) J. D Scholten, B. C. Leal and J. Dupont, *ACS Catal.*, 2012, **2**, 184; (b) J. Dupont and J. D. Scholten, *Chem. Soc. Rev.*, 2010, **39**, 1780.
115. (a) K. T. P. Charan, N. Pothanagandhi, K. Vijayakrishna, A. Sivaramakrishna, D. Mecerreyes and B. Sreedhar, *Eur. Polym. J.*, 2014, **60**, 114; (b) J. Yuan, S. Wunder, F. Warmuth and Y. Lu, *Polymer*, 2012, **53**, 43; (c) X. Yang, Z. Fei, D. Zhao, W. H. Ang, Y. Li and P. J. Dyson, *Inorg. Chem.*, 2008, **47**, 3292.

116. (a) K. L. Luska, P. Migowski and W. Leitner, *Green Chem.*, 2015, **17**, 3195; (b) K. L. Luska, P. Migowski, S. El Sayed and W. Leitner, *Angew. Chem., Int. Ed.*, 2015, **54**, 15750; (c) K. L. Luska, J. Julis, E. Stavitski, D. N. Zakharov, A. Adams and W. Leitner, *Chem. Sci.*, 2014, **5**, 4895.
117. M. I. Burguete, E. Garcia-Verdugo, S. V. Luis and J. A. Restrepo, *Phys. Chem. Chem. Phys.*, 2011, **13**, 14831.
118. (a) J. Restrepo, P. Lozano, M. I. Burguete, E. García-Verdugo and S. V. Luis, *Catal. Today*, 2015, **255**, 97; (b) J. Restrepo, P. Lozano, M. I. Burguete, E. García-Verdugo and S. V. Luis, *ACS Catal.*, 2015, **5**, 4743.
119. A. Dani, V. Crocellà, L. Maddalena, C. Barolo, S. Bordiga and E. Groppo, *J. Phys. Chem. C*, 2016, **120**, 1683.
120. H. Zhao, Y. Wang and R. Wang, *Chem. Commun.*, 2014, **50**, 10871.
121. N. M. Jiao, Z. L. Li, C. G. Xia and J. H. Liu, *ChemistrySelect*, 2017, **2**, 4545.
122. M. I. Burguete, E. García-Verdugo, I. García-Villar, F. Gelat, P. Licence, S. V. Luis and V. Sans, *J. Catal.*, 2010, **269**, 150.
123. Y. Yang, M. Ambrogi, H. Kirmse, Y. Men, M. Antonietti and J. Yuan, *Chem. Mater.*, 2015, **27**, 127.
124. V. B. Saptal, T. Sasakí and B. M. Bhanage, *ChemCatChem*, 2018, **10**, 2593.
125. S. V. Luis and E. Garcia-Verdugo, *Chemical Reactions and Processes under Flow Conditions*, RSC Green Chemistry No. 5, 2010.
126. E. García-Verdugo, B. Altava, M. I. Burguete, P. Lozano and S. V. Luis, *Green Chem.*, 2015, **17**, 2693.
127. N. Karbass, V. Sans, E. García-Verdugo, M. I. Burguete and S. V. Luis, *Chem. Commun.*, 2006, **29**, 3095.
128. V. Sans, F. Gelat, N. Karbass, M. I. Burguete, E. García-Verdugo and S. V. Luis, *Adv. Synth. Catal.*, 2010, **352**, 3013.
129. S. Martín, R. Porcar, E. Peris, M. I. Burguete, E. García-Verdugo and S. V. Luis, *Green Chem.*, 2014, **16**, 1639.
130. M. I. Burguete, H. Erythropel, E. García-Verdugo, S. V. Luis and V. Sans, *Green Chem.*, 2008, **10**, 401.
131. T. Wang, W. L. Wang, Y. Lyu, X. K. Chen, C. Y. Li, Y. Zhang, X. G. Song and Y. J. Ding, *RSC Adv.*, 2017, **7**, 2836.
132. P. Lozano, E. García-Verdugo, R. Piamtongkam, N. Karbass, T. De Diego, M. I. Burguete, S. V. Luis and J. L. Iborra, *Adv. Synth. Catal.*, 2007, **349**, 1077.
133. D. F. Izquierdo, J. M. Bernal, M. I. Burguete, E. García-Verdugo, P. Lozano and S. V. Luis, *RSC Adv.*, 2013, **3**, 13123.
134. P. Lozano, E. García-Verdugo, N. Karbass, K. Montague, T. De Diego, M. I. Burguete and S. V. Luis, *Green Chem.*, 2010, **12**, 1803.
135. P. Lozano, E. García-Verdugo, J. M. Bernal, D. F. Izquierdo, M. I. Burguete, G. Sánchez-Gómez and S. V. Luis, *ChemSusChem*, 2012, **5**, 790.
136. R. Porcar, P. Lozano, M. I. Burguete, E. Garcia-Verdugo and S. V. Luis, *React. Chem. Eng.*, 2018, **3**, 572.

137. M. Trobe and M. D. Burke, *Angew. Chem., Int. Ed.*, 2018, **57**, 4192.
138. R. Porcar, D. Nuevo, E. García-Verdugo, P. Lozano, J. Sanchez-Marcano, M. I. Burguete and S. V. Luis, *Chem. Commun.*, 2018, **54**, 2385.
139. (a) D. J. Wales, Q. Cao, K. Kastner, E. Karjalainen, G. N. Newton and V. Sans, *Adv. Mater.*, 2018, **30**, 1800159; (b) E. Peris, O. Okafor, E. Kulcinskaja, R. Goodridge, S. V. Luis, E. Garcia-Verdugo, E. O'Reilly and V. Sans, *Green Chem.*, 2018, **19**, 5345; (c) S. Montolio, G. Abarca, R. Porcar, J. Dupont, M. I. Burguete, E. García-Verdugo and S. V. Luis, *J. Mater. Chem. A*, 2017, **5**, 9733.

CHAPTER 14

CO_2 and Organic Carbonates for the Sustainable Valorization of Renewable Compounds

MAURIZIO SELVA*[a], ALVISE PEROSA[a], GIULIA FIORANI[a] AND LISA CATTELAN[b]

[a]Ca' Foscari University of Venice, Department of Molecular Sciences and Nanosystems, Via Torino 155, Venezia Mestre, 30175, Italy; [b]The University of Sydney, Laboratory of Advanced Catalysis for Sustainability, School of Chemistry, Eastern Avenue, Sydney, NSW 2006, Australia
*E-mail: selva@unive.it

14.1 Introduction

The transition from a fossil source-based economy to more environmentally virtuous options addressed at the use of renewable feedstocks is one the great challenges that mankind must cope with during the 21st century.[1] An analysis of this scenario includes not only economic and social reasons, such as the instability of oil prices and finite availability of non-renewable resources, even more at the present rate of world population growth, but also the increasing environmental/ethical awareness that is pushing consumers toward intrinsically safe daily-life products (plastics, materials, fuels, and even food) with reduced or neutral carbon footprint.[2] Although this contingency offers an opportunity for the development of

Green Chemistry Series No. 61
Green Synthetic Processes and Procedures
Edited by Roberto Ballini

Published by the Royal Society of Chemistry, www.rsc.org

sustainable chemicals and the management of the corresponding carbon cycle, the path is far from simple mostly because of the lack of chemical technologies for the recovery and conversion of renewable feedstocks, and the large number of products that are potentially accessible due to their chemical complexity.[3] To cite only one example on the relevance of this issue, a new plastic typically takes 2–3 years to establish early applications and 2–6 years to reach a platform position; thereafter, the market penetration of a plastic with enhanced technical functions on a complex supply chain may require even 20–40 years for production scales over 100 000 tons.[4] This perspective highlights how initial choices on the feedstock and technologies for the manufacturing of a material crucially reverberate not only in the economics, but also in the overall environmental impact of the production.

To combine sustainability with environmental protection and safety, the primary transformation of renewable feedstocks should focus on the most promising green technologies/transformations for the selective achievement of platform chemicals amenable to further valorization into high added-value functionalized molecules. The use of CO_2 and its derivatives, particularly organic carbonates, is playing an increasingly important role in this respect: looking at biomass as carbon dioxide removed from the atmosphere and converted into organic chemicals with sustainable (sunlight) energy, this approach aims at squaring the carbon cycle.[5]

14.2 Carbon Dioxide

The idea of using CO_2 to produce materials dates back to the 1970s, but the interest for such topic has progressively grown over time in parallel to issues on climate change and sustainability in the public debate.[6] Although environmental compatibility and intrinsic safety make carbon dioxide the natural candidate as a C1-building block for chemical synthesis, several aspects must be comparatively considered, including the energy demand of CO_2-mediated reactions and the overabundance of CO_2 generated by human activities.[7] A major concern is the high stability of carbon dioxide, implying the use of high-energy reaction partners: only hydrogen, unsaturated derivatives, or strained cyclic molecules are effective to steer chemical transformations of CO_2 toward low energy products, but the success of such processes can be undermined by the risk of emitting more CO_2 than that used and incorporated into new chemicals.[8] Another crucial aspect is the lifetime of the products of CO_2 conversion, which is variable from days or weeks (as in liquid fuels) to years (in plastics) or even decades to centuries (in cements). The longer this time, the more the release of CO_2 back to the atmosphere is prevented.[9] All these things being considered, the imbalance between current scales of emission and use of CO_2 remains impressively high. Recent estimates indicate that the total anthropogenic emissions of carbon dioxide were around 37 Gt in 2017,[10] which should be compared to CO_2 utilized for chemicals and fuels amounting approximately to 180 Mt and 2 Gt per year, respectively. This proves that options based on carbon capture and utilization (CCU) cannot provide a primary contribution to mitigating the CO_2 release, but they can be part of a larger strategy to complement other key enabling technologies, including carbon capture and storage (CCS) and exploitation of renewable (*e.g.*, solar, geothermal, wind) or decarbonated (*e.g.*, nuclear) energy sources.[5,7,11] A further added-value within the CCU context is the devising of routes where CO_2 is used for the valorization of renewable compounds. The following section describes some recent representative cases.

14.2.1 Catalytic Hydrogenation of CO_2

The synthesis of methanol by hydrogenation of CO_2 represents one of the best examples for the upgrading of renewables on condition that hydrogen is provided through a carbon-free technology reducing CO_2 emissions.[9,12] The first commercial demonstration plant based on this concept was implemented by Carbon Recycling International (CRI) in Iceland, in 2011:[13] the system integrates an electrolytic cracker producing renewable H_2 and a catalytic hydrogenation reactor, both taking advantage of the proximity of a geothermal power station that provides CO_2 as the reagent and low-cost energy for the electrolysis of water and the synthesis of MeOH (Scheme 14.1).

The hydrogenation is rather exothermic ($\Delta H_{25\,°C}$ = −49.5 kJ mol^{-1}) and proceeds with reduction in the mole number, meaning that the process is facilitated at high pressures and low temperatures. The typical catalysts comprised of doped metal oxides such as Cu/ZnO supported on Al_2O_3 are reported to operate at ~30 MPa and 125 °C to achieve high CO_2 conversion (>80%), although many other multicomponent systems are under investigation through the addition of ZrO_2 and Ga_2O_3.[14] The capacity of the plant sums up to 5 Mt_{MeOH} per year (2.5% of the total gasoline market in Iceland), and it is expected to reclaim 4500 t per year of CO_2. A similar, though smaller

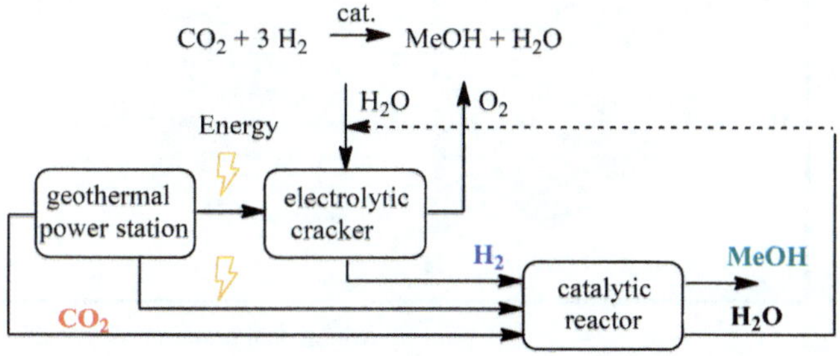

Scheme 14.1 Synthesis of MeOH from renewable H_2 and CO_2.

installation is operated by Mitsui Chemicals Inc., in Osaka, where 100 t_{MeOH} per year are synthesized from CO_2 recovered from factory emissions and H_2 from water photolysis.

14.2.2 Photocatalysis for Hydrogen Production and CO_2 Reduction

Photocatalysis is gaining increasing attention for the concurrent production of renewable H_2 from water splitting and reduction of CO_2 for the synthesis of organic building blocks as CH_3OH, HCOOH, HCHO, *etc.*, as well as CO.[15] Although these reactions use a green energy source such as sunlight and operate under milder conditions than the thermal-catalytic reforming of CO_2, the major challenges to make photocatalytic methods economically and industrially feasible are both efficiency and selectivity. These aspects have been recently reviewed by different authors, who pointed out the enormous efforts currently being undertaken in the design of new heterogeneous photocatalysts.[16,17] To cite only two significant examples among the many available, a recent procedure has been conceived by coupling Cu_2O with graphene and an array of TiO_2 nanotubes in a dual-chamber reactor for CO_2 photoreduction (Scheme 14.2).[18]

Graphene displays a multiple role for the protection of Cu_2O against photocorrosion, the promotion of charge separation and electronic transfer across the heterojunction interfaces, and the inhibition of backward reactions. Moreover, due to its enriched electron density, graphene allows the selective multi-electron reduction of CO_2 to methanol: the latter (MeOH) has been generated at a rate of 45 μmol cm^{-2} h^{-1} with a quantum

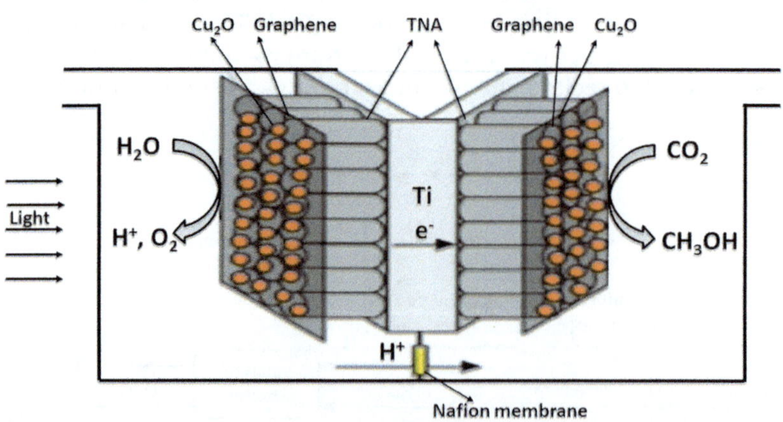

Scheme 14.2 CO_2 photoreduction for the synthesis of MeOH. Reproduced from ref. 18 with permission from Elsevier, Copyright 2016.

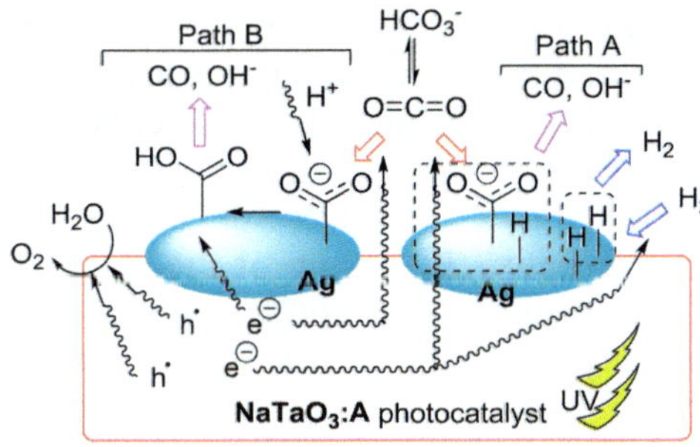

Scheme 14.3 CO_2 photoreduction for the production of CO.

efficiency of 5.71% at 420 nm. In a different configuration, A-doped $NaTaO_3$ (A = Mg, Ca, Sr, Ba, or La) has been used as a catalyst for CO_2 photoreduction in the presence of water as the electron donor and an Ag co-catalyst (Scheme 14.3).[19]

On the surface of the Ag co-catalyst, due to the conduction band level of A-doped $NaTaO_3$, the potential of photogenerated electrons is enough for the reduction of both CO_2 to $CO_2^{\bullet-}$(ad) and H_2O to H(ad), respectively. Then, the reaction of $CO_2^{\bullet-}$(ad) and H(ad) forms CO and OH^-, while part of H(ad) is consumed to form H_2 (path a). Alternatively, H^+ may react with $CO_2^{\bullet-}$(ad) to produce COOH(ad), which in turn captures an electron to release CO and OH^- (path b). By this arrangement, the selectivity for the formation of CO is around 90%, with an activity in the range of 88–176 μmol CO h^{-1}.

Notwithstanding the potential attractiveness of methods for artificial photosynthesis, an intrinsic issue is, at present, the capturing of CO_2 from air (where the average concentration is only 400 ppm), which costs approximately 500 € per ton of captured CO_2.[20] In the coming decades, the economic viability of photoreduction methods will have to focus on point sources of CO_2.

14.2.3 CH-carboxylation Reactions with CO_2

C–H bonds are virtually ubiquitous in hydrocarbons, natural products, drugs, and synthetic and biological polymers. Not surprisingly, vast research has been devoted to the direct reaction of C–H bonds with CO_2 to produce carboxylic functions, which are not only desirable motifs but often confer significant biological activity to organic compounds. This area has been initially developed for the functionalization of C(sp^2)–H bonds in both aromatic and heteroaromatic substrates. For these compounds, efficient catalytic systems based on metal (Au, Cu, Rh, Pd) complexes bearing a variety of ligands, such as N-heterocyclic carbenes (NHCs), acetate, phosphines, *etc.*, have been identified.[21–24] However, a new frontier is the more challenging carboxylation of C(sp^3)–H bonds, for which a promising approach is the implementation of light-driven CO_2-fixation processes.[25] It has been demonstrated that, under UV or sunlight irradiation, *o*-alkylphenyl ketones are able to react with CO_2 to generate the corresponding *o*-acylphenylacetic acids at room temperature (Scheme 14.4).[26]

The most plausible mechanism involves the photo-induced abstraction of a hydrogen from the *ortho*-methyl group of the reactant ketone, thereby providing a 1,4-biradical species. This spontaneously generates a reactive *o*-quinodimethane intermediate with long enough lifetime to trap CO_2 and afford the carboxylic acid product. Typical isolated yields are 60–95%. An even more recent photo-induced carboxylation reaction has been reported for allylic C–H bonds of alkenes with CO_2 in the presence of catalytic amounts of a ketone and a copper complex (Scheme 14.5).[27]

The photo-irradiation initially excites the ketone, which then abstracts the allylic hydrogen of the reagent, thereby generating a radical pair and, subsequently, a tertiary homoallyl alcohol. This is deprotonated by copper *tert*-butoxide (formed from the copper complex and *t*-BuOK) and the so-generated alcoholate species undergoes nucleophilic addition to CO_2 yielding the desired product as a carboxylate salt.

Another promising strategy in this field includes the selective α-carboxylation of amines *via* single-electron reduction of CO_2 for the photoredox-catalyzed continuous-flow synthesis of α-amino acids.[28]

14.2.4 Synthesis of Oleochemical Carbonates from CO_2

The insertion of CO_2 into epoxides derived from renewable fatty acid esters is a promising route to combine CO_2 utilization and biomass conversion for the synthesis of carbonated oleochemicals, which show a variety of applications for the synthesis of polymers and additives for both lubricants and

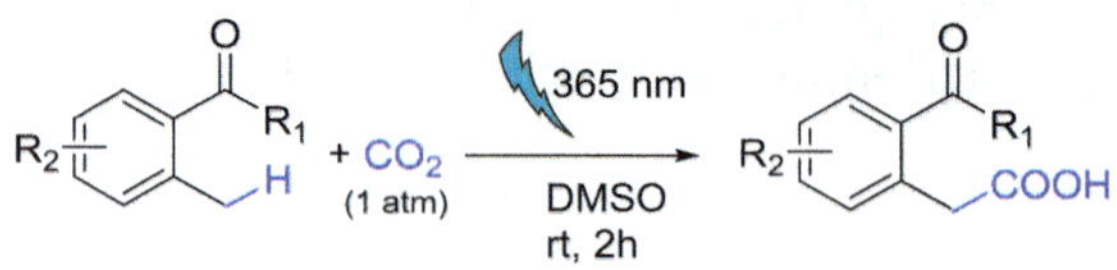

Scheme 14.4 Photochemical activation of C(sp^3)–H bonds with CO_2.

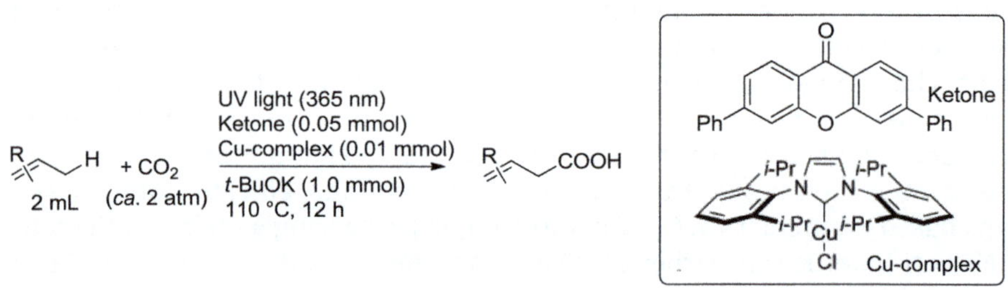

Scheme 14.5 Photo-induced carboxylation of allylic C–H bonds.

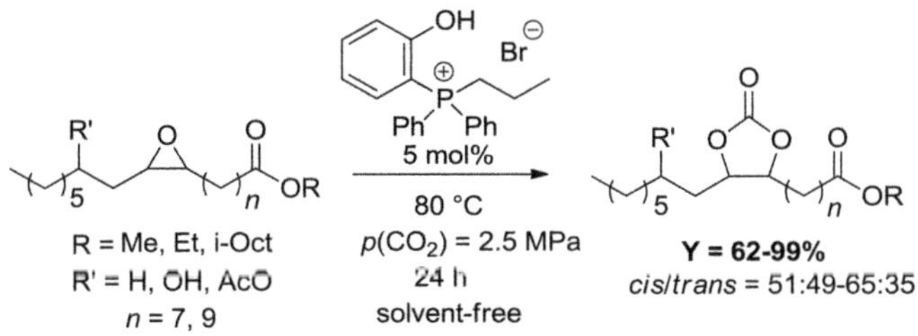

Scheme 14.6 Organocatalytic synthesis of carbonated oleochemicals from renewable epoxides.

fuels.[29,30] In this respect, the conversion of internal epoxides is a highly desired, but equally challenging transformation that typically requires temperatures and pressures of ≥100 °C and ≥5 MPa, respectively.[31] An elegant and efficient approach to this reaction has been recently proposed by Werner *et al.*, who introduced a new class of bifunctional organocatalysts based on phosphonium salts bearing phenolic moieties.[32] In the presence of the most active system of the series, the synthetic scope was extended to several oleochemical carbonates, which were isolated in excellent yields of up to 99% (Scheme 14.6).

The reaction proceeds under solvent-free and comparatively milder conditions (80 °C and 2.5 MPa) providing >99% CO_2 carbon-based products. Experiments have proven that the *ortho*-OH moiety in the catalyst structure is essential for high conversion rates, thereby indicating the role of H-bonding during the reaction. Moreover, of the different investigated halides, the bromide counterion offered the best compromise to reach both high conversions and product selectivity.

14.2.5 The Diagonal Approach

In 'horizontal and vertical approaches', CO_2 is used to generate CO and CN bonds without formal reduction of a carbon center or CO_2 is reduced to formic acid, formaldehyde, methanol, *etc.* By contrast, the so-called 'diagonal approach' combines both the reduction of CO_2 and formation of C–C, C–N, or C–O bonds. Former studies in this field examined a three-component system comprised of CO_2, a functionalization reagent, and a reductant: for example, it was demonstrated that, in the presence of an organic non-nucleophilic base (TBD, 1,5,7-triazabicyclo[4.4.0]dec-5-ene), the reaction of morpholine or dimethylamine with CO_2 yields carbamate intermediate salts, which undergo reduction to the corresponding formamides by cheap and non-toxic organosilanes [*i.e.*, $PhSiH_3$ or $(EtO)_3SiH$].[33] In this context, however, a breakthrough for potential large scale processes has been reported for the methylation of amines with CO_2 and H_2. This reaction was successfully described almost at the same time by the group of Klankermayer and Leitner and the group of Beller,[34] who claimed more than 50 successful cases of reductive mono- or

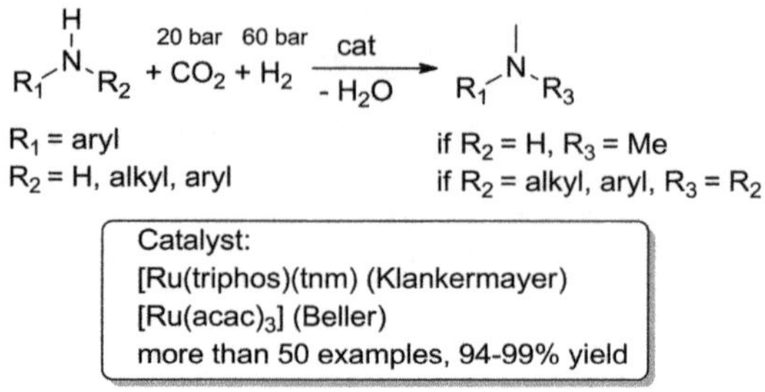

Scheme 14.7 Methylation of amines with CO_2 and H_2.

bis-*N*-methylations catalyzed by Ru-complexes (Scheme 14.7). The impact of these results to implement a viable technology for the transformation of CO_2 is immediate from both synthetic and economic standpoints: (i) methylations constitute the most used form of alkylation reactions in organic synthesis, life science applications, and materials science; (ii) considering only the simple, though representative, case of methylamines, such as $MeNH_2$, Me_2NH, and Me_3N, the market value of such platform chemicals exceeds 4000 € per ton.[35] Curiously, the reaction in Scheme 14.7 remained unknown until 2013.

14.3 Dialkyl Carbonates

Dialkylcarbonates (ROCOOR), potentially the product of dehydrative condensation of alcohols with CO_2, are safe reagents with tunable reactivity. The lighter homologues of the family – dimethyl and diethyl carbonate (R = Me, Et; DMC and DEC, respectively) – are key replacements of hazardous reagents such as phosgene and methyl- or ethyl-halides for alkoxy-carbonylation and alkylation reactions. The main bio-derived classes of bio-based substrates that can be targeted by reactions involving acyclic dialkylcarbonates were recently reviewed and include glycerol, triglycerides, fatty acids, polysaccharides, polysaccharide-derived platform molecules and lignin-based phenolics.[3]

The transesterification of dialkylcarbonates with glycerol yields glycerol carbonate (GC, Scheme 14.8), which is an excellent transcarbonation partner for further reactions with alcohols and phenols, aliphatic and aromatic amines, and carboxylates. In addition, the hydroxyl group of GC exhibits the typical nucleophilic reactivity toward carboxylic acids, anhydrides, acid halides, or esters. GC has also several potential direct and indirect applications as a solvent, electrolyte, additive, emulsifier, humectant, CO_2 adsorbent, and in the detergent/surfactant industry. Under basic catalysis and batch conditions, typical yields of GC are in excess of 90%, but at 300 °C and 20 MPa, thermal (catalyst-free) processes are also effective.[36]

Unlike conventional transesterification with methanol, one of the advantages of carrying out the transesterification of triglycerides using DMC as acyl acceptor to yield fatty acid methyl esters is that it is irreversible due to

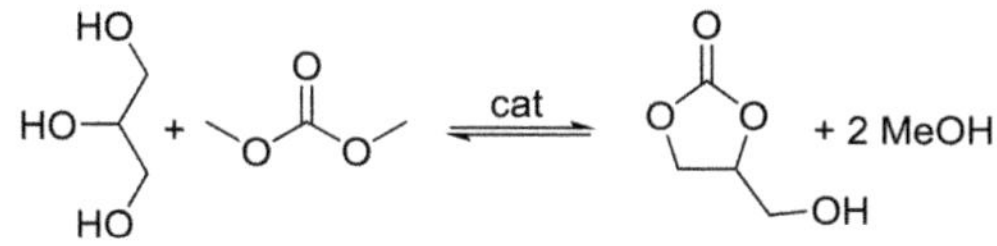

Scheme 14.8 Catalytic transesterification of DMC with glycerol for the synthesis of GCs.

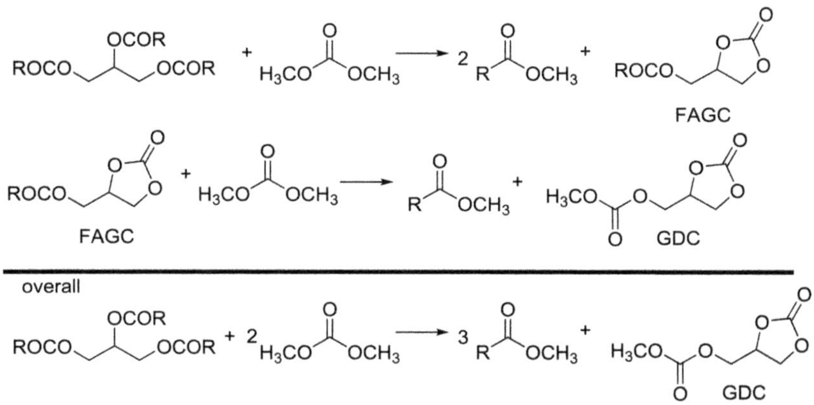

Scheme 14.9 Glycerol-free biodiesel by reaction of a triglyceride with supercritical DMC.[39] Reproduced from ref. 3 with permission from the Royal Society of Chemistry.

the formation of methanol and CO_2. In addition, the reaction can be engineered to combine the formation of biodiesel with the direct conversion of glycerol into glycerol carbonate, glycerol dicarbonate (GDC), and/or and fatty acid glycerol carbonates (FAGCs), which function themselves as fuels and need not be separated (Scheme 14.9).[37,38]

As far as polysaccharides are concerned, the direct reactions of cellulose, starch, and chitin with dimethylcarbonate have been rarely described in the literature.[40–43]

On the other hand, there is ample literature describing the reaction of dialkyl carbonates with organic platforms obtained by chemical or fermentative routes from polysaccharides, including acids (*i.e.*, lactic, levulinic, succinic, hydroxypropionic, malic, citric, adipic, glutamic, and itaconic), furanics (hydroxymethylfurfural (HMF)[44] and furfural) and lactones (butyro-, valero-, caprolactone).

Succinate and dimethyl-2,5-furandicarboxylate were converted into their methyl esters directly from their fermentative sodium salts, avoiding the stoichiometric production of waste salts.[45–47] The process integrates bio-based carboxylate production with direct downstream catalytic conversion by using a strong bicarbonate anion exchange resin that catalyzes the esterification with DMC (Scheme 14.10). Waste is avoided because DMC regenerates the bicarbonate anion resulting in only methanol as a by-product.

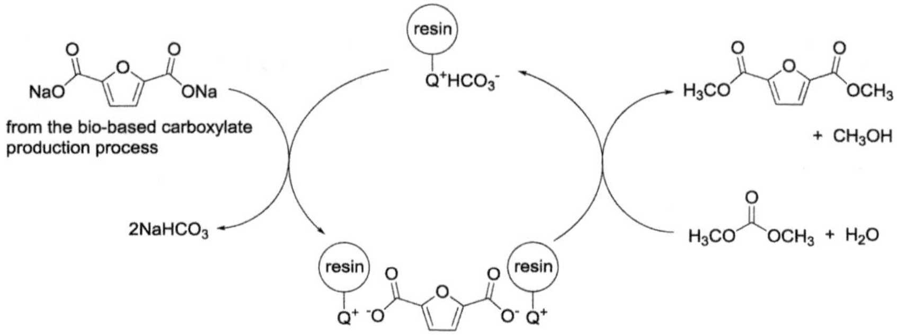

Scheme 14.10 Methylcarbonate esters by in-process integration of fermentative production with downstream catalytic conversion. Reproduced from ref. 3 with permission from the Royal Society of Chemistry.

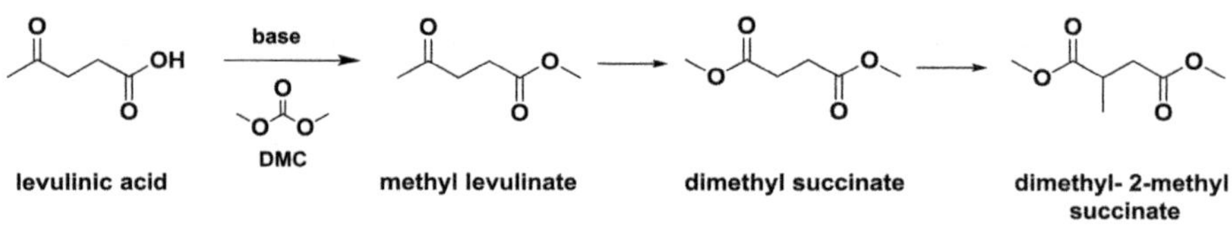

Scheme 14.11 Products of the reaction between levulinic acid and DMC under basic conditions.

We have described the reactivity of levulinic acid with DMC in detail,[48] illustrating the broad reactivity of levulinic acid with DMC to produce multiple outputs (Scheme 14.11).

The use of DMC and two higher homologues such as diethyl- and dibenzylcarbonate has also been described by us for the conversion of four bio-based lactones including γ-butyro-, γ-valero-, δ-valero-, and ε-caprolactone.[49] The reaction selectivity and conditions were investigated and rationalized. Of note was the base-catalyzed reaction of γ-valerolactone with dimethyl carbonate, which allowed the formation of the corresponding α-methylated product with excellent selectivity up to 90% (Scheme 14.12). A mechanistic hypothesis was formulated based on kinetic investigations carried out at different temperatures and reaction times.

The application of organic carbonates for the upgrading of lignin and its derivatives represents a promising goal, both because it could lead to more stable forms of lignin and to the formation of potentially interesting phenolic comopounds. The direct methylation of acetone soluble softwood kraft lignin (ASKL) with DMC at 150 °C indicated that the ASKL phenolics were extensively methylated, thereby improving the thermal stability and reducing self-polymerization.[50] The reactivity of organosolv poplar lignin (OPL) was investigated at high temperatures (300 °C) in MeOH/DMC with a Cu-doped porous metal oxide (PMO) catalyst. Under these conditions, OPL was depolymerized and *O*-methylated by DMC to

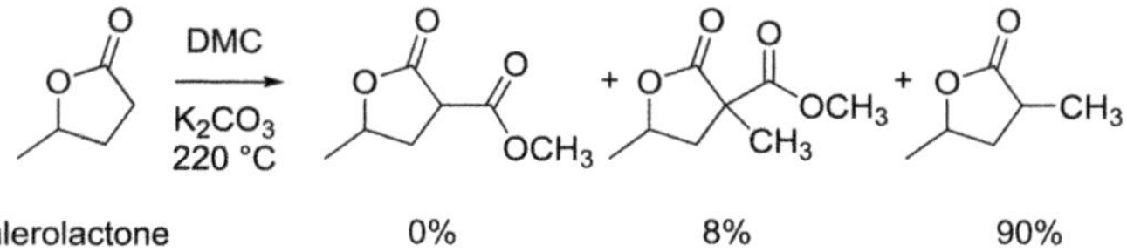

Scheme 14.12 Products of the base catalyzed reaction between γ-valerolactone and DMC.

yield alkylmethoxybenzenes.[51] More recently, free phenolics and carboxylic acids of lignins were hydroxyalkylated using ethylene carbonate as both the reagent and solvent, resulting in near complete conversion of phenolics and carboxylic acids into aliphatic hydroxyls and affording enhanced thermal stability.[52]

The use of DMC for the *O*-methylation of model phenolics was also investigated by our group. In one study, cinnamyl alcohol and 4-(3-hydroxypropyl) phenol (models of the major monolignols of lignin) were methylated to obtain the corresponding anisoles in 65–91% selectivity.[53] In another, microwave heating was implemented for the liquid-phase upgrading of catechol and hydroquinone, yielding the desired bis-*O*-methyl derivatives in >90% yield after 40 min.[54]

14.4 Organic Carbonates for the Synthesis of Biopolymers

The preparation of renewable-based plastics through the design of sustainable processes involving bio-based monomers represents an actual and ambitious challenge to obtain new materials replacing existing ones while maintaining the original properties.[55–57]

In the last 30 years, organic carbonates have been more and more used as monomers and co-monomers in both polymerization and co-polymerization processes. In particular, DMC, the lightest member of the organic carbonates series, has been extensively employed in the biopolymer field due to its eco-friendly nature, safety, and economics.[58,59]

14.4.1 DMC in the Synthesis of Bio-polycarbonates and Bio-polyurethanes

In recent years, DMC has been studied as a co-monomer in selected polycondensation reactions with inexpensive renewable diols to produce aliphatic polycarbonates (APCs), which are considered promising sustainable materials thanks to their bio-compatibility, low toxicity, and recyclability. These materials and their copolymers may find uses for biomedical, pharmaceutical, and environmental applications. Moreover, they are amorphous, display relatively low rates of hydrolysis in aqueous media, and remarkably they do

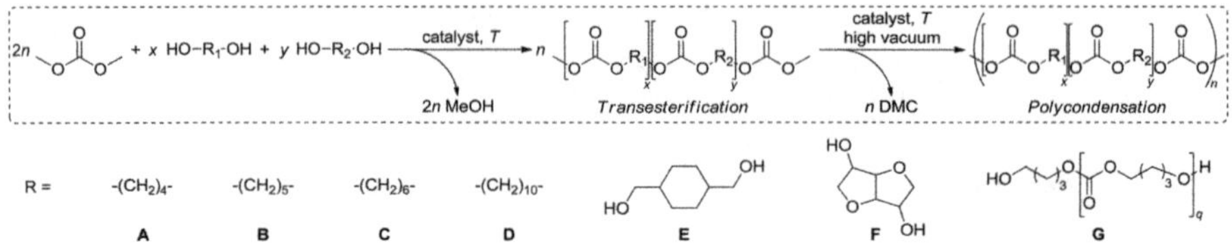

Scheme 14.13 Two-step polycondensation of DMC with aliphatic diol/aliphatic diol mixtures. A–G: structures of R groups in the reactant diols.

not produce the corresponding carboxylic diacids when subjected to *in vivo* degradation. This last aspect prevents any increase of local acidity in tissues, which is responsible for further degradation and the stimulation of inflammatory processes.[60]

Li and co-workers have surveyed and screened different homogeneous and heterogeneous catalysts for sequential transesterification/polycondensation of linear diols monomers with DMC. In the presence of a TiO_2/SiO_2-based catalyst (TS-44), narrowly dispersed high molecular weight polymers ($Đ \leq 1.86$; $M_w \geq 16.6 \times 10^4$ g mol^{-1}) were synthesized in good yield (>85%). No decarboxylation took place during the reaction, as confirmed by ^{1}H NMR spectroscopy. Moreover, the polymers showed thermal and mechanical properties (tensile strength of 34–40 MPa, and elongation at break) that were influenced by the amount of methylene groups present in the original diols and, overall, were compatible with the requirements for biodegradable plastics. Examples are illustrated in Scheme 14.13. Of note, the enzyme-catalyzed biodegradation of the material synthesized from monomer B was complete in less than 200 h.[61]

Under similar conditions, different catalysts were used to prepare copolymers (G in Scheme 14.13). In these products, the synthesis of the poly(butylene carbonate) (PBC) segment was carried out with $Ti(OBu)_4$ as the catalyst. The presence of PBC units further improved the impact strength of the materials without alteration of their thermal and biodegradability properties.[62] Renewable aliphatic polycarbonates were also obtained using isosorbide, which is a monomeric diol derived from sugars. In the presence of catalytic amounts of alkaline metal acac salts, the polycondensation of isosorbide with DMC yielded copolymers with M_w up to 2.8×10^4 g mol^{-1}.[63]

In 2013, Carbonate Metathesis Polymerization (CaMP) was proposed as a protocol to implement the catalytic carbonate interchange reaction between DMC and long chain alcohols (for example, 1,10-decanediol, a derivative of castor oil; Scheme 14.14). Accordingly, both symmetrical and unsymmetrical linear carbonates were prepared using $Zn(OAc)_2$ and K_2CO_3 as the catalysts. Aliphatic polycarbonates with M_w up to 8.8×10^4 g mol^{-1} were then synthesized in a two-step fashion by either isolating the α,ω-bis-carbonate intermediates or directly on α,ω-diols. Also, polyalkylene carbonates were obtained through the CaMP reaction (Scheme 14.14, top). The melting temperature of

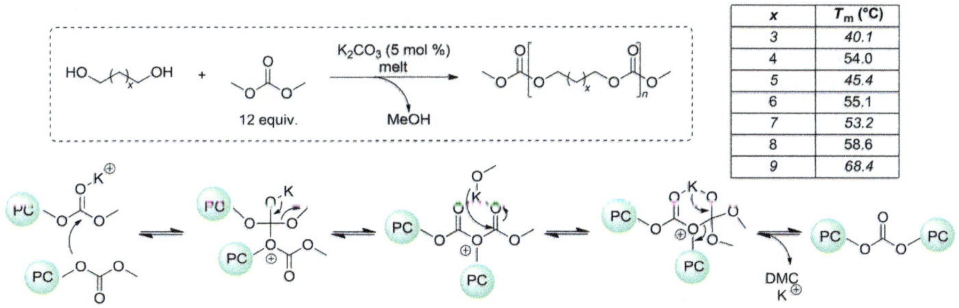

x	T_m (°C)
3	40.1
4	54.0
5	45.4
6	55.1
7	53.2
8	58.6
9	68.4

Scheme 14.14 Preparation of polycarbonates with different diol spacers (top) and proposed mechanism for CaMP (bottom). Reproduced from ref. 3 with permission from the Royal Society of Chemistry.

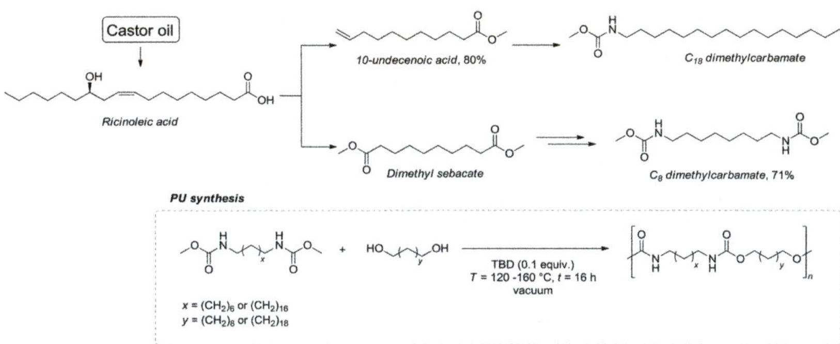

Scheme 14.15 Melt polymerization of dimethyl carbamate monomers from ricinoleic acid. Reproduced from ref. 3 with permission from the Royal Society of Chemistry.

such polymers increased with the increasing number of methylene groups, but the effect depended on odd (–$(CH_2)_x$–, where x = 5, 7, 9) and even (–$(CH_2)_x$–, where x = 6, 8, 10 and 12) numbers of such units.[64]

Promising routes for the synthesis of isocyanate-free bio-based polyurethanes (PUs) have been devised starting from plant oils. For example, thermal and chemical rearrangements were successfully reported to convert ricinoleic acid, a major component of castor oil, into the corresponding C_8 and C_{18} dimethyl carbamates (Scheme 14.15).[65]

The key synthetic step leading to the bis(carbamate) products was an environmentally benign modification of the Lossen rearrangement that required a mixture of a tertiary amine and methanol as catalysts, and DMC as the trapping agent. Then, operating at 120–160 °C and under vacuum, base (TBD)-catalyzed reactions of the resulting bis-carbamates with diols finally produced PUs with molecular weights (M_w) up to 2.5×10^4 g mol^{-1} and polydispersivity ($Đ$) values in the range of 1.24–2.14.

A different strategy for the preparation of non-isocyanate polyurethanes (NIPUs) was conceived starting from the base-catalyzed carbamation of renewable diamines $[R(NH_2)_2]$ with DMC to obtain diurethane monomers $[R(NHCO_2Me)_2]$. Then, again in the presence of a basic catalyst (TBD or K_2CO_3), a transurethanization process between such monomers and bio-based diols provided AA-BB type NIPUs that displayed T_g and T_m values in the range of 38–42 °C and 42–204 °C, respectively, and were stable at temperatures over 200 °C.[66] The same approach was used also to implement the reaction of dicarbamate monomers $[R(NHCO_2Me)_2]$ with diamines to obtain polyureas. To improve the thermal properties and stability of both NIPUs and polyureas, the terminal groups of such polymers were further modified through transesterification and amidation reactions. Final materials with crystalline melting temperatures (T_m) up to 113 °C were so achieved.[67]

A recent multi-step sequence was designed for the synthesis of renewable polyhydroxyurethanes (PHUs). An initial reaction of esterification/ring-aperture between 10-undecenoic acid (from castor oil) and glycidol (a cyclic synthon of glycerol) provided a monoglycide, which in turn underwent a base-catalyzed carbonate interchange with DMC to originate an ester derivative of glycerol carbonate. This species was the key intermediate to access PHUs *via* both metathesis and aminolysis processes. The resulting polymers showed low M_w ($(3.0–19.0) \times 10^3$ g mol^{-1}) and high $Đ$ (2.4–3.4) values. Moreover, as expected, PHUs with high T_gs were obtained starting from cyclic diamines as co-monomers.[68]

In 2016, Cramail and co-workers prepared renewable based polymers (polycarbonates and polyurethanes) characterized by low M_w ((6.7–20.4) × 103 g mol^{-1}) and good polydispersities ($Đ$ = 1.3–1.8).[69] The polymers were synthesized by homocondensation of carbamate and carbonate monomers, respectively, in the presence of TBD as the catalyst (5 mol%, Scheme 14.16). Starting from long chain bio-based unsaturated amines such as oleyl amine, the corresponding carbamate was prepared *via* a base-catalyzed reaction with DMC. Then, the internal double bond of methyl oleylcarbamate was subjected to a sequence of two transformations comprised of an isomerization

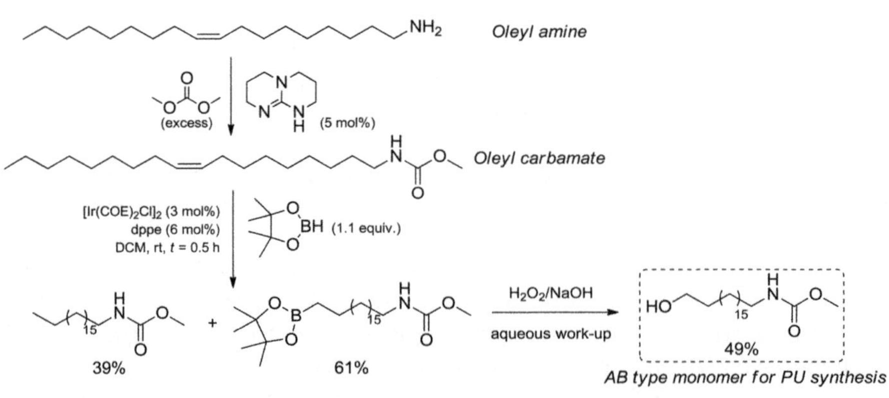

Scheme 14.16 Synthesis of a difunctional carbamate monomer starting from oleyl amine.

catalyzed by an Ir-based complex, chlorobis(cyclooctene)iridium (I) dimer (Ir(COE)$_2$Cl]$_2$), and an oxidative hydroboration, by which a terminal alcohol was obtained.

Vegetable tannins are mostly identified with two families of phenolic-based matrices, which include hydrolyzable tannins and polyflavonoid tannins.[70] Both these classes of compounds have been recently considered as an alternative to vegetable oils for the preparation of non-isocyanate bio-based aromatic polyurethanes (NIAPUs).[70,71] Starting from hydrolyzable tannins, particularly hydrolyzable chestnut (*Castanea sativa*) tannin, a two-step sequence based on the reaction with a water/DMC mixture, followed by condensation with hexamethylenediamine (HMD), produced a solid material consistent with the structure of NIAPUs. An array of techniques including MALDI-TOF spectrometry and ^{13}C NMR and FT-IR spectroscopies confirmed the occurrence of the reactions illustrated in Scheme 14.17 (top), which refer to the model case of gallic acid, the major component of hydrolyzable tannins.

At first, gallic acid undergoes a carbonate interchange reaction with DMC to produce a mixture of the corresponding mono-, di-, and tri-transcarbonated derivatives. Then, the reaction of HMD originates linear carbamate structures. The same two-step sequence was extended also to flavonoid tannins. In this case, however, due to the non-hydrolyzable nature of the starting materials, water was not used and DMC acted concurrently as a reagent and solvent. The formation of polyurethane solid products was coherent with the reactions shown in Scheme 14.17

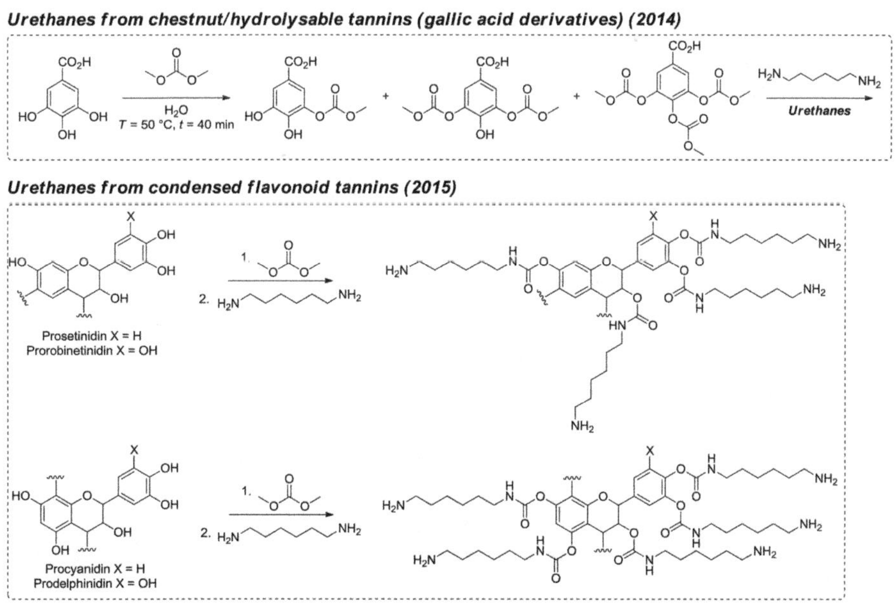

Scheme 14.17 Tannin-based urethanes. Reproduced from ref. 3 with permission from the Royal Society of Chemistry.

(bottom), which refer to the cases of the most abundant flavonoid structures in the tannin materials used by the authors, namely profisetinidin and prorobinetinidin (from Quebracho and Mimosa) and procyanidin and prodelphinidin (from Maritime and Radiata Pines). Transcarbonation and carbamation processes were responsible for the synthesis of highly crosslinked tridimensional polyurethanes, which showed promising uses as hydrophobic coatings.[71]

14.5 Co-polymers Based on CO_2 and Renewable-based Monomers

Polycarbonates are a family of polymers containing carbonate groups in their structure, typically comprising two classes: aromatic and aliphatic polycarbonates. Aromatic polycarbonates are the most common type of commercially available polycarbonates and are typically produced from bisphenol A according to Scheme 14.18 (bottom). Although Asahi Kasei Corporation developed a new sustainable manufacturing process for the production of bisphenol A-based polycarbonates (Scheme 14.18, top),[72] the use of this non-renewable monomer still poses serious environmental and health concerns.[73]

Aliphatic polycarbonates, on the other hand, are generally characterized by less favorable chemical properties (*i.e.*, low rigidity, brittleness, and moderate thermal stability). This latter class of polycarbonates, however, can be readily synthesized in a sustainable fashion from epoxides

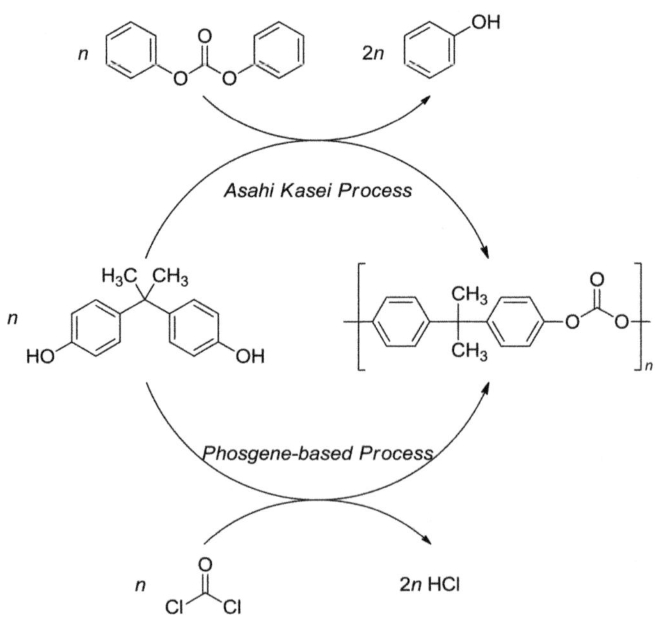

Scheme 14.18 Synthesis of aromatic polycarbonate from bisphenol A.

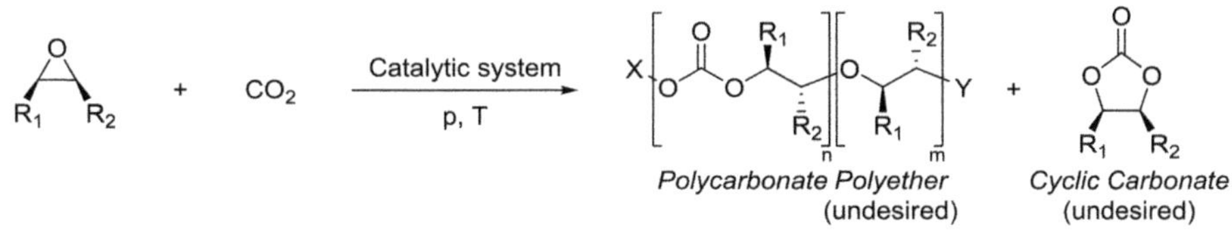

Scheme 14.19 Preparation of polycarbonates from epoxides and CO_2.

and CO_2 (Scheme 14.19) in the presence of a suitable catalytic system. In addition to establishing a sustainable value chain for the production of cyclic carbonates and new carbonate-based materials with highly tuneable properties, this process uses CO_2, an inexpensive and abundant starting material, as co-reactant (instead of toxic phosgene), which is embodied in the polymers in amounts as high as 50 mol%. Furthermore, in many cases, the epoxide can serve as the solvent, therefore reactions proceed in neat conditions.

Two of the most investigated epoxides are cyclohexene oxide (CHO, generating poly(cyclohexene)carbonate, PCHC, which contains up to 29 mol% of CO_2) and propylene oxide (PO, generating poly(propylene)carbonate, PPC, which contains up to 50 mol% of CO_2). Generally, copolymerization of CHO and CO_2 produces selectively the corresponding PCHC in a wide range of experimental conditions and in the presence of different metal-based catalysts, including salen and salophen based systems, β-diminidate complexes, amino(trisphenolate) based catalysts, and homo- and hetero-bimetallic catalysts.[74] PO can also react with CO_2 to form selectively the corresponding polycarbonate, PPC. However, in the latter case, competing formation of the cyclic by-products is an unavoidable problem, requiring careful catalyst and reaction condition selection. Notably, the in-air combustion of PCHC and PPC is clean, producing no harsh residues. Both PCHC and PPC can therefore be used as sacrificial binders in the production of materials for the electronic and ceramic industries. Due to its superior mechanical features, PPC is even more useful than PCHC, particularly for the preparation of degradable packaging materials and the synthesis of PUs. However, limitations for further applications of both these polycarbonates are the absence of functionalities and moderate glass transition temperatures (T_g = 37 °C for PPC, 115 °C for PCHC). Intriguingly, the development of functionalized co-polymers contributes to improving and/or tuning the mechanical properties of aliphatic polycarbonates.[75,76]

14.5.1 CO_2 Co-polymerization with Renewable-based Co-monomers

Co-polymerization reactions between CO_2 and epoxides derived from renewable feedstocks is one of the most promising strategies developed to improve the sustainability of this process, limiting the fossil fuel

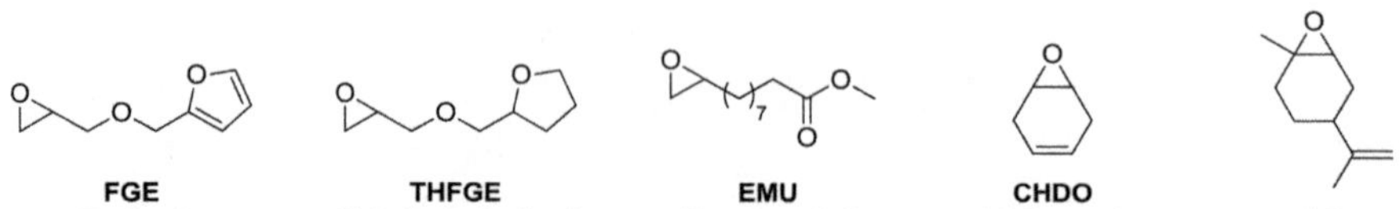

FGE	THFGE	EMU	CHDO	LO
Furfuryl glycidyl ether	*Tetrahydrofurfuryl glycidyl ether*	*Epoxy methyl 10-undecanoate*	*1,2-Epoxy-4-cyclohexadiene*	*Limonene oxide*

Figure 14.1 Examples of renewable-based epoxides active toward co-polymerization with CO_2.

dependency with the use of renewable-based epoxides. The majority of these monomers derive from lignocellulosic biomass, which is the most abundant and chemically complex renewable resource available (Figure 14.1).[77–79]

Besides developing bio-based routes to access commonly employed monomers, *e.g.*, glycerol-derived epichlorohydrin, which is widely employed for the preparation of epoxy resins,[78] various bio-based co-monomers were tested. These monomers originate from different renewable-based feedstocks, such as terpenes, vegetable oils, and carbohydrates. While the direct use of carbohydrates as co-monomers in CO_2 copolymerization processes is still an open challenge, promising results were obtained using carbohydrate derivatives. Furfural, in particular, is an established bio-based platform molecule, readily obtained from hemicellulose upon acid-catalyzed dehydration of pentose sugars.[80] Furfural, in turn, can be transformed into various synthetically useful C^4 and C^5 synthons by a range of complementary catalytic processes.[81] Two furfuryl-based epoxides, furfuryl glycidyl ether (FGE) and tetrahydrofurfuryl glycidyl ether (THFGE) (obtained *via* Williamson synthesis starting from epichlorohydrin and furfuryl alcohol or tetrahydrofurfuryl alcohol, respectively), have shown promising results in CO_2 copolymerization reactions in the presence of appropriate catalytic systems, including rare earth-based ternary catalysts[82] and Cr^{III} salen-based catalytic systems.[83] In the former example, co-polymerization of CO_2 and fully hydrogenated THFGE improved moderately the thermal properties of the resulting polycarbonate, increasing the T_g value from −10 °C for FGE/CO_2 copolymers[83] to −6 °C for THFGE/CO_2 copolymers.[82] The use of FGE as a co-monomer allowed the post-polymerization functionalization by a Diels–Alder (DA) reaction between *N*-phenylmaleimide and the pendant furan ring, improving the polycarbonate thermal properties upon formation of insoluble crosslinked networks characterized by high T_gs between 93 and 96 °C.[83] The crosslinking reaction was fully reversible *via* a retro-Diels–Alder reaction (occurring at T = 110 °C), effectively restoring the soluble, non-crosslinked polymer without any degradation of the polycarbonate backbone.[83]

Fatty acids derived from vegetable oils can also be transformed into polymerizable derivatives.[84,85] For example, ricinoleic acid, the main component of castor oil (a non-edible vegetable oil) accounting for >87% of the total mass, can be easily converted into the corresponding undecylic acid derivative[86] and, *via* a multi-step process, into epoxy methyl 10-undecenoate (EMU), which is suitable for CO_2 co-polymerization reactions.[87] Zhang and co-workers reported the first example of EMU/CO_2 co-polymerization.[87] The resulting polycarbonate was characterized by a high content of carbonate linkages (>99%); however, due to the long aliphatic pendant chains of EMU, the corresponding copolymers were characterized by low T_gs (−38 to −44 °C). The decomposition temperatures of this fully renewable-based co-polymer remained typical of other polycarbonates observing a $T_{d,\,5\%}$ = 240 °C. Additionally, the resulting fully alternating polycarbonate presented two hydroxyl end groups, which could be further employed as macroinitiator to form triblock co-polymers (*vide infra*). 1,4-Cyclohexadiene is a waste product of renewable-based polyunsaturated fatty acids (*e.g.*, linoleic acid) self-metathesis,[88] which can be selectively converted into the corresponding monoepoxide (cyclohexadiene oxide, CHDO) by carefully choosing the epoxidation conditions.[89,90] Williams and Meier and the group of Darensbourg reported independently the preparation of fully renewable-based poly(cyclohexadienecarbonate) (PCHDC) obtained by 1,4-cyclohexadiene oxide/CO_2 copolymerization in the presence of M^{3+}(salen)-based catalytic systems of Cr^{III},[90] Co^{III},[89] or dinuclear zinc or magnesium-based complexes.[90] Of importance, PCHDC is suitable for post-polymerization functionalization *via* thiol-ene click chemistry, affecting the bulk and thermal properties of the resulting functionalized co-polymers.[89]

Terpenes are another interesting class of renewable-based compounds with promising but largely unexplored potential toward polymerization reactions.[91,92] (*R*)-Limonene is a terpene-based platform chemical[93] that can be easily converted into the corresponding monoepoxide (limonene oxide, LO), particularly relevant in the context of epoxide/CO_2 copolymerization.[94,95] Indeed, poly(limonene)carbonate (PLC) still represents a breakthrough and, within the family of fully renewable-based polycarbonates, it displays thermal and physical properties suitable for practical applications.[95,96] Coates and co-workers reported the first successful example of alternating limonene oxide/CO_2 co-polymers in 2004.[97] Co-polymerization occurred selectively (>99%) in the presence of a zinc β-diiminate complex [(BDI)ZnX, X = OAc] under mild conditions (T = 25 °C, $p^0(CO_2)$ = 0.69 MPa), affording the corresponding regio- and stereo-regular PLC (>98% *trans* isomer incorporation) in 2 h (M_n = 9300 g mol^{-1}, PDI = 1.13). Scaling-up of the polymerization protocol was achieved starting from *trans*-LO-rich mixtures (>85%, easily prepared from (*R*)-limonene *via* a two-step bromohydrin/ring closure protocol): the corresponding high molecular weight PLC (>100 kDa) was obtained in large scale quantities (1.35 kg) in the presence of (BDI) Zn(OAc) as the catalyst (0.2 mol%). High-M_w *trans*-PLC displayed promising thermal (T_g = 130 °C) and optical properties (transmission 94%, haze 0.75%). In 2015, Kleij and co-workers prepared PLC using an Al^{III} amino(-trisphenolate) complex and a nucleophilic co-catalyst (PPNX, X = Cl, Br).[98] This catalytic system was also active under relatively mild conditions (T = 45

°C, $p^0(CO_2)$ = 0.5 MPa); however, the polymerization was much slower than with [(BDI)ZnX] catalysts (conversion = 53%, t = 48 h, M_n = 6700 g mol^{-1}, PDI = 1.55). Notably, this catalytic system reacted preferentially with *cis*-(*R*)-LO, producing selectively stereoregular perfectly alternating *trans*-PLC, whereas PLC with lower degrees of stereoregularity was obtained starting from pure *trans*-(*R*)-LO isomer and commercial *cis*/*trans*-LO. The residual C=C bond on limonene oxide can be converted either at the first[99,100] or after[95,101,102] polymerization, thereby providing PLC derivatives with an array of different properties (*e.g.*, modified solubility, improved hydrophilicity, higher T_g values).

In conclusion, although CO_2/epoxide co-polymerization is an industrially relevant process and commercially available PPC and PCHC are synthesized from CO_2 including Saudi Aramco, Empower Materials, Econic Technologies, and Covestro,[76] their applications are still limited to the formation of polycarbonate-polyols oligomers for PU synthesis.[78] The use of renewable-based co-monomers, however, could mitigate the environmental footprint of the process[103,104] and contribute to the development of new materials with practically relevant chemico-physical properties.

References

1. R. A. Sheldon, *Catal. Today*, 2011, **167**, 3.
2. N. M. P. Bocken and J. M. Allwood, *J. Cleaner Prod.*, 2012, **35**, 118.
3. G. Fiorani, A. Perosa and M. Selva, *Green Chem.*, 2018, **20**, 288.
4. E. de Jong, A. Higson, P. Walsh and M. Wellisch, *Biofuels, Bioprod. Biorefin.*, 2012, **6**, 606.
5. A. W. Kleij, M. North and A. Urakawa, *ChemSusChem*, 2017, **10**, 1036.
6. T. Bruhn, H. Naims and B. Olfe-Kräutlein, *Environ. Sci. Policy*, 2016, **60**, 38.
7. CO_2 Utilisation Today, Report 2017, ed. A. Zimmermann and M. Kant, 2017, DOI: 10.14279/depositonce-5806.
8. M. Peters, B. Kohler, W. Kuckshinrichs, W. Leitner, P. Markewitz and T. E. Muller, *ChemSusChem*, 2011, **4**, 1216.
9. E. A. Quadrelli, G. Centi, J.-L. Duplan and S. Perathoner, *ChemSusChem*, 2011, **4**, 1194.
10. https://www.carbonbrief.org/analysis-global-co2-emissions-set-to-rise-2-percent-in-2017-following-three-year-plateau, last access, October 21, 2018.
11. M. Pérez-Fortes, A. Bocin-Dumitriu and E. Tzimas, *Energy Procedia*, 2014, **63**, 7968.
12. M. Pérez-Fortes, J. C. Schöneberger, A. Boulamanti and E. Tzimas, *Appl. Energy*, 2016, **161**, 718.
13. http://carbonrecycling.is/, last accessed, October 22, 2018.
14. S. G. Jadhav, P. D. Vaidya, B. M. Bhanage and J. B. Joshi, *Chem. Eng. Res. Des.*, 2014, **92**, 2557.

15. M. Tahir and N. S. Amin, *Renewable Sustainable Energy Rev.*, 2013, **25**, 560.
16. K. C. Christoforidis and P. Fornasiero, *ChemCatChem*, 2018, **10**, 1–16.
17. L. Yuan and Y.-J. Xu, *Appl. Surf. Sci.*, 2015, **342**, 154.
18. F. Li, L. Zhang, J. Tong, Y. Liu, S. Xu, Y. Cao and S. Cao, *Nano Energy*, 2016, **27**, 320.
19. H. Nakanishi, K. Iizuka, T. Takayama, A. Iwase and A. Kudo, *ChemSusChem*, 2017, **10**, 112.
20. J. A. Martens, A. Bogaerts, N. De Kimpe, P. A. Jacobs, G. B. Marin, K. Rabaey, M. Saeys and S. Verhelst, *ChemSusChem*, 2017, **10**, 1039.
21. I. I. F. Boogaerts and S. P. Nolan, *J. Am. Chem. Soc.*, 2010, **132**, 8858.
22. L. Zhang, J. Cheng, T. Ohishi and Z. Hou, *Angew. Chem.*, 2010, **122**, 8852.
23. K. Sasano, J. Takaya and N. Iwasawa, *J. Am. Chem. Soc.*, 2013, **135**, 10954.
24. T. Suga, T. Saitou, J. Takaya and N. Iwasawa, *Chem. Sci.*, 2017, **8**, 1454.
25. Y.-Y. Gui, W.-J. Zhou, J.-H. Ye and D.-G. Yu, *ChemSusChem*, 2017, **10**, 1337.
26. Y. Masuda, N. Ishida and M. Murakami, *J. Am. Chem. Soc.*, 2015, **137**, 14063.
27. N. Ishida, Y. Masuda, S. Uemoto and M. Murakami, *Chem. - Eur. J.*, 2016, **22**, 6524.
28. H. Seo, M. H. Katcher and T. F. Jamison, *Nat. Chem.*, 2017, **9**, 453.
29. M. Bahr and R. Melhaupt, *Green Chem.*, 2012, **14**, 483.
30. L. Poussard, J. Mariage, B. Grignard, C. Detrembleur, C. Jerome, C. Calberg, B. Heinrichs, J. De Winter, P. Gerbaux, J. M. Raquez, L. Bonnaud and P. Dubois, *Macromolecules*, 2016, **49**, 2162.
31. A. Boyer, E. Cloutet, T. Tassaing, B. Gadenne, C. Alfos and H. Cramail, *Green Chem.*, 2010, **12**, 2205.
32. H. Bettner, J. Steinbauer, C. Wulf, M. Dindaroglu, H.-G. Schmalz and T. Werner, *ChemSusChem*, 2017, **10**, 1076.
33. C. Das Neves Gomes, O. Jacquet, C. Villiers, P. Thuery, M. Ephritikhine and T. Cantat, *Angew. Chem., Int. Ed.*, 2012, **51**, 187.
34. A. Tlili, X. Frogneux, E. Blondiaux and T. Cantat, *Angew. Chem., Int. Ed.*, 2014, **53**, 2543.
35. Y. Li, X. Cui, K. Dong, K. Junge and M. Beller, *ACS Catal.*, 2017, **7**, 1077.
36. Z. Ilham and S. Saka, *SpringerPlus*, 2016, **5**, 923.
37. K. T. Tan and K. T. Lee, *Renewable Sustainable Energy Rev.*, 2011, **15**, 2452.
38. S. Marx, *Fuel Process. Technol.*, 2016, **151**, 139.
39. Z. Ilham and S. Saka, *Lipid Technol.*, 2011, **23**, 10.
40. S. R. Labafzadeh, K. J. Helminen, I. Kilpeläinen and A. W. T. King, *ChemSusChem*, 2015, **8**, 77.
41. C. Hou, Y. Chen, W. Chen and W. Li, *Carbohydr. Res.*, 2011, **346**, 1178.
42. C. Hou, Y. Chen and W. Li, *Carbohydr. Res.*, 2012, **355**, 87.

43. P. S. Bangde, R. Jain and P. Dandekar, *ACS Sustainable Chem. Eng.*, 2016, **4**, 3552.
44. M. Krystof, M. Pérez-Sánchez and P. Domínguez de María, *ChemSusChem*, 2013, **6**, 630.
45. C. S. López-Garzón, L. A. M. van der Wielen and A. J. J. Straathof, *Chem. Eng. J.*, 2014, **235**, 52.
46. C. I. Cabrera-Rodríguez, L. A. M. van der Wielen and A. J. J. Straathof, *Ind. Eng. Chem. Res.*, 2015, **54**, 10964.
47. C. S. López-Garzón, L. A. M. van der Wielen and A. J. J. Straathof, *RSC Adv.*, 2016, **6**, 3823.
48. A. Caretto and A. Perosa, *ACS Sustainable Chem. Eng.*, 2013, **1**, 989.
49. A. Caretto, M. Noè, M. Selva and A. Perosa, *ACS Sustainable Chem. Eng.*, 2014, **2**, 2131.
50. S. Sen, S. Patil and D. S. Argyropoulos, *Green Chem.*, 2015, **17**, 1077.
51. J. A. Barrett, Y. Gao, C. M. Bernt, M. Chui, A. T. Tran, M. B. Foston and P. C. Ford, *ACS Sustainable Chem. Eng.*, 2016, **4**, 6877.
52. L.-Y. Liu, M. J. Cho, N. Sathitsuksanoh, S. Chowdhury and S. Renneckar, *ACS Sustainable Chem. Eng.*, 2018, **6**, 12251.
53. J. N. G. Stanley, M. Selva, A. F. Masters, T. Maschmeyer and A. Perosa, *Green Chem.*, 2013, **15**, 3195.
54. M. Y. Lui, K. S. Lokare, E. Hemming, J. N. G. Stanley, A. Perosa, M. Selva, A. F. Masters and T. Maschmeyer, *RSC Adv.*, 2016, **6**, 58443.
55. S. A. Miller, *ACS Macro Lett.*, 2013, **2**, 550.
56. A. Gandini and T. M. Lacerda, *Prog. Polym. Sci.*, 2015, **48**, 1.
57. A. Gandini, T. M. Lacerda, A. J. F. Carvalho and E. Trovatti, *Chem. Rev.*, 2016, **116**, 1637.
58. T. Matsuzaki and A. Nakamura, *Catal. Surv. Jpn.*, 1997, **1**, 77.
59. S. S. Fukuoka, I. Fukawa, M. Tojo, K. Oonishi, H. Hachiya, M. Aminaka, K. Hasegawa and K. Komiya, *Catal. Surv. Asia*, 2010, **14**, 146.
60. Y. Yamamoto, S. Kaihara, K. Toshima and S. Matsumura, *Macromol. Biosci.*, 2009, **9**, 968.
61. W. Zhu, X. Huang, C. Li, Y. Xiao, D. Zhang and G. Guan, *Polym. Int.*, 2011, **60**, 1060.
62. J. Wang, L. Zheng, C. Li, W. Zhu, D. Zhang, G. Guan and Y. Xiao, *Ind. Eng. Chem. Res.*, 2012, **51**, 10785.
63. Q. Li, W. Zhu, C. Li, G. Guan, D. Zhang, Y. Xiao and L. Zheng, *J. Polym. Sci., Part A: Polym. Chem.*, 2013, **51**, 1387.
64. R. Vanderhenst and S. A. Miller, *Green Mater.*, 2013, **1**, 64.
65. O. Kreye, S. Wald and M. A. R. Meier, *Adv. Synth. Catal.*, 2013, **355**, 81.
66. C. Duval, N. Kebir, A. Charvet, A. Martin and F. Burel, *J. Polym. Sci., Part A: Polym. Chem.*, 2015, **53**, 1351.
67. A. Martin, L. Lecamp, H. Labib, F. Aloui, N. Kébir and F. Burel, *Eur. Polym. J.*, 2016, **84**, 828.
68. S. Bigot, M. Daghrir, A. Mhanna, G. Boni, S. Pourchet, L. Lecamp and L. Plasseraud, *Eur. Polym. J.*, 2016, **74**, 26.

69. P. Sane, T. Lebarbe, E. Grau and H. Cramail, *Eur. J. Lipid Sci. Technol.*, 2016, **118**, 1620.
70. M. Thébault, A. Pizzi, S. Dumarçay, P. Gerardin, E. Fredon and L. Delmotte, *Ind. Crops Prod.*, 2014, **59**, 329.
71. M. Thébault, A. Pizzi, H. A. Essawy, A. Barhoum and G. Van Assche, *Eur. Polym. J.*, 2015, **67**, 513.
72. S. Fukuoka, M. Kawamura, K. Komiya, M. Tojo, H. Hachiya, K. Hasegawa, M. Aminaka, H. Okamoto, I. Fukawa and S. Konno, *Green Chem.*, 2003, **5**, 497.
73. N. Jalal, A. R. Surendranath, J. L. Pathak, S. Yu and C. Y. Chung, *Toxicol. Rep.*, 2018, **5**, 76.
74. For a comprehensive update review of the metal-based catalytic systems active towards epoxide/CO_2 co-polymerization, please refer to: C. M. Kozak, K. Ambrose and T. S. Anderson, *Coord. Chem. Rev.*, 2018, **376**, 565 and references therein.
75. M. Scharfenberg, J. Hilf and H. Frey, *Adv. Funct. Mater.*, 2018, **28**, 1704302.
76. Y. Wang and D. J. Darensbourg, *Coord. Chem. Rev.*, 2018, **372**, 85.
77. Y. Zhu, C. Romain and C. K. Williams, *Nature*, 2016, **540**, 354.
78. S. J. Poland and D. J. Darensbourg, *Green Chem.*, 2017, **19**, 4990.
79. Y.-Y. Zhang and X.-H. Zhang, in *Sustainable Polymers From Biomass*, ed. C. Tang and C. Y. Ryu, Wiley-VCH, Weinheim, 2017, ch. 12, pp. 279.
80. Y. Luo, Z. Li, X. Li, X. Liu, J. Fan, J. H. Clark and C. Hu, *Catal. Today*, 2019, **319**, 14.
81. X. Li, P. Jia and T. Wang, *ACS Catal.*, 2016, **6**, 7621.
82. Y. Hu, L. Qiao, Y. Qin, X. Zhao, X. Chen, X. Wang and F. Wang, *Macromolecules*, 2009, **42**, 9251.
83. J. Hilf, M. Scharfenberg, J. Poon, C. Moers and H. Frey, *Macromol. Rapid Commun.*, 2015, **36**, 174.
84. U. Biermann, U. Bormscheuer, M. A. R. Meier, J. O. Metzger and H. J. Shäfer, *Angew. Chem., Int. Ed.*, 2011, **50**, 3854.
85. E. B. Mubofu, *Sustainable Chem. Processes*, 2016, **4**, 11.
86. M. Van der Steen and C. V. Stevens, *ChemSusChem*, 2009, **2**, 692.
87. Y.-Y. Zhang, X.-H. Zhang, R.-J. Wei, B.-Y. Du, Z.-Q. Fan and G.-R. Qi, *RSC Adv.*, 2014, **4**, 36183.
88. H. Mutlu, R. Hofsäß, R. E. Montenegro and M. A. R. Meier, *RSC Adv.*, 2013, **3**, 4927.
89. D. J. Darensbourg, W.-C. Chung, C. J. Arp, F.-T. Tsai and S. J. Kyran, *Macromolecules*, 2014, **47**, 7347.
90. M. Winkler, C. Romain, M. A. R. Meier and C. K. Williams, *Green Chem.*, 2015, **17**, 300.
91. J. Zhao and H. Schlaad, *Adv. Polym. Sci.*, 2013, **253**, 151.
92. P. A. Wilbon, F. Chu and C. Tang, *Macromol. Rapid Commun.*, 2013, **34**, 8.
93. R. Ciriminna, M. Lomeli-Rodriguez, P. Demma Carà, J. A. Lopez-Sanchez and M. Pagliaro, *Chem. Commun.*, 2014, **50**, 15288.

94. O. Hauenstein, M. Reiter, S. Agarwal, B. Rieger and A. Greiner, *Green Chem.*, 2016, **18**, 760.
95. F. Parrino, A. Fidalgo, L. Palmisano, L. M. Ilharco, M. Pagliaro and R. Ciriminna, *ACS Omega*, 2018, **3**, 4884.
96. O. Hauenstein, S. Agarwal and A. Greiner, *Nat. Commun.*, 2016, **7**, 11862.
97. C. M. Byrne, S. D. Allen, E. B. Lobkovsky and G. W. Coates, *J. Am. Chem. Soc.*, 2004, **126**, 11404.
98. L. Peña Carrodeguas, J. González-Fabra, F. Castro-Gómez, C. Bo and A. W. Kleij, *Chem. - Eur. J.*, 2015, **21**, 6115.
99. C. Li, R. J. Sablong and C. E. Koning, *Angew. Chem., Int. Ed.*, 2016, **55**, 11572.
100. N. Kindermann, A. Cristòfol and A. W. Kleij, *ACS Catal.*, 2017, **7**, 3860.
101. C. Li, S. van Berkel, R. J. Sablong and C. R. Koning, *Eur. Polym. J.*, 2016, **85**, 466.
102. C. Li, M. Johansson, R. J. Sablong and C. R. Koning, *Eur. Polym. J.*, 2017, **96**, 337.
103. Life Cycle Assessment - Renewable and Sustainable Citrus Oils Final Report, Chicago, IL, 2012.
104. D. Zhang, E. A. del Rio-Chanona and N. Shah, *ACS Sustainable Chem. Eng.*, 2017, **5**, 4388.

CHAPTER 15

Transition Metal-catalysed Nucleophilic Additions of Terminal Alkynes in Water: Development and Synthetic Utility

ZOË HEARNE, SABRINA KEYS AND CHAO-JUN LI*

Department of Chemistry and FQRNT Centre for Green Chemistry and Catalysis (CCVC), McGill University, 801 Sherbrooke Street West, Montréal, Quebec, H3A 0B8, Canada
*E-mail: cj.li@mcgill.ca

15.1 Introduction

15.1.1 The Development of Transition Metal-catalysed Nucleophilic Addition of Terminal Alkynes in Water

The impetus to develop a suite of efficient methods to add terminal alkynes to a range of electrophiles lies in the utility of the reaction products, both as functional molecules in their own right and as useful synthetic intermediates. The products of these reactions come equipped with two functional handles: the alkyne and the remnant of the electrophile, be it an alcohol,

Green Chemistry Series No. 61
Green Synthetic Processes and Procedures
Edited by Roberto Ballini

Published by the Royal Society of Chemistry, www.rsc.org

carbonyl or amine. The alkyne group is a particularly versatile functionality, providing access to alkanes and alkenes *via* hydrogenation, aromatic compounds and heterocycles *via* cyclisation and cycloaddition, and ketones *via* hydration, to name a few.

Classical addition procedures of terminal alkynes to carbonyl and carbonyl-derived electrophiles involve the formation of metal acetylides using stoichiometric equivalents of a reactive metal such as zinc, magnesium or lithium. The synthesis and subsequent use of the unstable metal acetylide in the addition step requires large amounts of energy to cool the reaction and exclude oxygen from the apparatus. Furthermore, requirements for anhydrous organic solvents, additives, and stoichiometric quantities of metals leave much to be desired from the sustainability point of view.

Catalytic reactions offer a number of advantages over the method described above. Besides avoiding the use of stoichiometric metallic reagents, catalysts typically operate under mild and neutral conditions. This broadens the substrate scope compared to reactions operating under strongly basic conditions at extreme temperatures, and reduces the necessity for heavy use of protection/deprotection chemistry to preserve sensitive functional groups. Perhaps most interestingly, catalysts can unlock reactivity modes and achieve selectivity not possible using classical stoichiometric methods.

The catalytic addition of alkynes to electrophiles provides an extremely efficient way to establish carbon–carbon bonds and rapidly construct molecular complexity. Unlike methods utilising stoichiometric metallic reagents, these reactions generate very few by-products: in the case of alkyne addition to carbonyls, none are produced; in the case of alkyne addition to imine/iminium ions, one equivalent of water is produced; and in the case of alkyne addition to acid chlorides, one equivalent of protic acid is produced.

The stability of many transition metal catalysts in water provides yet another advantage, and a number of elements have demonstrable superiority in water, including the coinage metals gold, silver and copper, as well as palladium and zinc. The use of water as a reaction solvent is advantageous over organic solvents as it is renewable, non-toxic, cheap, abundant, and it provides a means for separating the hydrophilic metal salt catalyst from the organic reaction components. Furthermore, a ramification of being able to conduct catalytic reactions in water is that, if the reaction conditions can tolerate such a nucleophilic, protic and polar solvent, conventionally sensitive functional groups can often be tolerated without prior protection.

The beginnings of our group's involvement in this field dates back to 1991, when Li and Chan developed a Grignard-like allylation of aldehydes and ketones.[1] Stoichiometric in indium, the reaction could be conducted in water as the reaction solvent and had the advantage of wide reaction scope and mild reaction conditions. Despite relying on a stoichiometric quantity of

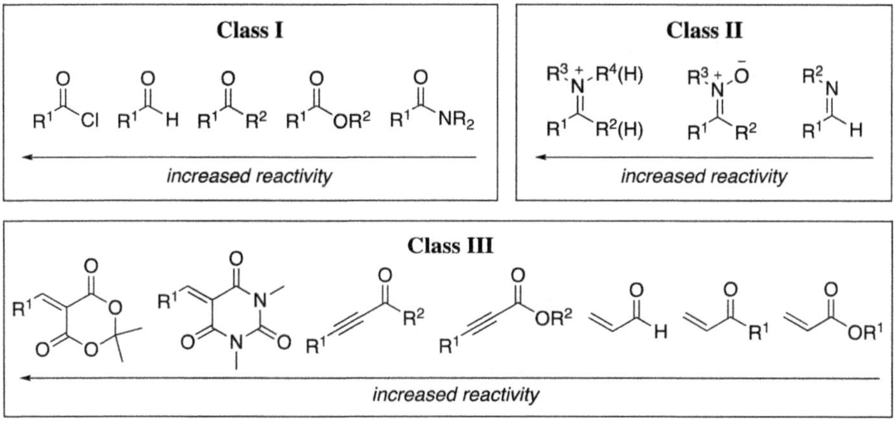

Figure 15.1 Relative reactivity of electrophiles towards acetylide nucleophiles.

a metal, this publication and the contribution of many others opened up a field of study, which has since become extensive in its scope and utility: transition metal-catalysed nucleophilic addition reactions in water.

15.1.2 Relative Reactivity of Various Electrophiles

In this review, we have organised the reactions into three different sections: the nucleophilic addition of terminal alkynes to carbonyl derivatives (class I), C=N bond derivatives (class II), and conjugated systems (class III). As the reactions involve the addition of a nucleophilic acetylide, the electrophilicity of the substrate is one relevant detail in the design and optimisation of catalytic methods. The relative reactivities of the different electrophile classes discussed in this review are illustrated in order of increasing reactivity† in Figure 15.1.[2]

15.1.3 Transition-metal Acetylides: Synthesis and Reactivity

The vast majority of reactions covered in this review are assumed to begin with the C–H activation of a terminal alkyne by a transition-metal catalyst to form a metal acetylide. This species is nucleophilic and can add to a variety of electrophiles. Another commonality between the reactions discussed in the following sections is the prevalence of reactions catalysed by the three coinage metals gold, silver and copper. Experimental evidence for the involvement of metal acetylides in reactions of this type is examined herein, focusing on these three metals.

†We have referred to Mayr's electrophilicity parameters for the relative reactivity of the selected electrophiles. Data were available for aldehydes, ketones, imines, iminium ions, nitrones, Meldrum's acid derivatives, barbituric acid derivatives, conjugated esters, and ketones. The remaining examples have been ordered based on extrapolation and experimental observations.

Despite extensive application of metal-catalysed additions of terminal alkynes, renowned in the A^3 and Sonogashira coupling reactions, the mechanism of metal acetylide formation still provides opportunities for study as some details remain unconfirmed. The most pertinent question is how the sp-proton of the terminal alkyne, with a pK_a of 23 in water (for phenylacetylene),[3] is deprotonated in solution by the weak bases present (amine, propargylamine product, or water). It is proposed that coordination of an electrophilic metal centre to the alkyne forms a π-metal-alkyne complex. The metal centre depletes the π-electron density on the alkyne and increases the acidity of the acetylenic sp-proton for subsequent deprotonation and conversion to a σ-metal complex: the nucleophilic acetylide.

For gold, the likelihood of this mechanism has been supported by a number of studies. Gold is known to readily form both π- and σ-complexes; some complexes have even been isolated with both bonding modes in the same complex. In 2011, Brown and Widenhoefer spectroscopically observed that gold N-heterocyclic carbene (NHC) complexes bearing a π-complexed alkyne could convert into σ,π-acetylide di-gold complexes at low temperatures (0 °C) in the absence of any external base. Another study observed that, at temperatures as low as −80 °C, the gold atoms in such complexes are fluxional and can undergo exchange.[4] This supports the proposal that switching from one bonding mode to another can occur in the same catalytic cycle in the absence of a strong base. In another study reporting the preparation of σ,π-acetylide di-gold complexes, the authors noted that the absence of a bulky, stabilising ligand bound to the gold centre was accompanied by formation of gold nanoparticles. This introduces the possibility that nanoparticle formation, whether it results in deactivation or generation of the active catalyst, should not be disregarded in mechanistic investigations involving metal complex pre-catalysts.[5]

For copper, computational studies on the copper-catalysed 1,3-dipolar cycloaddition of azides and terminal alkynes provided evidence that, in aqueous solution, π-coordination of copper(I) to an alkyne decreases the pK_a of the terminal proton by 10 units.[6] Furthermore, under these conditions, π-coordination by the metal is exothermic by 11.7 kcal mol^{-1} when water is available to act as a ligand. For the authors, these results were consistent with experimental results that the reaction proceeds more rapidly in aqueous solvent and does not require the addition of a base. Another study indicates that C–H activation of terminal alkynes is extremely efficient *via* copper(I)/copper(II) synergistic catalysis, introducing the possibility that the pre-catalyst can undergo redox processes to form more active catalytic combinations of oxidation states.[7]

Lu and co-workers used DFT calculations to elucidate the structure of active metal acetylides in the silver-catalysed carboxylation of terminal alkynes. They found that the anion-coordinated silver acetylide is more nucleophilic than the corresponding unligated neutral species.[8] Computational studies of the Sonogashira reaction mechanism also concluded that the anion-coordinated copper(I) acetylide is likely the active species.[9]

For copper-catalysed A^3 coupling reactions, studies have suggested that polymeric copper(I) acetylides with a ladderane structure can form under the reaction conditions in water, serving as an isolable reservoir of copper-acetylide nucleophiles. The polymeric material can be isolated and reused.[10] These experiments support the key involvement of copper acetylides in nucleophilic alkyne-addition chemistry. Other studies have revealed the structural diversity in polymeric metal(I) acetylides of copper, silver, and gold; structural elucidation by powder X-ray diffraction revealed that copper-acetylides can form catenanes, extended sheets and ladderanes, while silver forms ladderanes and gold forms extended honeycomb networks.[11]

15.1.4 Chapter Scope

This chapter presents a comprehensive review of transition metal-catalysed nucleophilic addition reactions of terminal alkynes to electrophiles (carbonyl, imine, iminium ion, and α,β-unsaturated carbonyl derivatives) in water as the solvent. Reactions with organohalide electrophiles and those that typically proceed by a cross-coupling mechanism are not included. The synthetic utility of the reviewed reactions is conveyed by highlighting relevant and interesting work for each reaction type. This component is not intended to be exhaustive, as that is beyond the scope of the review. Furthermore, many reactions are not conducted in water, but an effort has been made to choose examples that reflect the principles of green chemistry, such as the use of catalysis, benign solvents and additives, and those designed with high atom economy.

15.2 Addition of Terminal Alkynes to Carbonyl Derivatives (Class I)

15.2.1 Addition of Terminal Alkynes to Aldehydes and Ketones

15.2.1.1 Addition of Terminal Alkynes to Aldehydes

In 2002, Wei and Li reported the first transition metal-catalysed alkynylation of aldehydes in water, synthesising propargylic alcohols (Table 15.1a).[12] This work was on the heels of reports by the same group on the first catalysed coupling of internal alkynes with aldehydes for the synthesis of unsaturated ketones under acidic aqueous conditions.[13] For the synthesis of propargylic alcohols, the authors used a bimetallic catalytic system comprising ruthenium chloride and indium acetate, and conducted the reaction under basic conditions. The authors tentatively proposed a mechanism that takes place through carbonyl activation by the Lewis acidic indium salt and alkyne activation by ruthenium. These reaction conditions were amenable to the transformation of aromatic aldehydes and aliphatic aldehydes lacking an

Table 15.1 Addition of terminal alkynes to aldehydes for the synthesis of secondary propargylic alcohols.

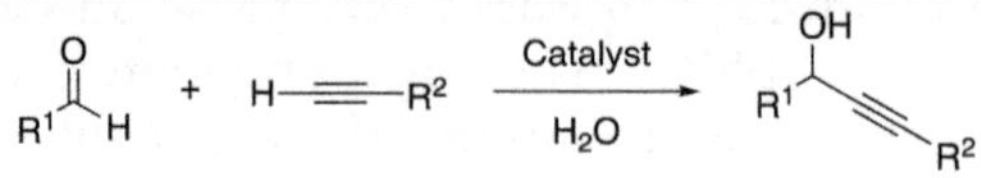

Entry	Conditions	Scope	Yield
a) Wei & Li, 2002	$RuCl_3$ (5 mol%) $In(OAc)_3$ (10 mol%) Morpholine (0.5 equiv.) K_2CO_3 (aq), 60–95 °C 24 or 48 h	R^1 = alkyl, Ph, Ar (–alkyl, –CN, –CF_3, –X), naphthyl R^2 = Ph	13 examples 27–94% yield
b) Yao & Li, 2005	Cy_3PAgCl (5 mol%) i-Pr_2NEt (20 mol%) r.t.–100 °C, air, 6–48 h	R^1 = H, alkyl, Ph, Ar (–Me, –CF_3, –OMe, –X), naphthyl, cinnamyl R^2 = alkyl, Ph	19 examples 35–98% yield
c) Yao *et al.*, 2012	Ag/TiO_2 (5 mol%) PPh_3 (10 mol%) NEt_3 (0.4 equiv.) 60 °C, air, 12 h	R^1 = alkyl, Ph, Ar (–CF_3, –X) R^2 = Ph, Ar (–Me, –OMe, –X)	15 examples 34–98% yield
d) Xu *et al.*, 2014	NanoAg@TiO_2@PMHSIPN (5 mol%), PPh_3 (10 mol%) $PhCOCF_3$ (10 mol%) 60 °C, 24 h	R^1 = Ar (–Ph, –OMe, –X), naphthyl R^2 = Ph, Ar (–Me, –X)	15 examples 53–99% yield

α-hydrogen, while electron-rich benzaldehydes performed more poorly. The modest reaction yields prompted the authors to investigate more active catalytic systems. To this end, in 2005, Yao and Li published the use of highly active silver-phosphine catalysts to perform this reaction (Table 15.1b).[14] This publication was discordant with the contemporary common knowledge that Group 1B acetylides (copper, silver and gold) cannot turnover catalytically and can only be used as stoichiometric reagents.[15] It is frequently observed that simple silver acetylides are too stable to add to carbonyls, but the authors hypothesised that complexation of silver with an electron-rich ligand would increase the electron-density on the metal sufficiently to destabilise the silver–carbon bond and make it more nucleophilic. A combination of tricyclohexylphosphine and silver chloride proved a highly reactive combination, in particular for the reaction of halo-substituted aromatic aldehydes and phenylacetylene (aliphatic aldehydes afforded moderate yields). Interestingly, the reaction was not sensitive to air, and water was not only a suitable reaction medium, but also the best one tested; any quantity of organic solvent decreased the reaction yield.

For the mechanism, the authors posited that an amine base facilitates the formation of a nucleophilic acetylide (**1**). This species adds to the carbonyl,

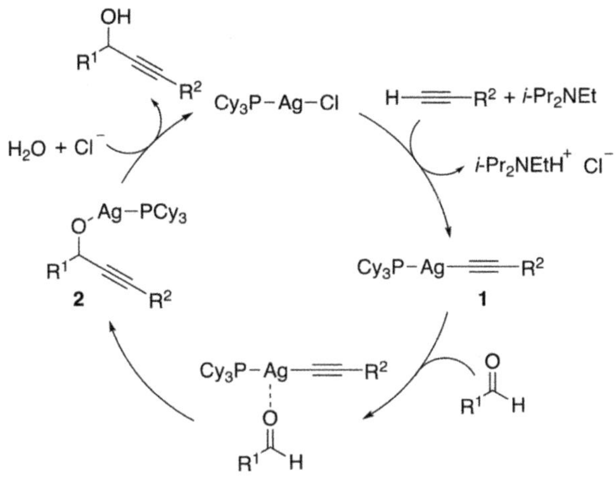

Scheme 15.1 Proposed catalytic cycle. Reproduced from ref. 14 with permission from American Chemical Society, Copyright 2005.

which is rendered electrophilic by coordination to the silver complex. The resulting silver alkoxide (**2**), upon protonation, releases the product and regenerates the catalyst (Scheme 15.1).

Yao and co-workers, interested in the construction of carbon–carbon bonds catalysed by silver nanoparticles, published an account in 2012 of their ability to catalyse the alkyne–aldehyde coupling reaction in water (Table 15.1c).[16] Silver nanoparticles are practical catalysts as they are easy to prepare, non-toxic, and stable in air and water. The authors found that silver nanoparticles, doped with catalytic quantities of triphenylphosphine, could give 89% yield of the propargylic alcohol addition product of 2-bromobenzaldehyde and phenylacetylene. However, the authors hypothesised that immobilising the nanoparticles on a support might enhance their catalytic activity by stabilising them against aggregation and providing additional Lewis acidic sites to activate the aldehyde. Accordingly, a titanium-oxide support increased the reaction yield to 97% and the catalyst could be recovered and reused twice. The reaction was optimised for halo-substituted substrates and these returned high yields, while cyclohexanal and phenylacetylene were poor substrates (45% and 34% yield, respectively).

In 2014, Xu and co-workers published a comprehensive study outlining the development and synthesis of a polymer/titanium oxide-immobilised silver nanoparticle catalyst, and its application to the direct addition of alkynes to aldehydes (Table 15.1d), and trifluoromethyl ketones (ketone addition is discussed in Section 15.2.1.2).[17] The synthesis of nanoparticles with high catalytic activity relies on the careful control of many physiochemical characteristics such as the particle size, dispersion, size distribution, and shape. In an effort to reproducibly prepare silver nanoparticle catalysts, the authors prepared a polymer support coated with an even film of titanium oxide and a dispersal of silver nanoparticles on the surface. The catalyst, dubbed 'NanoAg@TiO_2@PMHSIPN' was used in combination with triphenylphosphine to prepare propargylic alcohols from aromatic aldehydes and alkynes in moderate to excellent yield. Over the course of their optimisations, the authors observed an interesting phenomenon: adding a small quantity of a trifluoromethyl ketone boosted the yield of all the reactions by at least 13%. As will be discussed in the following section, trifluoromethyl ketones are very active substrates under these reaction conditions, and it was suggested that facile activation of the ketone leads to a nanosilver-ketone intermediate that might interchange with the less reactive aldehyde, lowering the activation energy for the initial step.

15.2.1.2 Addition of Terminal Alkynes to Ketones

Instances where ketone substrates are more active than aldehydes towards transition metal-catalysed alkynylation reactions are rare; in the majority of cases, the reduced electrophilicity and increased steric hindrance of ketones makes them more challenging substrates. However, harnessing ketones as substrates is highly desirable as a way of preparing quaternary carbon centres and tertiary propargylic alcohols. Unlike secondary propargylic alcohols, which can be prepared by direct addition of terminal alkynes to aldehydes (Section 15.2.1.1), or by the reduction of 2-alkynones, tertiary propargylic alcohols are inaccessible *via* reduction routes. This highlights the importance of developing methods to add terminal alkynes to ketones.

The catalytic alkynylation of trifluoromethyl ketones was reported by Shibasaki and Kanai in 2007 using a copper catalyst in THF or toluene.[18] Compared to unsubstituted aliphatic ketones, trifluoromethyl ketones are more reactive electrophiles due to the inductive electron-withdrawing effect of the trifluoromethyl group. In addition to having a favourable reactivity profile, the synthesis of fluorine-containing compounds is an active field of research due to their unique properties and disproportionately high representation in the pharmaceutical and agrochemical industries: it is estimated that 20% of currently administered drugs, and up to 30% of approved drugs, contain fluorine atoms.[19] Aqueous methods for the alkynylation of trifluoromethyl ketones were pioneered by Deng and Li the following year, who reported the high-activity of a silver-phosphine catalyst for such reactions (Table 15.2a).[20] Reaction conditions optimised for aldehyde–alkyne addition, namely the combination of silver chloride and tricyclohexylphosphine, gave only 12% yield. The authors found that switching the counterion to fluoride was key to unlocking high yields for the alkyne–ketone addition. The optimised conditions returned the highest yields of fluorine-containing tertiary propargylic alcohols using electron-rich aromatic alkynes and trifluoropyruvate substrates.

Table 15.2 Addition of terminal alkynes to ketones for the synthesis of tertiary propargylic alcohols.

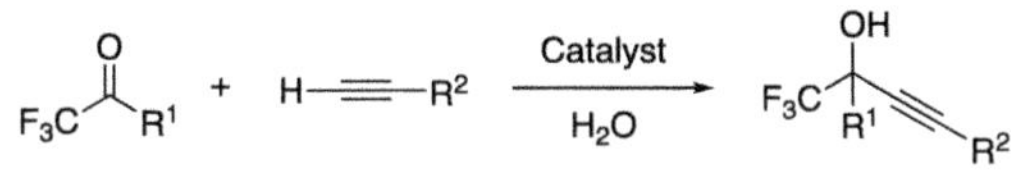

Entry	Conditions	Scope	Yield
a) Deng & Li, 2008	AgF (10 mol%) PCy_3 (10 mol%) r.t.–100 °C, N_2, 24–48 h	R^1 = CO_2R, Ph R^2 = alkyl, cyclohexenyl, Ph, Ar (–Me, –X, –OMe, –Ph, –CF_3), thiofuryl, naphthyl	15 examples 30–93% yield
b) Xu *et al.*, 2014	NanoAg@TiO_2@ PMHSIPN (5 mol%), PPh_3 (10 mol%) 60 °C, 2 h	R^1 = alkenyl, Ph, Ar (–Me, –Cl, –OMe) R^2 = Ph, Ar (–Me, –X, –OMe)	24 examples 88–98% yield
c) Michalak *et al.*, 2016	(IPr)CuCl (2 mol%) NEt_3 or TMG (20 mol%) 50–100 °C, 13–48 h	R^1 = alkyl (–CN, –P(O) OEt_2), alkenyl, Ph, Ar (Me, –C(O)R, –NMe_2, –F), naphthyl, fluorenyl, furyl, thiofuryl R^2 = alkyl (–NH_2, –OH, –OR, –Cl), Ph, thiofuryl, ferrocenyl	30 examples 28–98% yield
d) Cazin *et al.*, 2016	[Ag(IPr)(ICy)]BF_4 (1 mol%) 60 °C, air, 15 h	R^1 = Ph, Ar (–F) R^2 = alkyl, Ph, Ar (–alkyl, –CF_3, –OMe, –F)	13 examples 75–99% yield

In 2014, Xu and co-workers developed a catalyst composed of a titanium-oxide layered polymer scattered with catalytically active silver nanoparticles (80–120 nm, Table 15.2b).[17] The ability of this material to catalyse the addition of alkynes to aldehydes is summarised in Section 15.2.1.1; however, this catalyst is also highly active for the addition of alkynes to trifluoromethyl ketones. The addition of a catalytic quantity of triphenylphosphine in combination with the silver catalyst in water at 60 °C was sufficient to achieve complete conversion of the substrates. Electrophilic substrates with sp^2-substitution such as aromatic trifluoromethyl ketones and α,β-unsaturated trifluoromethyl ketones were ideal substrates, while aliphatic trifluoromethyl ketones and alkynes gave lower yields. Interestingly, when α,β-unsaturated trifluoromethyl ketones were used as substrates, smaller silver nanoparticles (10–30 nm) were required to achieve high yields of the 1,2-addition products.

In 2016, Michalak reported that NHC-copper complexes could catalyse the addition of trifluoromethyl ketones to alkynes in water (Table 15.2c).[21] This was a big advancement because it showed that copper, an inexpensive and abundant transition metal alternative, could be used to good effect. The optimised conditions of a low catalytic loading of [(IPr)CuCl] (0.1–2 mol%,

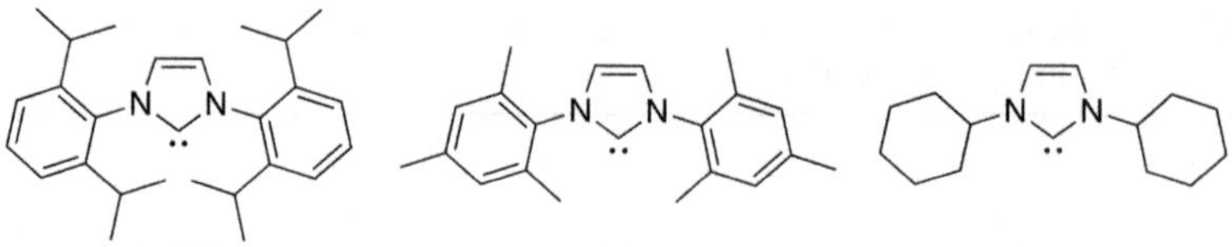

3 (IPr) 4 (IMes) 5 (ICy)

Figure 15.2 N-Heterocyclic carbenes used for the preparation of transition metal-complexes.

Figure 15.2) and a substoichiometric quantity of base converted a comprehensive scope of substrates in moderate to excellent yield. While previous methods were mostly suited to aromatic substrates, this method could be used for the addition of both aliphatic and aromatic starting materials with a variety of substituents such as amines, hydroxyls, phosphates, ferrocene, and thiofurans. After testing an assortment of NHC ligands, the researchers were able to correlate the reaction yield with the steric bulkiness‡ of the ligand.[22] The authors explained this with the support of DFT studies: in addition to the strongly electron-donating nature of NHC ligands, bulkiness also destabilises the metal–carbon bond of the acetylide and increases its nucleophilicity. For example, the energy barrier for the acetylide–ketone C–C bond-forming step was measured to be 26.1 kcal mol^{-1} for the [(IPr) CuCl] catalyst, *vs.* 29.4 kcal mol^{-1} for the comparatively less bulky [(IMes) CuCl] (Figure 15.2).

Cazin and co-workers, also interested in the high activity of metal-NHC complexes for the nucleophilic addition of alkynes, reported in 2016 that heteroleptic bis-NHC silver complexes could also catalyse the alkynylation of trifluoromethyl ketones (Table 15.2d).[23] With just 1 mol% of [Ag(IPr) (ICy)]BF_4, high yields using both electron-rich and poor aromatic alkynes could be obtained under air atmosphere and in water as the solvent. The authors also showed that the silver bis-NHC complex could catalyse the addition of aliphatic and aromatic terminal alkynes to *N*-protected-isatin compounds with quantitative conversion and yields ranging 90–99% (Table 15.3b). Like trifluoromethyl ketones, isatins are activated ketones; the electrophilicity is enhanced by an α-carbonyl group. To probe the active nucleophilic species, the bis-NHC complex was reacted with excess phenylacetylene, forming a silver acetylide ligated to a single IPr ligand (**3**). This complex was isolated and found to independently catalyse the addition reaction.

The silver-NHC catalysed addition of terminal alkynes to isatin compounds in water was first reported by Li and co-workers in 2011, as a strategy to create 3-hydroxyindolin-2-one scaffolds (Table 15.3a).[24] The

‡The researchers compared the percent buried volume, $\%V_{bur}$, a metric developed by Steven P. Nolan to specifically describe the steric bulkiness of NHC ligands. It is calculated from crystallographic data and defined as the percent of the total volume of a sphere occupied by the ligand, with the metal centre occupying the core.

Table 15.3 Addition of terminal alkynes to isatin derivatives.

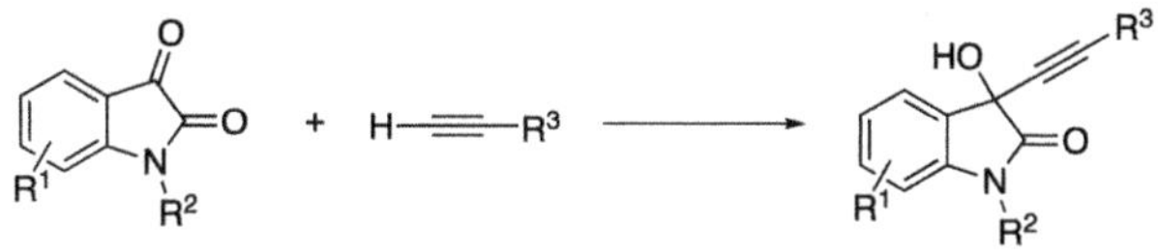

Entry	Conditions	Scope	Yield
a) Li *et al.*, 2011	(IMes)AgCl (5 mol%)	R^1 = H, OCF_3, X	18 examples
	DIPEA (10 mol%)	R^2 = H, Me, Bn	30–99% yield
	H_2O, 40–60 °C, air, 5–24 h	R^3 = alkyl, Ph, Ar (–Me, –NH_2, –OMe, –X), pyridyl, thiofuryl	
b) Cazin *et al.*, 2016	[Ag(IPr)(ICy)]BF_4 (1 or 2 mol%)	R^1 = H, OMe, X	18 examples
	1 : 1 MeOH–H_2O or H_2O	R^2 = Me, Bn	90–99% yield
	60 °C, air, 15 h	R^3 = alkyl (–NH_2, –OH), Ph, Ar (–alkyl, –CF_3, –OMe, –F), pyridyl	

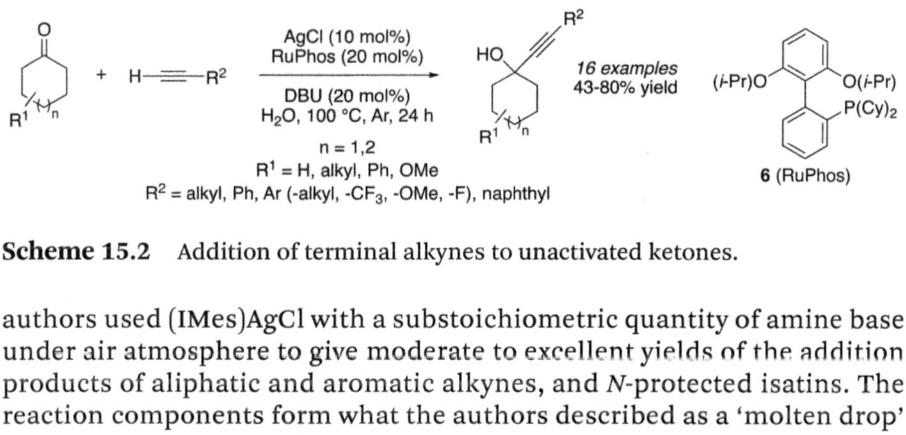

6 (RuPhos)

Scheme 15.2 Addition of terminal alkynes to unactivated ketones.

authors used (IMes)AgCl with a substoichiometric quantity of amine base under air atmosphere to give moderate to excellent yields of the addition products of aliphatic and aromatic alkynes, and *N*-protected isatins. The reaction components form what the authors described as a 'molten drop' on water.

Despite many advances in the alkynylation of electron-deficient ketones, it was not until 2012 that Li and co-workers devised a method capable of catalysing the addition of alkynes to unactivated ketones in water (Scheme 15.2).[25] This was realised by a continued effort to identify a strongly electron-donating and sterically bulky ligand to modify the metal centre of the catalyst to synthesise an unstable and highly reactive metal acetylide. The authors identified Buchwald ligand RuPhos (**6**) as a favourable choice. Compared to the use of activated ketones as substrates, relatively scrupulous conditions were required: the catalyst was pre-prepared from RuPhos (**6**) and silver chloride, the alkyne was gradually added with a syringe pump over the course of the reaction, and a strong base was necessary (DBU).

Nevertheless, after 24 h at 100 °C, a variety of aromatic and aliphatic alkynes, electron-rich and electron-poor, were added to cyclic alkynones in moderate to good yield.

15.2.1.3 *Synthetic Utility of Propargylic Alcohols*

The direct addition of terminal alkynes to aldehydes and ketones yields propargylic alcohols. Their dual-reactive nature, with a hydroxyl nucleophile and internal alkyne, enables their transformation into a variety of functional groups. The diversity of possible reactions makes selectivity a challenge, but catalysis can be used to guide the synthesis of one product over many possibilities under mild conditions. A common transformation of propargylic alcohols is direct substitution of the hydroxyl group *via* a cationic intermediate that exists in two resonance forms: the propargylic (7) and allenic cation (**8**, Figure 15.3). Nucleophilic addition to this intermediate yields the propargylic substitution product (Figure 15.3a) or the allene (Figure 15.3b), respectively. Propargylic alcohols can also isomerise to α,β-unsaturated carbonyl compounds. Isomerisation with oxygen transposition is known as the Meyer–Schuster rearrangement (Figure 15.3c), while isomerisation without transposition is simply a 'redox isomerisation' (Figure 15.3d).

The direct nucleophilic substitution of the hydroxyl group in propargylic alcohols is an efficient and atom economical approach to prepare propargylated molecules (Figure 15.3a). Under catalytic conditions, water is the only stoichiometric by-product and the reaction proceeds under redox neutral conditions (Figure 15.4a).

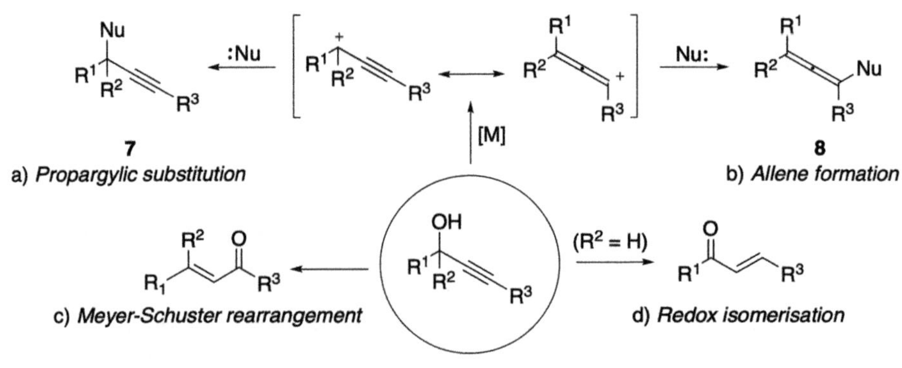

Figure 15.3 Synthetic utility of propargylic alcohols.

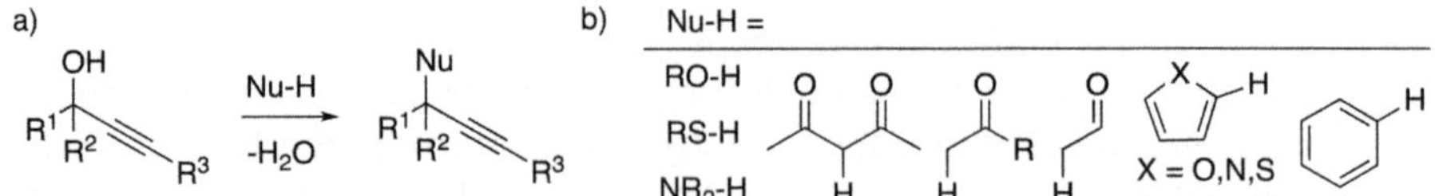

Figure 15.4 The direct substitution of propargylic alcohols (a) is a general reaction that can be carried out using a wide variety of nucleophiles (b).

The hydroxyl group of propargylic alcohols can be substituted with a wide variety of carbon, sulphur, nitrogen, and oxygen-centred nucleophiles (Figure 15.4b),[26] and catalysed by many transition metals including ruthenium,[27] rhenium,[28] gold,[29] iron[30] and copper,[31] to name a few. Some reactions are designed and carried out with di-nucleophilic species. One nucleophile substitutes with the hydroxyl group of the substrate, while the second can add intramolecularly to the internal alkyne to prepare cyclised products. This procedure has been used to prepare thiazoles,[32] furans,[33] isoxazoles and isoxazolines.[34]

The synthesis of allenes from propargylic alcohols (Figure 15.3b) is mechanistically related to the direct substitution by nucleophiles discussed in Section 15.2.1.3. The former can be considered the nucleophilic addition at the allenic carbocation, while the latter directly substitutes through its resonance form, the propargylic cation. Allenes are useful synthetic building blocks for the synthesis of complex molecular targets,[35] and are found in a number of natural products and biologically active compounds.[36] One of the earliest methods to access allenes from alkynes was through deprotonation by a strong base, while the use of transition-metal catalysts offers functional group tolerance, selectivity and mild conditions.

The Meyer–Schuster rearrangement (Figure 15.3c) offers an atom-economical way to prepare α,β-unsaturated carbonyl compounds, such as ketones and esters, from propargylic alcohols *via* 1,3-rearrangement and concomitant oxygen transposition.[37] α,β-Unsaturated carbonyls feature prominently in biologically active molecules[38] and as synthetic intermediates fated to undergo further transformations, such as nucleophilic Michael addition,[39] cyclopropanation[40] and Diels–Alder cycloaddition.[41] Typical reaction conditions for the Meyer–Schuster rearrangement dictate a strongly acidic environment and elevated temperatures, which are incompatible with many functional groups.[42] Utilising transition metal catalysts offers mild, neutral conditions. While propargylic alcohols with internal and terminal alkynes are compatible substrates, the focus of the following discussion is placed on the latter, as these can be prepared using the methods described in Section 15.2.1.

Nolan and co-workers have published extensively on the Meyer–Schuster rearrangement of propargylic alcohols, their particular interest being the ability of gold complexes to catalyse this transformation.[43] In 2010, the authors synthesised and characterised a catalytic gold complex with the formula $[((IPr)Au)_2(\mu\text{-}OH)]BF_4$, which they used to obtain 99% yield of the α,β-unsaturated ketone in 1 h at room temperature (Scheme 15.3).[44] The authors found that using this isolable and well-defined gold complex made unnecessary the use of the air, light and moisture-sensitive silver salts commonly employed in the preparation of active gold catalysts. In a rather impressive display of catalytic performance and researcher endurance, just 0.03 mol% of $[((IPr)Au)_2(\mu\text{-}OH)]BF_4$ achieved a turnover number of 3330 when the reaction was continuously fed substrate over a three week period. The substrate scope of three examples precludes drawing any conclusions

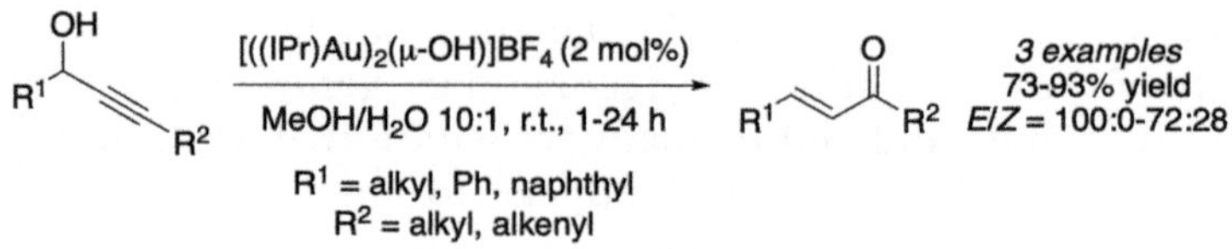

Scheme 15.3 Gold-catalysed Meyer–Schuster rearrangement.

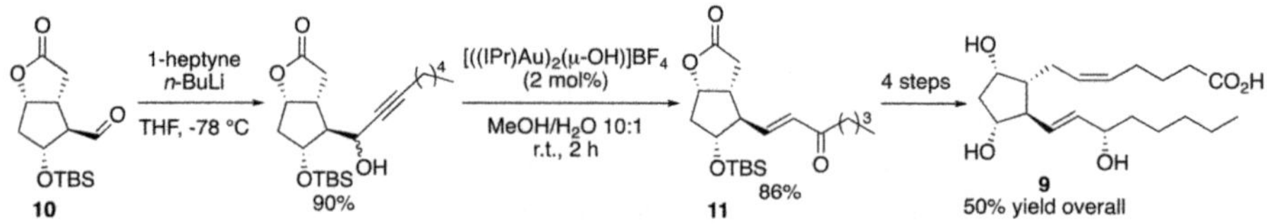

Scheme 15.4 Synthesis of prostaglandin F_2 (**9**) using the gold-catalysed Meyer–Schuster rearrangement at a key synthetic step.

about the versatility of these conditions, conveying instead that, with judiciously chosen substrates, the reaction can be fast and high yielding.

This was practically demonstrated using the gold-catalysed Meyer–Schuster rearrangement for the synthesis of prostaglandin F_2 (**9**, PGF_2), naturally produced by the human body and marketed as a pharmaceutical agent to induce labour. The researchers synthesised PGF_2 (**9**) in six steps from the commercially available (−)-Corey lactone aldehyde (**10**)[45] with 50% overall yield (Scheme 15.4). The gold-catalysed Meyer–Schuster rearrangement was employed at a crucial step to exclusively obtain the desired *E* isomer (**11**) in 86% yield.

One drawback of the Meyer–Schuster rearrangement is that it occurs with transposition of oxygen. Conversion of propargylic alcohols to α,β-carbonyl compounds without movement of oxygen should not be a leap of the imagination considering the two are isomers, related simply by hydrogen transfer. In 1995, Trost and Livingston reported that an indenyl ruthenium complex (**12**) could catalyse this 'redox isomerisation' (Figure 15.3d) to obtain good to moderate yields of the desired products.[46] More than a decade later, the same authors published a second generation method.[47] After extensive optimisation, it was found that addition of Lewis- and Brønsted acid co-catalysts broadened the reaction scope, and both α,β-unsaturated aldehydes and ketones could be obtained depending on whether primary or secondary propargylic alcohol substrates were used, respectively (Scheme 15.5).

15.2.1.4 Tandem Reactions Involving Alkyne Additions to Aldehydes and Ketones

In 2006, Yao and Li developed a highly efficient synthesis of the isochromene scaffold by nucleophilic addition of terminal alkynes to 2-alkynyl benzaldehydes, followed by cyclisation (Scheme 15.6).[48] In addition to being a novel

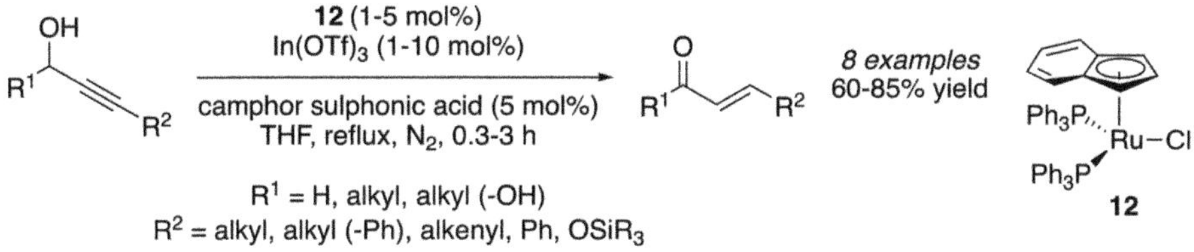

Scheme 15.5 Ruthenium and indium-catalysed redox isomerisation of propargylic alcohols.

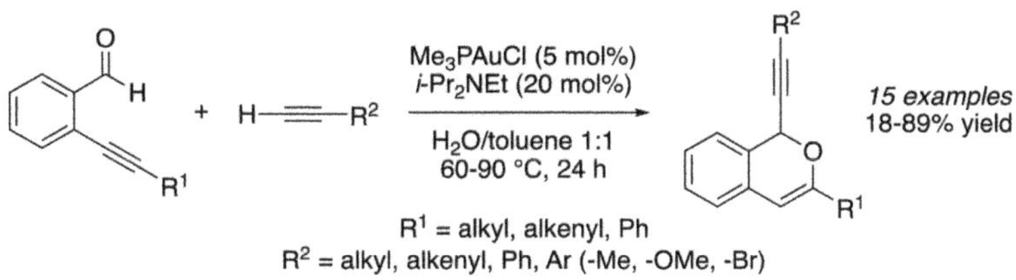

Scheme 15.6 Synthesis of isochromenes from 2-alkynyl benzaldehydes.

way of preparing isochromene derivatives, this was the first example of a gold-catalysed alkynylation of aldehydes. The authors discovered that the modification of gold(I) with an electron-rich phosphine ligand (PMe_3) gave a very active catalyst, and that a measure of water in the solvent mixture was crucial for obtaining synthetically useful yields. The addition of toluene as a co-solvent was necessary to solubilise the reaction components and facilitate efficient stirring. Benzaldehydes with sterically unhindered alkyl and phenyl substitution offered high yields when paired with aryl alkynes, while aliphatic terminal alkynes and benzaldehydes with bulky alkyl substituents gave poor yields.

The tentative mechanism in Figure 15.5 involves σ-activation of the terminal alkyne by gold to form an acetylide (**13**). This species coordinates to 2-alkynylbenzaldehyde to form a chelate (**14**), activating the carbonyl to nucleophilic attack by the acetylide (**13**). The resulting gold alkoxide intermediate (**15**) undergoes cyclisation to give a vinyl gold species (**16**). Protodeauration returns the final product (**17**) and regenerates the catalyst.

In 2009, Li and co-workers designed another reaction sequence involving the nucleophilic addition of a transition metal-acetylide, followed by cyclisation through a hydroxyl group upon an internal alkyne (Scheme 15.7).[49] In this example, hydrobenzofuranols (**18**) are the product, and these can be readily oxidised to the aurone scaffold, which is present in a number of biologically active natural products. Despite the mechanistic similarity between this report and the previously described synthesis of isochromenes (Scheme 15.6), gold complexes were not active for this transformation. Rather, silver salts paired with an electron-rich phosphine (PCy_3) in the presence of a tertiary amine base offered good yields of

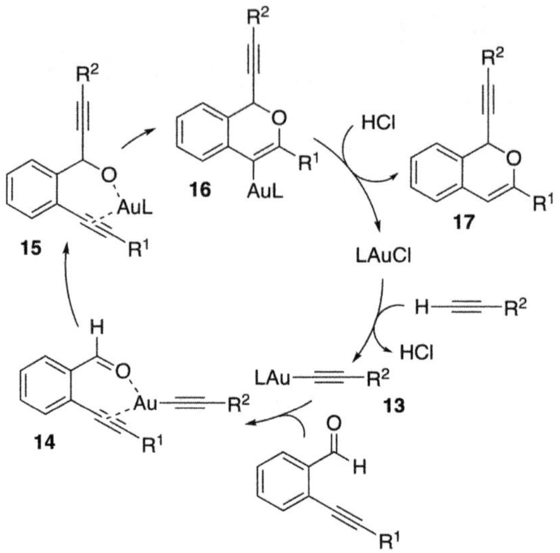

Figure 15.5 Proposed catalytic cycle. Reproduced from ref. 48 with permission from American Chemical Society, Copyright 2006.

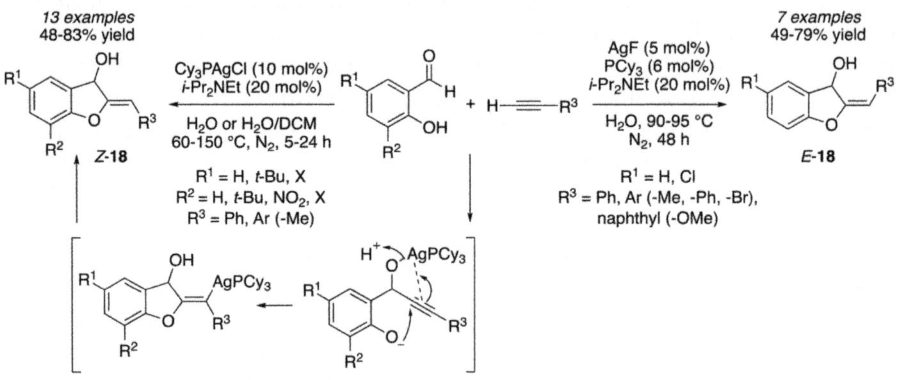

Scheme 15.7 Silver-catalysed synthesis of hydrobenzofuranols.

the desired product. The authors noted an interesting counter-ion dependence on the reaction stereoselectivity: the use of Cy_3PAgCl as the pre-catalyst favoured *Z*-stereoisomeric products (*Z*-**18**). For this, the authors attributed an organised transition state with coordination of the propargylic alkoxide and alkyne to the silver-complex, inducing a *trans*-addition by the phenolic nucleophile. However, when the pre-catalyst was changed to Cy_3PAgF (prepared *in situ* from AgF and PCy_3), the *E*-stereoisomer (*E*-**18**) was obtained with high selectivity. The authors tested the substrate scope under each set of conditions, and found that halogen- and nitro-substituted salicylaldehyde substrates gave the highest yields, while more forcing conditions were required and lower yields obtained for *tert*-butyl-substituted analogues.

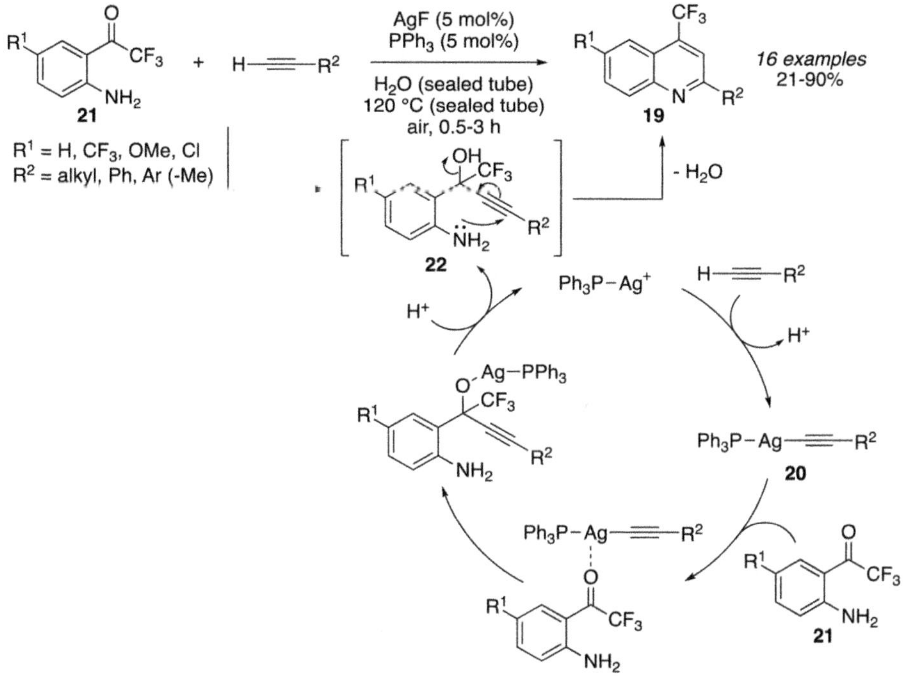

Scheme 15.8 Silver-catalysed synthesis of fluorinated quinolines (**19**) and proposed catalytic cycle.

In 2013, Zi-Gang Li and co-workers showed that fluorinated quinolines (**19**) could be prepared in one pot in a cascade reaction induced by the addition of a silver acetylide (**20**) to an aromatic trifluoromethyl ketone (**21**), generating a propargylic alcohol intermediate (**22**, Scheme 15.8).[50] The primary amine substituent on the aromatic ring then adds to the alkyne, followed by a condensation reaction to return the final product (**19**). An experiment was conducted to test the proposed mechanism; a trifluoromethyl ketone with a protected amine substituent incapable of adding to the alkyne was used. As expected, a high yield of the propargylic alcohol was obtained. Electron-deficient phosphines were not suitable to furnish efficient catalysts for this transformation, but the combination of silver fluoride and triphenylphosphine generated a catalyst capable of preparing a variety of quinolines in modest to good yield. Electron-poor ketones gave higher yields of the product than electron-rich ketones, and aromatic alkynes were superior to their aliphatic analogues.

15.2.2 Addition of Terminal Alkynes to Acid Chlorides

The direct addition of terminal alkynes to acid chlorides *via* transition metal catalysis in water yields ynones. Despite being a simple reaction on paper, in practice it is complicated by the high electrophilicity of acid chlorides and their propensity to react with water. This contrasts with other

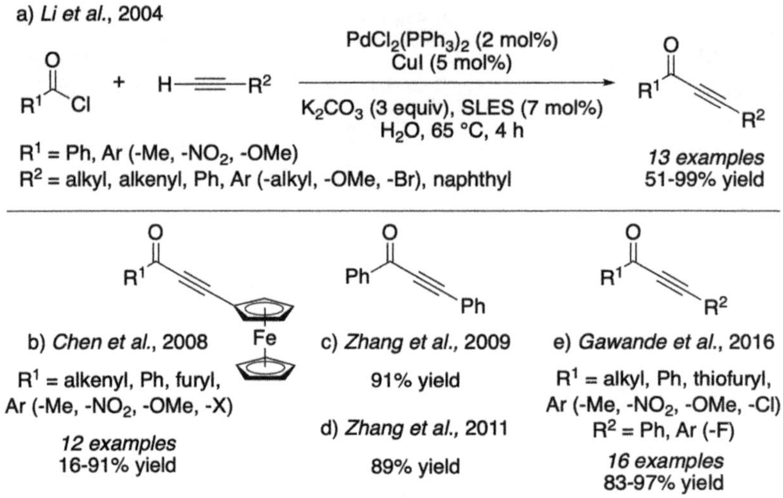

Scheme 15.9 Palladium-catalysed addition of terminal alkynes to acid chlorides.

electrophiles explored in this review, including aldehydes, ketones, and imines, that do not readily hydrolyse under the reaction conditions. The iminium ion intermediates of the A^3 coupling reaction (Section 15.3.2) are the exception, but the hydrolysis products are aldehydes and ketones, the original reaction substrates. Acid chlorides generate carboxylic acids upon hydrolysis that are not suitable substrates for the reaction and result in depleted reaction yields.

In 2004, Li and Chen were the first to overcome the challenge of acid chloride–alkyne coupling in water using a palladium and copper bicatalytic system in combination with a base (potassium carbonate) and a surfactant (sodium lauryl sulphate, SLES, Scheme 15.9a).[51] Conducting the reaction in aqueous medium offers the advantage of dissolving the hydrochloric acid generated over the course of the reaction, shifting the equilibrium to favour reaction completion, and using a surfactant prevents water from encountering acid chloride molecules. The importance of the surfactant was confirmed by conducting the model reaction, the addition of phenylacetylene to benzoyl chloride, without any surfactant. The yield dropped to 9%, compared to 98% when a surfactant was used. Aromatic and alkenic alkynes provided superior results to their aliphatic counterparts. This included an aryl bromide, which gave high yields and no trace of the Sonogashira coupling product, despite the similar reaction conditions.

In 2008, Chen and co-workers applied the conditions developed by Li to the synthesis of ferrocenyl-substituted ynones (Scheme 15.9b).[52] These exotic substrates have been used for the synthesis of uracil analogues *via* nickel-catalysed condensation and cyclisation chemistry, also in aqueous media.[53]

Zhang and co-workers have used the coupling of alkynes with acid chlorides, under the conditions originally reported by Li, as a model reaction to test the activity of novel heterogeneous catalysts. The authors studied materials consisting of periodic mesoporous silica frameworks containing covalently bonded diphenylphosphine functional groups ligated to palladium(II) ions. In 2009, the authors synthesised a material with diphenylphosphine ligands covalently tethered through an ethane bridge to the silica support (Scheme 15.9c).[54] Ligation of palladium(II) ions produced a catalyst that furnished 91% yield of the ynone product and could be recycled six times with only small losses in catalytic activity. In 2011, the researchers improved the synthesis of the support using a technique of 'aerosol-spraying assisted self-assembly'[55] in the presence of a surfactant, which creates spherical micellar templates and results in a material with uniform cavities (Scheme 15.9d).[56] Ligation of palladium(II) to the diphenylphosphine ligands embedded in the material generated an isolable and recyclable palladium catalyst for the synthesis of the ynone product in 89% yield.

Gawande and co-workers have also shown interest in the development of heterogeneous, silica-supported palladium catalysts to effect this transformation. In 2016, they synthesised silica-coated magnetic nanoparticles functionalised with a palladium-Schiff base complex (Scheme 15.9e). The magnetic core facilitates the facile catalyst recovery with an external magnet and the silica coating prevents iron leaching while also serving as an anchor to functionalise the material. The authors reported that acid chlorides could be alkynylated with just 0.2 mol% of the catalyst in the presence of an amine base under copper-free conditions; however, no explanation was given as to how acid chloride hydrolysis was inhibited in the absence of a surfactant in the miscible water/acetonitrile solvent mixture. The authors reported high yields within 2 h for a diverse selection of aliphatic and aryl acid chlorides with both electron-rich and poor groups. Even a nitro-substituted substrate gave high yields, opposite to the outcome of similar published reactions.[§]

15.2.2.1 Mechanistic Discussion

The palladium/copper-catalysed addition of terminal alkynes to acid chlorides is a distinct reaction amongst other reaction types explored in this review, where the focus has been the coupling of unactivated substrates. Acid chlorides are, arguably, activated substrates, and their coupling with terminal alkynes is considered a Sonogashira-like reaction in many publications.[57] The similarity between both reactions extends to the reaction conditions; both require a palladium(II) pre-catalyst that readily reduces to palladium(0) *in situ*, a copper(I) source, and a stoichiometric quantity of base. Investigation of the mechanism would reveal whether the carbon–carbon bond forming step proceeds by a Sonogashira-type

[§]Under both Li and Chen's conditions (ref. 51 and 52), the yields obtained for nitro-substituted aryl acid chlorides was in the range of 27–51% compared to Gawande's reported yield of 83%.

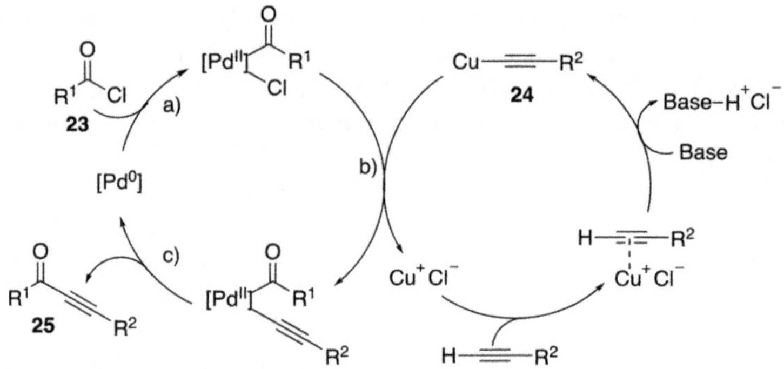

Scheme 15.10 Sonogashira-type mechanistic proposal for the addition of terminal alkynes.

mechanism,[58] namely oxidative addition of the acid chloride (**23**) to palladium (Scheme 15.10a), followed by transmetallation with a copper-acetylide (**24**, Scheme 15.10b) and reductive elimination to release the final product (**25**, Scheme 15.10c). Alternatively, the reaction might occur by nucleophilic addition of a metal acetylide to the acid chloride electrophile, as for other reactions in this review.

15.2.2.2 Synthetic Utility of 2-Alkynones

The multifunctional nature of 2-alkynones makes them ideal synthetic intermediates to access many diverse functional groups such as isoxazoles,[59] pyrimidines,[60] furans,[61] pyrazoles[62] and benzodiazepines.[63]

2-Alkynones also make suitable dienophiles for Diels–Alder reactions; the alkyne is rendered electron-poor through conjugation with the carbonyl group and can undergo cycloaddition with electron-rich dienes. This reactivity was utilised by Brummond and Kocsis, who prepared substrates (**26**) with a dienophile (2-alkynone) and diene (styrene) in the same molecule.[64] Microwave irradiation was used to induce an intramolecular Diels–Alder reaction, followed by oxidative aromatisation, giving naphthalene fluorophores (Scheme 15.11). Chloro-substituted naphthalenes were further transformed into amino-substituted analogues to generate a donor–acceptor system with the amine (donor) and ketone (acceptor) conjugated through the naphthalene ring. The authors measured the absorption maxima at 482–543 nm of the library of fluorescent dyes and observed solvatochromic behaviour.

Müller and co-workers also used 2-alkynones as precursors to fluorescent compounds (Scheme 15.12).[65] The authors developed a one-pot reaction to synthesise indolizines (**27**) commencing with the synthesis of 2-alkynones (**28**) under Sonogashira-coupling conditions, followed by addition of a pyridinium bromide salt. Triethylamine in the reaction mixture deprotonates the salt to yield a pyridinium ylide (**29**) *in situ*, which undergoes a 1,3-dipolar

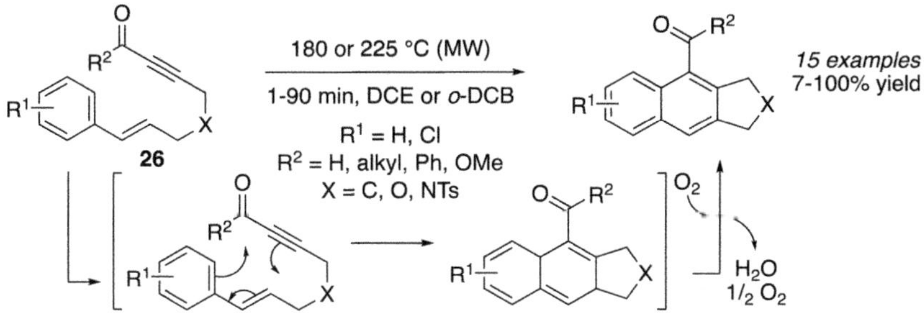

Scheme 15.11 Intramolecular Diels–Alder reaction of substrates with a 2-alkynone functional group.

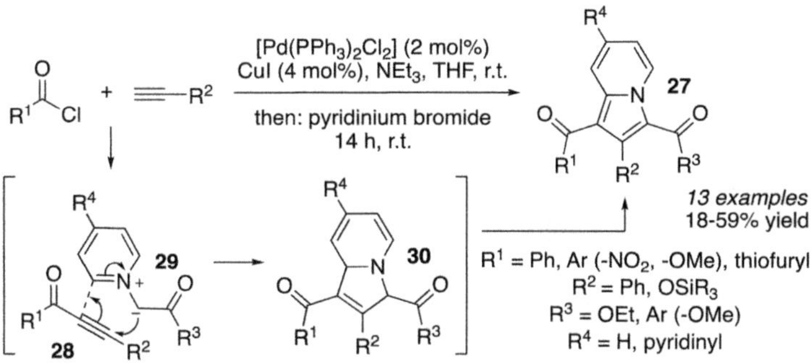

Scheme 15.12 Synthesis of indolizines from 2-alkynone intermediates.

cycloaddition with the 2-alkynone. The initial product, a dehydroindoline skeleton (**30**), spontaneously aromatises by expelling dihydrogen to generate a library of fluorescent products (**27**) with emission maxima at 471–510 nm. Bipyridine-derived ylides were also used to prepare more highly conjugated products.

15.3 Addition of Terminal Alkynes to Imines and Iminium Ions (Class II)

The transition metal-catalysed addition of terminal alkynes to imines and iminium ions is an atom economical way to prepare propargylamines. Credit for the first example of this reaction dates back to research conducted in the early 1950s by Guermont, who reported the coupling of terminal alkynes with the iminium ion condensation product of formaldehyde and a secondary amine.[66] Dax and co-workers reported a more contemporary version under the name of a 'Mannich condensation with alkynes' in organic solvents.[67] Such reactions were an advancement over chemical methods requiring the use of stoichiometric metal acetylides under inert, anhydrous conditions (Figure 15.6a). Three research groups are acknowledged for the simultaneous

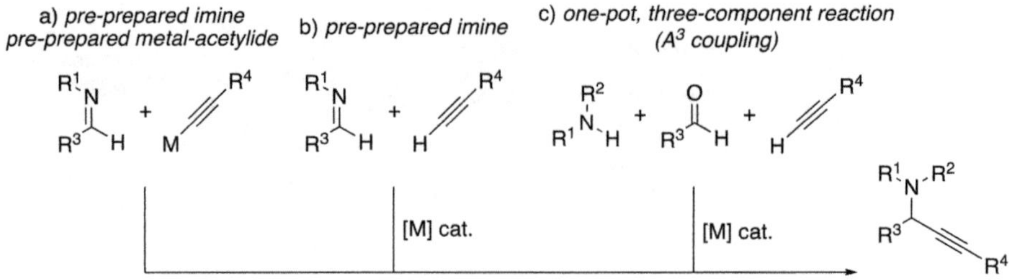

Figure 15.6 Evolution of methods for the addition of terminal alkynes to imines: the addition of metal acetylides to pre-prepared imines (a) was improved upon with the metal-catalysed addition of acetylenes to pre-prepared imines (b), which formed the groundwork for the development of metal-catalysed addition of acetylenes to *in situ* generated imines and iminium ions (c).

discovery and subsequent development of the modern coupling reaction of alkynes with imines and iminium ions: Ishii,[68] Carreira[69] and Li.[70] Of the three, the method by Li is the only example conducted in water with a broad scope of alkyne substrates. The timeline of the development of this reaction in water began with Li's account and use of pre-formed imines (Figure 15.6b). The reaction was further improved by the invention of a one-pot reaction, in which the imine and iminium ions are formed *in situ via* the condensation of aldehydes with primary and secondary amines, respectively (Figure 15.6c). This reaction was dubbed an A^3 coupling reaction (aldehyde–amine–alkyne coupling) by Li. The reduced number of reaction steps accompanied by performing a multicomponent reaction can reduce the quantity of by-products generated and minimise the use of solvent, purification and work-up materials, energy, and time. Numerous reviews have described its development and impact on chemical synthesis,[71] and herein we describe the relevant literature using water as the reaction solvent.

The challenges associated with this reaction include the hydrolysable nature of the imine and iminium ions. Furthermore, the resulting regeneration of the amine and aldehyde starting materials is favoured by equilibrium due to the excess of water present as the reaction solvent. Another challenge pertains to the multicomponent nature of A^3 reactions; reactivity must be balanced to favour acetylide addition to imine and iminium ions, rather than aldehydes, which are potent electrophiles in their own right.

15.3.1 Addition of Terminal Alkynes to Imines

Li and Wei's seminal report on the transition metal-catalysed preparation of propargylamines in water was published in 2002 (Scheme 15.13).[70,72] A bimetallic system consisting of copper bromide and ruthenium chloride proved to be an active catalyst for the coupling of alkynes and imines, which in many cases were synthesised in a separate thermal step from aldehydes and primary amines. However, the authors noted that, in some cases, the reaction could be carried out without pre-formation of the imine by using

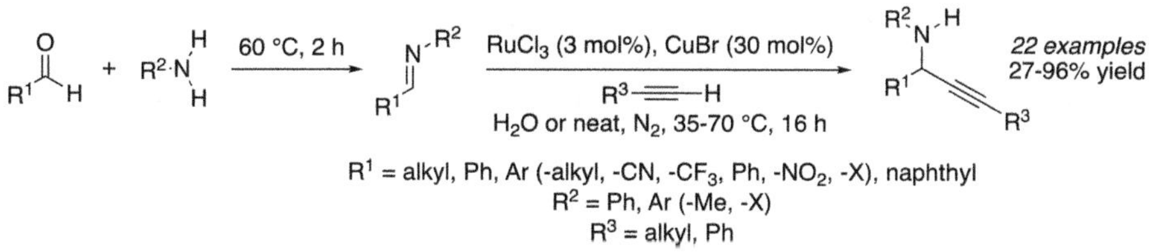

Scheme 15.13 Ruthenium- and copper-catalysed addition of terminal alkynes to imines.

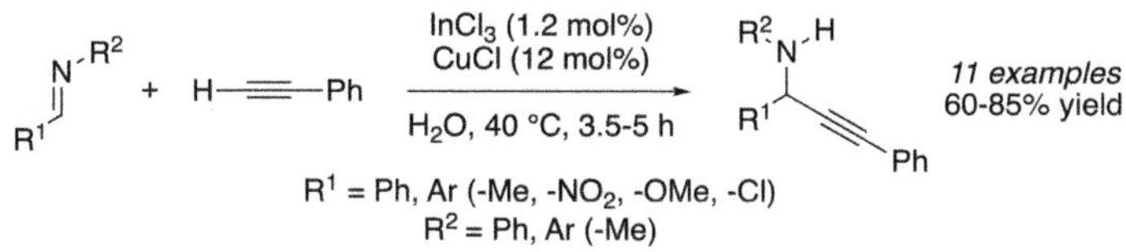

Scheme 15.14 Indium- and copper-catalysed addition of terminal alkynes to pre-formed imines.

a mixture of aldehyde, aniline and alkyne. This represented one of the first examples of the multicomponent reaction that would later be named the A^3 coupling reaction by Li. Excellent product yields of racemic propargylamines were obtained from aromatic aldehydes with either electron-withdrawing or donating groups; however, lower yields were obtained for aliphatic aldehydes (pivaldehyde: 64% yield). It was also noted that solvent-free conditions proved superior for imines susceptible to hydrolysis.

In 2011, Hu and co-workers used an indium salt in combination with copper to catalyse the alkynylation of pre-formed imines (Scheme 15.14).[73] The authors noted that the reaction could be catalysed by copper alone, but was accelerated by an additional quantity of indium. Under these conditions, quantitative conversion was attained within 5 h at 40 °C. Aryl imines with either electron-donating or withdrawing groups afforded good results; however, nitrogen-bound substituents with electron-withdrawing groups were not as well tolerated.

Van der Eycken and co-workers showed in 2015 that a ruthenium or indium co-catalyst was not necessary to achieve high product yields in a short timespan; rather, a combination of copper(I) and copper(II) catalysis under microwave irradiation in water can be used to give moderate to good product yields in just 25 min (Scheme 15.15).[74] The reaction conditions were optimised under one-pot, A^3 coupling conditions, without pre-formation of the imine. Whereas the work discussed previously was amenable to the synthesis of propargylic amines with aromatic substituents, this method is complementary insofar as aliphatic aldehyde and alkyne substrates are particularly high yielding.

The addition of alkynes to imines generates a stereogenic centre, which encouraged a number of groups to pursue the asymmetric variant of this reaction. In 2002, Wei and Li observed, while optimising the achiral reaction, that copper alone could catalyse the reaction (ruthenium alone could

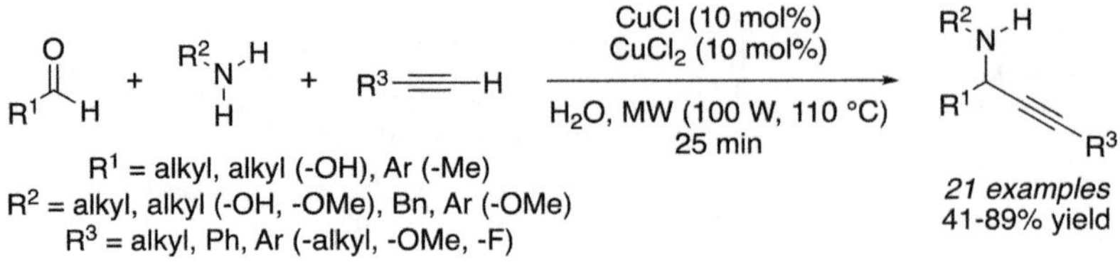

Scheme 15.15 Multicomponent synthesis of propargylamines from aldehydes, alkynes and primary amines.

Table 15.4 Copper-catalysed synthesis of chiral propargylamines.

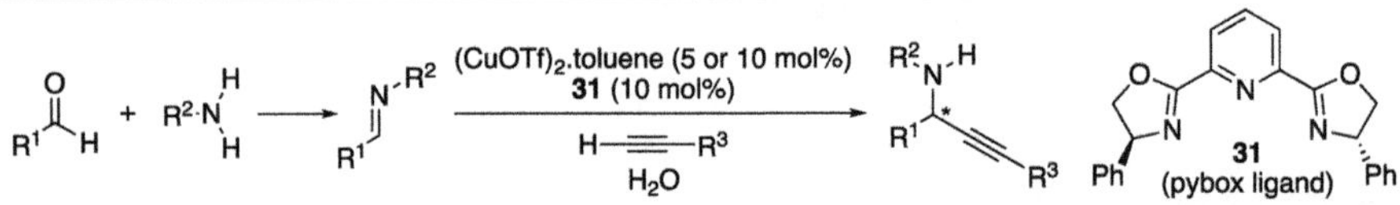

Entry	Conditions	Scope	Yield (enantio-selectivity)
a) Wei & Li, 2002	r.t. or 35 °C, 48 or 96 h	R^1 = Ph, Ar (–alkyl, –CF_3, –X), naphthyl, Ar = Ph, Ar (–alkyl, –X) R^2 = Ph	17 examples 48–86% yield (78–91% ee) o.r. = (+), (R)
b) Chan *et al.*, 2007	Stearic acid (1 mol%), r.t., 24 h	R^1 = Ph, Ar (–Me, –Cl) R^2 = Ph, Ar (–Me, –Cl)	12 examples 60–89% yield (35–97% ee)

not), albeit with long reaction times and low conversions (Table 15.4a).[75] It was hypothesised that the nucleophilicity of the copper acetylide could be enhanced by increasing the electron density on copper to weaken the copper–carbon bond. A survey of copper(I) complexes with electron-rich, chiral ligands was conducted to assess their ability to provide synthetically useful yields and chiral induction. Accordingly, a chiral pybox-copper(I) complex (pybox: bis(oxazolinyl)pyridine, **31**) was found to give high yields and enantioselectivities (slightly superior results were obtained using toluene as the solvent), and the authors named this reaction asymmetric A^3 coupling (AA^3 coupling). The best substrates were imines synthesised from benzaldehyde and alkyl-substituted aromatic aldehydes. All products exhibited positive optical rotations, and the absolute configuration was determined to be (*R*) based on X-ray crystallographic analysis of a representative bromine-containing product.[72]

In 2007, Chan and co-workers investigated whether the addition of surfactants might improve Wei and Li's conditions, with particular attention paid to reducing the reaction time required (Table 15.4b).[76] Following a survey of surfactants, the authors discovered that stearic acid ($CH_3(CH_2)_{16}CO_2H$)

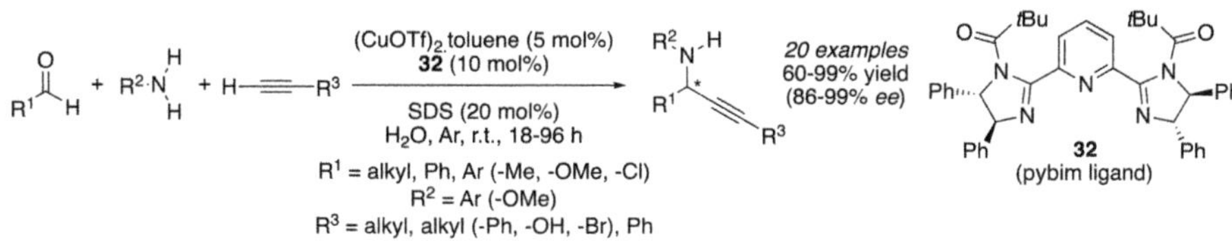

Scheme 15.16 Synthesis of chiral propargylamines using a copper-pybim ligand complex.

improved both the enantioselectivity and yield, and reactions reached completion in one day. For example, the addition of phenylacetylene to *N*-benzylideneaniline without a surfactant gave 77% yield and 80% ee in two days of reaction time. With a surfactant, these improved to 86% yield and 85% ee in one day.

In 2014, Nakamura and co-workers published an improved method for the copper(I)-catalysed asymmetric coupling of primary amines, aldehydes, and alkynes in water without pre-formation of the imine (Scheme 15.16).[77] The authors improved on the enantioselectivity possible using the pybox ligand (**31**) employed by Wei, Li and Chan (Table 15.4) by applying a sterically bulky bis(imidazoline) 'pybim' ligand (**32**). The combination of *p*-anisidine with aromatic aldehydes and aliphatic alkynes was extensively studied, and the authors found that addition of the surfactant sodium dodecyl sulphate (SDS) assisted in obtaining moderate to excellent product yields. The authors also tested tap water and sea water as reaction solvents. No appreciable loss of enantiomeric excess was observed, although a 12% and 26% reduction in yield, respectively, was recorded. The fact that synthetically useful yields can still be obtained demonstrates the robustness of this method and suggests that extensive water purification may be averted.

15.3.2 Addition of Terminal Alkynes to Iminium Ions

15.3.2.1 Gold-catalysed Methods for the A^3 Coupling Reaction

15.3.2.1.1 Homogeneous Gold Catalysts for the A^3 Coupling Reaction. In 2003, Wei and Li extended the scope of A^3 coupling reactions in water to include secondary amine substrates, which form iminium ions upon condensation with aldehydes and give tertiary propargylamine products (Table 15.5a).[78] Gold(I) and gold(III) salts emerged as the optimal catalysts for this transformation, the latter performing slightly better. Gold(III) bromide was chosen to test the reaction scope. In comparison to the A^3 coupling *via* imine intermediates, no co-catalyst was required, perhaps owing to the increased electrophilicity of the iminium ion. It was discovered that the optimised conditions were best suited to electron-poor aromatic aldehyde substrates; electron-rich aromatic aldehydes and aliphatic aldehydes gave decreased yields.

Table 15.5 A^3 coupling methods catalysed by homogeneous gold catalysts.

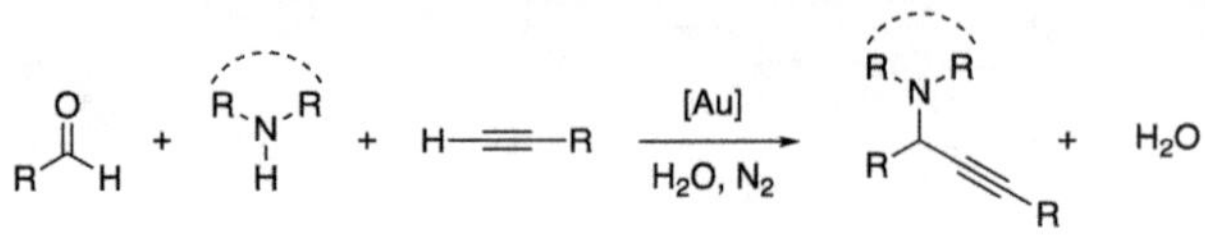

Entry	Conditions	Scope	Yield (diastereoselectivity)
a) Wei & Li, 2003	$AuBr_3$ (1 mol%) 100 °C, 12 h	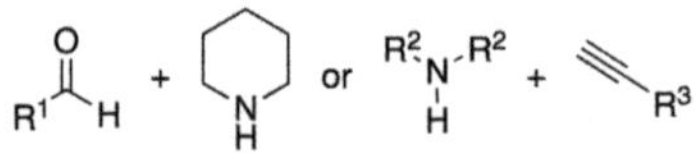 R = alkyl, alkyl (–Ph), Ph, Ar (–alkyl, –CF, –X)$^1{}_3$ R^2 = allyl, Bn; R^3 = alkenyl, Ph	16 examples 53–99% yield
b) Li *et al.*, 2006	AuI (5 mol%) r.t., 18 h	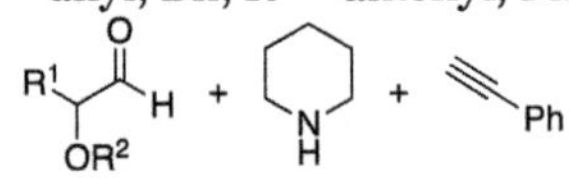 R = alkyl, Ph; R = H, Me, Bn[12]	6 examples 35–93% yield (dr = 73 : 27–50 : 50)
c) Che, Wong *et al.*, 2006	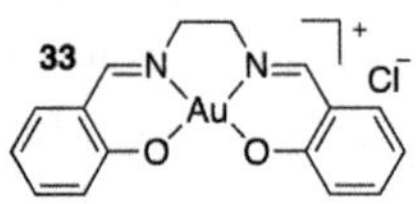 **33** (1 mol%) 40 °C, 24 h	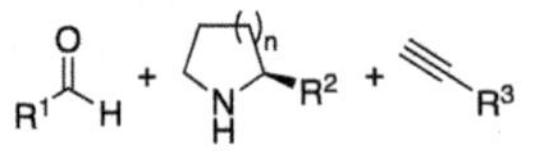 R^1 = alkyl, Ph R^2 = H, alkyl (–Ph, –OH, –OMe), CO_2Et R^3 = Ph, TMS; *n* = 1, 2	9 examples 67–99% yield (dr = 99:1–84 : 16)
d) Che, Wong *et al.*, 2013	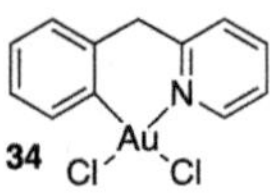 **34** (1 mol%) 40 °C, 24 h	R^1 = alkyl, Ph, Ar (–alkyl, –C(O) H, –OMe, –X) R^2 = H, alkyl (–Ph, –OH, –OMe, –pyrrole) R^3 = alkenyl, Ph, naphthyl, TMS; *n* = 1, 2	19 examples 62–99% yield (dr = 99:1–93 : 7)

In 2006, Li and co-workers studied the possibility of preparing α-oxygen-substituted propargylamines under A^3 coupling conditions beginning with α-oxygen-substituted aliphatic aldehyde precursors (Table 15.5b).[79] Despite their previous study suggesting that gold catalysis was inferior for coupling aliphatic aldehyde substrates, the authors found that a gold(I) catalyst provided the best yields.

The high activity of simple gold salts in catalysing A^3 coupling reactions was further explored by Srinivas and Koketsu in 2013.[80] The authors used gold(III) bromide to catalyse the synthesis of indole-based propargylamines from alkynes, secondary amines, and aldehyde-functionalised indoles.

The A^3 coupling reaction catalysed by simple gold salts has the benefit of broad substrate scope, mild conditions, and a convenient, readily available catalyst. However, the catalyst is not readily recovered and recycled, and cationic gold species readily reduce to elemental gold(0) in the presence of all manner of reductants, including alkynes, phosphines and alkenes.[81] This is a common route to catalyst deactivation. In 2011, Bowden and co-workers proposed a method to improve the catalytic efficiency of cationic gold salts in A^3 couplings in water.[82] They hypothesised that oxidising additives might inhibit the reduction to colloidal gold. Remarkably, the addition of a sub-stoichiometric quantity of TEMPO (60 mol%) and copper(II) chloride (0.25 mol%) allowed gold(III) chloride to be recycled 33 times with quantitative conversions, compared to a single turnover when no oxidising reagent was added.

Che, Wong and co-workers have shown a continued interest in the design and synthesis of gold(III) catalysts that resist reduction to elemental gold. In 2006, the authors developed a gold(III)-salen complex (**33**, Table 15.5c), which could still catalyse the A^3 coupling reaction after three successive cycles, albeit with a yield reduction of about 20%.[83] This catalyst could also be used to obtain high product diastereoselectivity using chiral-prolinol substrates. In 2009, the same authors developed another stable gold(III) complex capable of catalysing the A^3 coupling reaction, constructed with a pyridyl-phenylide mixed-donor bidentate ligand, with the formula $[Au(HC^\wedge N)Cl_2]$ ($HC^\wedge N$ = 2-phenylpyridine). This catalyst supersedes the stability of their former contribution, and remains active over ten consecutive reactions, corresponding to a 812 catalyst turnover.[84] The authors then went on to develop a variant of this catalyst using a benzylpyridine ligand (**34**, $HC^\wedge N$ = 2-benzylpyridine) and showed that it too was an efficient homogeneous catalyst for the A^3 coupling reaction (Table 15.5d). This catalyst could be used with low catalyst loadings and under mild conditions in air; however, recovery and recycling was not reported.[85] In addition to conventional A^3 coupling substrates, it was shown that protected galactose aldehydes made excellent starting materials, and unprotected D-raffinose aldehyde (a trisaccharide composed of galactose, glucose, and fructose) could be propargylated provided a stoichiometric quantity of the complex was used.[¶] [86]

In 2013, Wong and co-workers prepared a bis-cyclometallated variant $[Au(C^\wedge N)_2]BF_4$ ($HC^\wedge N$ = 2-phenylquinoline) with a distorted square planar geometry, again with high activity as A^3 coupling catalyst and high stability under the reaction conditions. The catalyst could be recycled seven times with marginal decreases in the yield (~10%, compared to ~70% yield loss for $KAuCl_4$ over the same number of cycles).[87]

Insight into the reactivity of homogeneous Au(III) complexes with the structure $[Au(C^\wedge N)Cl_2]$ ($HC^\wedge N$ = chiral benzylamine) in the A^3 coupling reaction was provided by Quayle and co-workers in 2017.[88] Whilst

¶When a catalytic (10 mol%) quantity of the gold complex was used with unprotected α-D-galactopyranose aldehyde as the substrate, only 10% yield of the product was obtained, suggesting stoichiometric conversion without catalyst regeneration.

attempting to achieve enantiomeric induction using chiral C^N chelating ligands, the authors found that, despite the high yields, there was a total lack of chiral induction and products were obtained as racemic mixtures. Furthermore, when kinetic studies were conducted with a variety of structurally distinct gold complexes, the kinetic data was strikingly similar. The researchers hypothesised that elemental gold particles were responsible for catalysis, and using TEM imaging they were able to confirm the presence of crystalline gold particles (5–50 nm diameter) in the spent reaction mixture. Although the isolation of gold nanoparticles does not constitute proof, the lack of stereocontrol and mirrored kinetic behaviour suggests they are the active species in some cases. This work contributes to the mounting body of evidence that many reactions assumed to be catalysed by homogeneous complexes are actually catalysed by metal nanoparticles.[89]

15.3.2.1.2 Heterogeneous Gold Catalysts for the A^3 Coupling Reaction. The focus of innovation in the A^3 coupling reaction catalysed by homogeneous catalysts has been in broadening the reaction scope, increasing the catalyst stability and improving the reaction conditions to favour shorter reaction times and lower temperatures. Unfortunately, homogeneous catalysis suffers intrinsic disadvantages: in many instances, the catalyst cannot be recovered and recycled, or catalyst separation is energy-intensive and requires large quantities of solvent and separation materials. Attempts to solve this problem have largely focussed on the immobilisation of transition-metal catalysts on heterogeneous supports, resulting in a large body of work prioritising high catalytic efficiencies and ease of catalyst recovery and reuse. Furthermore, many research groups have pursued the rational design of metal nanoparticle catalysts, a promising area of research due to their large surface area and unique size-dependent properties distinct from the bulk metal. Immobilisation of nanoparticles on a heterogeneous support can influence the reactivity and selectivity of the catalyst, prevent aggregation and coalescence of the particles, and provide a facile means for catalyst recovery.

Published in 2008, Corma and co-workers prepared a material composed of gold nanoparticles (~2 nm) supported on nanocrystalline cerium oxide (CeO_2, ~5 nm) to produce an air-stable and readily recoverable/reusable catalyst for A^3 coupling.[90] A number of nanocrystalline supports were trialled, including ZrO_2, TiO_2 and SiO_2, but the highest turnover number and conversion (788 and >99%, respectively, for the model reaction) was obtained using CeO_2. High catalyst activity is often associated with small nanoparticle size, but this did not seem to be a factor. Rather, the catalytic activity of each catalyst correlated with the concentration of gold(III) in the sample, inspiring the hypothesis that gold(III) ions stabilised on the CeO_2 surface are the active species. The comprehensive reaction scope revealed the broad utility of this method for both aliphatic and aromatic aldehydes and alkynes, as well as chiral prolinol derivatives and cyclic secondary amines.

Rhee and co-workers took a different approach to preparing a heterogeneous cationic gold catalyst. In 2015, they designed a thermosensitive polymer support functionalised with pyridyl groups to bind gold(III).[91] This polymer has an interesting property: it shrinks at temperatures higher than 47 °C, the temperature at which intramolecular hydrogen bonding overpowers intermolecular interactions with the water solvent. Shrinkage of the catalyst provides a highly concentrated and hydrophobic local environment, a phenomenon that may play a role in accelerating the reaction. The reaction conditions provided high yields, which did not decrease by an appreciable extent after recycling the catalyst ten times.

In 2016, researchers in Haruta's laboratory synthesised an A^3 coupling catalyst comprising an exotic titania-supported gold cluster with the general formula $Au_{25}(PPh_3)_{10}$(phenylacetylene)$_5$ X_2, which has the interesting property of containing gold atoms in many oxidation states: Au(0), Au(I) and Au(III).[92]

The immobilisation of gold nanoparticles on heterogeneous supports is a popular method to prepare A^3 coupling catalysts. A study in 2009 by Ying and co-workers reported the stabilisation of gold(I)-nanoparticles (1–2 nm) on semi-conducting lead-sulphide quantum dots,[93] and another study in 2016 by Nájera and co-workers showed how a polymer containing phosphinite and amine pendant ligands can immobilise gold nanoparticles (3–5 nm) to obtain a stable and recyclable catalyst.[94] In 2016, Feiz and Bazgir functionalised the silicon-hydroxyl groups on the surface of MCM-41 (a silicate with a regular arrangement of cylindrical mesopores) with an aliphatic thiol.[95] This ligand strongly coordinated and stabilised gold nanoparticles deposited in the silicate pores to prepare a high-surface area, reusable catalyst (at least five times) used to obtain high yields of propargylamines.

A common strategy to prepare gold-immobilised heterogeneous catalysts is to start from a support rich in hydroxyl groups such as silica, cellulose or graphene oxide. The hydroxyl groups provide a handle for functionalisation with an organic linker containing a terminal trialkoxysilane, which undergoes substitution with the surface hydroxyl groups forming strong silicon–oxygen bonds. The opposite end of the organic linker can be customised with a variety of functional groups capable of ligating gold nanoparticles and immobilising them on the surface of the material. This strategy was applied by He-Xing Li and co-workers in 2011 (Table 15.6a). The authors dispersed small gold nanoparticles (1–2 nm) within the pores of periodic mesoporous silica through an organic linker with coordinating thiol groups. In addition to being an efficient A^3 coupling catalyst, the authors demonstrated the material's broad utility for a number of organic transformations including alkyne hydration, intramolecular hydroamination, and styrene oxidation.[96] A collaboration between Li's research group and Gray in 2013 studied nanocrystalline cellulose as a hydroxyl-rich support (Table 15.6b).[97] A thiol moiety was similarly used to coordinate small gold nanoparticles (2–3 nm), and good yields were maintained through 11 iterations of catalyst recycling. The

Table 15.6 Functionalisation of heterogeneous supports for the synthesis of A^3 coupling catalysts.

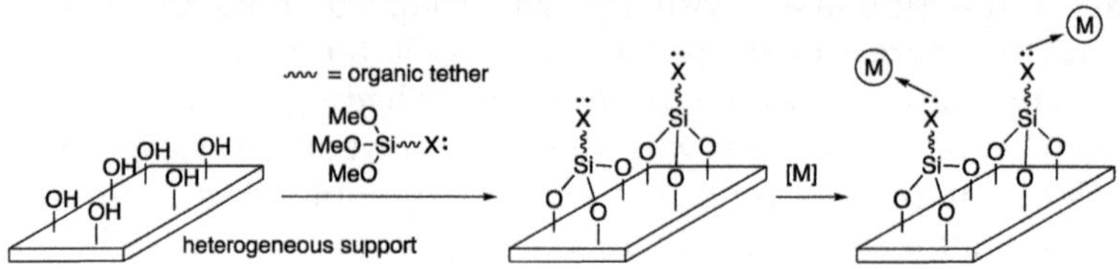

Entry	Heterogeneous support	X =	(M) =
a) Li *et al.*, 2011	Periodic mesoporous silica	–SH	Au nanoparticles
b) Li *et al.*, 2013	Cellulose nanocrystals	–SH	Au nanoparticles
c) Pourjavadi *et al.*, 2015	Magnetite/silica core–shell nanoparticles	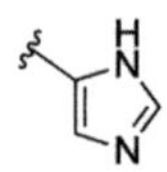	Au(III) ions
d) Zohreh *et al.*, 2017	Magnetite/silica core–shell nanoparticles	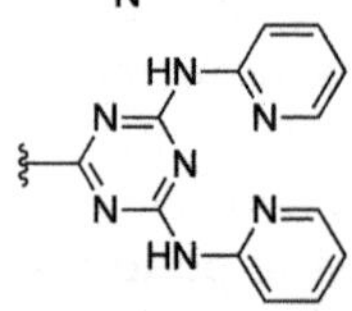	Au nanoparticles
e) Najera *et al.*, 2018	Magnetite/silica core–shell nanoparticles	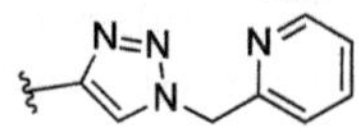	Au nanoparticles
f) Movahedi *et al.*, 2014	Zn-oxide nanoparticles	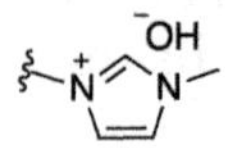	Ag nanoparticles
g) Heravi *et al.*, 2017	Hollow magnetite/silica core–shell nanoparticles	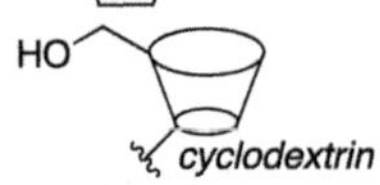	Ag nanoparticles
h) Heravi *et al.*, 2018	Halloysite nanotube	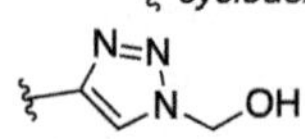	Ag nanoparticles
i) Tajbakhsh *et al.*, 2015	Magnetite/silica core–shell nanoparticles	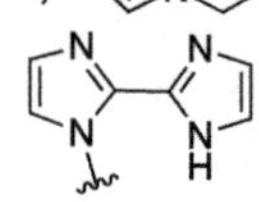	Cu(I) ions
j) Pathak *et al.*, 2016	Graphene oxide	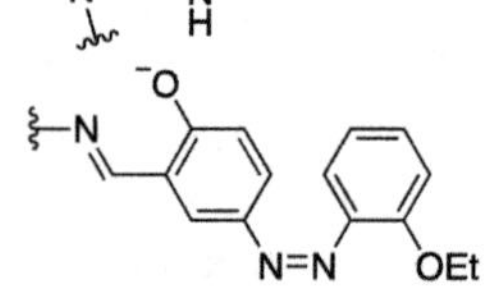	Cu(II) ions
k) Heravi *et al.*, 2017	Halloysite nanoclay	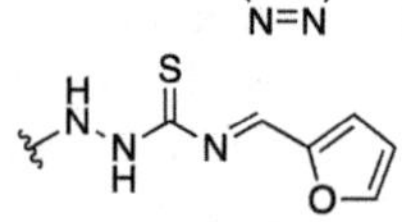	Cu(II) ions

catalyst performed well in water, but the authors discovered that neat reaction conditions were superior for the synthesis of un-functionalised aromatic and aliphatic substrates.

In an effort to simplify catalyst recovery, a number of research groups have reported the synthesis of catalytic magnetic nanoparticles, which can be separated by applying a magnet to the exterior of the reaction vessel and decanting the remaining solution. The first instance of magnetically separable catalysts for the A^3 coupling reaction was reported by Li in 2010.[98] The authors used non-functionalised magnetite nanoparticles, although using an aqueous reaction medium gave only modest yields and an organic solvent was necessary to achieve high yields of the propargylamine products. To prepare magnetically separable catalysts with high activity in water, a number of research groups have coated magnetic nanoparticles in silica, functionalised, then decorated them with active gold species (nanoparticles or ions). In 2015, Pourjavadi used this method to synthesise magnetic particles functionalised with a tether containing a terminal vinyl group to cross-link with vinyl-substituted imidazole and imidazolium monomers (Table 15.6c).[99] This constructed a second polymer coating rich in nitrogen-containing ligands, which were used to bind gold(III) ions. The catalyst was used for the A^3 coupling reaction with 1% catalyst loading, and was reused 10 times with 10% yield reduction. Zohreh and co-workers functionalised magnetic particles with an NNN-pincer ligand comprising a triazine core and two coordinating pyridyl groups, providing three-point coordination of gold nanoparticles (4–6 nm, Table 15.6d).[100] The authors found that the conditions were best suited to aryl aldehydes and alkynes; however, aliphatic examples still gave moderate yields. Nájera and co-workers immobilised gold nanoparticles on magnetic particles through covalently bound pyridyl-triazole mixed-donor ligands (Table 15.6e).[101] The authors reported that extremely low catalyst loadings were sufficient (0.01 mol%) and high catalytic activity in A^3 coupling could be maintained over 10 runs.

15.3.2.2 *Silver-catalysed Methods*

In addition to gold, many silver salts have emerged as highly active catalysts for the A^3 coupled synthesis of propargylamines. An interesting aspect is that silver salts offer complementary reactivity: where gold is the judicious choice for aromatic aldehyde substrates, silver is superior for aliphatic aldehydes. This was first discovered in 2003, when Wei and Li published the first silver-catalysed A^3 coupling reaction (Scheme 15.17).[102] The reaction is conducted in water, reaching completion in 2 h. A follow-up study on the ability of silver iodide to catalyse A^3 coupling reactions using α-branched aliphatic aldehydes was reported in 2006.[79] Moderate diastereoselectivity (71:29–52:48) and reasonable yields (57–83%) were achieved.

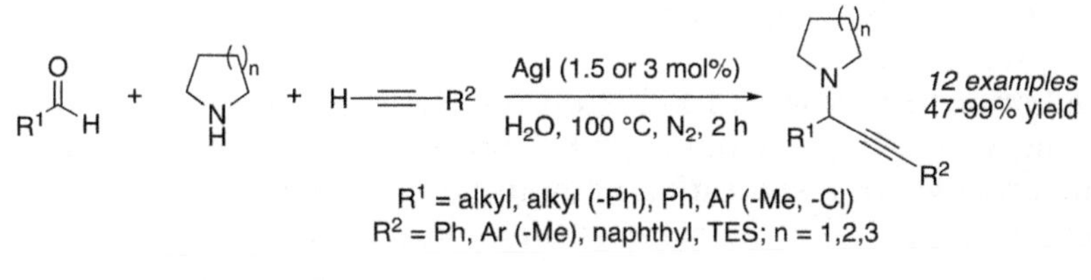

Scheme 15.17 A^3 coupling reaction catalysed by a simple silver salt.

Heterogeneous catalysts have also been prepared with silver as the catalytically active species. Wang and co-workers supported silver-oxide nanoparticles (Ag_2O, 6–8 nm) on alumina and demonstrated their catalytic utility for the A^3 coupling reaction.[103] Consistent with Li's findings, aliphatic aldehydes provided the best product yields under the reaction conditions. In 2014, Islam and co-workers prepared a polymeric network from melamine (a triazine substituted with three amine groups), which forms a Schiff base with a rigid organic strut functionalised with aldehydes at each end.[104] Silver nanoparticles (15–40 nm) were fixed to the surface of the resulting nitrogen-rich, porous assembly. The catalyst could be recycled at least five times, and electron-rich aromatic aldehyde substrates gave the best yields. Pitchumani and co-workers synthesised a heterogeneous A^3 coupling catalyst using montmorillonite clay, a naturally occurring silicate with a layered structure.[105] Interspersed between the silicate sheets are cations, and these were replaced by silver(I) cations. The authors optimised the A^3 coupling conditions with formaldehyde and diisopropylamine, and 86% of the desired product was obtained after 8 h at room temperature in water. This is in contrast with the conditions required for aromatic aldehyde substrates, namely neat conditions at 100 °C.

The strategy of grafting silver to a hydroxyl-rich support through an organic linker, previously discussed as applied to gold catalysis (Section 15.3.2.1), has also been used to prepare heterogeneous silver catalysts. In 2014, Movahedi and co-workers prepared zinc-oxide nanoparticles (25–35 nm), and methyl-imidazolium hydroxide moieties were covalently bound to the surface, resulting in an ionic liquid layer (Table 15.6f).[106] Silver nanoparticles (6–10 nm) were embedded in this layer and the material proved to be an active A^3 coupling catalyst.

15.3.2.3 Copper-catalysed Methods

In 2004, Fan and co-workers showed that, under the right conditions, copper represents a cheap and abundant alternative to silver and gold as an A^3 coupling catalyst (Table 15.7a).[107] Using microwave irradiation, a broad scope of substrates were transformed into propargylamines in high yield within 30 min (the model reaction required five days under thermal conditions). The reactivity profile of copper was found to be unique compared to those of silver and gold: electron-poor aromatic aldehydes were the best substrates, but their electron-rich aromatic and aliphatic counterparts also gave good yields. Also of note, two examples of primary amines were tested with satisfactory results, showing that copper can also catalyse A^3 coupling through imine intermediates.

In addition to microwave irradiation, other energy sources have shown utility in the A^3 coupling reaction. In 2005, Sreedhar and co-workers used ultrasound, provided by a sonication bath, to promote the A^3 coupling catalysed by copper(I) iodide.[108] The same authors, in a continued effort to develop

Table 15.7 A^3 coupling procedures catalysed by homogeneous copper catalysts.

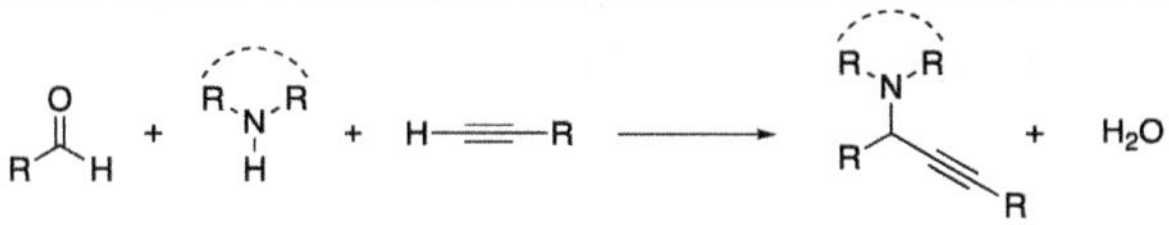

Entry	Conditions	Scope	Yield
a) Fan *et al.*, 2004	CuI (15 mol%) Ar, H_2O MW, 5–30 min	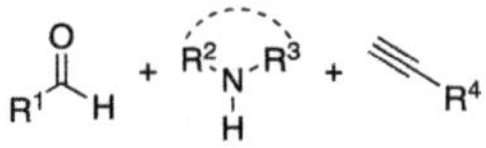 R^1 = alkyl, Ph, Ar (–Me, –NO_2, –X), naphthyl, 2-furyl R^2/R^3 = morpholine, pyrrolidine, piperidine, H, alkyl, Ph, Bn R^4 = alkyl, alkyl (–OTBS), Ph, TMS	25 examples 41–93% yield
b) Jain *et al.*, 2016	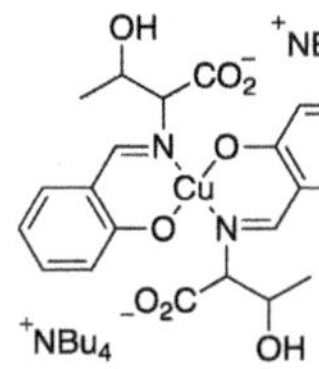 35 **35** (2.5 mol%) H_2O, 100 °C, 6 h	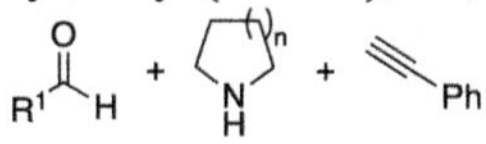 R^1 = Ph, Ar (–Me, –OMe, –Cl, –NH_2, –NO_2) n = 1, 2	12 examples 88–96% yield

innovative A^3 coupling procedures, developed a recoverable and reusable (at least five times) heterogeneous copper catalyst by tethering copper chloride to silica gel through a sulphonate ligand.[108]

Efforts to develop new catalysts, in particular heterogeneous approaches that can be readily separated and reused, has been the focus of many groups seeking to improve the efficiency of the A^3 coupling reaction in water. In particular, heterogeneous-catalysts that bind copper through Schiff base ligands are a popular approach. In 2016, Jain and co-workers synthesised a copper(II) complex with Schiff base ligands constructed from salicylaldehyde and the amino acid threonine (**35**, Table 15.7b).[109] Deprotonation of the amino acid, countered with a tetrabutylammonium cation, generated a thermally stable and readily recoverably catalyst. The authors demonstrated the application of this catalyst in the A^3 coupling reaction with a small scope of electron-rich and poor aromatic aldehydes and excellent yields were reported for all cases.

In 2015, Tajbakhsh functionalised magnetite nanoparticles (Fe_3O_4) with 2,2′-biimidazole ligands ligated to copper ions, and used this catalyst for the A^3 coupling reaction in water (Table 15.6i).[110] In 2016, Pathak and co-workers prepared a recyclable A^3 coupling catalyst composed of a copper(II) Schiff-base complex tethered to graphene oxide through the surface hydroxyl groups (Table 15.6j).[111] Heravi and co-workers used a similar approach using halloysite nanoclay (an aluminosilicate with pendant hydroxyl groups) as the heterogeneous support (Table 15.6k)[112] and, in 2016, Islam and co-workers decorated polystyrene beads with an organic tether incorporating a Schiff base, this time using copper(I) as the catalytic metal.[113]

15.3.2.4 *KA² Coupling*

The coupling of a ketone, alkyne and amine, known as KA^2 coupling, can be used to prepare propargylamines with a quaternary carbon centre. Compared to A^3 coupling, KA^2 coupling has not been extensively explored; no doubt due to the reduced electrophilicity of ketones *versus* aldehydes. In the rare instances that A^3 coupling catalysts were applied to ketone substrates, yields were very low, as observed by Ramón and co-workers in 2010, who prepared a copper(II)-magnetite catalyst for A^3 coupling under neat conditions.[114] With aldehyde substrates, yields in excess of 80% were attained for all entries within 3 h at 120 °C. The reaction time increased to seven days for aliphatic ketone substrates to achieve yields below 40%.

In 2010, Van der Eycken and co-workers were the first to achieve synthetically useful yields (up to 82%) for KA^2-coupling, using copper(I) iodide under neat, microwave irradiation conditions.[115] Many follow-up reports have similarly used neat conditions to perform this transformation, catalysed by either copper[116] or gold.[117]

Two recent reports by Heravi and co-workers have outlined the use of heterogeneous silver catalysts to execute KA^2-coupling in aqueous solvents. In 2017, the authors prepared magnetite/silica core–shell nanoparticles functionalised with a cyclodextrin moiety (Table 15.6g).[118] Then, silver nanoparticles were prepared using hollyhock flower extract, which served to both reduce and stabilise the synthesised particles with an organic matrix. Immobilisation of silver on the heterogeneous support, facilitated by interactions with the pendant cyclodextrin groups, culminated in a catalyst capable of catalysing both the A^3 and KA^2 coupling reaction using ultrasonic irradiation as the energy source. The authors reported a similar strategy in 2018; halloysite nanotubes (a naturally occurring sheet silicate) scattered with covalently bound triazole ligands provided an heterogeneous support for silver nanoparticles (Table 15.6h).[119] The nanoparticles were bio-reduced using an extract from a species of burdock, and the catalyst was used to provide yields of 80–92% of the KA^2-coupling products in water.

15.3.3 Addition of Terminal Alkynes to Nitrones

Nitrones are the *N*-oxide of imines, and their chemistry is most commonly associated with cycloadditions as they are a 1,3-dipole. For example, the Kinugasa reaction proceeds by the cycloaddition of terminal alkynes with nitrones, forming an isoxazolidine intermediate that rearranges and offers a convenient route to β-lactams.[120] In 2015, Michalak reported reaction conditions to avert the cycloaddition pathway and instead undergo the direct addition of terminal alkynes to nitrones (**36**) for the synthesis of propargylic *N*-hydroxylamines (**37**, Scheme 15.18).[121] To achieve this aim, the authors prepared a copper-NHC complex, as these ligands display strong σ-donating but weak π-accepting properties. It was anticipated that loading the copper centre with electron-density would destabilise the copper-acetylide bond and increase the nucleophilicity of the alkyne. Furthermore, the steric shielding provided by the ligand favours the direct addition over the cycloaddition pathway. High yields were obtained when both aromatic and aliphatic alkynes were used; however, a stronger base (tetramethylguanidine, TMG) was required for the latter. When chiral substrates were tested, a high level of stereoinduction was observed, with diastereomeric ratios exceeding 95 : 5 for all relevant examples.

The authors showcased their novel reaction by embarking on the formal synthesis of (−)-lentiginosine (**39**), an alkaloid natural product isolated from the desert shrub *Astragulus lentiginosus*, which holds academic interest for its potential chemotherapeutic effects against viral and bacterial infections (Scheme 15.19). The first step of the synthesis utilised the copper-catalysed direct addition of terminal alkynes to nitrones to yield an acetylated pyrrolidinol (**40**). This species was hydrogenated, which served to reduce the nitrogen centre, triple bond and remove the benzyl protecting group. The resulting

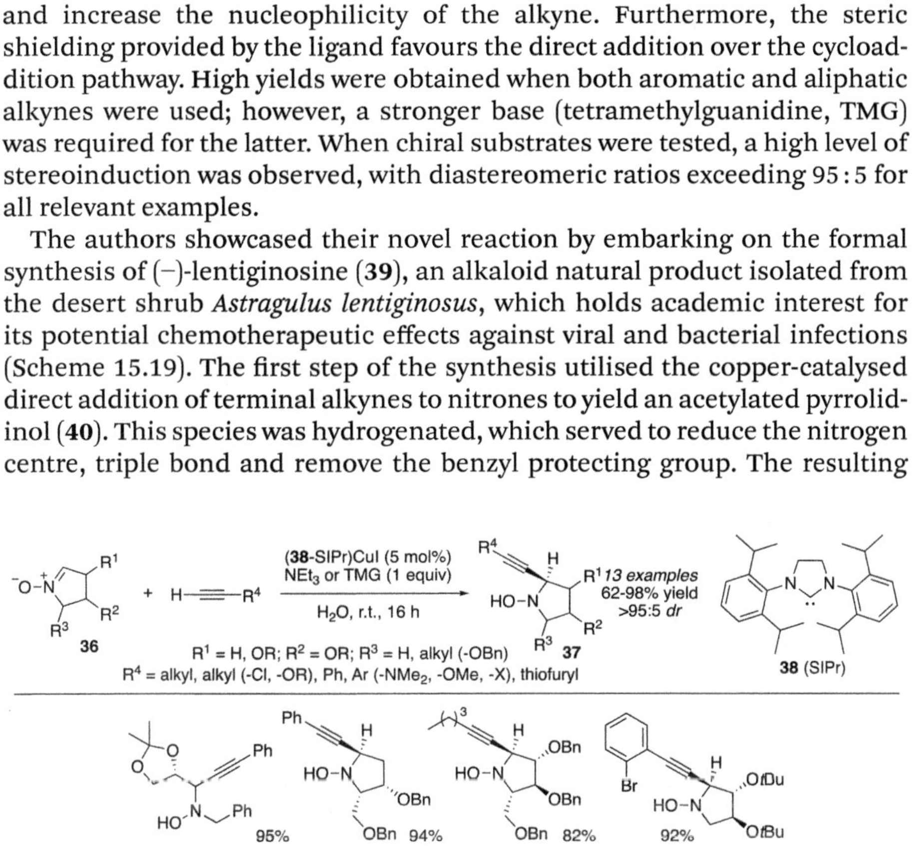

Scheme 15.18 Addition of terminal alkynes to nitrones catalysed by a copper-NHC complex.

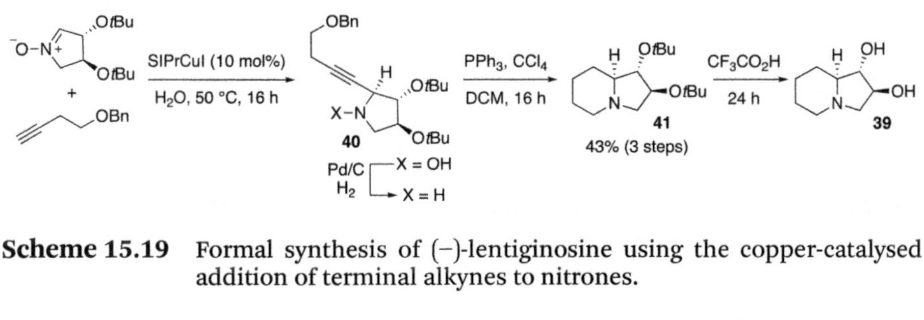

Scheme 15.19 Formal synthesis of (−)-lentiginosine using the copper-catalysed addition of terminal alkynes to nitrones.

intermediate underwent cyclisation in the presence of triphenylphosphine and carbon tetrachloride to yield the indolizidine core (**41**), *en route* to **39** upon alcohol deprotection.[122]

15.3.4 A^3 Couplings with Multiple Additions

In 2007, Bonfield and Li sought an A^3 coupling strategy for the preparation of di-propargylated amines (Scheme 15.20).[123] This was achieved by coupling formaldehyde with a primary amine and two equivalents of a terminal alkyne, using copper bromide and ruthenium bromide, the catalyst combination that had proved effective in Li's seminal A^3 coupling publications.[70] The reaction was conducted at room temperature under 'semi-neat' reaction conditions and efficient stirring required the addition of some water. The optimised reaction conditions were applied to aliphatic and electron-rich aromatic amines and alkynes, while electron-poor substrates were not tested.

In 2012, Uhlig and Li demonstrated how the di-propargylation reaction could be applied to the site-selective functionalisation of amino acids and peptides (Scheme 15.21).[124] Introducing an alkyne functionality into biomolecules provides a useful handle for further bioconjugation, for example *via* the azide–alkyne click reaction. The authors developed a procedure to prepare di-propargylamines from formaldehyde, terminal alkynes and amino acid salts (or their corresponding esters) in the presence of catalytic copper chloride. Moderate to excellent yields were

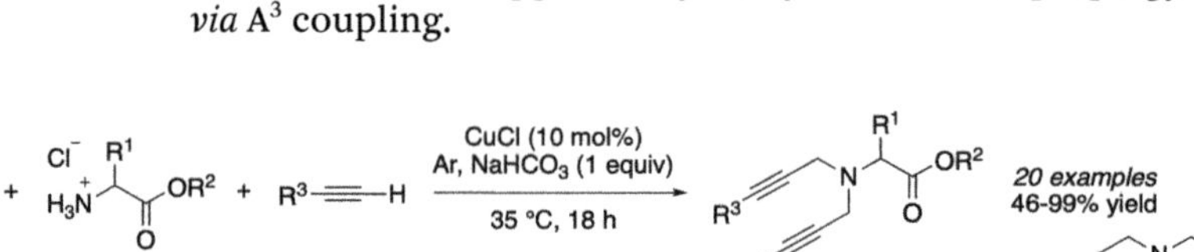

Scheme 15.20 Ruthenium and copper-catalysed synthesis of di-propargylamines *via* A^3 coupling.

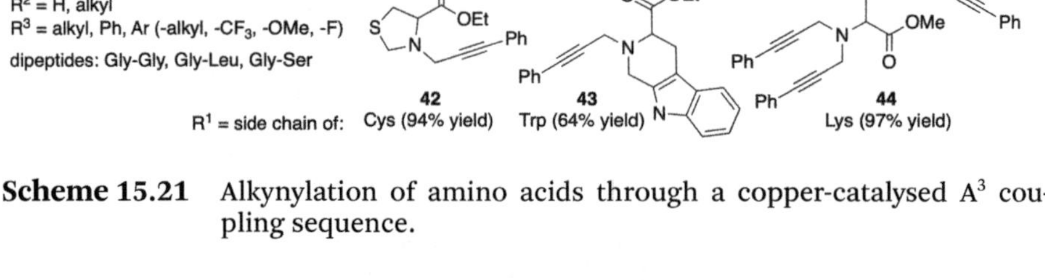

Scheme 15.21 Alkynylation of amino acids through a copper-catalysed A^3 coupling sequence.

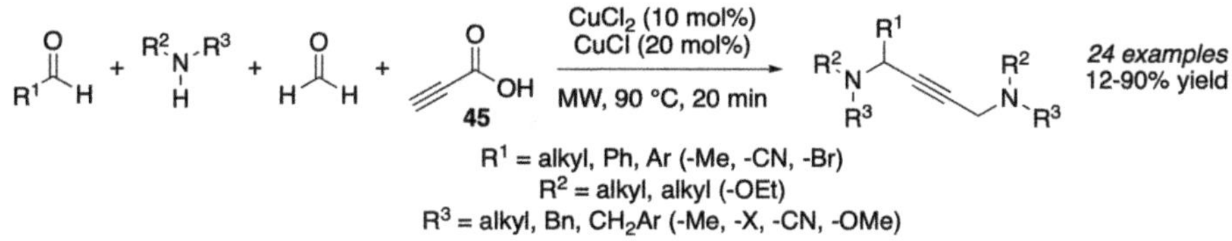

Scheme 15.22 Decarboxylative A^3-type multicomponent coupling for the synthesis of 1,4-diamino-2-butynes.

obtained under 'semi-neat' conditions: the quantity of water from the formaldehyde solution is sufficient. Some substrates gave unexpected products; cysteine and tryptophan underwent nucleophilic attack of the side-chain to the reactive intermediate (rather than addition of a second nucleophilic acetylide species), giving thiazolidine (**42**) and N-heterocyclic products (**43**), respectively, while lysine's side chain took part in a double A^3 coupling sequence (**44**).

In 2017, Tong and co-workers developed a multicomponent coupling reaction, described by the authors as a decarboxylative A^3-type coupling under copper(I)/copper(II) catalysis (Scheme 15.22). Four substrates were selectively combined to form unsymmetrical 1,4-diamino-2-butynes.[125] The reaction was designed using propiolic acid (**45**), which acts as an alkyne surrogate under copper catalysis. Over the course of the reaction, both of the acetylenic sites are functionalised: one reacts with an aminal formed *in situ* from the primary amine and formaldehyde, and the second adds to the iminium-ion condensation product of the amine and aldehyde (conventional A^3-type reactivity).

15.3.5 Tandem Reactions of Terminal Alkynes with Imine and Iminium Ions

Sometimes referred to as domino, cascade or one-pot reactions, tandem reactions combine a number of sequential reactions in a single reaction vessel, and are designed to maximise synthetic efficiency by minimising the number of reaction steps.[126] When other synthetic iterations are combined with A^3 coupling, already a highly efficient multicomponent reaction, a remarkable level of molecular complexity can be constructed in a single operation.

Yan and Liu showed in 2007 that a cationic gold(III) salt catalyses the A^3 coupling of pyridine-2-carboxaldehyde (**46**), secondary amines, and terminal alkynes, followed by cyclisation of the resulting propargylic pyridines (**47**, Scheme 15.23).[127] The final products are fused heterocycles: 1,3-disubsituted indolizines. The optimised conditions were amenable to a variety of substrates, including aromatic and aliphatic amines, terminal alkynes, and a selection of optically pure α-amino acid esters.

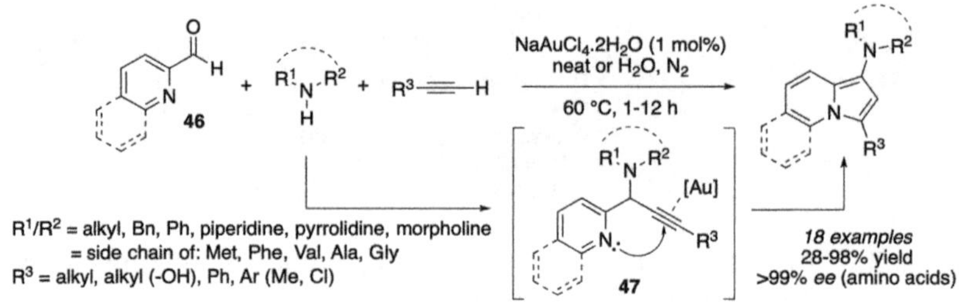

Scheme 15.23 Gold-catalysed tandem synthesis of 1,3-disubstituted indolizines.

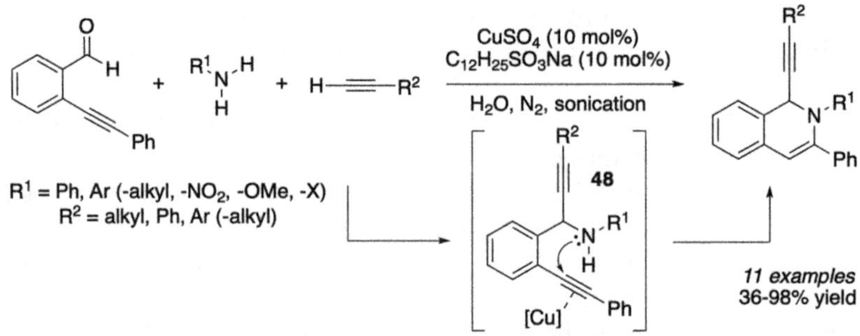

Scheme 15.24 Copper-catalysed tandem synthesis of 1,2-dihydroisoquinones.

Moderate to high yields were attained and no loss in enantiomeric purity was observed for the relevant examples. It should be noted that, while the reaction can be conducted in water, neat conditions gave superior yields in most cases.

In 2008, Wu and co-workers carried out a tandem reaction to prepare a library of 1,2-dihydroisoquinoline compounds (Scheme 15.24).[128] Primary amines, terminal alkynes and 2-alkynylbenzaldehydes were transformed *via* a copper-catalysed A^3 coupling reaction into propargylamines (**48**), which underwent intramolecular cyclisation. The researchers emulsified the reaction media with a Lewis acid-surfactant combined catalyst, which contained a metal cation and a dodecylsulphonate anion. Ultrasonic irradiation was critical to achieve high reaction yields, and aromatic primary amines and terminal alkyne substrates provided the best yields.

15.3.6 Mechanistic Discussion of the A^3 Coupling Reaction

The generally accepted mechanism for A^3 coupling begins with C–H activation of the terminal alkyne (Scheme 15.25a) to form a metal acetylide (**49**), a nucleophilic species.[71e] The metal acetylide adds to an electrophilic imine or iminium ion (Scheme 15.25b), formed *in situ* by the condensation of an aldehyde with a primary or secondary amine, respectively. This returns the propargylamine product and regenerates the metal catalyst.

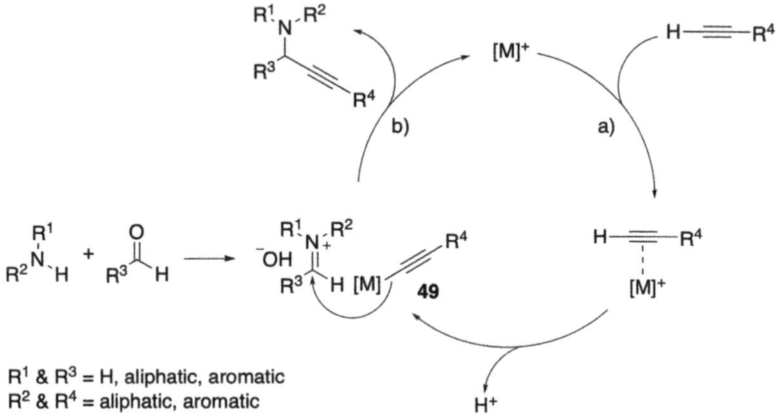

Scheme 15.25 Proposed catalytic cycle for the A^3 coupling reaction.

15.3.7 Synthetic Utility of Propargylic Amines

The propargylamine synthon, prepared by the multicomponent coupling of alkynes, amines and aldehydes, is a versatile functional group found in the core structures of organic compounds with diverse utility. Of note are the structurally related pharmaceuticals rasagiline and selegiline, which are used to treat both Parkinson's disease and depression by inhibiting the breakdown of L-DOPA. Propargylamines can also take part in further synthetic operations that profit from the presence of both an alkyne (electrophilic or nucleophilic depending on the reaction conditions) and a nucleophilic amine present in the same molecule. Propargylamines are particularly suited to the synthesis of nitrogen heterocycles[129] such as pyrroles,[130] pyrrolines,[131] pyrrolidines,[132] indole derivatives,[133] tetrahydropiperidines,[134] quinolines,[135] polycyclic quinolines,[136] oxazolidinones,[137] and oxazoles.[138]

Cacchi and Fabrizi showed that *N*-propargylic β-enaminones are practical starting materials to prepare either pyrroles (**50**) or pyridines (**51**) depending on the reaction conditions (Scheme 15.26).[139] The route to both products proceeds by nucleophilic addition of the carbonyl's α-carbon to the alkyne. Under basic conditions, an anion is formed that attacks the proximal carbon of the triple bond, closing a five-membered ring and forming a library of substituted pyrroles (**50**) upon rearrangement. Under copper catalysed conditions, the carbon attacks the farthermost carbon of the triple bond, closing a six-membered ring and returning pyridine final products (**51**) after oxidation.

Van der Eycken and co-workers developed a one-pot synthesis of thiazolidine-2-thiones commencing with initial A^3 coupling using primary amines, aliphatic aldehydes, and aromatic alkynes, followed by incorporation of carbon disulphide (CS_2, Scheme 15.27).[140] This strategy is reminiscent of prior syntheses of oxazolidinones,[141] replacing CO_2 with CS_2. The authors

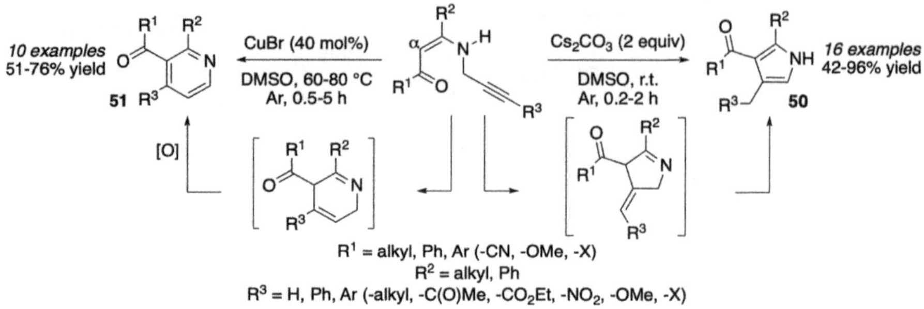

Scheme 15.26 Divergent synthesis of pyrroles and pyridines from *N*-propargylic β-enaminones.

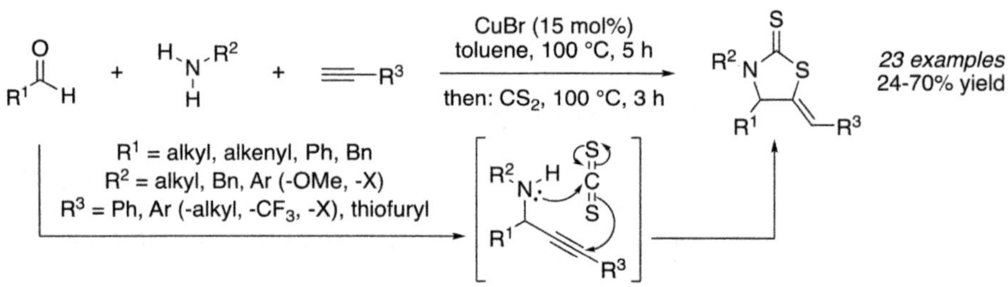

Scheme 15.27 Copper-catalysed one-pot synthesis of thiazolidine-2-thiones.

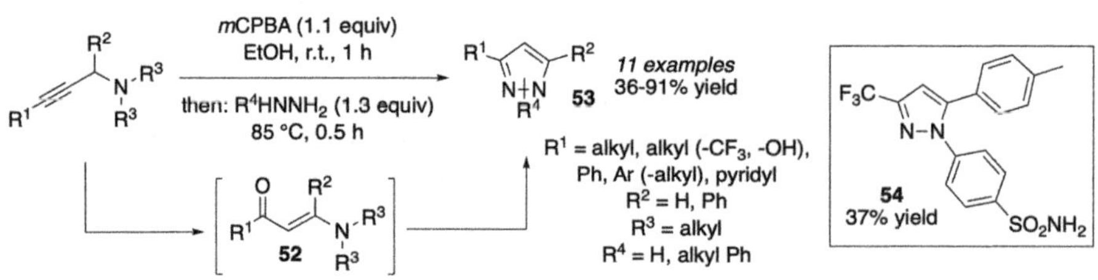

Scheme 15.28 Copper-catalysed one-pot synthesis of pyrazoles.

optimised the reaction conditions and found that, under copper catalysis with excess CS_2, moderate to good yields of the desired products could be obtained.

Chen and co-workers at the Pfizer research laboratories published a metal-free, one-pot synthesis of pyrazoles *via* enaminone intermediates (Scheme 15.28).[142] The propargylamine starting materials were first oxidised at room temperature, generating an unstable *N*-oxide. The reaction temperature was then increased to 85 °C, which promoted a thermal rearrangement and enaminone formation (**52**). If these are the desired products, they may be isolated and used, but the authors further showed that addition of a hydrazine during the thermal step resulted in the formation of aliphatic or aromatic pyrazoles

(**53**) in moderate to good yield. Attesting to its utility, this transformation was used as the key step to prepare Celecoxib (**54**), an anti-inflammatory drug developed and marketed by Pfizer.

15.4 Addition of Terminal Alkynes to Conjugated Electrophiles (Class III)

The conjugate (Michael) addition is the reaction of nucleophiles with olefins, rendered electrophilic by conjugation to electron-withdrawing functional groups (such as carbonyl functional groups). This reaction offers an efficient method for the construction of carbon–carbon bonds and offers synthetic flexibility to the suite of direct addition reactions by enabling functionalisation at the γ-position with preservation of the electrophilic carbonyl.

The first report of a transition metal-catalysed conjugate addition of terminal alkynes was published in 1990 by Nikishin and Kovalev. An electron-rich rhodium catalyst [$RhCl(PMe_3)_3$] was used to effect the conjugate addition of non-functionalised phenyl- and aliphatic alkynes to alkyl vinyl ketones in acetone at room temperature. Although reaction times were long (48–108 h), yields were modest to good (20–91%) and the ability to conduct the reaction at ambient temperatures was a significant development.[143] Nearly a decade later, Dixneuf published a ruthenium catalysed variant.[144] The ruthenium complex [$Ru(O_2CH)(CO)_2(PR_3)$]$_2$ (R = Me, Ph) catalysed the conjugate addition of an assortment of carbonucleophiles to methyl vinyl ketone in acetonitrile at 100 °C. Substrates included classical coupling partners (β-diketones, malonates, β-keto-, nitro- and cyano-esters) and a small scope of aryl-, vinyl-, and silyl-substituted alkynes (46–74% product yield) and aliphatic alkynes (20–29% product yield). These seminal publications have been supplemented by many examples in organic solvents,[145] and a discussion of relevant examples in water follows.

15.4.1 Addition of Terminal Alkynes to α,β-Enones

In 2004, Chen and Li reported the first palladium-catalysed conjugate addition of terminal alkynes to α,β-conjugated enones, with the added benefit that it could be performed in water under air (Scheme 15.29).[146]

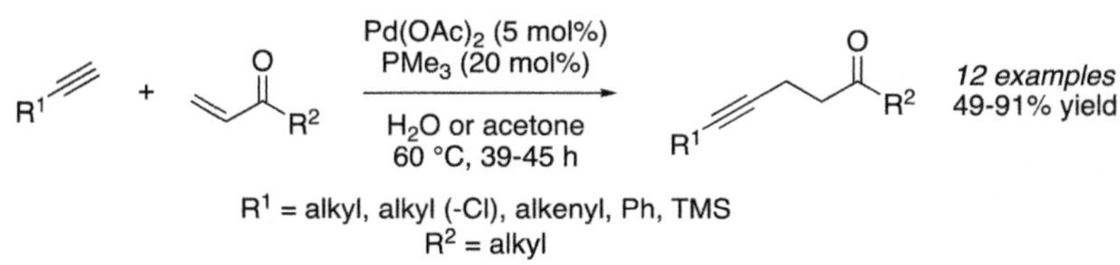

Scheme 15.29 The first palladium-catalysed conjugate addition of terminal alkynes to α,β-enones.

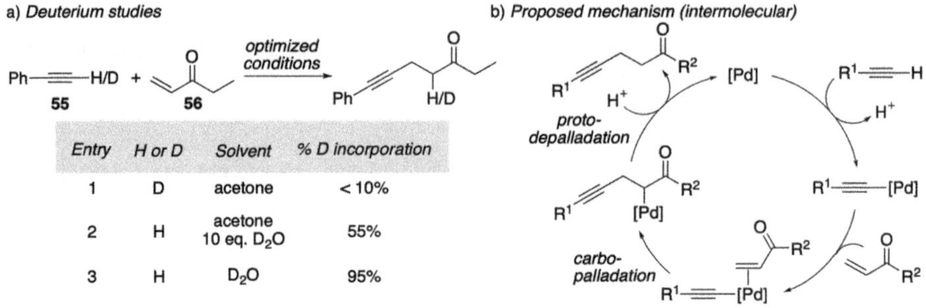

Entry	H or D	Solvent	% D incorporation
1	D	acetone	< 10%
2	H	acetone 10 eq. D_2O	55%
3	H	D_2O	95%

Scheme 15.30 Deuterium-labelling experiments (a) and proposed mechanism (b) for the conjugate addition of terminal alkynes to enones.

The authors attributed the lack of precedent reactions to the stability of the palladium–alkyne bond, and the proclivity of alkynes to participate in palladium-catalysed di- and trimerisation reaction.‖[147] During early attempts to accomplish the reaction, phenylacetylene and ethyl vinyl ketone were combined in water with a catalytic system comprised of palladium(II) acetate and triphenylphosphine (PPh_3), but only trace yields of the 4-alkynone were obtained. Applying wisdom from earlier studies by Trost, who used electron-rich ligands to increase the reactivity of σ-alkynyl palladium intermediates,[148] the authors replaced PPh_3 with the electron-rich ligand trimethylphosphine (PMe_3). This increased the yield to 77%, and doubling the ligand quantity had an additional benefit (91% yield). In addition to phenylacetylene, a variety of aliphatic, alkenic, and silyl-substituted alkynes gave moderate to good yields of the enone products. Both water and acetone could be used as the solvent, but in the majority of instances acetone was superior.

The reaction mechanism was studied by conducting deuterium-labelling experiments. Interestingly, the combination of deuterium-labelled phenylacetylene (**55**) and ethyl vinyl ketone (**56**) under the optimised conditions in dry acetone gave minimal deuterium-incorporation at the α-position of the product (Scheme 15.30a, entry 1). In contrast, the reaction of non-labelled phenylacetylene in a solvent with varying quantities of deuterium oxide gave high percentages of deuterium incorporation (Scheme 15.30a, entries 2 and 3), implying that the α-hydrogen derives from an external proton source. These experiments provided evidence against the intramolecular reaction, where the α-hydrogen is installed following reductive elimination of the palladium-bound addition product with a hydride ligand, originally installed by oxidative addition of the terminal alkyne. Considering the deuterium-labelling data, a mechanism was proposed beginning with η^2-alkyne coordination and followed by deprotonation to

‖This reaction has been developed as a synthetically useful method for the preparation of enynes and benzene derivatives.

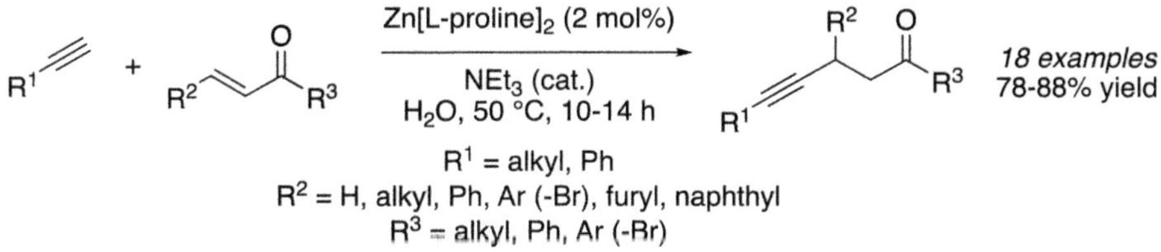

Scheme 15.31 Zinc-catalysed conjugate addition of terminal alkynes to enones.

generate an alkynyl-palladium species. Coordination of the enone, followed by carbopalladation and protodepalladation from an external proton source, returns the product (Scheme 15.30b).

As a follow up to this work, the authors modified the reaction conditions to extend the reaction scope to the comparatively less-electrophilic acrylate esters. This was possible using even more electron-rich ligands: NHCs. Good yields of the desired products were possible if electron-rich alkynes were used (41–85% yield), but the reaction was not possible in water and could only be conducted in acetone.[49]

In 2011, Bhardwaj and co-workers reported that very low loadings of a zinc-proline catalyst could be used to realise the conjugate addition of phenylacetylene and ethynylcyclohexane to a variety of α,β-unsaturated ketones bearing aromatic and aliphatic substituents in high yields (Scheme 15.31).[149] Of particular note is the ease with which sterically hindered β-substituted substrates are purported to undergo addition.

15.4.1.1 *Synthetic Utility of α,β-Enones*

Like many of the substrates discussed in this review, the bifunctional nature of 4-alkynones makes them convenient substrates for the preparation of many diverse functional groups such as α,β-cyclic enones *via* gold-catalysis,[150] aryl-substituted 3-cyclohexenols *via* rhodium catalysis,[151] carbazoles *via* gold copper tandem catalysis,[152] and diketones *via* base- and transition metal-catalysed alkyne hydration.[144,153] 4-Alkynones are also common synthetic intermediates *en route* to functional molecules, such as in the industrial scalable synthesis of a Pfizer pharmaceutical candidate to treat osteoarthritis, asthma and neuropathic pain,[154] and in the four-step total synthesis of the natural product (+)-lauthisan.[155]

The value of 4-alkynones for the synthesis of oxygen heterocycles is of particular note. In such syntheses, the bifunctional nature of the substrates is used to its full potential: cyclisation occurs following nucleophilic attack by the ketone to the triple bond, rendered electrophilic by coordination to a transition metal. This strategy was used by Iwasawa and co-workers, who investigated the reactivity of 4-alkynones under platinum catalysis, for the synthesis of 8-oxabicyclo[3.2.1]octanes (**56**, Scheme 15.32a). This transformation is instigated by coordination of platinum to the internal alkyne of the substrate, activating the alkyne towards

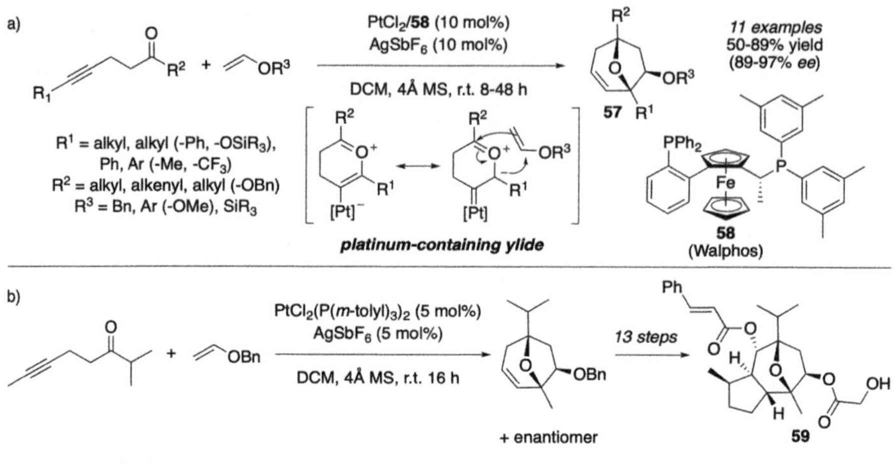

Scheme 15.32 General scheme and proposed mechanism for the platinum-catalysed preparation of 8-oxabicyclo[3.2.1]octanes from α,β-enones (a) and application of this method to the synthesis of the natural product englerin A (**59**, b).

nucleophilic attack by the ketone in a 6-*endo* fashion. The resulting species, a cyclic ylide bound to platinum, is a 1,3-dipole that reacts with electron-rich alkenes *via* a [3 + 2] cycloaddition reaction. A sequence of hydrogen and alkyl migrations concludes with regeneration of the catalyst and return of the product.

In their seminal work, the authors reported that platinum chloride alone could catalyse the reaction at room temperature, but long reaction times (up to five days) and the opportunity to investigate an asymmetric variant inspired further study.[156] The authors found that cationic platinum(II)-phosphine complexes were efficient catalysts, and could be synthesised *in situ* from the dichloro complex *via* chloride abstraction by silver. Furthermore, the use of a chiral Walphos ligand (**58**) accomplished an asymmetric reaction, returning products in moderate to high yield with high enantiomeric excess within two days at room temperature (Scheme 15.32a).[157] The level of molecular complexity possible in one reaction step is an impressive feat. Further, the 8-oxabicyclo[3.2.1]octane skeleton **57** appears in a number of natural products, such as cortistatins[158] and englerin A (**59**), both of which have demonstrated anti-tumour activity. The authors developed a 14-step total synthesis of **59** using platinum-catalysed 1,3-dipolar cycloaddition as the first step, obtaining 16% overall yield of the final product (Scheme 15.32b).[159]

Belting and Krause have further extended the scope of heterocycles accessible from 4-alkynones using a cationic gold catalyst to transform these substrates into tetrahydrofuranyl ethers (**60**), 4*H*-pyrans (**61**) or furans (**62**, Scheme 15.33).[160] The mechanism for all cases involves nucleophilic attack through the ketone to the gold-activated alkyne, but the exact product obtained depended on the substrate substitution pattern and the reaction solvent. In alcohol, the tetrahydrofuranyl ether (**60**) was obtained regardless

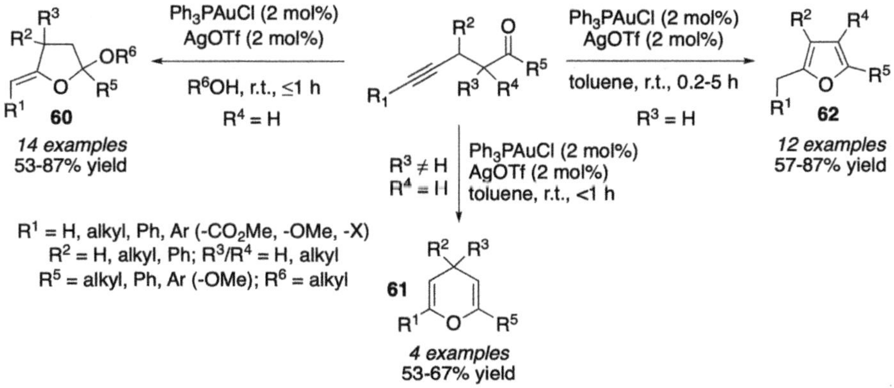

Scheme 15.33 Preparation of tetrahydrofuranyl ethers 4*H*-pyrans and furans from α,β-enones under gold-catalysed conditions.

of the substitution pattern. When toluene was used, substrates doubly substituted at the β-carbon with respect to the carbonyl gave 4*H*-pyrans (**61**), while other substitution patterns yielded the furan (**62**).

15.4.2 Addition of Terminal Alkynes to Meldrum's Acid Diesters

In 2003, Carreira's laboratory published the first copper-catalysed conjugate addition of terminal alkynes (Table 15.8a).[161] This was a notable addition to the field, turning conventional wisdom on its head. Previously, copper-acetylides had been judged unreactive, so much so that they only appeared in conjugate addition chemistry as 'dummy ligands', namely non-transferable organocuprate ligands. Carreira found that copper-acetylides synthesised under aqueous conditions were sufficiently nucleophilic to add to Meldrum's acid-derived electrophiles (**63**). Water was discovered to be a crucial component, and any diminished water content due to the addition of organic solvents caused the yield to drop. Under the optimised conditions, sodium ascorbate was used to reduce copper(II) to copper(I) for the *in situ* preparation of copper-acetylides, which add to Meldrum's acid-derived electrophiles with aliphatic, aryl, and vinyl-substituents in a water/*tert*-butanol solvent system at room temperature. The benefits of using Meldrum's acid derived electrophiles are their easy preparation *via* the condensation of aldehydes with Meldrum's acid, and the fact that their conjugate addition products can be thermally converted to β-alkynal acids.

In 2007, Carreira and co-workers extended the substrate scope of this reaction to propiolates and barbituric acid derivatives (**64**, Table 15.8b).[162] It can be considered that barbituric acid derivatives (**64**) contain a diimide-activated alkene that acts as an electrophilic acceptor in the same way that Meldrum's acid derivatives (**63**) contain a diester-activated alkene. The use of propiolate nucleophiles (Table 15.8b, $R^2 = CO_2R$) is challenging with regards

Table 15.8 Addition of terminal alkynes to Meldrum's acid and barbituric acid derivatives.

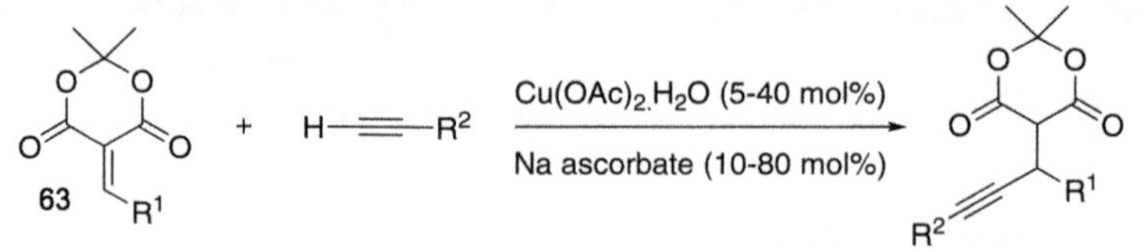

Entry	Conditions	Scope	Yield (diastereo-selectivity)
a) Carreira *et al.*, 2003	H_2O/*t*-BuOH (10 : 1) r.t., 3–42 h	R^1 = alkyl, Ph, Ar (–OMe, –NO_2, –Cl), furyl R^2 = Ph, Ar (–Br), furyl	10 examples 55–98% yield
b) Carreira *et al.*, 2007	H_2O, 0 °C–r.t., 2–48 h	R^1 = alkyl, alkyl (–OR), acetal R^2 = CO_2Et	12 examples 54–82% yield (>19 : 1 dr)

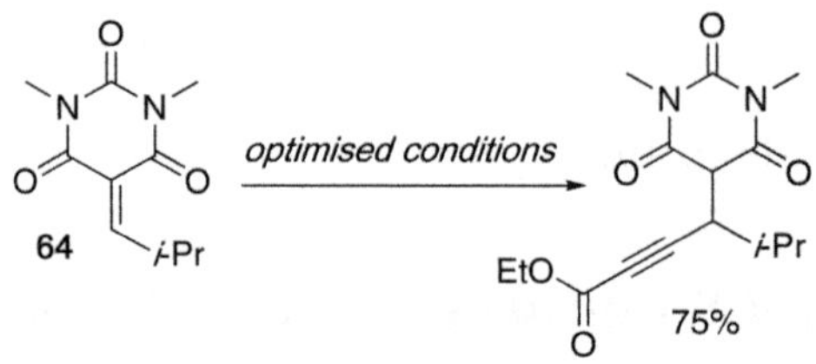

to chemoselectivity, as there are a number of different electrophiles present in the reaction mixture. In addition to the desired addition to the Meldrum's or barbituric acid-derived electrophiles, the propiolate can add to another equivalent of propiolate substrate or product alkynoate, as ester conjugated alkynes are themselves excellent acceptors for conjugate additions, as covered in Section 15.4.3. Gratifyingly, with the right conditions, the reaction was completely selective. A combination of a high loading of copper acetate and sodium ascorbate in water transformed alkyl substituted electrophiles in moderate to good yield. Electrophilic substrates with a γ-oxygen substituent (1,2- and 1,3-dioxolanes, and chiral ethers) reacted rapidly, and most reached completion at 0 °C in short times. Asymmetric induction was observed when chiral substrates were used, resulting in preferential formation of the *syn* diastereomer.

In 2005, the same group achieved the first transition metal-catalysed asymmetric conjugate addition of terminal alkynes (Table 15.9a).[163] Focussing on the copper-catalysed addition of Meldrum's acid derivatives (**63**) previously developed, the authors investigated the effect of a variety of chiral ligands on the yield and enantioselectivity of the reaction. The best results were obtained using a new class of ligand similar in structure to QUINAP (**65**), developed in-house and termed 'PINAP' (**66**). Both **66** and **65** exhibit atropisomerism: the biaryl rings sterically constrain

Table 15.9 Asymmetric addition of terminal alkynes to Meldrum's acid derivatives.

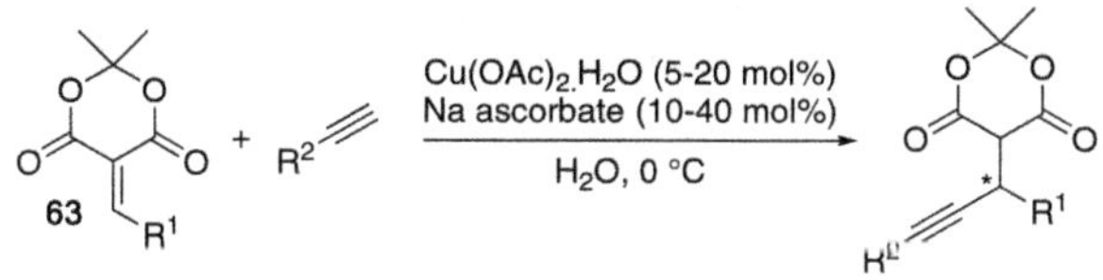

Entry	Conditions	Scope	Yield (enantio-selectivity)
a) Carreira *et al.*, 2005	(*R,M*)-PINAP (5–20 mol%), 14–66 h	R^1 = alkyl, Ph, Ar (–Me) R^2 = Ph	7 examples 64–94% yield (82–94% ee)
b) Carreira *et al.*, 2012	(*R,M*)/(*R,P*)-PINAP 60 : 40 (100 mol%), 14 h	R^1 = alkyl R^2 = Ph	80% yield (94% ee)
c) Aponick *et al.*, 2017	(*S*)-Me-Stackphos (5.5 mol%), 24 h	R^1 = Ph, Ar (–OMe, –Cl), naphthyl, furyl R^2 = alkyl (–OR, –NR_2, –Br, –Ph), alkenyl, Ph, Ar (–Me, –F), TMS	22 examples 45–92% yield (76–98% ee)

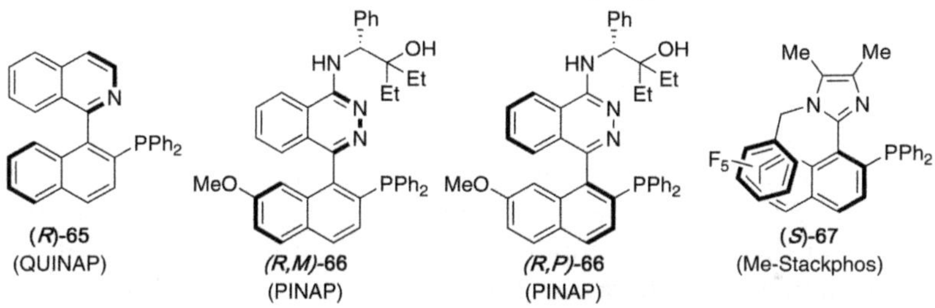

rotation of the connecting carbon–carbon bond, resulting in axial chirality. In the synthesis of **66**, obtaining an optically pure sample is achieved through diastereomeric resolution, due to an additional fixed stereogenic centre in the structure. This is comparably easier than the difficult enantiomeric resolution required for **65**.[164] Studies of matched/mismatched effects using PINAP in the conjugate addition reaction revealed that both the stereogenic axis and stereogenic centre influence the enantioselectivity. The authors applied a 1 : 1 ratio of copper/ligand in water at 0 °C to a range of Meldrum's acid acceptors (**63**), and the best results were obtained for substrates with γ-branched alkyl substituents such as cyclohexyl or isopropyl (79–94% yield, 94–95% ee). Other acceptors (R = aliphatic, aryl) gave slightly worse results (64–87%, 82–90% ee). Further investigations showed that an optically impure mixture of PINAP diastereomers (60 : 40 *R,M*-**66**/*R,P*-**66**) could be used to obtain the conjugate addition product in up to 94% ee (Table 15.9b).[165] This is an example of a non-linear effect, whereby a catalyst with low enantiomeric purity can be used to synthesise reaction products with a high level of enantiopurity. In fact, when a highly pure PINAP ligand (**66**) was used for the same reaction, the enantiomeric excess was barely improved to 95% ee. This has practical implications, as the authors found that enriched ligand mixtures with different diastereomeric ratios could be synthesised and used without resolution to achieve high enantiomeric excesses in the conjugate addition reaction. The authors hypothesised that the basis for such non-linear effect would be the formation of thermodynamically stable and catalytically inactive aggregates in the reaction mixture, with the formula [(Cu$_2$(*R*,*M*-**66**)(*R*,*P*-**66**)$_n$)]. These may serve as a reservoir, enriching the diastereomeric purity of the remaining mixture.

In 2017, Aponick demonstrated that another atropisomeric ligand, (*S*)-Me-Stackphos (**67**), is a valuable and high-performing ligand for the asymmetric alkynylation of Meldrum's acid derivatives (**63**, Table 15.9c).[166] The copper/**67** complex gave moderate to excellent yields and high enantioselectivities of conjugate addition products, synthesised from a diverse array of alkyl, aryl, alkenyl, and silyl-substituted substrates, including Meldrum's acid alkylidenes lacking γ-branching. This is noteworthy as these performed poorly under the copper/PINAP conditions developed by Carreira and coworkers (Table 15.9, entries a and b).

15.4.2.1 Synthetic Utility of α,β-Enoates

Compared to their aldehyde and ketone analogues, 4-alkynoates do not commonly partake in reactions where the carbonyl and aldehyde react in a single, connected reaction. Nevertheless, 4-alkynoates are convenient synthetic intermediates and the two functional groups can be individually modified. The ester is often used as a surrogate for the corresponding carboxylic acid, and the alkyne can be used as a handle for further functionalisation. For example, in the synthesis of the cytotoxic natural product (−)-amphidinolide V (**68**, Scheme 15.34), the 4-alkynoate-derived ester is deprotonated at a key step, forming an acid which cyclises to a lactone macrocycle. On the other hand, the alkyne is converted by enyne cross-metathesis to a [2]dendralene

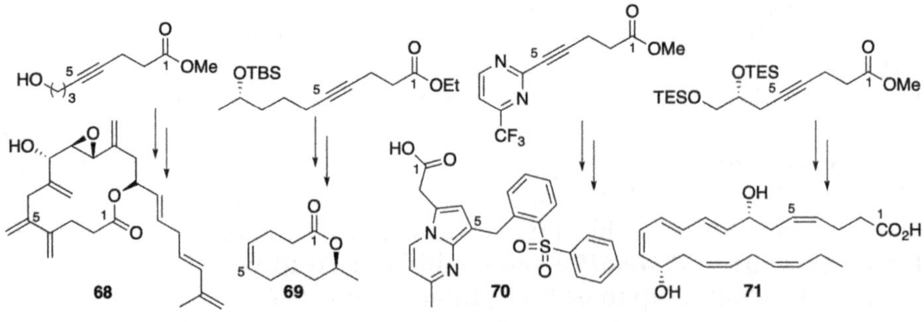

Scheme 15.34 Synthesis of a variety of biologically relevant compounds from α,β-enoates.

motif.[167] A similar strategy was used for the synthesis of phoracantholide J (**69**, Scheme 15.34),[168] where deprotection of the ester to the acid induces ring closure to form the lactone and the alkyne is stereospecifically converted into a *cis*-alkene. In the synthesis of pyrrolopyrimidine compounds (**70**, Scheme 15.34), 4-alkynoates were used as a protected form of the acid present in the final product, while the alkyne participates in a copper-catalysed cyclisation with the pyrimidine.[169] In maresin 1 (**71**, Scheme 15.34), a carboxylic acid is also present in the final product, and deprotection of the ester occurs in the final step.[170] This stereospecific conversion of the alkyne to a *cis*-alkene is a crucial step in faithfully replicating the stereochemistry in this alkene-rich compound.

15.4.3 Addition of Terminal Alkynes to α,β-Ynoates

The variety of publications exploring the addition of terminal alkynes to activated alkene electrophiles in water contrasts with the scarce reports using activated alkyne electrophiles. Reactions of this type were first reported by Trost in 1987, with the homodimerisation and cross-coupling of terminal alkynes using palladium acetate and an electron-rich triarylphosphine in benzene.[147a] In 2004, Chen and Li showed that the cross-coupling of terminal alkynes with activated internal alkynes, generating conjugated enynes, could be catalysed by a copper/palladium bimetallic system in water (Scheme 15.35).[171] The success of the bicatalytic system was hypothesised to lie in the ability of copper to activate the C–H bond of the terminal alkyne, and the ability of palladium to catalyse the carbon–carbon bond-forming step. The authors showed that internal alkynes conjugated to electron-withdrawing ketones and esters made convenient substrates. Regarding the scope of terminal alkynes, phenylacetylene could be used under the mildest conditions, and increasing the temperature enabled the addition of alkenyl-, aliphatic-, and silylated alkynes in moderate to good yield.

15.4.4 Addition of Terminal Alkynes to α,β-Enals

The conjugate addition of terminal alkynes to unsaturated aldehydes in water remains relatively unexplored. Worth noting is the work by Jana and Tunge, who used a recyclable rhodium-phosphite catalyst to effect the conjugateaddition of a small number of substrates in a toluene–water mixture.[172]

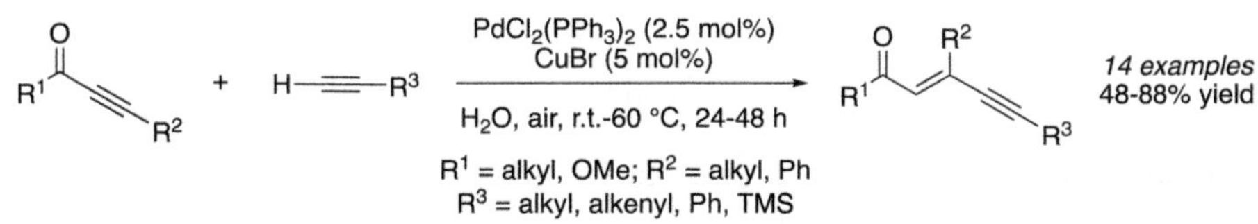

Scheme 15.35 Palladium- and copper-catalysed conjugate addition of terminal alkynes to ynoates.

In 2015, Wang and co-workers reported a general method to effect the conjugate addition of terminal alkynes to acrolein (Scheme 15.36).[173] The authors found that pre-formation of the palladium/phosphine catalyst at temperatures above 110 °C was key to achieve high yields and reducing the incidence of acrolein polymerisation, which can occur spontaneously in the presence of phosphines. A diverse range of terminal alkynes were transformed to 4-alkynals, with both electron-rich and poor, aromatic and aliphatic groups. Moderate to good yields of the addition products of aromatic alkynes and acrolein (42–67% yield) were achieved in water, despite aliphatic substrates performing poorly (5–24% yield). For the cases where water fell short, acetone could be used to obtain satisfactory yields.

The proposed mechanism, supported by DFT calculations and isotope-labelling experiments, is hypothesised to involve four steps. Firstly, coordination of the alkyne to the palladium centre forms a π-complex and, secondly, acetate-assisted deprotonation results in the formation of a palladium acetylide (Scheme 15.37a). Thirdly, coordination of acrolein to the metal centre (Scheme 15.37b) is followed by migratory insertion of the alkynyl group (Scheme 15.37c) and, finally, protonation extrudes the product and regenerates the catalyst (Scheme 15.37d).

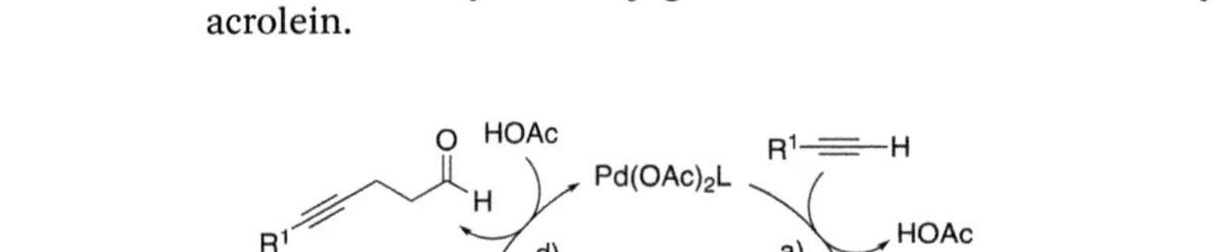

Scheme 15.36 Palladium-catalysed conjugate addition of terminal alkynes to acrolein.

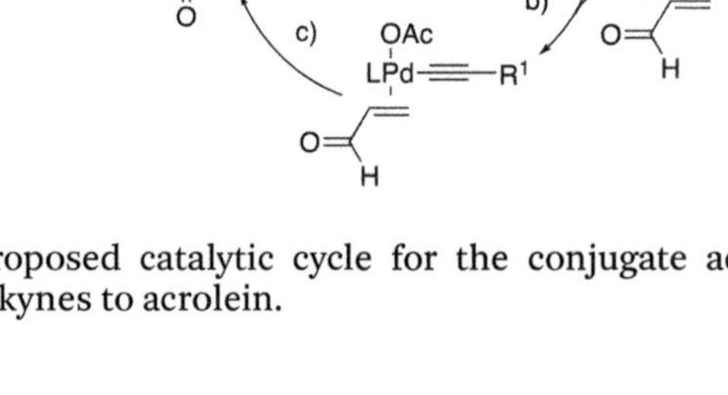

Scheme 15.37 Proposed catalytic cycle for the conjugate addition of terminal alkynes to acrolein.

15.4.4.1 *Synthetic Utility of α,β-Enals*

4-Alkynals are practical and versatile precursors to a number of functional groups such as cyclic alkenyl ethers under palladium-catalysed conditions,[174] and methylene pyrrolidines *via* cyclisation of the corresponding Mannich adducts.[175] 4-Alkynals feature in the published routes to a number of natural products, such as (+)-8-*epi*-xanthatin,[176] echinopine A and B,[177] leinomycin analogues,[178] and pestalotioprolide C.[179]

Tanaka has published extensively on reactions of 4-alkynals to prepare cyclic products using cationic rhodium catalysts. The earliest report was the hydroacylation of 4-alkynals to synthesise cyclopentenones (**72**, Scheme 15.38).[180] The reaction mechanism is proposed to begin with oxidative addition of the aldehydic carbon–hydrogen bond to the rhodium(I) centre. The resulting rhodium(III)-hydride adds across the internal alkyne to generate a rhodacycle (**73**). Two intermediates are possible: the six-membered rhodacycle (**73**), which returns cyclopentenones (**72**) upon reductive elimination, or the five-membered rhodacycle (**74**). The authors discovered that the five-membered rhodacycle could be intercepted by alcohol nucleophiles to synthesise *cis*-4-alkenoates (**75**),[181] electron-poor alkenes to prepare cyclohexanones (**76**),[182] alkynes to prepare cyclohexenones (**77**),[183] isocyanates to prepare piperidinediones (**78**),[184] and carbodiimides to prepare imine-substituted piperidones (**79**).[185]

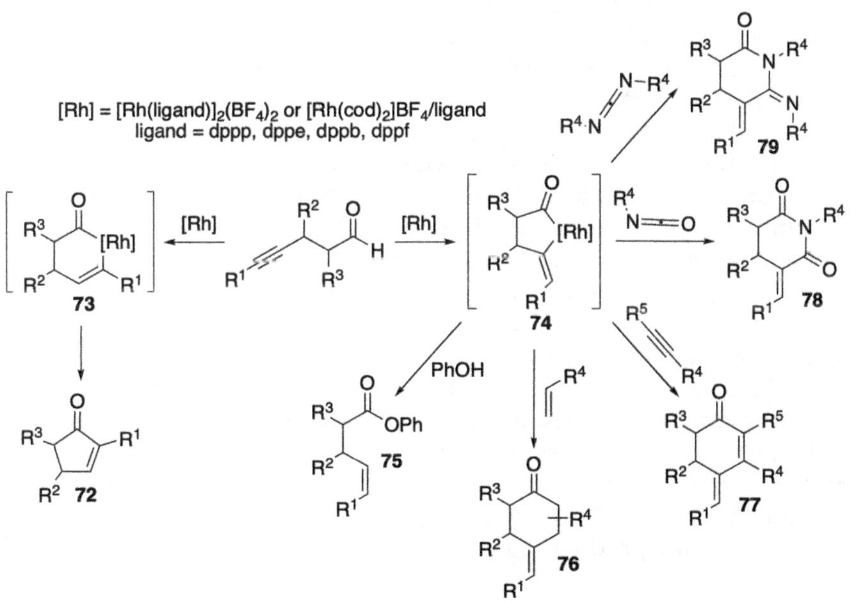

Scheme 15.38 Rhodium-catalysed transformations of α,β-enals.

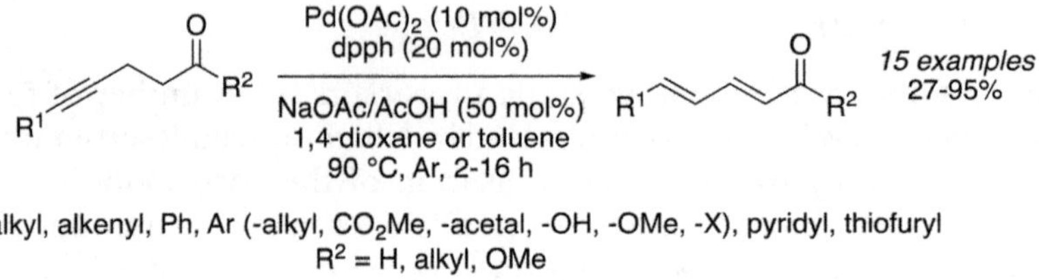

Scheme 15.39 Palladium-catalysed isomerisation of α,β-enals for the preparation of dienals.

In 2017, Hearne and Li used 4-alkynals to prepare (*E,E*)-dienals *via* the palladium-catalysed isomerisation of the internal alkyne (Scheme 15.39).[186] Aryl- and alkenyl-substituted substrates gave high product yields with high diastereoselectivities. Aliphatic substrates performed more poorly due to a lack of electronic bias; the alkyne can isomerise in either direction, giving multiple products and lower yields. The internal alkyne in 4-alkynone and 4-alkynoate substrates also isomerised under the same conditions to give high yields of dienones and dienoates, respectively.

15.4.5 Tandem Reactions of Terminal Alkynes with Conjugated Electrophiles

Jiao and co-workers adapted Carreira's procedure for the conjugate addition of terminal alkynes to Meldrum's acid derivatives to the one-pot preparation of γ-alkylidene lactones (**80**) using two inexpensive and abundant metal catalysts, copper and iron (Scheme 15.40).[187] This molecular scaffold is found in a variety of natural products,[188] and is commonly prepared *via* the intramolecular hydroacyloxylation of 4-alkynoic acids in several steps.[189] Under the optimised conditions, aryl-substituted alkynes could be combined with Meldrum's acid derivatives substituted with alkyl, aryl and heteroaryl groups in low to good yield. In this report, the authors performed a copper-catalysed conjugate addition (Scheme 15.40a), followed by addition of iron(III) chloride to induce a cascade of steps. The author's proposed mechanism** proceeds by dual activation of the conjugate addition product by copper and iron. The hard Lewis acid iron activates the oxygen atom of the intermediate, while the soft Lewis acid copper activates the alkynic π system. Cyclisation (Scheme 15.40b), followed by hydrolysis, decarboxylation and extrusion of acetone (Scheme 15.40c), generates a copper alkylidene (**81**). In the final step, protodecupration (Scheme 15.40d) returns the lactone product (**80**).

**Please note that some components of the catalytic cycle have been edited for this review to preserve atom and charge balance.

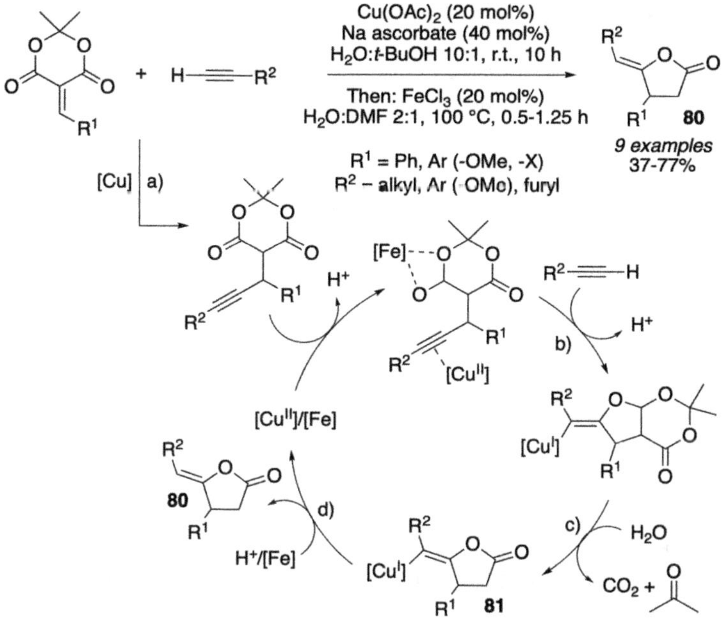

Scheme 15.40 One-pot synthesis of γ-alkylidene lactones under iron- and copper-catalysed conditions.

15.5 Conclusions and Outlook

In this chapter, we have shown the breadth of transition metal-catalysed methods in water for the addition of terminal alkynes to a variety of electrophiles, providing an efficient strategy for the preparation of alkynylated molecules. The chemical space traversed is expansive, and is further extended when considering the synthetic utility of the alkynylated products. It is our hope that this chapter will encourage chemists to consider the nucleophilic addition of terminal alkynes as an accessible method to prepare practical intermediates or synthetic targets. With regards to outlook, the exploration of transition metal-catalysed alkyne addition reactions is imbalanced; the majority of methods pursue the synthesis of propargylamines *via* A^3 coupling strategies. However, reflecting on how A^3 coupling has developed over time is instructive when considering the potential of other addition reactions. A^3 coupling methods have increasingly turned to heterogeneous catalysts prepared from abundant transition metals that can be recovered and recycled, provide broad substrate scopes and operate under mild conditions. We anticipate that the preparation of propargylic alcohols, alkynones and 4-alkynyl carbonyl derivatives will advance towards similar aims.

Acknowledgements

We acknowledge the hard work of many past and present members of Professor C.-J Li's laboratory, whose work is outlined in this review. We are also grateful to the Canada Research Chair (Tier 1) foundation (to C.-J. Li), the Fonds de Recherche du Québec Nature et Technologies, Natural Sciences and Engineering Research Council of Canada, CFI and L'Oréal R&I for their support of our research.

References

1. C.-J. Li and T. H. Chan, *Tetrahedron Lett.*, 1991, **32**, 7017.
2. (a) H. Mayr and M. Patz, *Angew. Chem., Int. Ed. Engl.*, 1994, **33**, 938; (b) H. Mayr and A. Ofial, Mayr's Database of Reactivity Parameters, http://www.cup.lmu.de/oc/mayr/reaktionsdatenbank/, accessed October 15, 2018; (c) R. Appel, S. Chelli, T. Tokuyasu, K. Troshin and H. Mayr, *J. Am. Chem. Soc.*, 2013, **135**, 6579.
3. D. H. Ripin and D. A. Evans, Evans pK_a Table, evans.rc.fas.harvard.edu/pdf/evans_pKa_table.pdf, accessed September 15, 2018.
4. A. Grirrane, H. Garcia, A. Corma and E. Álvarez, *ACS Catal.*, 2011, **1**, 1647.
5. A. Grirrane, H. Garcia and A. Corma, *Chem. - Eur. J.*, 2013, **19**, 12239.
6. F. Himo, T. Lovell, R. Hilgraf, V. V. Rostovtsev, L. Noodleman, K. B. Sharpless and V. V. Fokin, *J. Am. Chem. Soc.*, 2005, **127**, 210.
7. R. Bai, G. Zhang, H. Yi, Z. Huang, X. Qi, C. Liu, J. T. Miller, A. J. Kropf, E. E. Bunel, Y. Lan and A. Lei, *J. Am. Chem. Soc.*, 2014, **136**, 16760.
8. C. Liu, Y. Luo, W. Zhang, J. Qu and X. Lu, *Organometallics*, 2014, **33**, 2984.
9. X. Wang, Y. Song, J. Qu and Y. Luo, *Organometallics*, 2017, **36**, 1042.
10. B. B. Buckley, A. N. Khan and H. Heaney, *Chem. - Eur. J.*, 2012, **18**, 3855.
11. S. S. Y. Chui, M. F. Y. Ng and C.-M. Che, *Chem. - Eur. J.*, 2005, **11**, 1739.
12. C. Wei and C.-J. Li, *Green Chem.*, 2002, **4**, 39.
13. (a) G. S. Viswanathan and C.-J. Li, *Tetrahedron Lett.*, 2002, **43**, 1613; (b) G. S. Viswanathan, M. Wang and C.-J. Li, *Angew. Chem., Int. Ed.*, 2002, **114**, 2242.
14. X. Yao and C.-J. Li, *Org. Lett.*, 2005, **7**, 4395.
15. (a) R. H. Pouwer, C. M. Williams, A. L. Raine and J. B. Harper, *Org. Lett.*, 2005, **7**, 1323; (b) S. Dillinger, P. Bertus and P. Pale, *Org. Lett.*, 2001, **3**, 1661.
16. M. Yu, Y. Wang, W. Sun and X. Yao, *Adv. Synth. Catal.*, 2012, **354**, 71.
17. H. Wang, K.-F. Yang, L. Li, Y. Bai, Z.-J. Zheng, W.-Q. Zhang, Z.-W. Gao and L.-W. Xu, *ChemCatChem*, 2014, **6**, 580.

18. R. Motoki, M. Kanai and M. Shibasaki, *Org. Lett.*, 2007, **9**, 2997.
19. D. O'Hagan, *J. Fluorine Chem.*, 2010, **131**, 1071.
20. G.-J. Deng and C.-J. Li, *Synlett*, 2008, **10**, 1571.
21. P. Czerwiński, E. Molga, L. Cavallo, A. Poater and M. Michalak, *Chem. - Eur. J.*, 2016, **22**, 8089.
22. H. Clavier and S. P. Nolan, *Chem. Commun.*, 2010, **46**, 841.
23. F. Lazreg, M. Lesieur, A. J. Samson and C. S. J. Cazin, *ChemCatChem*, 2016, **8**, 209.
24. X.-P. Fu, L. Liu, D. Wang, Y.-J. Chen and C.-J. Li, *Green Chem.*, 2011, **13**, 549.
25. Z. Jia, X. Li, A. S. C. Chan and C.-J. Li, *Synlett*, 2012, **23**, 2758.
26. (a) D.-Y. Zhang and X.-P. Hu, *Tetrahedron Lett.*, 2015, **56**, 283; (b) E. B. Bauer, *Synthesis*, 2012, **44**, 1131.
27. (a) Y. Nishibayashi, Y. Inada, M. Yoshikawa, M. Hidai and S. Uemura, *Angew. Chem., Int. Ed.*, 2003, **42**, 1495; (b) Y. Inada, Y. Nishibayashi, M. Hidai and S. Uemura, *J. Am. Chem. Soc.*, 2002, **124**, 15172; (c) Y. Nishibayashi, A. Shinoda, Y. Miyake, H. Matsuzawa and M. Sato, *Angew. Chem., Int. Ed.*, 2006, **17**, 4835.
28. (a) B. D. Sherry, A. T. Radosevich and F. D. Toste, *J. Am. Chem. Soc.*, 2003, **125**, 6076; (b) M. R. Luzung and F. D. Toste, *J. Am. Chem. Soc.*, 2003, **125**, 15760.
29. M. Georgy, V. Boucard and J.-M. Campagne, *J. Am. Chem. Soc.*, 2005, **127**, 14180.
30. Z.-P. Zhan, J.-L. Yu, H.-J. Liu, Y.-Y. Cui, R.-F. Yang, W.-Z. Yang and J.-P. Li, *J. Org. Chem.*, 2006, **71**, 8298.
31. H.-H. Hui, Q. Zhao, M.-Y. Yang, D.-B. She, M. Chen and G.-S. Huang, *Synthesis*, 2008, 191.
32. X. Gao, Y.-M. Pan, M. Lin, L. Chen and Z.-P. Zhan, *Org. Biomol. Chem.*, 2010, **8**, 3259.
33. X. Feng, Z. Tan, D. Chen, Y. Shen, C.-C. Guo, J. Xiang and C. Zhu, *Tetrahedron Lett.*, 2008, **49**, 4110.
34. O. Debleds, E. Gayon, E. Ostaszuk, E. Vrancken and J.-M. Campagne, *Chem. - Eur. J.*, 2010, **16**, 12207.
35. (a) R. W. Bates and V. Satcharoen, *Chem. Soc. Rev.*, 2002, **31**, 12; (b) A. S. K. Hashmi, *Angew. Chem., Int. Ed.*, 2000, **39**, 3590; (c) S. Yu and S. Ma, *Angew. Chem., Int. Ed.*, 2012, **51**, 3074.
36. A. Hoffmann-Röder and N. Krause, *Angew. Chem., Int. Ed.*, 2004, **43**, 1196.
37. V. Cadierno, P. Crochet, S. E. García-Garrido and J. Gimeno, *Dalton Trans.*, 2010, **39**, 4015.
38. (a) S. Amslinger, *ChemMedChem*, 2010, **5**, 351; (b) T. Shiraki, N. Kamiya, S. Shiki, T. S. Kodama, A. Kakizuka and H. Jingami, *J. Biol. Chem.*, 2005, **280**, 14145; (c) J. M. Padrón, P. O. Miranda, J. I. Padrón and V. S. Martín, *Bioorg. Med. Chem. Lett.*, 2006, **16**, 2266.

39. (a) J.-W. Xie, W. Chen, R. Li, M. Zeng, W. Du, L. Yue, Y.-C. Chen, Y. Wu, J. Zhu and J.-G. Deng, *Angew. Chem.*, 2007, **119**, 393; (b) G. Bartoli, M. Bartolacci, M. Bosco, G. Foglia, A. Giuliana, E. Marcantoni, L. Sambri and E. Torregiani, *J. Org. Chem.*, 2003, **68**, 4594.
40. (a) M. G. Edwards, R. J. Paxton, D. S. Pugh, A. C. Whitwood and R. J. K. Taylor, *Synthesis*, 2008, **2008**, 3279; (b) E. J. Corey and M. Chaykovsky, *J. Am. Chem. Soc.*, 1965, **87**, 1353.
41. (a) R. P. Singh, K. Bartelson, Y. Wang, H. Su, X. Lu and L. Deng, *J. Am. Chem. Soc.*, 2008, **130**, 2422; (b) A. B. Northrup and D. W. C. MacMillan, *J. Am. Chem. Soc.*, 2002, **124**, 2458.
42. K. H. Meyer and K. Schuster, *Chem. Ber.*, 1922, **55**, 819.
43. (a) R. S. Ramón, N. Marion and S. P. Nolan, *Tetrahedron*, 2009, **65**, 1767; (b) N. Marion, P. Carlqvist, R. Gealageas, P. d. Frémont, F. Maseras and S. P. Nolan, *Chem. - Eur. J.*, 2007, **13**, 6437; (c) S. Gaillard, A. M. Z. Slawin and S. P. Nolan, *Chem. Commun.*, 2010, **46**, 2742; (d) S. Gaillard, J. Bosson, R. S. Ramón, P. Nun, A. M. Z. Slawin and S. P. Nolan, *Chem. - Eur. J.*, 2010, **16**, 13729; (e) A. Gómez-Suárez, Y. Oonishi, S. Meiries and S. P. Nolan, *Organometallics*, 2013, **32**, 1106.
44. R. S. Ramón, S. Gaillard, A. M. Z. Slawin, A. Porta, A. D'Alfonso, G. Zanoni and S. P. Nolan, *Organometallics*, 2010, **29**, 3665.
45. E. J. Corey and R. Noyori, *Tetrahedron Lett.*, 1970, **11**, 311.
46. B. M. Trost and R. C. Livingston, *J. Am. Chem. Soc.*, 1995, **117**, 9586.
47. B. M. Trost and R. C. Livingston, *J. Am. Chem. Soc.*, 2008, **130**, 11970.
48. X. Yao and C.-J. Li, *Org. Lett.*, 2006, **8**, 1953.
49. L. Zhou, L. Chen, R. Skouta, H.-f. Jiang and C.-J. Li, *Org. Biomol. Chem.*, 2008, **6**, 2969.
50. B. C. Zhao, Q.-Z. Zhang, W.-Y. Zhou, H.-C. Tao and Z.-G. Li, *RSC Adv.*, 2013, **3**, 13106.
51. L. Chen and C.-J. Li, *Org. Lett.*, 2004, **6**, 3151.
52. Q.-R. Lu, X. Meng, J.-S. Wu, Y.-J. Gao, C.-L. Li, Q.-Q. Zhu and B.-H. Chen, *Catal. Commun.*, 2008, **9**, 2127.
53. I. Arellano, P. Sharma, J. L. Arias, A. Toscano, A. Cabrera and N. Rosas, *J. Mol. Catal. A: Chem.*, 2007, **266**, 294.
54. C. Kang, J. Huang, W. He and F. Zhang, *J. Organomet. Chem.*, 2009, **695**, 120.
55. Y. Lu, H. Fan, A. Stump, T. L. Ward, T. Rieker and C. J. Brinker, *Nature*, 1999, **398**, 223.
56. F. Zhang, C. Kang, Y. Wei and H. Li, *Adv. Funct. Mater.*, 2011, **21**, 3189.
57. X. Yao and C.-J. Li, in *Metal-Catalyzed Reactions in Water*, ed. P. H. Dixneuf and V. Cadierno, Wiley-VCH, Weinheim, 2013.
58. R. Chinchilla and C. Nájera, *Chem. Soc. Rev.*, 2011, **40**, 5084.
59. Y. Jeong, B.-I. Kim, J. K. Lee and J.-S. Ryu, *J. Org. Chem.*, 2014, **79**, 6444.
60. D. M. D'Souza and T. J. J. Müller, *Nat. Protoc.*, 2008, **3**, 1660.
61. A. V. Kel'in and V. Gevorgyan, *J. Org. Chem.*, 2002, **67**, 95.

62. D. B. Grotjahn, S. Van, D. Combs, D. A. Lev, C. Schneider, M. Rideout, C. Meyer, G. Hernandez and L. Mejorado, *J. Org. Chem.*, 2002, **67**, 9200.
63. B. Willy, T. Dallos, F. Rominger, J. Schönhaber and T. J. J. Müller, *Eur. J. Org. Chem.*, 2008, **2008**, 4796.
64. K. M. Brummond and L. S. Kocsis, *Acc. Chem. Res.*, 2015, **48**, 2320.
65. A. V. Rotaru, I. D. Druta, T. Oeser and T. J. J. Müller, *Helv. Chim. Acta*, 2005, **88**, 1798.
66. J. P. Guermont, *Bull. Soc. Chim. Fr.*, 1953, 386.
67. M. A. Youngman and S. L. Dax, *Tetrahedron Lett.*, 1997, **38**, 6347.
68. S. Sakaguchi, T. Kubo and Y. Ishii, *Angew. Chem., Int. Ed.*, 2001, **40**, 2534.
69. C. Fischer and E. M. Carreira, *Org. Lett.*, 2001, **3**, 4319.
70. C.-J. Li and C. Wei, *Chem. Commun.*, 2002, 268.
71. (a) C. Wei, Z. Li and C.-J. Li, *Synlett*, 2004, 1472; (b) W.-J. Yoo, L. Zhao and C.-J. Li, *Aldrichimica Acta*, 2011, **44**, 43; (c) W.-J. Yoo, L. Zhao and C.-J. Li, in *Science of Synthesis*, ed. T. J. J. Müller, Thieme Stuttgart, 2013; (d) N. Uhlig, W.-J. Yoo, L. Zhao and C.-J. Li, in *Modern Alkyne Chemistry: Catalytic and Atom-Economic Transformations*, ed. B. M. Trost and C.-J. Li, Wiley-VCH Weinheim, 2014; (e) V. A. Peshkov, O. P. Pereshivko and E. V. Van der Eycken, *Chem. Soc. Rev.*, 2012, **41**, 3790.
72. C. Wei, J. T. Mague and C.-J. Li, *Proc. Natl. Acad. Sci. U. S. A.*, 2004, **101**, 5749.
73. D. Prajapati, R. Sarma, D. Bhuyan and W. Hu, *Synlett*, 2011, **5**, 627.
74. T. T. T. Trang, D. S. Ermolat'ev and E. V. Van der Eycken, *RSC Adv.*, 2015, **5**, 28921.
75. C. Wei and C.-J. Li, *J. Am. Chem. Soc.*, 2002, **124**, 5638.
76. J. Liu, B. Liu, X. Jia, X. Li and A. S. C. Chan, *Tetrahedron: Asymmetry*, 2007, **18**, 396.
77. M. Ohara, Y. Hara, T. Ohnuki and S. Nakamura, *Chem. - Eur. J.*, 2014, **20**, 8848.
78. C. Wei and C.-J. Li, *J. Am. Chem. Soc.*, 2003, **125**, 9584.
79. B. Huang, X. Yao and C.-J. Li, *Adv. Synth. Catal.*, 2006, **348**, 1528.
80. V. Srinivas and M. Koketsu, *Tetrahedron*, 2013, **69**, 8025.
81. X. Zhang and A. Corma, *Chem. Commun.*, 2007, 3080.
82. T. A. Graf, T. K. Anderson and N. D. Bowden, *Adv. Synth. Catal.*, 2011, **353**, 1033.
83. V. K.-Y. Lo, Y. Liu, M.-K. Wong and C.-M. Che, *Org. Lett.*, 2006, **8**, 1529.
84. V. K.-Y. Lo, K. K.-Y. Kung, M.-K. Wong and C.-M. Che, *J. Organomet. Chem.*, 2009, **694**, 583.
85. K. K.-Y. Kung, V. K.-Y. Lo, H.-M. Ko, G.-L. Li, P.-Y. Chan, K.-C. Leung, Z. Zhou, M.-Z. Wang, C.-M. Che and M.-K. Wong, *Adv. Synth. Catal.*, 2013, **355**, 2055.
86. K. K.-Y. Kung, G.-L. Li, L. Zou, H.-C. Chong, Y.-C. Leung, K.-H. Wong, V. K.-Y. Lo, C.-M. Che and M.-K. Wong, *Org. Biomol. Chem.*, 2012, **10**, 925.

87. H.-M. Ko, K. K.-Y. Kung, J.-F. Cui and M.-K. Wong, *Chem. Commun.*, 2013, **49**, 8869.
88. G. A. Price, A. K. Brisdon, S. Randall, E. Lewis, D. M. Whittaker, R. G. Pritchard, C. A. Muryn, K. R. Flower and P. Quayle, *J. Organomet. Chem.*, 2017, **846**, 251.
89. J. A. Widegren and R. G. Finke, *J. Mol. Catal. A: Chem.*, 2003, **198**, 317.
90. X. Zhang and A. Corma, *Angew. Chem., Int. Ed.*, 2008, **47**, 4358.
91. S. Shabbir, Y. Lee and H. Rhee, *J. Catal.*, 2015, **322**, 104.
92. Y. Chen, C. Liu, H. Abroshan, Z. Li, J. Wang, G. Li and M. Haruta, *J. Catal.*, 2016, **340**, 287.
93. L. L. Chng, J. Yang, Y. Wei and J. Y. Ying, *Adv. Synth. Catal.*, 2009, **351**, 2887.
94. M. Gholinejad, F. Hamed and C. Nájera, *Synlett*, 2016, **27**, 1193.
95. A. Feiz and A. Bazgir, *Catal. Commun.*, 2016, **73**, 88.
96. F.-X. Zhu, W. Wang and H.-X. Li, *J. Am. Chem. Soc.*, 2011, **133**, 11632.
97. J.-L. Huang, D. G. Gray and C.-J. Li, *Beilstein J. Org. Chem.*, 2013, **9**, 1388.
98. T. Zeng, W.-W. Chen, C. M. Cirtiu, A. Moores, G. Song and C.-J. Li, *Green Chem.*, 2010, **12**, 570.
99. F. M. Moghaddam, S. E. Ayati, S. H. Hosseini and A. Pourjavadi, *RSC Adv.*, 2015, **5**, 34502.
100. N. Zohreh, S. H. Hosseini, M. Jahani, M. S. Xaba and R. Meijboom, *J. Catal.*, 2017, **356**, 255.
101. M. Gholinejad, F. Zareh and C. Najera, *Appl. Organomet. Chem.*, 2018, **32**, e4452.
102. C. Wei, Z. Li and C.-J. Li, *Org. Lett.*, 2003, **23**, 4473.
103. Y. Zhou, T. He and Z. Wang, *ARKIVOC*, 2008, **13**, 80.
104. N. Salam, S. K. Kundu, R. A. Molla, P. Mondal, A. Bhaumik and S. M. Islam, *RSC Adv.*, 2014, **4**, 47593.
105. M. Jeganathan, A. Dhakshinamoorthy and K. Pitchumani, *New J. Chem.*, 2014, **2**, 781.
106. F. Movahedi, H. Masrouri and M. Z. Kassaee, *J. Mol. Catal. A: Chem.*, 2014, **395**, 52.
107. L. Shi, Y.-Q. Tu, M. Wang, F.-M. Zhang and C.-A. Fan, *Org. Lett.*, 2004, **6**, 1001.
108. B. Sreedhar, P. S. Reddy, C. S. V. Krishna and P. V. Babu, *Tetrahedron Lett.*, 2007, **44**, 7882.
109. M. Varyani, P. K. Khatri and S. L. Jain, *Catal. Commun.*, 2016, **77**, 113.
110. M. Tajbakhsh, M. Farhang, S. M. Baghbanian, R. Hosseinzadeh and M. Tajbakhsh, *New J. Chem.*, 2015, **39**, 1827.
111. S. Kumari, A. Shekhar and D. D. Pathak, *RSC Adv.*, 2016, **6**, 15340.
112. S. Sadjadi, T. Hosseinnejad, M. Malmir and M. M. Heravi, *New J. Chem.*, 2017, **41**, 13935.
113. M. M. Islam, A. S. Roy and S. M. Islam, *Catal. Lett.*, 2016, **146**, 1128.
114. M. J. Aliaga, D. J. Ramón and M. Yus, *Org. Biomol. Chem.*, 2010, **8**, 43.
115. O. P. Pereshivko, V. A. Peshkov and E. V. Van der Eycken, *Org. Lett.*, 2010, **12**, 2638.

116. (a) C. J. Pierce and C. H. Larsen, *Green Chem.*, 2012, **14**, 2672; (b) C. J. Pierce, M. Nguyen and C. H. Larsen, *Angew. Chem., Int. Ed.*, 2012, **51**, 12289; (c) U. Gulati, U. C. Rajesh and D. S. Diwat, *Tetrahedron Lett.*, 2016, **57**, 4468.
117. M. Cheng, Q. Zhang, X.-Y. Hu, B.-G. Li, J.-X. Ji and A. S. C. Chan, *Adv. Synth. Catal.*, 2011, **353**, 1274.
118. S. Sadjadi, M. Malmir and M. M. Heravi, *RSC Adv.*, 2017, **7**, 36807.
119. M. Malmir, M. M. Heravi, S. Sadjadi and T. Hosseinnejad, *Appl. Organomet. Chem.*, 2018, **32**, e4291.
120. M. Kinugasa and S. Hashimoto, *J. Chem. Soc., Chem. Commun.*, 1971, 466.
121. Ł. Woźniac, O. Staszewska-Kcajewska and M. Michalak, *Chem. Commun.*, 2015, **51**, 1933.
122. F. Cardona, A. Goti, S. Picasso, P. Vogel and A. Brandi, *J. Carbohydr. Chem.*, 2000, **19**, 585.
123. E. R. Bonfield and C.-J. Li, *Org. Biomol. Chem.*, 2007, **5**, 435.
124. N. Uhlig and C.-J. Li, *Org. Lett.*, 2012, **14**, 3000.
125. H. Feng, P. Zhao, L. Huang, Z. Sun and M. Tong, *Asian J. Org. Chem.*, 2017, **6**, 161.
126. Y. Hayashi, *Chem. Sci.*, 2016, **7**, 866.
127. B. Yan and Y. Liu, *Org. Lett.*, 2007, **9**, 4323.
128. Y. Ye, Q. Ding and J.-S. Wu, *Tetrahedron*, 2008, **64**, 1378.
129. K. Lauder, A. Toscani, N. Scalacci and D. Castagnolo, *Chem. Rev.*, 2017, **117**, 14091.
130. A. Saito, T. Konishi and Y. Hanzawa, *Org. Lett.*, 2010, **12**, 372.
131. A. Kinoshita, N. Sakakibara and M. Mori, *Tetrahedron*, 1999, **55**, 8155.
132. I. Ojima, A. T. Vu, S.-Y. Lee, J. V. McCullagh, A. C. Moralee, M. Fujiwara and T. H. Hoang, *J. Am. Chem. Soc.*, 2002, **124**, 9164.
133. G. Hattori, Y. Miyaki and Y. Nishibayashi, *ChemCatChem*, 2010, **2**, 155.
134. T. Imahori, H. Ojima, Y. Yoshimura and H. Takahata, *Chem. - Eur. J.*, 2008, **14**, 10762.
135. B. Roy, I. Ansary, S. Samanta and K. C. Mujumdar, *Tetrahedron Lett.*, 2012, **53**, 5119.
136. Q. Yang, T. Xu and Z. Yu, *Org. Lett.*, 2014, **16**, 6310.
137. S. Hase, Y. Kayaki and T. Ikariya, *ACS Catal.*, 2015, **5**, 5135.
138. A. Saito, K. Iimura and Y. Hanzawa, *Tetrahedron Lett.*, 2010, **51**, 1471.
139. S. Cacchi, G. Fabrizi and E. Filisti, *Org. Lett.*, 2008, **10**, 2629.
140. A. A. Nechaev, A. A. Peshkov, K. Van Hecke, V. A. Peshkov and E. V. Van der Eycken, *Eur. J. Org. Chem.*, 2017, **2017**, 1063.
141. (a) X.-T. Gao, C.-C. Gan, S.-Y. Liu, F. Zhou, H.-H. Wu and J. Zhou, *ACS Catal.*, 2017, **7**, 8588; (b) W.-J. Yoo and C.-J. Li, *Adv. Synth. Catal.*, 2008, **350**, 1503.
142. J. Chen, R. Properzi, D. P. Uccello, J. A. Young, R. G. Dushin and J. T. Starr, *Org. Lett.*, 2014, **16**, 4146.
143. G. I. Nikishin and I. P. Kovalev, *Tetrahedron Lett.*, 1990, **31**, 7063.
144. M. Picquet, C. Bruneau and P. H. Dixneuf, *Tetrahedron*, 1999, **55**, 3937.

145. B. M. Trost and A. H. Weiss, *Adv. Synth. Catal.*, 2009, **351**, 963.
146. L. Chen and C.-J. Li, *Chem. Commun.*, 2004, 2362.
147. (a) B. M. Trost, C. Chan and G. Ruhter, *J. Am. Chem. Soc.*, 1987, **109**, 3486; (b) B. M. Trost and A. E. Harms, *Tetrahedron Lett.*, 1996, **37**, 3971.
148. B. M. Trost, M. T. Sorum, C. Chan, A. E. Harms and G. Rühter, *J. Am. Chem. Soc.*, 1997, **119**, 698.
149. M. Kidwai, A. Jain and S. Bhardwaj, *Catal. Lett.*, 2011, **141**, 183.
150. T. Jin and Y. Yamamoto, *Org. Lett.*, 2007, **9**, 5259.
151. B. M. Partridge, M. Callingham, W. Lewis and H. W. Lam, *Angew. Chem., Int. Ed.*, 2017, **56**, 7227.
152. S. Choi, V. Srinivasulu, S. Ha and C.-M. Park, *Chem. Commun.*, 2017, **53**, 3481.
153. X.-T. Bai, Q.-Q. Zhang, S. Zhang, D.-Y. Chen, J.-Y. Fu, J.-Y. Zhu, Y.-B. Wang and Y.-T. Tang, *Eur. J. Org. Chem.*, 2018, **2018**, 1581.
154. P. G. M. Wuts, S. W. Ashford, B. Conway, J. L. Havens, B. Taylor, B. Hritzko, Y. Xiang and P. S. Zakarias, *Org. Process Res. Dev.*, 2009, **13**, 331.
155. G. Hernández-Torres, J. Mateo, A. Urbano and M. C. Carreño, *Eur. J. Org. Chem.*, 2013, **2013**, 6259.
156. H. Kusama, K. Ishada, H. Funami and N. Iwasawa, *Angew. Chem., Int. Ed.*, 2008, **47**, 4903.
157. K. Ishada, H. Kusama and N. Iwasawa, *J. Am. Chem. Soc.*, 2010, **132**, 8842.
158. S. Aoki, Y. Watanabe, M. Sanagawa, A. Setiawan, N. Kotoku and M. Kobayashi, *J. Am. Chem. Soc.*, 2006, **128**, 3148.
159. H. Kusama, A. Tazawa, K. Ishida and N. Iwasawa, *Chem. - Asian J.*, 2016, **11**, 64.
160. V. Belting and N. Krause, *Org. Biomol. Chem.*, 2009, 7, 1221.
161. T. F. Knöpfel and E. M. Carreira, *J. Am. Chem. Soc.*, 2003, **125**, 6054.
162. S. Fujimori and E. M. Carreira, *Angew. Chem., Int. Ed.*, 2007, **46**, 4964.
163. T. F. Knöpfel, P. Zarotti, T. Ichikawa and E. M. Carreira, *J. Am. Chem. Soc.*, 2005, **127**, 9682.
164. T. F. Knöpfel, P. Aschwandem, T. Ichikawa, T. Watanabe and E. M. Carreira, *Angew. Chem., Int. Ed.*, 2004, **43**, 5971.
165. P. Zarotti, T. F. Knöpfel, P. Aschwanden and E. M. Carreira, *ACS Catal.*, 2012, **2**, 1232.
166. S. Mishra, J. Liu and A. Aponick, *J. Am. Chem. Soc.*, 2017, **139**, 3352.
167. (a) A. Fürstner, S. Flügge, O. Larionov, Y. Takahashi, T. Kubota and J. Kobayashi, *Chem. - Eur. J.*, 2009, **15**, 4011; (b) I. Volchkov and D. Lee, *J. Am. Chem. Soc.*, 2013, **135**, 5324.
168. K. F. Avocetien, J. J. Li, X. Liu, Y. Wang, Y. Xing and G. A. O'Doherty, *Org. Lett.*, 2016, **18**, 4970.
169. M. A. Buil, M. Calbet, M. Castillo, J. Castro, C. Esteve, M. Ferrer, P. Forns, J. González, S. López, R. S. Roberts, S. Sevilla, B. Vidal, L. Vidal and P. Vilaseca, *Eur. J. Med. Chem.*, 2016, **113**, 102.
170. A. R. Rodriguez and B. W. Spur, *Tetrahedron Lett.*, 2012, **53**, 4169.

171. L. Chen and C.-J. Li, *Tetrahedron Lett.*, 2004, **45**, 2771.
172. (a) R. Jana and J. A. Tunge, *Org. Lett.*, 2009, **11**, 971; (b) R. Jana and J. A. Tunge, *J. Org. Chem.*, 2011, **76**, 8376.
173. H. Wang, Z. Hearne, T. Knauber, M. Dalko, J. Hitce, X. Marat, M. Moreau and C.-J. Li, *Tetrahedron*, 2015, **71**, 5866.
174. N. Asao, T. Nogami, K. Takahashi and Y. Yamamoto, *J. Am. Chem. Soc.*, 2002, **124**, 764.
175. A. Jean, J. Blanchet, J. Rouden, J. Maddaluno and M. De Paolis, *Chem. Commun.*, 2013, **49**, 1651.
176. D. A. Kummer, J. B. Brenneman and S. F. Martin, *Tetrahedron*, 2006, **62**, 11437.
177. P. A. Piexoto, R. Severin, C.-C. Tseng and D. Y.-K. Chen, *Angew. Chem., Int. Ed.*, 2011, **50**, 3013.
178. A. Szilágyi, F. Fenyvesi, O. Majercsik, I. F. Pelyvás, B. Bákckay, P. Fehér, J. Várad, M. Vecsernyés and P. Herczegh, *J. Med. Chem.*, 2006, **49**, 5626.
179. K. S. N. Reddy and G. Sabitha, *Tetrahedron Lett.*, 2017, **58**, 1198.
180. K. Tanaka and G. C. Fu, *J. Am. Chem. Soc.*, 2001, **123**, 11492.
181. K. Tanaka and G. C. Fu, *Angew. Chem., Int. Ed.*, 2002, **41**, 1607.
182. K. Tanaka, Y. Hagiwara and M. Hirano, *Eur. J. Org. Chem.*, 2006, **2006**, 3582.
183. K. Tanaka and G. C. Fu, *Org. Lett.*, 2002, **4**, 933.
184. K. Tanaka, Y. Hagiwara and M. Hirano, *Angew. Chem., Int. Ed.*, 2006, **45**, 2734.
185. K. Tanaka, M. Mimura and D. Hojo, *Tetrahedron*, 2009, **65**, 9008.
186. Z. Hearne and C.-J. Li, *Chem. Commun.*, 2017, **53**, 6136.
187. S. Li, W. Jia and N. Jiao, *Adv. Synth. Catal.*, 2009, **351**, 569.
188. (a) X.-P. Fang, J. E. Anderson, C.-J. Chang and J. L. McLaughlin, *Tetrahedron*, 1991, **47**, 9751; (b) R. Kazlauskas, P. T. Murphy, R. J. Quinn and R. J. Wells, *Tetrahedron Lett.*, 1977, **18**, 37; (c) X. Yang, Y. Shimizu, J. R. Steiner and J. Clardy, *Tetrahedron Lett.*, 1993, **34**, 761.
189. I. Tellitu, S. Serna, M. T. Herrero, I. Moreno, E. Domínguez and R. SanMartin, *J. Org. Chem.*, 2007, **72**, 1526.

Subject Index